THE CITY

GREAT CITIES OF THE WORLD

POLITICS OF THE CITY

GREAT CITIES OF THE WORLD

Their Government, Politics and Planning

Edited by

WILLIAM A. ROBSON

LONDON AND NEW YORK

First published in 1954

This edition published in 2007
Routledge
2 Park Square, Milton Park, Abingdon, Oxon, OX14 4RN
Simultaneously published in the USA and Canada by Routledge
711 Third Avenue, New York, NY 10017
Routledge is an imprint of Taylor & Francis Group, an informa business

Transferred to Digital Printing 2007

First issued in paperback 2013

British Library Cataloguing in Publication Data
A CIP catalogue record for this book
is available from the British Library

ISBN13: 978-0-415-41763-1 (volume hbk)
ISBN13: 978-0-415-41930-7 (subset)
ISBN13: 978-0-415-41318-3 (set)
ISBN13: 978-0-415-86041-3 (volume pbk)

Routledge Library Editions: The City

LONDON

The Thames, showing the Houses of Parliament and Westminster Abbey in the foreground.
Almost facing, on the opposite bank, is County Hall, the headquarters of the London County Council.

[Photo: Fox Photos

GREAT CITIES
OF THE WORLD

THEIR GOVERNMENT, POLITICS
AND PLANNING

EDITED BY

WILLIAM A. ROBSON

Professor of Public Administration at
the London School of Economics and Political
Science (University of London)
Past President of the International
Political Science Association

GEORGE ALLEN AND UNWIN LTD
RUSKIN HOUSE MUSEUM STREET LONDON

*Printed in Great Britain
in 12 point Bembo type
by Unwin Brothers Limited
Woking and London*

To
Elaine, Philip
and Ronald

It now only resteth, having brought our city to that
dignity & greatness, which the condition of the Scite
and other circumstances afford unto it: that we labor
to conserve, to maintaine and uphold the dignitie and
greatnes of the same.

Giovanni Botero: *A Treatise Concerning
the causes of the Magnificencie and
greatness of Cities*, done into English
by Robert Peterson (London, 1606).

CONTRIBUTORS

Rafael Bielsa

F. A. Bland

Ralph H. Brookes

Keith B. Callard

Brian Chapman

Giuseppe Chiarelli

Winston W. Crouch

Lorenz Fischer

L. P. Green

Peter van Hauten

Gunnar Heckscher

Axel Holm

Per Holm

Maurice Hookham

Max Imboden

Dean E. McHenry

José Arthur Rios

William A. Robson

Masamichi Royama

Lady Shena D. Simon

Roger Simon

Rexford G. Tugwell

M. Venkatarangaiya

Robert A. Walker

J. P. Wildschut

Editorial Assistant

Arch T. Dotson

M.A., Ph.D. (Harvard); Teaching Fellow, Harvard University, 1946–47; Assistant Professor of Political Science, University of Virginia, 1947–49; engaged in post-doctoral research, London School of Economics and Political Science, 1950–51; Assistant Professor of Political Science, Cornell University, 1951–56; Associate Professor since 1957.

PREFACE

WHEN I was writing *The Government and Misgovernment of London*,[1] I was struck by the remarkable dearth of books dealing with the government of great cities in foreign countries or within the Commonwealth. 'There is a tendency,' I wrote in the preface to that book, 'for the government of the principal metropolitan cities to be either ignored by social scientists, or else treated as part of the general subject of local government. It is obvious, however, that giant conurbations such as London are in a category by themselves and demand separate and individual treatment. Such treatment has so far been lacking or else inadequate in depth and extent, in view of the magnitude and complexity of the subject. I have not so far discovered any satisfactory studies of Paris, New York or Berlin.'

Several books have been published dealing with metropolitan areas in the United States, such as *The Government of Metropolitan Areas* by Paul Studenski (1933); *The Metropolitan Community* by R. D. McKenzie (1933), and the report on *Urban Government* of The National Resources Committee (1939). But these are general in character and do not attempt the intensive study of a particular city. An exception is the Chicago region, on which a good deal of research was carried out under the guidance and stimulus of the late Professor Charles E. Merriam. Among the valuable results of Merriam's impulse and interest were his own book *Chicago: A more intimate view of urban politics* (1929), and the monograph on *The Government of the Metropolitan Region of Chicago* by Merriam, Parratt and Lepawsky (1933). In 1942 Dr. Victor Jones published his *Metropolitan Government*, but this again was quite general in its scope and outlook; and, like other American books on this subject, it dealt only with U.S. cities. In recent years a number of monographs have been published dealing with individual cities of the United States. These afford welcome evidence of the increased attention being paid to the problems of metropolitan areas, but they are related entirely to American experience.[2]

---

[1] George Allen and Unwin. First published in 1939. Second revised edition 1948.

[2] Geddes, William Rutherford: *Administrative Problems in a Metropolitan Area: the National Capital Region* (Public Administration Service. Special Publications No. 61, Chicago, 1952); Betty Tableman: *Governmental Organization in Metropolitan Areas*. This deals mainly with the State of Michigan, and particularly Detroit. A whole series of studies of Los Angeles have been produced by the Haynes Foundation. See Edwin A. Cottrell and Helen L. Jones: *Characteristics of the Metropolis*; Richard Bigger and James D. Kitchen: *How the Cities Grew: a century of municipal independence and expansion in metropolitan Los Angeles*; Judith Norvell Jamison: *Regional Planning*; Robert F. Wilcox: *Law Enforcement*; Winston W. Crouch and others: *Sanitation and Health*; James Trump and others: *Fire Protection*; all published in Los Angeles in 1952.

A*

It was my original intention to try to persuade a few political scientists in other countries to remedy the situation by producing full length studies of one or more of their own metropolitan cities. But the Second World War broke out a few months after the publication of my book on London, and this drove all thoughts of the matter out of my mind. For the next five or six years great cities were being destroyed or defended rather than studied, and I was myself fully engaged on war work. It was not until 1949 that I came to consider the matter again in a serious way.

By that time I had formed the opinion that a book containing a series of fairly long essays dealing with the government of individual cities might be of interest, not only to teachers and students of political science, but also to politicians and ministers, councillors and officials, town-planners and journalists, and perhaps also the intelligent citizen and the traveller.

I had a twofold purpose in mind in planning the present volume. First, that it should describe the local government of a representative group of great cities. Second, that it should show whether, and if so how far, these vast communities are confronting common problems; and what steps, if any, they are taking to overcome them. It was clearly desirable that each chapter dealing with an individual city should be written by a contributor who had lived and worked in that city, or at least made a special study of it on the spot.

In the early stages of my work I thought the range of choice open to me was as extensive as the wide world itself. All the great cities of our planet, I fondly believed, were waiting to be plucked by my discriminating editorial hand. Alas! these illusions were quickly dispelled. I soon learned that in many countries political science is in so undeveloped a condition that it is impossible to find anyone who is able to give an accurate account of the government of a great city regarded as a living organism. Officials are often neither competent nor willing to write essays of this kind; and although there is an abundant supply of lawyers who might have been invited to contribute, I was particularly anxious to avoid having essays of the purely legalistic type, which would provide a detailed account of the statutes, enactments, regulations and other legal documents relating to a municipality, without the least regard to the political forces or the administrative realities of city government. I sought contributors who would explain the actual working of the great municipality, how its elective and executive organs are organized, the kind of political forces which motivate their activities, the scope and character of the municipal services, how they are financed, what are the relations between the great city and the state or national government, to what extent the machinery is adequate or obsolete, and what effect town and country planning is having on the great metropolis.

With such an object in mind, the field of choice soon began to dwindle rapidly. I must not disclose the secrets of the editorial office, but I have myself learned a great deal about foreign countries from this experience. Whatever else reviewers or critics may wish to say about this volume, I hope they will refrain from pointing out that I have omitted Prague, Madrid, Lisbon, Cairo, Brussels and many other great cities. I can assure them that there are very good reasons both for what is included and what is omitted. I should perhaps explain that Berlin, Vienna and Tokyo were omitted because, owing to the presence of occupying forces, they were (or are) in an abnormal condition.[1] I am fully aware of the fact that some of the cities which are included in the work are smaller than some of those which have been omitted.

I have endeavoured to obtain a substantial degree of consistency of treatment in the essays. With this in view a detailed synopsis of the scope, outlook and purpose of the book was sent to each contributor with an indication of the various matters with which he should deal in his essay Not all the contributors followed these directions in all respects, and deviations and omissions in some chapters could not always be remedied by subsequent additions or amendments. The degree of consistency which has been achieved is certainly less than I expected; but this is partly due to the differences of education, background and outlook of the authors. Nevertheless, the contributors generally have given me their fullest and most cordial co-operation, and I should like to thank them for their help in making it possible to produce the book. This is a somewhat new type of multi-national research project; and it is in some ways a pioneering effort on a modest scale. I hope it will be followed by more ambitious and intensive full-length studies of these and other cities.

The preparation, writing, editing and production of this book has taken nearly five years. The first manuscript was received in October 1950 and the final revised version of the last one in July 1953. During this period of nearly three years I was in constant communication with contributors scattered all over the world, and in many instances the typescripts were sent backwards and forwards several times in order to dispose of difficulties of one kind or another. As the final versions of the chapters began to reach me, it became evident that there would be some disparity between

---

[1] Vienna and Tokyo are now free of occupying forces. A chapter on Tokyo and Osaka is included in the present edition.

their several dates of completion. It also became clear that it would have been impossible to arrange for all the authors to deal with the position in their respective cities as at a uniform point of time—e.g. the end of 1953. If I had attempted this the book would probably never have been published at all.

Most of the contributors were good enough to insert the most recently-available factual information on the proofs. In one or two cases, however, this could not be done. The chapters on Manchester, Moscow and Wellington, for example, present the state of affairs existing in 1951.

While every effort has been made to present a contemporary picture of the great cities, small changes of detail which occur from year to year in the finance, population, or similar matters relating to a great city do not affect the general picture which this book seeks to provide. The object of the work is to show the problems of government, of politics and of planning which confront the great cities of the world, and how they are trying to deal with them. This object has, I think, been substantially achieved.

I am specially indebted to Mr. Arch T. Dotson, now associate professor of political science in Cornell University, for his immensely valuable help as editorial assistant. Dr. Dotson came to the London School of Economics and Political Science during the sessions 1949 to 1951 to do post-graduate work on local government. He was working under my supervision and I was so much impressed by his outstanding ability, scholarship, and judgment that I invited him to act as my editorial assistant on this book. He occupied this post for a year, including a period of three months when I was abroad. He helped to edit the manuscripts, corresponded with authors on my behalf, and was an invaluable aid in every way. We discussed together at length many of the problems dealt with in the essays.

My warm thanks are due also to Mr. H. A. Cartledge of the British Council, who has been unbelievably generous with his help and counsel in regard to the essays on Latin American cities. I must also thank Mr. John Hampden of the British Council for obtaining valuable information for me from the Council's overseas representatives; while yet another officer of the British Council, Mr. M. V. Kitchin, of the Cambridge office, kindly gave me much help with the translation of Professor Bielsa's essay.

I also wish to thank Professor R. A. Humphreys of University College, London; Professor Lashley Harvey of Boston University; Mr. W. Eric Jackson of the London County Council for information used in my own chapter on London; my colleague Professor R. O. Buchanan of the

Geography Department of the London School of Economics and Political Science, for help and advice on the maps which appear in the book and Miss S. V. Webb, a member of his staff, for cartographic work; Dr. M. Plant and Dr. H. Schurer, of the British Library of Political and Economic Science, for help in compiling the select bibliography; my colleague, Professor David Glass, for advice on urban statistics; Mr. A. T. Williams, of Victoria, and Mr. S. Haviland, of Sydney, Australia; Mrs. E. Wistrich, B.Sc.(Econ.), former research assistant in the Government Department of the London School of Economics and Political Science, and last but not least, Mrs. D. W. Cleather, my secretary, who helped me to conduct voluminous correspondence with the contributors and kept the mass of papers, illustrations and documents in marvellous order. Mrs. Cleather also kindly offered to read through the final page proofs of the book.

WILLIAM A. ROBSON

London School of Economics and
   Political Science
July 1954

PREFACE TO THE SECOND EDITION

THE substantial demand for this book both in Britain and in overseas countries has made it necessary for the publishers to issue a second edition within a comparatively short space of time for a work of this character.

In preparing the present edition I have taken advantage of the occasion to correct a number of errors which managed to creep into the original text. These included the misdescription of two of the illustrations, which resulted from incorrect information being supplied to me with the photographs.

The most important feature of the second edition is the addition of three new chapters dealing with Cologne, Johannesburg, Tokyo and Osaka. These supplementary studies fill serious gaps to which I drew attention in the preface to the original edition. I there explained that Berlin, Vienna and Tokyo were omitted because at the time when the book was being prepared, they were in an abnormal condition owing to the presence of occupying forces. The resumption of political freedom in Japan has enabled me to include a study of Tokyo and Osaka. I am particularly indebted to Professor Masamichi Royama, who contributed this chapter with the assistance of the Tokyo Institute for Municipal Research. Members of the Tokyo Institute are engaged in translating the book into Japanese. I should like to express my thanks to Professor Royama, Dr. Tanabe, the Managing Director of the Institute, and others who are participating in this enterprise.

The first edition also lacked a study of a South African city owing to the inability of two authors in that country to produce the chapter they had in turn agreed to write. I am very glad that this unfortunate gap has now been filled by Dr. Green's essay on Johannesburg.

It is still impracticable to include a chapter on Berlin, but I have managed to obtain an essay on Cologne; so that a representative metropolitan area of Western Germany is now presented. Dr. Lorenz Fischer, who undertook the task, had done much of the work but had not completed the chapter when, to my deep regret, he died in October 1956. I am grateful to Dr. Peter van Hauten for completing the chapter.

These three supplementary studies add to the comprehensiveness, and hence to the value of the book; but they do not affect in any significant respect my analysis of the problems of the great city, and the conclusions at which I arrived about those problems. Partly in order to save the very heavy additional expense which would have been incurred in resetting the type, and partly because the result would not have justified the labour involved, Part One has not been revised so as to take into account the supplementary chapters.

Similarly, with one exception the chapters in Part Two have not been revised by the authors. The reasons for this are threefold. First, I did not believe that it would have been possible to get all the chapters revised promptly enough to ensure early publication of the second edition. Second (as I pointed out in the original preface), small changes of detail which occur from year to year do not affect the general view of the great city and its problems which this book endeavours to provide. Third, it was necessary from a commercial point of view for the publishers to use the standing type in producing the present edition. The only exception is the chapter on Bombay and Calcutta, which Professor Venkatarangaiya has revised rather considerably in order to eliminate certain errors in the original text.

The International Political Science Association included *The Government of Great Cities* as one of the subjects for discussion at its Third World Congress held in Stockholm in August 1955. I acted as rapporteur-general at the discussions devoted to this theme. On opening the subject I presented an analysis of the main problems of the great city on the lines indicated in Part One of this book. A lively and illuminating discussion took place by political scientists from many parts of the world. Ultimately the Congress agreed that the problems I had emphasized (summarized herein on pages 99–102) represent the major problems which demand attention at the present time in regard to the great metropolitan community. I was much encouraged by this support and confirmation of my analysis.

My friend Dr. Luther Gulick, then City Administrator of the City of New York, informed the Congress that more than twenty research projects are being carried out in the United States into metropolitan areas, and many millions of dollars have been allocated for this purpose by the great foundations. So far as I am aware, no funds have been granted for any international studies in this subject. In my opinion effort and money might usefully be directed to promote further study and research in the problems of the metropolitan community on an international scale. There is ample need for more intensive and widespread work in this field. Above all we need to produce more men and women who are trained to understand the political, economic, social, administrative, and technological aspects of metropolitan areas.

W. A. R.

February, 1957

CONTENTS

SUPPLEMENTARY STUDIES

ILLUSTRATIONS

Plate

SUPPLEMENTARY ILLUSTRATIONS

MAPS AND DIAGRAMS

PART ONE

The Great City of Today

THIS book is concerned with the government, politics, and planning of a selected group of great cities in Europe, America, Asia, and Australasia. Its purpose is not only to gain knowledge of a neglected field but to see how far the great city produces problems of more or less universal significance. I believe that the studies will provide some valuable comparative material on matters of general interest.

I. A MODERN PHENOMENON

Men have lived in towns for thousands of years, but the great city of today is a very modern phenomenon. The vast aggregations of population, numbering millions, to be found in the largest towns could not have been supplied with enough food to keep them alive for a single day before the advent of modern means of transport. The railway, the motor car, the cargo ship driven by steam or oil, are the veritable founders and supporters of the great modern city. We are equally dependent on modern technology for the conveyance, filtration and distribution of the enormous quantities of water required by the citizens of every large city for domestic, industrial and commercial uses. The Pont du Gard in Provence is a superb relic of a Roman aqueduct which successfully carried water a considerable distance; but the Romans would have been totally unable to provide London or Paris with the quantities of water they consume today.

Much the same is true of the garbage collection and sewage disposal services of the great modern city. They too are the result of modern technology. And without power-driven factories and mills, mechanized offices, telecommunications, and swift local transport, it would be impossible for such vast populations to earn their living in concentrated urban centres.

The cities described in this volume cover a very wide range of size, both as regards population and area. The following table shows the population and the territory contained within the legally defined boundaries of the cities. These boundaries bear little relation to the real size of the city regarded as a social and economic aggregation.

There is, of course, no definition of a great city, in terms of its minimum size. The conception will usually have some relation to the total population of the country in which the city is situated. Zürich, for example,

TABLE I
Population and area of cities

	Population	Date	Area in sq. miles
Amsterdam	846,000	1951	26·75
Bombay	2,839,270	1951	94
Buenos Aires	3,000,371	1947	118·53
Calcutta	2,548,677	1951	36
Chicago	3,620,962	1950	206·7
Copenhagen	970,100	1952	28·83
London	3,348,336	1951	117
Los Angeles	1,970,358	1950	453·4
Manchester	715,000	1951	42·6
Montreal	1,021,520	1951	50·39
Moscow	5,100,000	1950	125
New York	7,891,957	1950	315
Paris	2,869,000	1953	41
Rio de Janeiro	2,377,451	1950	450
Rome	1,735,354	1950	20·7
Stockholm	740,000	1950	70·31
Sydney	212,360	1949	11
Metropolitan Toronto	1,134,000	1953	245
Wellington	120,064	1951	25·47
Zürich	390,020	1950	36

ranks as a great city in a small country like Switzerland, whereas it would be regarded as of only medium size in the United States. Wellington is a much more extreme case, and could only be regarded as a large city in a country with a very small population. Size is, however, only one of the characteristics of the great city. Its economic significance, either in regard to manufacturing industry, commerce, or financial power, is generally a dominant factor. A third feature is its cultural eminence. A city which is merely large and wealthy lacks the quality of greatness unless it can show signs of distinction in the arts and sciences. This explains the urge on the part of public authorities or public spirited citizens to establish or develop a university; to have a symphony orchestra, some fine museums, art galleries, and other displays; to have an opera house, a leading theatre, and other manifestations of culture, even though a public subsidy is needed to support them.

Lastly, there is the political importance of the great city. Professor Mackenzie regards the potential dominance of the great city in the sphere of national politics as one of the chief problems calling for consideration.[1]

[1] W. J. M. Mackenzie: *The Government of Great Cities*, p. 2 (Percival Lecture, 1952); 93 Memoirs and Proceedings of the Manchester Lit. and Phil. Society, No. 5.

He points out that the domination of great cities has been accepted as one of the marks of the Western way of life; and yet the political problem which they present has been largely ignored, despite its urgency.

The political importance of the great city does not necessarily derive from its position as a national capital. Indeed, only nine of the twenty cities dealt with in this book are national capitals. It arises rather from the massive concentration of highly organized power which the great city is able to exert. When that concentration of power is associated with the prestige and authority of a capital, the result is formidable in the extreme.

The great city may not be a capital, but it is almost certain to be a metropolitan area, in the modern sense of the term. This connotes a great commercial, industrial, cultural or governmental centre surrounded by suburbs, housing estates, dormitory towns or villages. Large numbers of people who work in the city reside in these outlying areas beyond its boundaries. They and their families use the shops, hospitals, colleges or schools, theatres, concert halls, and other institutions provided by the 'mother-city.'[1] Conversely, the outlying areas supply the city with market garden and dairy produce; and they offer those who dwell in the city open spaces, playing fields, camping grounds, and many different kinds of rural amenity. There is also sometimes a centrifugal movement of people who live in the city and work in the outlying areas, but this is seldom of substantial proportions.

2. THE METROPOLITAN COMMUNITY

The economic and social reality we have to bear in mind is that of a metropolitan community living in a metropolitan area of the kind mentioned above. This again is the result of modern improvements in transportation and communication. Never before has the townsman been able to live a considerable distance from his work. The age-long distinction between the way of life of the town-dweller and the country-man has almost ceased to exist in the typical metropolitan area. People who live in the country receive frequent deliveries of goods by motor van from the great city; they enjoy electric light, gas supply, main drainage, motor cars, the telephone, radio, television, daily

---

[1] Dr. Victor Jones defines a metropolis as 'a mother-city from and toward which people move to establish suburban aggregations on the periphery. Many of the people who live in the suburbs work in the central city; and they and their families use the cultural, recreational, trade, professional and commercial facilities of the mother-city almost as freely as do those who live within its boundaries.' *Metropolitan Government* (University of Chicago Press, 1942), p. 2.

postal deliveries, laundry and cleaning services, metropolitan news-papers, travelling libraries, refrigerators, local cinemas—all this and heaven too.

Despite the growth of suburbs and increasingly distant housing estates, despite a relative and sometimes absolute decline of population in the inner core compared with the outer belt, the original city has not merely held its own but even increased its dominance as a commercial and financial centre, a magnet which attracts those in search of amusement and culture. 'The whole increasing population,' wrote Dr. Thomas H. Reed, '—the overtaken farmer as well as the invading suburbanite—more and more seek the stores, churches and professional services of the centre to the detriment of the crossroads store, the wayside church and the country doctor. A nation has been put on wheels, and its mobility has created metropolitan regions with common political problems and no adequate machinery to solve them.'[1]

The growth of metropolitan areas has occurred in every part of the world. One finds metropolitan cities in the West and in the Orient; in all the continents; in developed and underdeveloped countries; under capitalism and under communism; in new countries and in old. It is a movement peculiar to our age and has now assumed vast quantitative importance. Thus, in 1950, 44,440,496 people, or 29·5 per cent of the total population of the United States, were living in 14 metropolitan districts each containing over a million inhabitants. A further 12,398,635 persons were living in 19 metropolitan areas with populations of between half a million and a million. In the same year there were 14,594,878 people living in 44 metropolitan areas in the United States whose populations lay between 250,000 and 500,000 each. Thus in all nearly 71½ million persons were living in 1950 in these 77 metropolitan districts.[2] In England and Wales, 17 million people, or about two-fifths of the nation, live in London and five other metropolitan areas recognized as conurbations by the Registrar-General.[3] The position is similar in Germany, while Paris embraces about one-eighth of the population of France.[4]

In Australia, the metropolitan communities have been increasing in population at a more rapid rate than either the separate states or the Commonwealth. Between 1933 and 1947, the population of Australia increased by 14 per cent, while that of Australian metropolitan com-

---

[1] *Metropolitan Areas.* Encyclopaedia of the Social Sciences, X, p. 396.
[2] *United States Census of Population*, 1950, Vol. I, p. 33, Washington, 1952.
[3] Census of England and Wales, 1951. Preliminary Report, Table IV, p. 39.
[4] Victor Jones: *op. cit.*, p. 97; W. J. M. Mackenzie: *loc. cit.*

munities increased by 24 per cent. During this period the population of the metropolitan area of Perth grew by 31 per cent, while the population of Western Australia increased by the national figure of 14 per cent. In 1901 the Perth urban complex contained 38 per cent of the population of Western Australia. By 1947, it comprised 54 per cent of the state population. Yet in that year, two other Australian states had a higher degree of 'conurbation' than Western Australia.[1]

The proportion of the metropolitan community living within the city boundaries is indicated in some of the studies presented in this book. Thus, the population of Greater Buenos Aires was 4,644,000 in 1947, of which 3,000,371 was contained in the city of Buenos Aires. The population of metropolitan Chicago was 5,495,364, according to the United States census of 1950, of which nearly 35 per cent lives outside the city limits. The city of Copenhagen has a population of 764,100, while the capital, consisting of the city and two adjoining municipalities of Frederiksberg and Gentofte, contains 970,100 persons. The metropolitan area of Copenhagen comprises this inner core and 19 neighbouring municipalities, which have a total population of 1,230,000. The metropolitan population outside the city is thus 465,900, or 259,900 outside the capital. The population of Montreal in 1951 was 1,021,520, compared with 1,395,400 for Greater Montreal. This means that 373,980 people in Greater Montreal live outside the city limits. The population of the Ville de Paris is 2,725,000, while that of the whole metropolitan area is said to exceed 5 millions. In 1950, out of a total population of 980,000 in Greater Stockholm, 740,000 lived inside the city and 240,000 outside. The recently constituted Metropolitan Municipality of Toronto contained in 1951 1,117,470 persons, of whom 675,754 lived in the city and 441,716 in the suburbs. The city of Los Angeles, according to the 1950 census, has a population of 1,970,358, compared with a total population for the metropolitan area of 4,367,911.

These figures are symptomatic of the increasing tendency for large numbers of persons to live outside the central core of the old city, whose boundaries remain fixed for long periods of time.

It is, however, impossible to present the position of metropolitan cities in its proper quantitative significance because there are no areas which adequately represent them for census purposes. For example, the area known as Greater London is used by the Registrar-General for population statistics, but the metropolitan region extends far beyond this. The same is doubtless true of Greater Stockholm, Greater Buenos Aires, and all the

[1] J. R. H. Johns: *Metropolitan Government in Western Australia*, pp. 4–6.

other metropolitan areas. The metropolitan area of Manchester, defined as the immediate region of which Manchester is the centre, contains 1,296,000 persons; but Professor Mackenzie, of the University of Manchester, says that he considers the real city in which he lives, a city which has neither a name nor boundaries, to have a population of nearly 2,250,000.[1] The people of the Perth-Fremantle area in Western Australia are becoming 'metropolitan-conscious and aware of metropolitan needs.' But an Australian scholar reports that few of them could give a precise answer to the question: What is our metropolitan area?[2] The same would doubtless be equally true in Rome, Boston, Brussels, Cairo, Madrid, or Johannesburg.

If the typical metropolitan city consisted of a continuous built up area, it would be far easier to define and to describe in terms of population and territory; but actually no metropolis is composed of a single compact urban concentration.[3] And just because the metropolitan community is so attenuated, spreading further and further afield in ever-widening circles like the ripples on a pond, it becomes increasingly hard to define, to measure, to comprehend, and to govern. In the United States a good deal of research has been devoted to metropolitan areas; and the usual means of defining such areas is by laying down a minimum figure of aggregate population combined with a density minimum for the surrounding districts.[4] So crude a method is clearly unsatisfactory.

The present method of defining 'standard metropolitan areas' used by the United States census in 1951 is to take an area containing one, and perhaps more than one, city of 50,000 or more. Each standard metropolitan area has as its nucleus the county or counties containing this central city or cities. Adjacent counties are included in the metropolitan area if they satisfy certain criteria. First, the county must have 10,000 non-agricultural workers, or 10 per cent of the non-agricultural workers in the standard metropolitan area, or at least half of its population residing in minor civil divisions with a population density of 150 or more per square mile. Second, non-agricultural workers must constitute at least two-thirds of the total employed persons of the county. Third, there must be evidence of social and economic integration of the county with the central city as shown by the proportion of persons living in the county and working in the city, or vice versa, and the number of telephone calls between the two areas.[5]

[1] J. R. H. Johns: *op. cit.*, p. 2.   [2] *Ibid.* p. 1.   [3] Thomas H. Reed: *op. cit.*, p. 397
[4] Paul Studenski: *The Government of Metropolitan Areas in the United States*, p. 9.
[5] *United States Census of Population*, 1950, Vol. II, p. 27.

These criteria are satisfied by a very large number of areas which cannot claim to be metropolitan centres in a proper sense of the term. The figure of 50,000 for the 'core' city is so small that it robs the word 'metropolitan' of any sociological and political significance; and the definition takes no account whatever of the functions which should be performed by a metropolitan area worthy of the name. It should be either a great political and governmental centre, or a commercial and industrial centre, or a cultural centre, or perhaps all of these, or at least two of them.

The effect of the United States census definition is shown by the fact that by applying it in the 1950 United States census no fewer than 168 standard metropolitan areas were enumerated, of which only 42 contained populations exceeding 400,000; while more than 77 contained fewer than 200,000 persons. The populations of 28 standard metropolitan areas lay between 56,000 and 110,000. They included such insignificant places as Gadsden, Alabama, with a central city of 55,725 and a total population of 93,892, and Ogden, Utah, with a central city of 57,112 and a total population of 83,319.[1]

In our view, in a country as large and highly developed as the U.S.A., only metropolitan areas with a central city of not less than 300,000 and a total population of at least 400,000 should be regarded as possessing metropolitan status. There were in 1951 about 33 such areas. They would all fulfil the functional conditions referred to above. To place a vast mass of small fry in the same category as the great metropolitan communities of New York, Chicago, Los Angeles, Philadelphia, Detroit, Boston and so forth does not assist the serious study or understanding of the problem of the metropolitan city.

Nowhere has any serious attempt been made to provide the metropolitan community with a system of government designed to satisfy its present and future needs in regard to organization, services, finance, co-ordination, planning, or democratic control. The great city is struggling along with an out-of-date structure vainly trying to grapple with mounting difficulties and to solve problems which cannot be overcome without drastic reforms. Meanwhile, the population washes outward over the irrelevant administrative boundaries like the ocean tide washing over submerged breakwaters. The inhabitants of the outlying suburbs, dormitory towns, communes, housing estates, and villages flood into the central city each day, scarcely aware when they cross the boundary. The outward growth of the metropolis is not checked by the

[1] *United States Census of Population*, 1950, Vol. I, p. 33, Washington, 1952.

frontiers of counties, states, or even countries. Metropolitan New York sprawls over much of New Jersey, Connecticut and other states. Detroit is partly in the United States and partly in Canada. The development of civil aviation, and particularly the helicopter, will accentuate this problem. It is already possible to imagine businessmen living on the French coast and flying daily to and from their work in London. And if the housing situation is difficult in Copenhagen, why should one not live in Norway or Sweden and fly every day to the Danish capital?

The picture we have tried to paint of the great city outgrowing its boundaries should be regarded as prolegomena to an understanding of the government and administration, the politics and planning, of the modern metropolis. We should think of the metropolitan community which lives, works, and plays in a widespread metropolitan area as being the basic social and economic reality. This is the great city of today, not the narrowly confined part of it which happens to live within the obsolete legal and administrative boundaries laid down in a former age.

We shall return to the political, governmental and administrative implications of this problem after we have considered the present methods of local government.

The task of governing a very large city is fundamentally different from that of governing one of small or medium size. After a certain stage is reached, quantitative differences produce qualitative differences. The great city of today is so large that it is difficult to evoke from its citizens or inculcate in them the spirit of community. People usually do not know their neighbours, and so the sense of neighbourhood, of belonging to the same place and sharing common interests, is lost. The city council and its executive organs often seem remote and aloof from the daily life of the citizens, with whom they are not in close contact. In consequence, apathy is often shown towards municipal policies and municipal elections. It is difficult for local government to flourish in an atmosphere of indifference, suspicion or distrust.

3. SPECIAL CONSTITUTIONAL FEATURES

The great city is sometimes, but not always, distinguished from other towns either by reason of its constitution or because of its relations with the national government, or both. Thus, the Ville de Paris differs from all other French communes by virtue of the fact that it has no mayor, all executive power being in the hands of either the Prefect of the Seine or the Prefect of Police. Moreover, the former has acquired many powers which are usually possessed by the municipal council in a

French commune. His powers in regard to the departmental council of the Seine are greater than is accorded elsewhere to the prefect.

In Moscow, as in London, there is a two-tier structure specially designed for the capital city. The Moscow City Soviet has direct relations with the Union Soviet on certain matters, owing to its position as capital of the Soviet Union as well as of the Russian Socialist Republic. The London County Council is in almost the same position as any other major local authority in Britain as regards its relations with the national government. Manchester has the same form of government as eighty-two other county boroughs, which combine the powers of a municipal corporation with those of a county council. Stockholm also has a county borough status. It is, however, peculiar in having city directors, who resemble municipal ministers in charge of executive functions. Rio de Janeiro and Buenos Aires, like Washington, D.C., have been separated from their respective states and placed in a special district under the federal legislature. In both cities the mayor is appointed by the President of the Republic, instead of being elected by the local community or by its city council.

The city government of Rome differs surprisingly little from the local government of other Italian towns, except for the requirement (often found in the countries of continental Europe) that the financial estimates must be approved by the central government instead of by the provincial authority. Moreover, the Minister of the Interior exercises control over all matters of wide economic importance affecting Rome. In Copenhagen, which also belongs to the county borough species, the executive organs are rigidly separated from the city council, whereas in other Danish towns the mayors and aldermen are chosen from among the councillors. The separation of powers is also reflected in the provision that committees of the council do not participate in administration.

Among the cities which are not national capitals, one finds that Montreal has been the subject of special legislation passed by the province of Quebec, while the province of Ontario has recently created a new form of metropolitan government for Greater Toronto. Sydney also depended on special legislation until the Local Government (Areas) Act, 1948, brought the city into line with the ordinary local government system of the state. New York City has a constitution peculiar to itself which was adopted in 1938; but in the United States the principle of 'home rule' is so widely applied to local government, that this is a very commonplace occurrence.[1] Zürich differs from the general run of communes in German-speaking Switzerland mainly by reason of the fact that supreme

[1] See Harold Zink: *Government of Cities in the United States*, pp. 121–30.

B

power is shared between the communal council and the electorate, which decides certain questions by means of a referendum. The usual Swiss practice is for supreme communal authority to be vested in the gathering of the voters known as the communal assembly—an obviously impractical arrangement in a city as large as Zürich.

From this brief survey it appears that the tendency is for capital cities[1] to have a special form of local government different from that of other cities in the country. Great cities which are not capitals usually have the system of local government applied in their country to substantial towns. Both these propositions are subject to exceptions.

It is undoubtedly true that a capital city tends to enjoy a lower degree of self-government than other cities in the same country. The explanation is not far to seek. A capital city is the seat of the national government; it contains the diplomatic representatives of foreign countries whose security is a national responsibility; and possession of the capital is often the key to political control over the entire country. For these reasons the national government often exercises a much greater degree of control over the capital than over other large towns. Paris is the classic example of this; but one might equally well cite Buenos Aires, Rio de Janeiro or Washington, D.C. In London the police force, elsewhere administered by the watch committee of the municipal corporation, is administered by the Metropolitan Commissioners of Police who are under the control of the Home Secretary.

4. THE ELECTED COUNCIL

In local government, as in government at higher levels, we normally look for both the representative element and the bureaucratic element. The representative element is found in the city council. The degree to which a municipality can be regarded as endowed with democratic local government depends on the composition and functions of the council, and the extent to which the executive organs are subject to popular control either by the council or by the electorate.

All but one of the great cities comprised in this volume have an elected council. The exception is Buenos Aires, which was entitled to a city council under an Act of 1882. The council was from time to time replaced by a citizens' committee nominated by the national government, and in 1943 it was definitely abolished by the Peron régime. There is thus no longer any local self-government in Buenos Aires.

---

[1] I refer here to national capitals, and not to state capitals in countries with a federal form of government.

The city councils vary greatly in size, the qualifications required of candidates, tenure of office, remuneration, and also the franchise. The size of the council ranges from 15 persons in Los Angeles and Wellington, to 1,392 in Moscow. Apart from Moscow, which is in a class by itself, the councils fall into the following three groups in regard to size:—

Number of Councillors

Below 50	*Between 50 and 100*	*Between 101 and 150*
Amsterdam 45	Chicago 50	Bombay 135
Los Angeles 15	Copenhagen 55	Manchester 144
New York 27	Paris 90	London 144
Toronto 25	Montreal 100	Zürich 125
Wellington 15	Rio de Janeiro 50	
Sydney 30	Rome 80	
	Stockholm 100	
	Calcutta 86	

The size of the council is a matter of great importance. It affects not only the effectiveness of the deliberative body but also the relations between the city council and the citizens. In New York City, where a councilman represents about 75,000 voters, it must be much harder for the elected representative to keep in close touch with his constituents than in Amsterdam, for example, where the ratio is rather less than 1 to 20,000 voters.

One of the most serious disadvantages usually associated with the great city is, indeed, the steep reduction in the number of citizens who participate directly in the process of government. Manchester has a representative on the council for every 5,000 citizens; while in neighbouring South-east Lancashire there is an elected councillor for every 1,500 citizens.[1] The Ville de Paris has one councillor for every 30,000 citizens, but in the Department of the Seine and its communes the ratio is one to fewer than 1,000 persons. Hence one can infer that the amalgamation of outlying areas with the city will usually cause a decrease in the ratio of councillors to citizens, thereby reducing the number of men and women taking an active part in local government.

This problem has been solved by two methods. One is the establishment of a two-tier structure which provides, in addition to the major council, a lower tier of elected councils exercising minor functions in their own districts. There are twenty-eight such councils in the administrative county of London, and as a result there is one representative for every 2,250 persons. This is a higher rate of participation than in the outlying

[1] W. J. M. Mackenzie: *op. cit.*, pp. 8–9.

parts of the metropolis which have not been amalgamated with the county.

Moscow also has a lower tier of twenty-five district soviets, which provide about 3,000 elected councillors. The Soviet capital has, however, approached the problem of representation from the other end. One councillor sits on the city soviet for every 3,000 electors. Three consequences flow from this approach. First, the size of the soviet fluctuates in accordance with the size of the population. Second, the assembly is nearly ten times larger than that of any other great city outside the U.S.S.R. Third, the proportion of persons participating in local government is much higher in Moscow than elsewhere, for in addition to the 1,392 members of the city soviet there are the 3,000 district councillors, together with about 10,000 so-called 'activists' serving as co-opted members of standing committees.

There are obvious disadvantages in having so large an assembly as the principal governing body of the municipality. It is comparable in size to the annual conference of the British Labour Party; and even in a one-party organ of this size, detailed discussion is impracticable, debate is difficult, and oratory offers the most effective way to influence. Members can reveal streams of opinion in such an assembly, voice the discontent of their constituents, and express popular needs. By such means policy can be influenced and administrative action determined. But the business of very large assemblies is almost always decided 'off-stage'; and the agenda and resolutions are usually managed by the executive committee. We may conclude, therefore, that while Moscow has apparently solved the problem of popular participation to a greater extent than elsewhere, in doing so it has transformed the city council from a council of city fathers into a mass meeting. Much more thought and study must be given to this aspect of metropolitan government if genuine democracy is to be attained in the great city.

The two-tier structure possesses definite advantages for a great city quite apart from the increase of popular representation which it makes possible. In a city with a population of more than, say, half-a-million, the city council finds great difficulty in maintaining a close contact with the citizens. The town hall, the councillors and officials, seem to become remote and aloof: and the citizens see the city government as 'they' rather than 'we.' However good the intentions of the governors may be, the bureaucratic element tends to prevail as the size of the city becomes really large. This tendency is assisted by the increasing scope, magnitude, and complexity of municipal functions.

The introduction of a lower tier of secondary councils is a method of decentralizing power in respect of purely local matters to district organs. These organs are in a position to know much more than the city council about their own districts; and they should be able to keep in close and intimate touch with their local affairs. If well organized, and free from corruption and nepotism, they should be able to conduct purely local functions in a more efficient and democratic manner than the major council.

I have in an earlier book criticized the arrangements prevailing in London, especially the insufficient co-ordination and control exercised over the metropolitan borough councils by the London County Council.[1] Nevertheless, I believe a two-tier structure to be right in principle as a form of government for the largest cities, and it is somewhat surprising that Moscow is the only one which has so far followed the example of London.[2]

A two-tier form of organization is especially necessary in a metropolitan area. For this consists of an extensive region based on a great city, which forms the commercial, industrial, political, and cultural centre, surrounded by suburbs, dormitory towns, housing estates, villages, market-garden areas, and so forth. By whatever means the need for a unified government over the whole metropolitan area may be met—and we shall discuss this later—the need for the purely local government of all these diverse places will remain. It can best be satisfied by means of a two-tier organization conferring the greatest possible amount of local autonomy consistent with the well-being, good government, co-ordination and planned development of the whole metropolitan area.

A step in this direction has recently been taken in Toronto, although it falls short of a fully articulated two-tier system. The municipality of Metropolitan Toronto, which began to operate on January 1, 1954, consists of a metropolitan council responsible for certain services in an area comprising the city of Toronto and twelve contiguous municipalities. The services include the supply and wholesale distribution of water to the component municipalities, trunk sewers and sewage disposal plants,

---

[1] W. A. Robson: *The Government and Misgovernment of London* (2nd edition), pp. 94-8.

[2] The *arrondissement* in Paris is not comparable to the metropolitan borough in London, for it has only a mayor appointed and dismissible by the Minister of the Interior, and no elected council. The *arrondissement* is in no sense an area of self-government. The same applies to the districts into which Rome is divided, each with its *delegazione*. They are merely branch offices of the commune of Rome itself. In New York City, each of the five constituent boroughs has a borough president, who exercises certain functions within his respective area, but the boroughs do not have separate councils.

and major highways. Public transportation and town planning will also be dealt with on a metropolitan scale by *ad hoc* bodies. The twelve outlying municipalities have been separated from the county of York in which they were formerly situated, and the services previously carried out by the county have been transferred to the metropolitan council. These include the provision and upkeep of court houses and jails, the hospitalization of indigent patients, and post-sanitorium care of tubercular patients.

These changes have left untouched the existing local authorities, although they have been deprived of some of their functions. They will continue to carry out all the residual local services. The metropolitan council is constituted on the federal principle. Its twenty-five members consist of a chairman appointed by the provincial government, the mayors or reeves of the twelve smaller municipalities, and twelve representatives of the city of Toronto. The latter comprise the mayor, two of the controllers elected at large, and an alderman from each of the nine wards who gained the largest number of votes at the last election.

The Toronto reform is clearly a compromise. It makes no attempt to rationalize the lower tier of local authorities. The principal city remains an undivided unit of 675,754 population, while the outlying municipalities have populations ranging from 8,000 to nearly 100,000. Representation on the major authority bears no relation to population.

The allocation of functions is not based on scientific principles. Services requiring large-scale organization, such as the police and fire brigades, are left with the minor authorities in order to avoid increasing the tax rates in the outlying municipalities. On the other hand, some of the functions which the metropolitan council has inherited from York county, such as the provision and maintenance of homes for the aged, the maintenance of neglected children and of women committed to industrial refuges, could quite well be carried out by secondary authorities of reasonable size and resources.

Finally, metropolitan government cannot rest entirely on indirect election if it is to be satisfactory from a democratic standpoint, although a combination of direct and indirect election may have certain advantages.

It has been said of the Toronto experiment that it is 'not the prized product of the political scientist but rather, a victory in practical politics.'[1] In a sphere where so much is needed and so little has been done, the best should not be the enemy of the good; and the reform is undoubtedly

[1] 'Metropolitan Area Merges,' by Eric Hardy, in *National Municipal Review*, New York, July 1953, p. 363.

a move in the right direction. We may hope that in course of time
further improvements will take place in the reorganization of Metropolitan Toronto. Meanwhile, we shall watch with interest the new
system at work.

To return to the council. We will not trouble the reader with all the
minor variations in regard to such matters as the qualifications of candidates, the franchise, tenure of office, remuneration of councillors, and so
forth. These are not without interest; but little would be gained from a
summary of the different practices.

In order to understand the role of the city council we must consider
its relation to the executive power. There are fundamental differences
to be found in different city constitutions, and they are of crucial importance. We may preface our remarks by pointing out that in many
countries the separation of powers is applied in city government inasmuch
as the executive power is entrusted to an organ which is separate from
the council. Where this occurs, the council is frequently regarded as a
deliberative body which discusses policy, legislates by enacting ordinances
or by-laws, passes the budget, approves proposals submitted to it, and is
responsible for approving or making certain appointments. This is a
typical situation which is subject to many variations.

The constitutions of the great cities dealt with in this volume can be
divided into five main classes as regards the executive power and its
control.

5. THE CITY COUNCIL AS EXECUTIVE

This is the universal principle to be found among all local authorities
in Great Britain. It applies to the London County Council, to the metropolitan borough councils, and also to Manchester. The essence of the
system is that the council possesses full executive power within the limits
laid down by law, and has complete control over administration. This
control is exercised in practice by committees of the council dealing
with specified functions such as education or public health. The committees report to the full council, which can either accept their proposals,
reject them, or refer them back for reconsideration. The council will,
however, often delegate to a committee the power to decide specific
matters without referring them to the council for prior approval. Under
this system the council elects the mayor (or in a county the chairman of
the council) who presides over the council and occupies a position of
great dignity and civic prestige. But he possesses no executive authority
and does not control the administration.

In Manchester the city council elects the mayor annually. It appoints and dismisses the town clerk, the chief professional, salaried, non-political officer of the city. It appoints representatives to various joint boards such as the Port of Manchester committee. It also appoints co-opted members of the committees.

Every member of the council is expected to serve on at least three committees, and this may also involve service on sub-committees. The council has to approve the executive proposals of each committee. It passes the minutes of every committee at its monthly meetings, and the volume of business is so large that the council inevitably tends to become a report-receiving body rather than a place where policy originates. The standing committees have immense potential power both to control administration and to determine administrative policy subject to the over-riding authority of the council. In Manchester, we are told, 'the initiation of policy is now mostly in the hands of the officials, but the committee approve and take responsibility for it.' This is certainly not true in London and in many other cities in Britain, although officials rightly exercise a considerable influence on committees.

Finance is the prime responsibility of the finance committee, to which all other committees must submit their proposals both for revenue expenditure and capital projects. The ultimate authority over finance, as over all other matters, is the city council.

The system described above has been adopted in some of the Commonwealth countries. We find it applied in Sydney, except that there is no co-option on the municipal committees. The system also exists in Wellington, except that the mayor is directly elected every three years by the city at large. His position is, however, similar to that of an English mayor, and executive control is exercised largely through a number of functional committees.

6. AN EXECUTIVE APPOINTED BY THE COUNCIL

The nearest thing to the British committee system is the method of entrusting executive power to a committee appointed by the city council. The simplest case exists in Moscow, where there is no mayor. In that city an executive committee elected by the soviet is the executive and administrative organ. It contains sixty members, including a chairman, vice-chairman, and secretary. It is primarily responsible for planning and directing the work of the municipality, and full executive powers have been delegated to it by the soviet.

In Montreal, the city council appoints an executive committee, consisting of six of its members, and a chairman. This body is in charge of the city's administration, although certain powers of supervision are accorded to the mayor. It reports to the council on all matters within its jurisdiction; submits by-laws to the latter for approval; and prepares the annual budget and capital development programme. It appoints all heads of departments and approves the appointments of most other officials. It is responsible for enforcing the law and many administrative duties.

The mayor of Montreal is directly elected on a popular vote for a three-year term of office. He receives a salary of $10,000 a year. He is the first magistrate of the city, and represents the city on ceremonial occasions. He presides over the council and can submit recommendations to the executive committee and to the council. He is required to superintend, investigate, and control all departments of the municipality to ensure the laws are properly and impartially enforced; and he can suspend municipal employees. But his executive authority would appear to be circumscribed, not only by the executive committee but also by the director of departments, who is appointed by the council and who is a non-voting member of the committee. The director is responsible for city administration and keeps the departments in touch with the executive committee.

Copenhagen also comes into this category. In the Danish city the executive council (*magistraten*) is composed of a chief mayor (*overborgmester*), five other mayors (*borgmestre*), and five aldermen (*radmaend*). All of them are elected by proportional representation for eight years by the city council. Anyone is eligible who is qualified to vote for the lower house of Parliament. The present practice is to choose mayors from among members of the city council. A councillor who is elected to be a mayor or alderman must vacate his seat on the council.

The mayors and aldermen are the repositories of executive power in municipal affairs. They act partly through the executive council, over which the chief mayor presides, and partly through their individual departments or jurisdictions. The chief mayor and each of the other mayors is at the head of a department. There are six departments, and the municipal statutes lay down the distribution of functions among them. Thus, to the chief mayor's department belong municipal accounts, audits, wages, pensions, and local rates; he also acts as city treasurer. The third department, to take another example, embraces public assistance, social insurance, and the child guidance service. An alderman is allotted to each department.

B*

Each mayor has charge of all matters referred to his department. He can act independently and on his own responsibility and no appeal lies from his decision to the executive council. Questions affecting the municipality as a whole, such as by-laws or statutes and the annual budget; or matters which concern several branches of the city's administration; or which are of vital importance to the city or its economy, go to the executive council for decision. The executive council also deals with the appointment and discharge of officials, except the chief officers, who are appointed by the town council on the nomination of the executive council.

The mayors possess great executive power, both individually and collectively. It will be noticed that they are not tied to committees appointed by the elected council, which happens in some cities. The aldermen act as the link between them and the council, and seek to reconcile the bureaucratic and the democratic points of view. In the executive council the aldermen are in a position of equality with the mayors and their votes are of equal weight. The aldermen are largely occupied with committees on which they serve as representatives of the executive council.

Despite this concentration of power in the hands of the mayors and the executive council, Dr. Holm assures us that 'the centre of gravity in the administration of Copenhagen lies not only nominally but actually in the town council.'[1] The city council elects the mayors and aldermen. It appoints ten standing committees and any others which may be required. It appoints the highest municipal officials on the nomination of the executive council. It alone can grant money or approve the raising of loans.

Stockholm is another Scandinavian example of the same species, though it differs in several important respects from Copenhagen. In the Swedish capital, the city council elects a board of aldermen (*stadskollegiet*) consisting of twelve members of the council. Their duty is to prepare all business going to the city council and also to exercise general control over the municipal administration. This town board is the central authority of the city and wields greater power than might appear from the city charter, by reason of the fact that all the party leaders on the council are members of it. In consequence, its proposals are generally accepted by the council. The aldermen, like the councillors, are unpaid.

Stockholm has no mayor, but has instead eight city directors elected by the council for four years. They must not be aldermen but they may

[1] *Post*, p. 234.

be councillors. They receive the highest salaries paid in the Swedish public service—considerably more than those paid to Ministers of the Crown.

Each director is in charge of a municipal department. The directors are not, however, simply officials but occupy a politico-administrative position more akin to that of municipal ministers. They participate in the meetings of the city council and of the aldermanic board, to which they report on departmental matters. They also form a board of their own, presided over by the director in charge of finance.

The city directors possess great power, but they do not exercise undivided authority over their departments. The most important decisions are taken by separate administrative boards elected by the city council to take responsibility for the various spheres of municipal activity. Each of these administrative boards is made up of members nominated by the political parties in proportion to their relative strengths on the council. Members of a board may be councillors, but a great proportion of them are drawn from outside occupations. They bear some resemblance to the committees of a British municipality, except that a director generally acts as chairman of the board or boards[1] which deal with his department. The point to notice is that a director cannot dictate the decision of an administrative board, if the members are opposed to his opinion. On the other hand, an administrative board has nothing like the political influence of the aldermanic board, and therefore cannot rely on the city council accepting its recommendations. Nevertheless, an administrative board exercises a considerable amount of independent administrative power, as we should expect in Sweden, which is accustomed to give a large measure of autonomy to boards at all levels of government. Hence, the administrative boards can take decisions which neither the city council nor the board of aldermen can affect, except in the long run by the replacement of members whose term of office has expired, or by control over finance.

This structure of innumerable boards is typical of Swedish local government, and by no means clear or coherent. Underlying the system, we are told, lies a deep distrust of municipal bureaucracy and the desire to secure the participation of a considerable number of citizens in the administration of local affairs.

The intention is that powers should be divided between the city council on the one hand, and the aldermanic board, the city directors and the

---

[1] There are usually several boards for each department. Thus, in the public welfare department there is a poor law board, a child welfare board, and an unemployment relief board.

administrative boards on the other. In practice, the board of aldermen acts largely as a committee of the council. In theory, policy should be decided by the city council and administrative matters determined by the other aut orities. In fact, many of the council's decisions, including the budget, are based on the advice of the administrative boards.

The system of city directors, which is peculiar to Stockholm, has been under discussion from time to time. There is some controversy about whether the combination of political and full-time administrative posts is desirable in a city government. A similar combination existed in Germany under the former *bürgermeister* system, which in our opinion had many disadvantages.[1] It has now been replaced by a system which renders the burgomaster a purely political and civic personage. His former administrative responsibilities have been transferred to a *stadtdirektor*. It is also found in another form in the American 'strong mayor' towns which we shall presently describe. In Britain, local government is based on the assumption that a sharp dividing line should be drawn between the political or democratic element, consisting of the elected representative, and the bureaucratic or administrative element, consisting of the salaried professional officials; but this is difficult to establish elsewhere where a different tradition prevails. Another question which has been raised in Stockholm is whether it would not be better to replace the present method of electing city directors according to political party representation on the council by a responsible cabinet system, whereby the political party with a majority on the council would nominate all the directors.

The municipality of Rome falls under the same heading, although the organization is far simpler than in Stockholm. The city council is the fundamental organ of the municipality; and its approval is required for all important questions involving policy. It passes the budget, and must approve the estimates of all administrative agencies to which the commune contributes money. It supervises all institutions created for the benefit of the citizens.

Executive power is confided to an executive committee or *giunta*, composed of the mayor and 12 councillors (or *assessori*), together with 2 substitute members, who are elected by the council. The mayor is also elected by the council from among its members. He is head of the municipality and presides over both the council and the *giunta*. He signs all municipal contracts, when they have been authorized by the council,

[1] See William A. Robson: 'Local Government in Germany,' in *Political Quarterly*, Vol. XVI, No. 4 (October–December 1945), p. 277.

and is responsible for enforcing all municipal regulations and by-laws. He inspects municipal offices and institutions. But he does not exercise power independently of the *giunta* and the council. Except in case of an emergency, the mayor is always subordinate to these bodies. In a few matters he acts as an officer of the central government, e.g. as registrar for births, deaths and marriages in the municipality.

Amsterdam has a municipal executive which combines an aldermanic element elected by the city council with a mayor appointed by the central government. The system contains some features displayed by the munici-palities dealt with in this section—i.e. those in which the executive is appointed by the city council; but on balance we consider that Amsterdam falls into the category of cities whose executive organ is appointed by the central government. We shall therefore discuss it under that heading.

7. THE ELECTED MAYOR AS EXECUTIVE

A third type of constitution to be found among great cities is that in which a directly-elected mayor possesses the executive power. This system embodies the twin doctrines of the separation of powers, and popular choice of the executive. It is much favoured in the United States, particularly in the larger cities which have not adopted the city manager plan.

New York is an example of the 'strong mayor' form of city govern-ment. The mayor is elected on a popular vote and is the chief executive officer. He appoints and can remove all heads of departments, commis-sioners, and non-elective officials, and is regarded as responsible for their acts. He is responsible for enforcing the law within the city limits. The administration of the city is under his control.

The powers of the council were greatly reduced when the present charter was adopted in 1938. To the council belongs legislative power of the city, but legislation must also be passed by the mayor and the Board of Estimate. In consequence, the council's function tends to be chiefly that of imposing a veto.

The mayor has to work with the Board of Estimate. This body consists of the mayor, comptroller (also a directly elected official), the president of the council, and the presidents of the five boroughs which make up New York City.[1] The voting is unequal, for the mayor (or in his absence the deputy mayor), the comptroller, and the president of the council have three votes each, while the presidents of Manhattan and Brooklyn

[1] The presidents are elected in their respective boroughs for a four-year term of office.

boroughs have two votes each, and the presidents of Queens, Bronx and Richmond boroughs have only one vote each. According to the charter, the Board of Estimate may exercise all the powers vested in the city which are not specifically allocated by law to other authorities. In practice it acts as a general purposes committee for determining policy in regard to the city's affairs, local assessments, etc. It fixes the remuneration, pensions and retirement of municipal employees and lays down the personnel establishment. It prepares the city's annual revenue budget. The mayor has usually been able to dominate the Board of Estimate, sometimes through the force of his personality, which was notable in the case of Mayor La Guardia, but more often through sheer voting strength. The comptroller and the president of the council usually run with the mayor in the municipal elections, and as the leading political figure in the city he is naturally able to exert considerable influence on his party associates. If the mayor can secure the votes of the president and the comptroller, he is in a position to control the Board of Estimate. He is often able to do this through his ability to build up a strong political machine.

The constitution of the city does not concentrate executive power exclusively in the mayor. Thus, fiscal affairs are supervised by the comptroller, who is separately elected. The charter also provides for a planning commission, consisting of the chief engineer to the Board of Estimate and six members to be appointed by the mayor for staggered terms of eight years. The mayor designates the chairman, who becomes head of the department of city planning. The principal function of the commission is to prepare a comprehensive master plan for the city. It also prepares and submits to the mayor each year a budget of capital expenditure by the city. Furthermore, no improvement contrary to its recommendations may be included in the general revenue budget, prepared by the Board of Estimate, except by a three-quarters majority in the council.[1] These constitutional provisions do not, however, prevent the government of New York City from being centred in, and dominated by, the mayor. It therefore falls into a different class from cities like Copenhagen, Montreal or Rome, where the executive power is in the hands of a committee of some kind appointed by the council, and of which the mayor is a member or chairman.

The two other United States cities dealt with in this volume have constitutions which differ from that of New York, but both in Chicago and

[1] See W. A. Robson: *The Government and Misgovernment of London* (second edition), pp. 465-8.

Los Angeles the position of the mayor is essentially similar to that of the mayor of New York City. In Chicago, however, the mayor is not only the chief executive and head of the municipal administration: he is also the presiding officer of the council. His constitutional powers are reinforced by his political power as party boss. In Los Angeles the mayor possesses very large ostensible authority, but in practice he exercises little direct power over the departments, which are headed (as in Chicago) by boards or commissions of citizens. The power of any one individual to control effectively a large and complex organization like a city government must always be limited by the sheer impossibility of his having sufficient time and energy to keep his eye on more than a few of the multitudinous matters to be dealt with each day. This is especially true in a democratic country where serious blunders caused by hasty or ill-considered intervention from the top executive will not be lightly tolerated and will evoke immediate criticism and resentment. In Los Angeles, it is said, the council is inclined to meddle overmuch in the administration. This assumes that the council should keep aloof from the executive side of the city government—an assumption which would be entirely unacceptable in Britain and other countries which have adopted the British committee system.

8. AN ELECTED COMMITTEE AS EXECUTIVE

A fourth type of city constitution is one in which, instead of an elected mayor wielding executive power, we find a small group of individuals being elected to the highest executive position by the direct vote of the citizens.

In the city of Toronto[1] (not to be confused with the recently created municipality of Metropolitan Toronto), a board of control has general responsibility for the administration. This board consists of the mayor, who is elected each year by the city at large, together with four other members elected in a similar manner. The board of control prepares the annual estimates, supervises revenue, expenditure and investment. It awards municipal contracts and considers by-laws. It recommends to the council appointments to the heads of departments and sub-departments; it can also recommend the dismissal of chief officers. The city council in Toronto possesses power to over-ride the board of control in several respects by a two-thirds vote: for example, by such means it can make appropriations or authorize expenditure not recommended by the

---

[1] For the new metropolitan authority in Greater Toronto see *post*, pp. 376-7.

board; and it can refuse to sanction the dismissal of a senior official. A two-thirds vote in the council is necessary to confirm the removal of a municipal auditor, or to award a municipal contract. In Toronto we see a mixture of the American predilection for a separation of powers and a directly elected executive, with the British preference for entrusting authority to a committee rather than to an individual.

A similar system is found in Zürich. There a board of commissioners (*stadtrat*), consisting of nine members, is directly elected for a term of four years, and this constitutes the executive organ. The mayor, who is also directly elected for a similar term, presides over the board of commissioners. The other members, but not the mayor, are given individual authority in specific fields of activity as heads of departments. But in principle all important decisions must be made by the *stadtrat*. Where individual members act independently as heads of departments, they are often assisted by expert committees whose advice they can seek.

9. AN EXECUTIVE APPOINTED BY THE CENTRAL GOVERNMENT

The several forms of government which we have so far considered are those in which executive power is given either to the elected city council, or to an organ elected by the council, or to a person or body elected by the citizens. These are all different types of local self-government.

We now turn to the great cities in which executive power is given, wholly or partly, to an officer or organ appointed by a higher authority. This higher authority is either the central government, or where a federal constitution is in force, the state or national government. It is obvious that cities which are governed wholly by such means do not enjoy local self-government; while those which are governed partly by such methods do not enjoy full self-government. For this reason they must be distinguished from those which we have examined above. It will be found, however, that in some instances an elected council is associated with a centrally-appointed municipal executive. Where this occurs, the character of the régime depends on the balance of forces between these two elements.

We have already explained that the executive system in Amsterdam is of a mixed type.[1] The constitutional feature which has caused us to classify Amsterdam with the great cities whose executive is appointed

---

[1] *Ante,* p. 45.

by the central government is the arrangement whereby the mayor or burgomaster is appointed by the Crown for a period of six years, on the recommendation of the Cabinet council. He can be re-appointed for a further period of office. The mayor presides over the city council, and he can in law be a member of it, but in practice this rarely happens. His right to vote depends on whether he is a member of the council, but he can always participate in the deliberations of the council.

The mayor is chairman of a body known as the Board of Burgomaster and Aldermen, which controls the executive machinery of the city. This board consists of the burgomaster, who presides, and six aldermen, who are elected by the councillors from among their own ranks. The aldermen are chosen from the political parties in accordance with their relative strengths on the council. Thus, all the members of the board except the mayor are appointed by the municipal council, and must be members of the council.

The aldermen receive full-time salaries, and it is customary to entrust each of them with the direction of an important branch of the administration, such as education, finance, or public works. Strictly speaking, the aldermen have no legal authority either individually or collectively, except when they are sitting with the mayor as the Board of Mayor and Aldermen.

It would be wrong to regard the division of powers between the council and the board as based simply on the distinction between 'legislative' or deliberative functions on the one hand and administrative functions on the other. 'Both in the field of legislation and in that of administration, the council's powers are primary,' writes Mr. Oud, a former Mayor of Rotterdam and President of the International Union of Local Authorities.[1] The law confides all authority in these two fields to the council, except where the Municipal Act expressly confers powers either on the Board of Burgomaster and Alderman, or on the mayor personally. The council may, of course, delegate some of its powers to the board or to the mayor. The board is chiefly concerned with day to day administration, and in particular with executing the decisions of the council. Moreover, the law has conferred on the Board of Burgomaster and Aldermen numerous powers in the exercise of which it is legally independent of the council. The entire machinery of administration is under the control of the Board of Burgomaster and Aldermen. Neverthe-

[1] In a paper on 'The Netherlands Burgomaster' prepared for the Congress of the International Political Science Association held at The Hague, September 8–12, 1952. Printed in *Public Administration*, Vol. XXXI, Summer 1953.

less, as Mr. Oud explains, the board is answerable to the council for all they do or fail to do.[1]

The mayor is at once the representative of the city and an agent of the central government. Apparently no one, not even a minister, can give him orders, though he can be dismissed by the Crown. He is in no sense an official either of the state or of the municipality, but occupies rather the position of a public officer who takes his instructions only from the law itself. What is important is that he possesses a large and increasing range of powers which have been conferred upon him in his individual capacity, and for these he is not accountable to the council. He is head of the municipal police force; he has many administrative functions vested exclusively in him, especially those relating to public order. Even in regard to some matters in which the council retains certain regulatory powers, such as theatres, public houses, etc., it is the mayor alone who is charged with the task of enforcing them. The office of the mayor of Amsterdam thus partakes of that 'dual capacity' which is found so frequently in the countries of continental Europe. A surprising feature is the way in which this centrally appointed mayor is regarded as an instrument of 'self-government' even when he acts in his independent capacity quite apart from the council. In conferring such powers upon the mayor, Mr. Oud declares, matters have been carried much further than is consistent with the aim and nature of so-called 'self-government.'

It must not be thought that the mayor, though appointed by the Crown, is in effect the council's nominee. This is quite an erroneous supposition. The municipal council is excluded from exercising any influence on the appointment; and it happens not infrequently that the council is displeased either with the choice of the individual or dissatisfied with his political views. The mayor is not politically neutral; and his views may upset the political balance on the Board of Burgomaster and Aldermen.

In Rio de Janeiro the mayor is also appointed by the national government. His appointment and dismissal rests with the president of the republic, subject to approval by the senate. The mayor of Rio is described as 'the president's man.' He wields executive power in the city, and he also has authority to promulgate laws and to enact decrees, regulations and instructions. He prepares the budget for submission to the council.

The mayor of Rio de Janeiro has much greater personal power in his official capacity than the mayor of Amsterdam. Unlike the latter, he does not have to share his power with a collegiate body such as the Board of

[1] *Ibid., Public Administration*, Vol. XXXI, Summer 1953.

Burgomaster and Aldermen. The only exceptions to this are in the realm of finance. The council must approve the budget, and there is a body called the Financial Board of Review, which supervises certain aspects of finance. This board consists of seven councillors appointed for life by the mayor. Its duties include the examination of contracts which impose financial burdens on the city, or which involve municipal loans, etc.

The mayor is in a position of superiority to the council, for he can veto the latter's bills and decisions. In 1946 the council lost its former power to over-ride the mayor's veto, and in consequence its responsibility and authority have visibly diminished. A veto by the mayor can now be referred on appeal to the senate for a final decision. The mayor has to report to the council, to render accounts to them, and to provide any information they may require. In the councillors' last year of office they fix the mayor's salary for the forthcoming period.

An essential feature of the situation is that the mayor, as a presidential appointee, feels himself to be in a sufficiently strong position to govern the city without paying much attention to the city council, and he does not hesitate to veto the council's decisions. His executive powers are very extensive. He directs and supervises the public services; appoints, promotes, dismisses or retires public servants; collects the revenues of the municipality and applies them on expenditure authorized by law; conserves and manages the public property belonging to the city; promotes town planning and submits planning schemes to the council; requisitions property; borrows money in anticipation of revenue; enters into contracts on behalf of the city; and represents it in the courts. He controls the administration through a secretary-general and five general secretaries who are in charge of the various departments.

In Buenos Aires the system of government is entirely undemocratic, for in the Argentine capital there is no elected municipal assembly at all. The mayor is appointed, as in Rio de Janeiro, by the president of the republic; but he exercises unconditional power and is neither elected nor responsible to an elected body. An organic law of 1882 provided for the constitution of a municipal council with considerable powers of deliberation, policy-making, legislation and finance. The council, after many vicissitudes, was abolished in 1943.

Bombay has a representative city council which each year elects the mayor from among the councillors. The mayor presides over the council meetings; he is the first citizen of the city and performs many ceremonial functions. But, like his counterpart in Britain, he has no special powers and does not control the executive. The régime in Bombay provides for executive power to be concentrated in the municipal commissioner, an official appointed by the state government for a three-year term of office, subject to renewal. The commissioner controls the municipal civil service; he attends meetings of the council, though he cannot vote; he appoints all junior officials—those earning over a specified amount are appointed by the council.

The city council can remove the municipal commissioner from office by a five-eighths vote of all the members of the council. This provision is the key to understanding the relationship between the state-appointed commissioner and the municipal assembly, for it gives the city council a supremacy over the commissioner in the last resort, if friction develops and an *impasse* is reached. Nevertheless, though the representative organ ultimately prevails in a conflict of wills, of policy, or of personalities, we cannot regard Bombay as fulfilling the essential conditions to qualify as a self-governing city: namely, that not only the deliberation of policy, the passing of ordinances, and the control of finance, shall be within the ambit of an elected council, but also that executive power shall belong either to the council, or to an organ appointed by the council, or to officers directly elected by the citizens.

In Calcutta a similar system is in existence. The entire executive power is in the hands of a municipal commissioner, who is appointed by the state government on the recommendation of the State Public Service Commission. He holds office for a period of five years, but the corporation can by a majority pass a resolution asking for his removal. Apart from that they have no control over him.

In Paris, executive power is vested in the prefect of the Seine and the prefect of police, who are administrative agents of the central government. Like all other prefects in France, they are appointed and dismissed by the Minister of the Interior, and are answerable to him. Their power is enormous and the Ville de Paris has neither an elected mayor nor any other executive organ. The prefect of the Seine exercises powers which in other French towns would belong to the municipal council.

As a consequence of all these centralizing tendencies, the municipal council of Paris is in a relatively weak position. Its decisions are binding on the prefect only when they deal with such matters as municipal property, or the contribution of the capital to state work of interest to the city. In regard to a host of questions, any resolutions which the council may pass are liable to be over-ridden by the prefect. For instance, on such matters as the budget, loans, the organization of municipal services, or highway development, decisions of the council can be vetoed by the prefect of the Seine, the Minister of the Interior, and perhaps also by the Minister of Finance. Even these conditional decisions apply only if the proposed expenditure is above a certain figure. Below this minimum, the prefects determine their own administrative policy, and the role of the council is confined merely to asking questions and demanding information. Thus, almost any project put forward by the municipal council can be turned down by the central authorities. On the other hand, the latter cannot substitute their own proposals for those which they have disapproved, and hence the initiative still rests to some extent with the elected council.

The prefect of police controls a police force which is almost autonomous. The prefect of the Seine controls all the municipal services, draws up the budget, settles the council's agenda, and prepares proposals for debate in the council. He can insert in the budget items of expenditure which the council has refused to pass, if they are required by law; but he cannot compel the council to vote money for permissive services. Most overwhelming of all is his power in certain circumstances to suspend the meetings of the council for a period up to three months, or to suspend its activities altogether, or to dismiss councillors from office. The prefect would not, of course, exercise such arbitrary powers except in the most serious circumstances, and no doubt only with the support of the ministers concerned; but such considerations touch only the question of expediency, not of the distribution of power between the central government and the municipality. The distribution is, indeed, vastly unequal, and Paris cannot be regarded as in any real sense a self-governing city. The bureaucratic and centralized character of prefectoral power is, indeed, so manifest, that the municipal council can sometimes influence the prefects by mobilizing the force of public opinion in support of their views. Hence, though from a formal standpoint the municipal council is little more than a consultative body, the very weakness of its legal or constitutional powers sometimes enables it to invoke the force of public opinion against the government and the prefects.

It is a disturbing fact that in no fewer than six of our great cities the executive organ is appointed wholly or partly by the central government. This group of metropolitan areas, comprising Paris, Rio de Janeiro, Buenos Aires, Bombay, Calcutta and Amsterdam, is the largest category in the classification we have attempted above. Here we have a group of famous cities, situated in Western Europe, in South America, and in Asia, where the executive authority is not entrusted to councillors elected by the local body of citizens, or to a person or persons chosen by them and responsible to them, or to a mayor or other high officer voted into office. How comes it that in these vast metropolitan cities with their millions of inhabitants, their high cultural attainments, their relative economic prosperity, their proud history and traditions, their busy industrial and commercial life, the democratic spirit burns at so low an ebb that the aspiration to govern itself which has inspired every great city since the days of ancient Greece, has not been achieved? This indeed affords food for reflection, especially if one believes, as the writer does, that without successful self-government in the local sphere, a country is unlikely to attain a satisfactory level of self-government at the national level.

We should be fully aware of the fact that the cities referred to above present vast differences not only in the formal machinery of government but also in the spirit which informs its working. In Amsterdam, for example, the city council exercises a most important influence on the entire organization of the municipal administration, and everyone is familiar with the political maturity and democratic outlook of the Dutch people. Moreover, the mayor is associated with a group of aldermen appointed by the city council to the Board of Burgomaster and Aldermen. There is a vast contrast between Amsterdam, whose constitution contains many of the elements of self-government, and which operates within a democratic national régime, and Buenos Aires, where power is asserted in arbitrary and degraded forms and the true spirit of democracy is at present lacking at all levels of government—federal, state and local.

But when full allowance has been made for these differentiations, it is impossible to regard a city in which the executive power is possessed by an organ appointed by higher authority, as enjoying democratic local government.

The explanation of the position lies partly in the immense increase in the importance of the great city both in the national economy and in the national polity. The sheer size of the modern megalopolis—especially if it is a capital—its preponderance of resources of all kinds, has made the state regard it as a potential threat to the central government.[1] Indeed, the possession of Paris has in the past often determined the fate of the French régime; while Mussolini's so-called 'march on Rome' was a decisive act in his seizure of power, despite the fact that he made the journey in a railway sleeping compartment. Historical facts and contemporary fears combine to make national governments anxious to keep control of the government of the capital city; and this desire expresses

[1] W. J. M. Mackenzie: *op. cit.*, p. 13.

itself most frequently in the direct or indirect retention of executive authority. The city council, if there is one, may be permitted to do the talking; but a mayor appointed by a minister or the head of the state, better still a prefect, will in time of emergency do what he is told by the government of the day.

10. MUNICIPAL SERVICES

For what functions is the municipal government of the great city responsible? This is obviously a question of large importance to our enquiry, and one would like to be able to give a reply which would be at once comprehensive and simple. Unfortunately, it is almost impossible to give such an answer. One reason is that the contributors of the individual chapters, despite urgent and repeated entreaty by the editor, have not always produced the necessary information, or they have not given it in sufficient detail. A second reason is that in many of the cities some functions are carried out by independent or semi-independent boards which are often linked in one way or another with the city government, and it is exceedingly difficult to decide whether such functions should or should not be regarded as performed by the city government. To rule out all such functions would often be to give a totally misleading picture of the city government. A third reason is that while about eighteen or twenty services are commonly carried out by the municipalities of great cities, and about another eight or nine are found in several of them, there is a wide range of functions which are peculiar to a particular city, and are not to be found elsewhere.

The most widespread services administered by the municipalities of great cities are public health, hospitals, city planning, water supply, sewerage and sewage disposal, public cleansing, schools, highway construction, street maintenance and lighting, public assistance, welfare, police forces, fire fighting brigades, the provision of public housing and housing regulation, parks and playgrounds. These services form the central core of local government, but all of them are not everywhere entrusted to the municipal government of the great city.

A second group of services comprises those of a public utility character; or, as they are often called in Britain, municipal trading services. They include the supply of water, gas and electricity; the operation of street transport services within the city and its environs; wholesale food markets, municipal slaughterhouses or abattoirs; a port or harbour undertaking where one exists; with sometimes an airport as a modern addition. These services are less frequently the responsibility of the city government than are those of the first group. They are, or can usually be made to be, financially self-supporting or even profitable. In consequence, municipal enterprise in the trading services is often rejected in favour of a commercial company operating for profit and regulated by public utility legislation or bound by the terms of a concession. Another alternative to municipal enterprise is an *ad hoc* authority, particularly the public corporation, which has become extremely popular in one form or another for public utility undertakings requiring management of a commercial or industrial kind.[1]

Direct municipal ownership and administration is in my opinion the most satisfactory method of operating public utility services in the great city, if certain conditions are satisfied. First, the local government must cover a sufficiently large part of the population of the metropolitan area to enable the services to be provided on a basis of economy and efficiency. Second, the city government must be democratic, honest and able to provide capable management. Third, public utility services should not be run primarily for profit to relieve taxation, but in order to provide good service at the lowest cost to the citizens. Well run municipal enterprise is far better than the concessionaire company, which all too often is unwilling or unable to replace obsolete plant and equipment because of the uncertainty of its future position in relation to the undertaking. Whenever one sees in a city antiquated, obsolete tramways or motor buses, one can usually assume that they are owned by a company which is nearing the end of its franchise. Municipal enterprise, at its best, can be superior to the public corporation running a nationalized industry. For the opportunity for effective democratic control over the undertaking is direct and immediate when it is run by the municipality. The voters, who are also the consumers, know that responsibility for the service rests with a body (the city council, the mayor, etc.) which they can influence not only by their votes but also by their protests and representations. A public corporation, independent board, or commission, is a much more remote body to the ordinary citizen. The man in the street, and still more the woman in the home, is scarcely aware of its existence; and its non-elective directors are remote from any living contact with the common man.

These two groups of services do not exhaust the functions of the city government. There is, in addition, a great variety of functions which are peculiar to individual cities and which scarcely reveal a common pattern. In the sphere of recreation and entertainment, for example, one

[1] See W. A. Robson: *Problems of Nationalized Industry* (1952), London, George Allen and Unwin.

finds municipal art galleries, museums, theatres, opera houses, concerts, playing fields, tennis courts, and swimming pools. One finds municipal universities in some cities; municipal pawnshops, savings banks, laundries, restaurants and hotels in others.

II. THE NEED FOR INTEGRATION

The task of providing a great city with the major services required by a civilized community has produced immense problems of a technical, administrative, and financial character. The teeming millions of Paris, New York or London expect to be provided with an adequate water supply, an efficient main drainage and sewage system, highways adapted to modern traffic needs, and housing accommodation on a continually rising standard. It is becoming increasingly difficult to provide these and other basic services like education on a scale which will satisfy their ever-growing demands.

A special difficulty which confronts the great city is the need to provide services for the army of workers of all classes and occupations which pours into the city on every week-day. These hordes of workers in offices, shops and factories require water, drainage, police protection, highways, fire brigades, traffic and transport facilities, food regulation and inspection, and many other costly services; but they reside outside the boundaries of the city, and in consequence escape wholly or in large part the taxes levied by the city government, in particular, those which fall on the houses or flats in which they live. This phenomenon has intensified the financial problem of the great municipality. It has also aroused the desire of the city government to extend its boundaries in order to incorporate the outlying suburbs, towns, villages, housing estates, and other places within the metropolitan area.

There are many technical or administrative reasons connected with municipal functions which support and reinforce the desire of the great city on financial grounds to absorb the neighbouring areas outside its boundaries. Town and country planners, for example, are agreed that today a great city should not be planned in isolation from the surrounding suburban, rural, or semi-rural hinterland. Accordingly, a development plan should cover the whole metropolitan area.

Technical education is a service which requires for a high standard of achievement, a wide range of differentiated institutions containing expensive equipment and highly trained specialist staffs. These can only be provided by a large local authority disposing of great resources. Where the great city provides technical education, people living in the outlying districts will seek to benefit from the superior facilities afforded by the large municipality, particularly if they work in the city. Where they are permitted to do so, the result is usually to the financial disadvantage of the large city, since such services are often subsidized out of local taxation. Moreover, technical education should be carefully related to the needs of local industry or commerce; and the requirements of outlying areas may differ from those of the central municipality. They may not be taken into consideration unless the entire metropolitan community is treated as a whole for this purpose.

To supply a great city with water often means bringing supplies from long distances at great cost; while to provide it with main drainage will often mean transporting vast quantities of sewage to large disposal plants or outfalls. Sewage works and main drainage systems require high capital expenditure, great engineering works, and skilled personnel far beyond the reach of small local authorities. The technical reasons for treating the entire metropolitan community as a single entity in respect of such services are very strong. Similar considerations apply to main highways and bridges. The system of public transport services within the region should clearly be organized as a whole. It is not simply the travelling needs of their own citizens which must be considered by the government of the great city, but also those of the entire metropolitan community of which it forms the traffic focus and radiating centre. There are several other services which should also be planned and administered over the entire metropolitan area, such as hospital services, fire brigades, public libraries, parks and large open spaces, the distribution of gas and electricity, and certain health services requiring highly specialized organization.[1]

We have already pointed out that nowhere, with the possible exception

[1] For other examples, see Albert Lepawsky: *Urban Government* (National Resources Committee, Washington, 1939), Vol. I, pp. 29–30. Some of these refer to regulatory and police functions. Thus, Professor Lepawsky writes: 'While metropolitan life overflows the artificial network of urban boundary lines, each little bailiwick of government preserves its independent island of authority, with odd results. The city of Evanston, Chicago's suburban neighbor on the north, finds that its own automobile inspection system cannot solve all of its traffic accident problems, since almost all of its automobile fatalities and half of its other motor accidents involve drivers or vehicles from places outside the city. Metropolitan mobility is such that the majority of defendants in misdemeanor cases tried before some of the most important suburban courts are Chicagoans, not local residents. Criminals hop over jurisdictional lines which local police dare not cross without elaborate devices for administrative co-ordination which are only now beginning to develop. Germs and contagious diseases recognize no borders, but only seldom are local officers free to pursue the exigencies of public health outside their own jurisdiction.' See also Thomas H. Reed: *Metropolitan Areas*, in Encyclopaedia of the Social Sciences, X, p. 397.

of Toronto, has a serious attempt been made to provide the metropolitan community with a system of government designed to satisfy modern needs in regard to organization, planning, co-ordination, finance, etc. The clearest evidence of this is to be seen in the fact that in no instance have we been able to discover a municipal government covering, for the major services, the metropolitan area, or serving the metropolitan community living in that area.

The nearest attempt in recent times is the Municipality of Metropolitan Toronto, which provides a major authority for the City of Toronto and a dozen neighbouring municipalities outside its boundaries. But in 1949, a committee set up to study the problem of Greater Toronto defined the metropolitan area as including ten municipalities outside the area covered by the new metropolitan municipality.[1] Nevertheless the situation in Toronto is greatly improved and is much better than in Montreal, where there are nearly 60 municipalities in the metropolitan area. The Chicago metropolitan area contains no fewer than 821 governmental units, of which 593 are school districts, 115 municipalities, 42 townships, and 5 counties. Greater New York contains about 550 cities, towns and villages in the metropolitan area. In Sydney there are 33 municipalities and 5 shires in the metropolitan area. In the region centred on Manchester there are 14 outlying local authorities, and 72 local authorities within a radius of 15 miles of the city. Outside the Ville de Paris there are 80 communes in the department of the Seine; and the metropolitan area of the French capital spills over into many other communes in the departments of Seine-et-Oise and Seine-et-Marne. Rome is surrounded by about 100 suburbs each of which is a commune with its separate organ of local government. In Denmark, the adjoining municipalities of Copenhagen, Frederiksberg and Gentofte are recognized as 'the capital.' But the true metropolitan area includes a total of 19 suburban municipalities administered by parish councils. Greater Stockholm comprises, outside the city proper, 5 other independent city councils and 6 or 7 boroughs and townships, each with their separate local authorities. There are in Outer London, outside the area of the London County Council, 3 county boroughs, 35 municipal boroughs, 30 urban districts, 4 rural districts and 6 parish councils, together with several counties, making a total of over 80 local authorities. In Los Angeles the city is largely coterminous with Los Angeles County and Orange County. Within the former are 44 incorporated cities and within the latter 13 incorporated

---

[1] 'Metropolitan Area Merges,' by Eric Hardy, in *National Municipal Review*, New York, July 1953, p. 330.

cities; there are also 30 unincorporated urban areas dependent on the county for services, and numerous suburban special districts partly providing their own services and partly receiving those provided by the county. In Wellington, Johnsonville Town District is the only ancillary authority close to the main city; but there are 7 other authorities scattered over the neighbouring countryside which can be regarded as within the metropolitan region. These examples serve to show how widespread is the situation we have described.

12. THE ATTEMPT TO EXPAND

The expansion of the principal municipality to cover the metropolitan area is a development which might appear at first sight to be the most obvious solution to the problem of metropolitan government. As a general proposition, the principle that political and governmental institutions should expand in order to keep pace with the enlarged scale of human activity, is incontestable. Among smaller units of local government, the principle often prevails. When, however, attempts are made by great cities to extend their territories, their efforts meet with such fierce resistance that this method has proved of small use as a means of providing metropolitan areas with appropriate organs of local government. New York City has annexed no territory for more than forty years; Philadelphia is only one-tenth of a square mile larger than it was in 1854; San Francisco is the same size as it was in 1856; Chicago occupies the same territory as it did in 1889.[1] The London County Council has remained since its creation in 1888 within the area assigned to the Metropolitan Board of Works in 1855.

There are only a few exceptional instances of success in this direction. In Bombay, for example, the policy of boundary extension, first advocated in 1918, was adopted in 1950 by the amalgamation with the city of a number of neighbouring communities. This assures Bombay of sufficient land to meet all its housing requirements that can at present be foreseen. Manchester, after repeated efforts, at last induced Parliament in 1930 to extend the city's boundaries so as to include land now occupied by the garden suburb of Wythenshawe. Sydney, by invoking statutory powers enacted in 1948, secured the amalgamation of eight municipalities in the industrial suburbs adjoining the city. A similar number of communes were brought within the City of Zürich in 1934. Los Angeles was an outstanding case of large-scale territorial expansion, especially in the years

[1] Albert Lepawsky: *Urban Government* (National Resources Committee, Washington, 1939), Vol. I, p. 33.

between 1913 and 1925; but this was due to the difficulty of obtaining
water in a semi-arid region. Los Angeles was drawing its water supply
from areas more than 150 miles away; and this involved expenditure
on a scale beyond the resources of any save a very large and wealthy
city. In order to share in the benefits of this water supply, outlying areas
were ready and even anxious to come within the territory of the city;
and many of them actually applied for annexation. Nevertheless, more
than 2 million people lived outside the city limits in 1950.[1] Moreover,
there are in the Los Angeles area some thirty islands of unincorporated
territory which are surrounded by the city of which they do not form
part. These unintegrated fragments defy all attempts at amalgamation
with the city. (See the relief map reproduced on plate number 21.)

There are only a few rare instances where a great city has succeeded
in expanding its boundaries to include the metropolitan community. 'The
large American cities,' writes Dr. Victor Jones, 'have never been able
to keep pace, by means of annexation or consolidation, with the accumulation of population on the margin of the city.'[2] The prospects today are
less favourable even than in the past that the process of amalgamation
will bring any substantial part of the metropolitan population under the
jurisdiction of a true metropolitan government. There is, indeed, a
continued proliferation of small local authorities in the outlying parts of
the principal metropolitan areas of the United States. In 1940, there were
1,103 incorporated places within the 17 larger metropolitan areas, an
increase of 71 over 1930, and 240 more than existed in 1920.[3]

A similar situation is reported from Australia. 'It is unlikely,' observes
Dr. J. R. H. Johns, 'that the problems of metropolitan local government
in Western Australia will be solved, at least in the near future, by either
a Greater City Movement or the adoption of Regionalism.'[4] Local
parochialism, the refusal of the wealthier areas to share their rating
resources with the poorer local authorities, and the indifference of elected
councillors towards projects which may result in their loss of office, are
among the reasons which make amalgamation of local government areas
with the central city almost impossible to achieve.[5]

In view of this it is not surprising to learn that the distribution of the
metropolitan population in Western Australia has occurred without any

---

[1] Edwin A. Cottrell and Helen L. Jones: *Metropolitan Los Angeles: I. Characteristics of the Metropolis*, Table II, p. 4.

[2] Victor Jones: *Metropolitan Government* (1942), p. 129; see also Lepawsky: *op. cit.*, p. 33.

[3] Victor Jones: *ibid.*, pp. 16–20.

[4] J. R. H. Johns: *Metropolitan Government in Western Australia*, p. 73.

[5] *Ibid.*, p. 59.

corresponding changes in local government boundaries or fiscal resources. Since 1900 only one municipality and four road board districts have been incorporated. 'Changes in the status of local authorities have been rare and boundary adjustments few. Structural changes in local government have lagged behind changes in function and fiscal need.'[1]

It is obvious that a large municipality, surrounded by a multiplicity of small local authorities of various kinds, cannot hope to meet the social, political or administrative needs of a great metropolitan area. A medley of scattered and disintegrated local authorities cannot provide the unity required for a coherent scheme of development. Many small areas will be unable to obtain the range or standard of services which demand large-scale administration or substantial resources of money, population, specialized institutions, and highly trained personnel.[2] It has been truly remarked that 'disintegrated local government in metropolitan areas results in unequalized services, in a disparity between need and fiscal ability to meet the need, and in a dispersion and dissipation of political control of the development of social, economic, and political institutions.'[3]

Even where the government of a metropolis is largely or even wholly controlled by the central government, there has been no tendency to extend the boundaries of the city. The administration of Paris, for example, has for long been dominated by the prefect of police and the prefect of the Seine. Yet nothing whatever has been done to extend the boundaries of the Ville de Paris so as to make them reflect the growth of Paris; nor have the outlying communes been amalgamated with the central city.

Opposition to the expansion of the great city is essentially political. It is often founded on a sense of separateness on the part of the inhabitants of the outlying suburban areas, who do not identify themselves completely with the metropolitan community, although they may work in the great city, use it for shopping, recreation, and many other purposes. The local politicians, councillors, and officials usually try to fan the flames of such resistance and their attitude may be quite sincere in that they themselves fail to identify their district with the larger metropolis of which it forms a part. On the other hand, their attitude is sometimes due to self-regarding motives springing from a desire to retain an official appointment or to remain a councillor at all costs, regardless of the wider public interest involved. When a full account has

[1] J. R. H. Johns: *Metropolitan Government in Western Australia*, p. 6.

[2] This was emphasized by the Ontario Municipal Board in its decisions and recommendations on the Toronto metropolitan question. See report dated January 20, 1953, p. 14.   [3] Victor Jones: *op. cit.*, p. 52.

been taken of such considerations, the stark fact remains that amalgamation usually results in the absorbed unit ceasing to be a separate entity, and thereby losing its local government institutions. Instead of having its own elected council, mayor, chairman or city manager, it becomes an insignificant fraction of a vast city governed from a remote centre with which it has little contact. Is it surprising that, faced with such a prospect, the small towns, urban and rural districts, villages, etc., should resist what appears to them to be the lethal encroachments of an advancing tide? From their point of view, it is a fight for life.

Technical, financial, and administrative considerations concerning efficiency or economy, the equalization of services and resources, the broadening of the incidence of taxation: all these count for little compared with the primitive emotions which are aroused by the urge to survive.

All kinds of arguments both true and false, and all manner of accusations, are employed by the spokesmen of suburban or outlying areas to oppose absorption by the great municipality. These arguments and allegations vary from place to place and from time to time.[1] They are for the most part, however, mere surface phenomena, and as such scarcely warrant detailed study. They usually represent an attempted rationalization of underlying currents of feeling which are often subconscious.

By far the best way of meeting these objections is by means of a two-tier system of local government. There are substantial advantages in establishing a major authority for the planning, co-ordination and administration of large-scale functions, while leaving all the purely local services to a lower tier of minor authorities. The arguments for a two-tier system in a great metropolitan area are overwhelming, for only by such a method is it possible for the suburban and outlying districts to retain their institutional identity and communal life whilst becoming part of the metropolitan area for the larger governmental purposes.

Only by this means, moreover, can we hope to find a solution to the problem of providing the metropolitan area with a democratic system of local government while also giving the citizen a smaller and more easily comprehensible unit of community life in whose government he can participate. It is perfectly feasible and logical to aim simultaneously at both larger and smaller units of local government in metropolitan

[1] For an analysis of the reasons formulated by politicians and various organizations in outlying areas in U.S.A. in opposing amalgamation, see Victor Jones: 'Politics of Integration in Metropolitan Areas,' in *The Annals of the American Academy of Political and Social Science*, Vol. 207, January 1940, pp. 161–4. For a recent example drawn from Canada see the Report of the Ontario Municipal Board dated January 20, 1953, summarizing the objections by outlying municipalities to amalgamation with the City of Toronto.

areas; and to evoke in the citizens a sense of civic interest in both the larger community and the smaller. It may even be worth while establishing a federal type of metropolitan council, in which a proportion of the members would be elected by the minor authorities in the area, although we would insist that not less than half of them would be directly elected by the electors for the metropolitan area, in order to ensure an adequate representation of the regional outlook.

13. THE AD HOC AUTHORITY FOR SPECIAL PURPOSES

The English-speaking countries have made considerable use of the *ad hoc* authority set up to perform a single function, or sometimes more than one, over a larger area than that in which the city government exercises power. In London, for example, there is the Port of London Authority, the Metropolitan Water Board, the London Electricity Board, the North Thames Gas Board, the London Transport Executive (now part of the British Transport Commission) and several others. In Los Angeles there is the Metropolitan Water District, the Water Pollution Board, the County Air Pollution Control District, the Flood Control District, and one or two other 'special districts' as these bodies are called in the United States. Metropolitan Toronto has a Metropolitan School Board, the Toronto and York Planning Board, and the Toronto Transit Commission. In Montreal we find a Transportation Authority controlling all surface transport facilities throughout the metropolitan area, and the Metropolitan Commission, set up to control the finances of all the municipalities on the Island of Montreal. New York has its Port Authority, and also the Triborough Bridge and Tunnel Authority. Sydney has had a Water and Sewerage Board since 1888; a Main Roads Commission; the Sydney County Council, created to generate and distribute electric light and power to the entire metropolitan area; and many other *ad hoc* bodies to provide transport, fire protection, abattoirs, hospitals and public health services, maritime services and housing. In Wellington, there is a Fire Protection Board serving Wellington and Johnsonville; a Harbour Board consisting of representatives elected from a wide area served by the port; a City and Suburban Water Board; and the Wellington Regional Planning Council. In metropolitan Chicago there are several *ad hoc* bodies, of which the Sanitary District of Chicago is the best known. When it was set up in 1889, its area covered 185 square miles. Today the territory extends to 442 square miles.

The city which leads all others in America in the number of functions which are entrusted to special *ad hoc* authorities is Boston. The movement began in 1889 with the creation of a metropolitan sewerage district, followed soon afterwards by the setting up of a metropolitan park commission and a metropolitan water board. In 1919, the Metropolitan District Commission was constituted, and absorbed the special authorities dealing with sewers, water supply, parks and planning. Its jurisdiction now extends over more than forty cities and towns in the Boston metropolitan area. In 1929 a further *ad hoc* body was set up to deal with local passenger transport.[1]

In the 26 largest cities of the United States there were, in 1940, no less than 61 *ad hoc* bodies, of which 25 have power to levy general property taxes. They included 17 housing authorities, 13 independent school boards, 5 park authorities, 5 library authorities, 5 sewer authorities, 2 public works authorities, and others dealing with water supply, highways and bridges, ports, local passenger transport, and planning.[2] There has been, moreover, a notable increase in the number of these special purpose authorities.

The Port of New York authority is a leading example of a vigorous *ad hoc* authority exercising functions in a district covering about 1,500 square miles and containing about twelve million persons. It was established in 1921 by a compact between the states of New York and New Jersey to deal with the complex transport problems which have arisen in the metropolitan area. Six unpaid commissioners are appointed by the governor of each state with the approval of the state senate. The authority has provided and administers three bridges linking Staten Island with New Jersey; one bridge and two tunnels linking Manhattan with New Jersey; four marine terminals for shipping; four air terminals, including La Guardia and International airports; and four land terminals for motor bus services, motor truck transport, and railway goods.

In Australia a similar trend can be observed. In Western Australia alone there are nineteen statutory authorities concerned with the provision or regulation of metropolitan services. Among them are *ad hoc* bodies carrying on business enterprises such as sawmills, hotels, brick works, meat processing, and insurance.[3]

In Britain the movement towards the *ad hoc* authority has swung

---

[1] Victor Jones: *Metropolitan Government*, pp. 93–5.

[2] William T. R. Fox and Annette Baker Fox: 'Municipal Government and Special Purpose Authorities,' *The Annals of the American Academy of Political and Social Science*, January 1940, pp. 176–7, 182. [3] J. R. H. Johns: *op. cit.*, pp. 67–8.

C

forward in the course of nationalizing several major industries and services. There are now numerous area gas boards and area electricity boards, regional hospital boards, and licensing authorities for road transport vehicles, operating over very large regions and exercising functions which were in many instances formerly carried on by city councils.[1]

There are several different types of *ad hoc* body among those which have been or might be mentioned. Dr. Thomas H. Reed distinguishes the following four main categories:[2]

(1) Those appointed by and responsible to the central government, or, in the case of a federal constitution, to the state government. Examples of this type are the Metropolitan Police Commissioners in London, who are appointed and controlled by the Home Secretary; and the Metropolitan District Commission of Greater Boston, whose members are appointed by the Governor of Massachusetts with the advice and consent of the state senate.

(2) Those composed of representatives of the constituent bodies, such as the London Metropolitan Water Board, the Montreal Metropolitan Commission, and the Metropolitan Water District of South California.

(3) Those directly elected by the voters. In this category come the Sanitary District of Chicago, and the East Bay Municipal Utility District of California.

(4) Those where the governing body represents chiefly the interests concerned in its activities. Examples of this type are the Port of London Authority and some of the boards set up in Australia which contain members representing producers, consumers, employees, and other interests.

The *ad hoc* authority has been widely adopted as an easy solution to the problem of metropolitan government. In this connection I will venture to quote the following passage on the subject from my book, *The Government and Misgovernment of London:*[3]

The attractiveness of the *ad hoc* idea is not difficult to understand in a situation such as that which exists in London. The ground is littered with multifarious elected authorities possessing jurisdiction over utterly inadequate areas. Each one of those authorities is a centre of potential opposition to any rational scheme of reform. On the other hand,

---

[1] See W. A. Robson: *Problems of Nationalized Industry*; J. W. D. Grove: *Regional Administration. An Enquiry into the Regional Organization of the Central Government,* Fabian Society Research Series, No. 147.

[2] 'Metropolitan Areas,' *Encyclopaedia of the Social Sciences,* Vol. X, p. 398.

[3] Second edition, pp. 333–4.

the technical needs of a service—water, transport or whatever it may be—are easily ascertained and strongly urged by responsible administrators or independent experts who at least desire to promote the efficiency of that service. Hence the wary politician, the timid civil servant and the technical specialist readily turn to the *ad hoc* authority as the easiest way out of their difficulties. 'Ministers have almost ceased to apologise,' writes Mr. Herbert Morrison, 'for creating Greater London authorities for purposes which, if local government were rationally organised in the area, could have been discharged under normal local government auspices. Indeed, some enthusiasts with specialist minds occasionally bob up demanding yet another special Greater London authority in respect, for example, of housing or town planning. There are people who believe that the establishment of a special authority will solve most problems for co-ordination, whereas it may have done little more than create a salary list.' Sometimes the more pressing technical difficulties may be assuaged for a time. But ultimately the *ad hoc* body gives rise to as many problems as it solves.

The most serious drawback of the *ad hoc* body is that there is no method of co-ordinating its work with related activities carried out by other bodies. It has one, and only one, object in view; and it is in a sense failing to discharge its duty if it attempts to take a comprehensive view of things. Yet the services of a great modern city are becoming more interrelated every day, and even their efficiency is determined to no small extent by the degree of co-ordination that is attained. Housing, planning, transport, highways —how can one separate such a group as this? And housing in turn involves education, drainage, public health, gas and electricity and many other services. There are hundreds of other points of contact between the various public and social services where 'the single eye' is needed to obtain the best result.

Professor Reed[1] agrees that the metropolitan problem as a whole cannot be solved by the creation of special districts, though he rightly points out that the services rendered by many *ad hoc* authorities are of immense value. He considers that where several important functions have been entrusted to a single *ad hoc* authority, the result is a close approach to what he calls 'the federated city.' He cites as an example the Massachusetts Metropolitan District Commission, which has been made responsible for sewers, water supply, parks and planning in Greater Boston. But this commission is in no sense an organ of local self-government, for its members are appointed by, and responsible to, the governor of the state. In consequence, it has the overwhelming disadvantage of failing to satisfy the basic criterion of local emocracy. On this ground alone we must reject such a solution of the problem of metropolitan government.[2]

[1] *Encyclopaedia of the Social Sciences: op. cit.*

[2] See J. R. H. Johns: *Metropolitan Government in Western Australia* (1950), p. 67. 'A widening range of metropolitan services is rendered by statutory boards which are subject to diverse directions in the composition of the controlling agency and have varying degrees of fiscal and administrative autonomy. Since 1930 three major authorities have been created for transport co-ordination, the regulation of milk supplies and electricity generation. In other instances, local government services and functions have been transferred to "line"

The same objection would not apply to a directly elected, or even an indirectly elected, *ad hoc* authority gradually acquiring the power to perform a wide range of functions, and thereby becoming in effect the principal organ of metropolitan government. But is there any reason to believe that such a movement would not meet with fierce opposition on the part of all the minor local authorities, and in addition the chief municipality, whose interests would be jeopardized by such a development? They would, we believe, resist this movement for the same reasons as the former resist the expansion of the great city. Indeed, from their point of view the ultimate result would be almost identical.

The *ad hoc* authority operating within a special district which is larger than the municipality attempts to solve the technical, administrative, and financial problems arising in connection with a particular service by isolating it from the complex of municipal services of which it forms a part. It may solve that problem, but only at the cost of weakening the general structure of local government in the great city and its environs, whereas the real need is to strengthen it.

The *ad hoc* authority appears to have emerged chiefly, or perhaps only, in the English-speaking countries as a method of dealing with metropolitan areas. The reason for this is not clear. But whatever the reason may be, one finds no attempt in Paris, Rio, Buenos Aires or Rome, to solve their local government problems by such means.

It is possible that a prefect or other local representative of the central government may sometimes act as a unifying influence in an urban centre divided among numerous local authorities. It is suggested, for example, that in the Scandinavian countries the governor or intendant provides a central point of reference which fosters local unity.[1] Similarly the principle of 'democratic centralism' which governs the relations between the Moscow soviet and the higher authorities in the U.S.S.R. may provide an explanation of why it is not necessary to establish *ad hoc* organs in the Soviet capital. In our view the disadvantages and dangers to local self-government which the prefect or intendant brings in his train outweigh any possible advantages which the system may have; and we would actively oppose a suggestion recently put forward in Britain by Sir Geoffrey Hutchinson, Q.C., M.P., that there should be a Minister for

State departments. The future good government of the Metropolitan Area and indeed of the State demands an examination of the functions which have been removed from local control and the selection of those which local government is competent and willing to administer.'

[1] W. J. M. Mackenzie: *The Government of Great Cities* (University of Manchester, 1952), p. 11. 93 Memoirs and Proceedings of the Manchester Lit. and Phil. Society No. 5.

London to effect co-ordination and direction in Greater London. Certainly such co-ordination and direction are needed; but we should avoid that method at all costs.

14. OTHER ATTEMPTS TO SOLVE THE PROBLEM

There are many other devices which either exist or can be introduced to overcome the difficulties arising from the lack of an integrated metropolitan government. One such device is for the great city to provide the outlying districts or towns with some of its services. For example, the London County Council has admitted many of the outlying areas into its main drainage system, and the same thing has occurred in Los Angeles, although there is a physical limit to what can be done in this direction. It is quite common for large cities which own and operate motor bus, trolley bus and tramway undertakings to provide services to points far beyond the city boundaries. Manchester City Council brings water to the city from a distance of 100 miles, and supplies in bulk twenty-six other districts and also the neighbouring town of Salford. In Buenos Aires, municipal services are usually extended to suburbs outside the city limits as the metropolis expands. Legal permission has usually to be obtained from higher authorities before activities of this kind can be carried out.

A similar result is attained when the great city permits residents from outside areas to use its municipal services, on payment of an appropriate sum by the outlying local authority or by the person concerned. This arrangement is easily applied to services which are rendered to specific individuals, like technical or higher education, hospital treatment and so forth. Arrangements of this kind must usually be negotiated and agreed between the local authorities concerned.

Large cities also sometimes acquire property outside their boundaries, such as water reservoirs, hospitals, greenhouses, parks and playing fields. Sometimes they are even authorized to exercise regulatory or police powers in outlying areas.[1]

A further development of more recent date is the provision of housing to take the overspill of the great city's population in outlying areas within the metropolitan region. Where insufficient land is available in the great city to enable slum clearance, the abolition of overcrowding, and good housing standards to be achieved, arrangements are sometimes made whereby the municipality provides or finances housing accommodation for its citizens in outlying areas. The London County Council has built

[1] Albert Lepawsky: *op. cit.*, Vol. I, p. 34.

a great many housing estates outside its own boundaries. In 1939, about 42,000 out of 71,634 dwellings provided by the London County Council were situated outside its territory; and this policy has been continued on a big scale since 1945. The largest housing estate in the world was constructed by the council at Becontree in Essex, several miles outside the administrative county. It is estimated that the rehousing of the congested parts of Manchester will require 150,000 people (nearly 20 per cent of the population) to be accommodated outside the city's limits.

Where this occurs the great municipality becomes a colonizing power, planting its citizens in alien lands beyond its own borders. In the past, these colonizing activities have often been unwelcome to the 'receiving' local authorities, who were obliged to provide schools, hospitals, police forces, fire brigades, and many other services for the overspill. This not only cast a financial burden on the receiving local authority, but also created a problem of assimilation, where large numbers of families were uprooted from a slum district in which they had long resided and were suddenly transplanted to an area which was quite new to them and in which they had neither friends nor relatives. A recent English law, the Town Development Act, 1952, which seeks to encourage town development in county districts in order to relieve congestion or overpopulation elsewhere, should remove some of these causes of friction. The statute enables the central government to make exchequer contributions to the council of a receiving district towards the cost of providing or extending such services as water supply, main sewerage and sewage disposal, housing, etc. The 'exporting' local authority can also assist the receiving local authority with financial contributions.

These diverse arrangements, whether legal, financial, or administrative, are essentially devices for overcoming the lack of unity and coherence in the government of the metropolitan area. They are not solutions of the problem; but rather substitutes for a solution. Regarded merely as expedients, they have much to commend them; but if they are regarded as attempts to cope with some of the greatest problems of modern urban life, they are puny and insignificant. Moreover, they bring great disadvantages in their train. It is not obvious wherein lies the good sense or justice in permitting or encouraging the London County Council or Manchester City Council to impose taxes on their citizens in order to subsidize the accommodation of their fellow-citizens in housing estates in other towns. Such activities do not strengthen local government in the metropolitan area; and it is doubtful if they even enhance the sense of community among the people in the metropolitan region.

One other method of attempting to solve the problem which we may note is by amalgamating the city with the county in which it is situated. This approach is familiar in the United States, where proposals to consolidate the county with the city have been put forward in recent years in a number of great cities, including Pittsburgh, Detroit, Cleveland, Milwaukee and St. Louis. This usually means reorganizing the county, which in the United States is a unit of state government, and making it the basis of a metropolitan municipality. Some important cases of city-county unification occurred last century in New York City, Washington, D.C., New Orleans and other large American cities,[1] but in general the movement is a rural rather than an urban one.

15. POLITICS IN THE GREAT CITY

The growth of party politics in the great city can be observed all over the world. This phenomenon is an aspect of the general movement towards the increase of political parties and party organization which has occurred during the twentieth century. Municipal politics have become the subject of intense organization and propaganda in many countries during the last twenty or twenty-five years to a much greater degree than formerly. Moreover, the political party machines have become much more closely integrated than hitherto at the municipal, provincial, and national levels.

These tendencies are particularly strong in the great cities, for reasons which are easy to understand. Political control of a great municipality—especially a capital city—is in itself a considerable prize, a substantial asset in the struggle for power. Moreover, such control possesses great propaganda value and prestige in the wider sphere of national politics. Indeed, so important are these influences in the municipal election campaigns which take place in great cities, that one often feels the main object of the conflict is to demonstrate that public opinion is strongly in favour of, or opposed to, the party which controls the national (or state) government rather than to gain control of the city council. These secondary effects can become the primary objects of the political contest within the municipality.[2]

This tendency is fortified by the close relationship which exists in most countries between the great municipality and the central (or the state)

---

[1] Lepawsky: *op. cit.*, p. 31.

[2] An example of this occurred in Rio de Janeiro, where the Brazilian Labour Party secured a victory in 1950 because they represented a movement for the return of Vargas to power and had the support of his political machine.

government. This relationship takes various forms and many different threads are woven into the pattern; but the general trend is towards increased control of the city by higher authority, and greater reliance by the municipality on grants of money contributed by the national or state government. One result of this has been to assimilate municipal and national politics, especially in regard to such matters as the social services, full employment, and fiscal policy.

The emerging picture is one in which the national political parties participate in the municipal elections, and usually nominate the candidates. The 'independent' candidate for municipal office without specific party ties has become a rarity; and it is extremely difficult, if not impossible, for any-one to get elected to the council without the support of a powerful party. All or most of the main political parties are nowadays represented on the city council; and executive organs, committees, commissions or boards appointed by the council usually reflect the relative strengths of the various parties on the council. Where the mayor is directly elected by the citizens, he will usually be a party candidate, and his executive powers will be immensely reinforced by his political influence as a party chief.

There is nothing inherently wrong with these tendencies. They are merely the result of the drive for better party organization to cope with the enormous size of the modern electorate, and for more coherence to deal with the greater scope and complexity of municipal activity. Political parties can be as useful to the local government of a great city as they are to democracy at the national or provincial level. The value of political parties can be tested by what happens in their absence. In Los Angeles, and in the adjoining county, municipal government is conducted on a non-partisan basis. The result, we are told, is that all kinds of pressure groups representing commercial, vocational and other interests exert an influence on the conduct of city affairs, and the newspapers have a dominant voice in determining policy. This may well be a less desirable state of affairs than a straightforward party system, provided the parties are not corrupt.

There are, however, some very undesirable consequences which some-times flow from the intrusion of party politics in city affairs. First, there is the tendency to sacrifice questions of genuine municipal impor-tance to spurious national controversies. Municipal elections should be fought on municipal issues; and where the intervention of national parties has the effect of substituting national questions the quality of political life within the city is lowered. It is absurd to conduct a municipal election campaign about such matters as foreign policy, rearmament, or the national budget, for the city council exercises no power in such matters, which are outside its jurisdiction. Yet this is quite a common occurrence. In Stockholm, we are told, municipal elections are fought on national issues and little attention is given to real municipal policy. This is largely because Stockholm and other Swedish municipalities elect members to the upper house of the parliament, but the same thing occurs elsewhere where this function is not performed. In Wellington, to take another instance, elections are decided on irrelevant issues and many votes are cast on national and not local programmes. Election campaigns are seldom fought on questions of principle, and the voters are not given an opportunity to approve or reject a policy applicable to the city. Another trouble in Wellington seems to be an over-emphasis on the merits or demerits of the respective candidates at the expense of any serious discussion about their programmes. In Amsterdam, by contrast, the practice is to vote for parties rather than for persons. Each party chooses its candidates, and it is the party lists which are placed before the electors.

A second and more serious defect of party politics in city government is the tendency in some places to treat every municipal question, great or small, as a party question, even when no political principle is involved. The rebuilding of a century-old cross-river bridge in London, which was no longer capable of carrying the traffic, was made a burning political party issue on the London County Council for many years, the Labour Party advocating a completely new bridge while the Conservatives favoured reconditioning the old bridge. A purely technical matter of this kind should never have been treated as a political question. The result was to postpone action for several years.

In Buenos Aires the party system appears to have degraded municipal government to a point at which it became almost unworkable. Not only was the mental and moral calibre of the councillors unequal to their task, but the introduction of unsound parliamentary methods and obstructionism discredited local self-government in the city. The party zealots, Professor Bielsa complains, made the deliberative council a miniature parliament rather than a corporate organ for active administration. The elected municipal council was abolished in 1942, and there appears to have been no widespread demand for the re-establishment of local self-government in the Argentine capital.

The charge levelled against the municipal politicians of Buenos Aires is not only that their excessively partisan attitude led them to put the interests of party before the well-being of the city, but also that they

c*

were incompetent and corrupt. It must be admitted that corruption is an ever-present danger in great cities, and many of them have suffered from this evil at one time or another. The politics of Chicago are described by Professor Walker as 'bawdy, corrupt, and unashamed.' The political life of the city has been dominated by political bosses with a strong personal following, sometimes resembling a mercenary army. It has also been marked by 'amazing loyalty to party organizations,' with equally unfortunate results. In Chicago the maintenance of a political machine has involved great waste and inefficiency in municipal administration. Party supporters are often appointed to public offices regardless of their incompetence or dishonesty. Patronage has discouraged the recruitment of capable, well-trained and honest officials, and so a vicious circle is created. The conclusion is that party politics in Chicago city government has meant 'incompetence, favouritism, distortion of the public interest, and contempt for the public service.' An improvement is nevertheless in sight and there are forces at work which will make it harder in future for party politicians to exploit the need and the ignorance of the less fortunate citizens.

In New York City the worst abuses of Tammany Hall have been corrected; but it has proved beyond the power of even the best-intentioned mayor to keep the vast network of municipal departments honest and efficient, and political scandals still occur. A recent innovation is the appointment of a city administrator of high status who will be directly responsible to the mayor for the effective management of the municipal administration. This appointment follows the recommendation of a New York State Commission set up to study the organization and functioning of the city government. The first holder of the new office is Dr. Luther Gulick, a well-known authority on public administration who has devoted many years to the study and improvement of New York's city government.

It is easy to blame all these evils on political parties and to assume that if party politics were abolished municipal government would become miraculously transformed and everywhere exhibit qualities of truth, goodness and beauty. In fact, however, political parties reflect the civic qualities—or the lack of them—to be found in the electorate. Where the citizens are indifferent to the public weal, politically immature, not averse to corruption or patronage, and ready to take advantage of every weakness in the city government, the party politicians will possess the characteristics familiar in Chicago or Buenos Aires.

Municipal administration in Rio de Janeiro, for example, is gravely injured by nepotism and other maladies derived from 'machine' politics. The boss-ridden city uses a corrupt bureaucracy to disseminate propaganda among the electors and to bring pressure to bear on them. This is made possible, however, by the lack of popular interest in democratic institutions, and the irresponsible outlook of the wealthier classes. The situation, in short, is the outcome of a lack of political education among the citizens. 'Only after a period of thorough education,' observes Dr. Rios, 'of daily practice in the democratic process, will our people be able to discriminate between candidates and hold them responsible.'

16. DIVIDED CIVIC AND POLITICAL INTEREST

The impact of the social, psychological and educational background on the political régime is a matter of which we are all aware both in a general way and through our individual experiences. This is the realm of political sociology, which has thrown much light on the subject. What has not been sufficiently studied are the reasons which cause apathy towards city government and an undeveloped sense of community among the denizens of the great city, as compared with those who live in smaller towns.

The most important factor in our view is the inadequacy of the local government institutions in metropolitan areas. Nowhere do the local authorities correspond to the social, economic and political realities of the area. In consequence, large numbers of men and women live in one municipal area and work in another. Their interest, their loyalty and their allegiance are divided. They suffer from a kind of political schizophrenia which weakens their desire to participate actively either as citizens or as councillors. The sense of community among those who live in the great city is diluted by the presence of the invading hordes of workers who swarm in each day. Those who live in the suburbs beyond the city are, in a sense, escapists: they no longer feel any sense of responsibility for the well-being of the great city nor do they participate in its government. On the other hand, their interest in the local government of the area in which they dwell cannot be very strong and vital, because they spend so much of their time and energies elsewhere.

In a recent biographical sketch of her parents Lady Chorley has given an interesting description of the effect which 'moving out' of Manchester to the exclusive suburb of Alderley Edge had both on the well-to-do business men who could afford to go there and on the corporate life of Manchester. The exodus spoilt the appearance of Manchester by vacating the fine Regency houses in the older residential parts of the city. These quarters soon degenerated in the usual way when they were abandoned by the better-off families. But the 'moving out' process had deeper effects by depriving the city of many vigorous and capable leaders. When they lived in Manchester, the city was the centre of their lives; they had a civic pride which was more than mere philanthropic zeal. They founded colleges and a university, picture galleries and libraries; they established the Hallé symphony orchestra and supported its famous concerts. They sought to make the city proud and beautiful. 'But when the sons of these nineteenth century citizens moved out they could no longer carry on the tradition. The city became the place they worked in by day and abandoned in the evening as quickly as might be. Their leisure interests and recreations were elsewhere and the time they gave to civic duties dwindled. The city was no longer the centre of their cultural lives and though . . . they still contrived to run a host of charities, they tended to withdraw their services from the city council. They grumbled enough about its quality . . . but it never occurred to them that they as individuals were perhaps to blame.'[1] The history of Lady Chorley's own family typified the process. Her grandfather had been an alderman and mayor, but none of his sons served on the council.

There was clearly an element of personal irresponsibility in thus evading the problems of the city which provided the labour, the wealth, and the property which made it possible to build the large and luxurious houses at Alderley Edge. But part of the blame must also rest on the governments and parliaments which have refused for so long to recognize the problem which is epitomized in the story of Manchester and Alderley Edge. A similar process has taken place, and is now taking place at an enhanced tempo, in nearly all the great cities of the world. And almost nothing is being done to counteract the social and political disintegration which results. If a proper system of metropolitan government were introduced, the Alderley Edges would become part of the Greater Manchesters to which they belong, and the 'moving out' process would not necessarily produce such unfortunate results.

The expedients which have been introduced in metropolitan areas to overcome the difficulties of local government without drastic reform have produced an extraordinary tangle of areas and authorities. The medley of unco-ordinated units which exists outside the narrow boundaries of the city proper, the welter of *ad hoc* authorities set up to administer a whole series of services, add to the sense of confusion, incoherence and

[1] Katharine Chorley: *Manchester Made Them*, p. 139.

disharmony in the mind of the average citizen when he thinks about the government of the great city in which he lives. How can we expect to find in the minds of men and women that sense of unity on which community is founded if we do nothing to develop and express it through appropriate political institutions?

The very size of the great city today makes the task of creating or enhancing a sense of community among its inhabitants much more difficult than in a small town. The sheer scale of the metropolitan area, in terms both of population and territory, is in some respects a handicap to good government. For although the resources of the great city make it possible to employ the best and most highly qualified officers, to undertake developments requiring immense sums of capital expenditure, and to provide services (in the spheres of education and health, for example) which are quite beyond the power of smaller authorities, there are countervailing disadvantages. It is extremely difficult, if not impossible, for the city government to keep in close touch with the mass of citizens, to be aware of their attitude towards the services which are provided, or of their unsatisfied needs.

The Moscow soviet attempts to bridge the gap between the governors and the governed by enlisting the help of a large number of voluntary workers to assist the deputies on the various committees. These so-called 'activists' are co-opted to a particular committee, such as that dealing with health, housing, or education; and their duty presumably consists largely of visiting municipal institutions such as schools, colleges, blocks of flats, hospitals or clinics, and ascertaining the standard of efficiency and also the public attitude towards them. These activists, together with the deputies whom they assist, number some 4,000 persons—a substantial political force for maintaining public relations and communications between the principal city soviet and the people of Moscow. This figure does not include the district soviets, whose 3,000 deputies are assisted by a further 10,000 activists. No other metropolis tries to solve the problem of the giant city by multiplying to so large an extent the political elements engaged in the city government. The activists are doubtless chosen by more or less rule of thumb methods, and it must be exceedingly difficult to imbue them with a consistent viewpoint, a sense of discretion, and sufficient responsibility to make them useful, while at the same time conscious of their own limited powers. Nevertheless, the attempt represents an experiment which may be of value.

17. RELATIONS WITH HIGHER AUTHORITIES

The relations between the government of a great city and higher authorities reflect to a considerable degree the general relations between central and local government prevailing in the country concerned. There may also be, however, certain special factors which can influence the position. The most important of these exists where the great city is a national capital. For this may lead to a much greater degree of central control, as in Paris, or to the supersession of the democratic organs of local government, as in Buenos Aires. It may result, in a federal constitution, in a special district being established for the capital under the direct control of the national legislature, as in Rio de Janeiro, Washington, D.C., or Buenos Aires. It may result, as in Moscow, in direct relations between the executive of the U.S.S.R. and the city soviet, and the by-passing for certain purposes of the Russian Socialist Federative Soviet Republic in which the capital is situated. It may result, as in London, in a special organization of local government not found elsewhere.

Even where the great city is not a capital, or at any rate not a national capital, it may occupy a special position. This occurs particularly in federal types of constitution, where the higher authority for local government is the state or province. Frequently there is a conflict between the city and the state government which arises from the under-representation of the former in the state legislature. New York City, Montreal, and Chicago are examples of cities located in provinces or states in which a rural minority is able to dominate an urban majority on account of the over-representation of the former in the legislature. A typical result of this situation is that rural areas are favoured in matters of taxation, in the administration of federal aid through state or provincial governments, or in the calculation of subventions by the latter.

Pure autonomy for any municipality, however large, is of course impossible; for unlimited freedom would be tantamount to national sovereignty and we should then have the city-state of ancient Greece. Hence, the autonomy of a city is always limited, and the most usual method of determining or controlling it is by legislation emanating from the higher authority (national or provincial). State law and state constitutions thus impose limitations on even the 'home rule' cities of the U.S.A., which enjoy an exceptionally high degree of freedom. Under the Soviet system there are no specific limitations on the powers of the local authorities in Moscow; but the latter are subordinate to the higher authorities and must conform to their plans and directives in every respect, and this is the all-important fact.

The general tendency seems to be for the higher authorities to exercise, or at least to acquire, increasing powers of control over the great city. In India, the administrative departments of the state government exercise a large amount of control over the municipal government of Bombay. The state must approve all its by-laws and loans. The state appoints and controls certain senior officials; it can intervene in case of default on the part of the corporation; and it has power to conduct a special audit when it thinks fit. Similar powers are possessed by the state in respect of Calcutta, quite apart from the power which the state government exercises of appointing a commissioner to control municipal administration. In Australia, the state can regulate the organization of local government in Sydney. It lays down rules governing the procedure at council meetings, determines the qualifications and status of local government officers, and settles many other details. The municipality cannot make its own ordinances, regulations, or by-laws. It must obtain the approval of the State Department of Local Government for loans and bank credits; the state examines the annual statement of accounts, the qualifications and integrity of local government officers. A somewhat similar situation exists in Canada in regard to the relations between Toronto and the provincial government. The Department of Municipal Affairs and the Ontario Municipal Board are the provincial organs responsible for supervising the city government. The department is concerned with the general operation of the law. It has power to regulate methods of accounting, estimates, and auditing. The Municipal Board must approve loans and zoning ordinances; it hears appeals against assessments for city taxes, and arbitrations regarding the acquisition of land by the city. It supervises the operation of the city's public utility undertakings. The appointment and dismissal of some of the chief municipal officers are subject to the approval of the provincial government. In New Zealand the pattern of relationship is not fundamentally different, except that the national government is the higher authority.

Turning to the countries of Northern Europe, we find that although in principle Amsterdam can legislate on all matters of local interest which have not been regulated by the provincial or central government, there is nevertheless a considerable amount of central control. The approval of the provincial government ('deputed states,' as they are called) is required for the budget, for decisions of the city council concerning municipal property and various other matters. The Crown—that is, the central government—can suspend or annul decisions of the city which it considers to be illegal or not in accordance with the public interest.

The city's housing quota is allotted by the provincial government. In future, Amsterdam is likely to be bound even more closely than at present by the policy of the central government, largely owing to its increasing financial dependence on the centre. The loss of financial independence is certain to be followed by diminished municipal autonomy.

From Denmark comes a similar report that the influence of the central government on municipal administration is steadily increasing. The formal relations between Copenhagen and the state are determined by constitutional law, which declares that the law shall establish the right of municipalities to govern their own affairs under the supervision of the state. This means that the rules according to which Copenhagen is governed are contained in acts of the Danish Parliament, and municipal regulations must be approved by the Home Office. These regulations include the creation or abolition of all the higher posts in the city. Home Office approval must be obtained for any loans required for capital expenditure, and also for the purchase or disposal of municipal property.

The general duty of the state to supervise the municipality is carried out in Copenhagen by the lord lieutenant (*overpraesident*) who acts as superintendent on behalf of the central government. He can attend the meetings of both the city council and the executive committee, and reports to the Home Office any matter which he regards as illegal or undesirable. 'The local administration,' observes Dr. Holm, 'is in practice largely dependent on the central administration.' The principal reason for this is to be found in the realm of finance. The municipality of Copenhagen, it appears, has no real authority as regards taxation. The provisions which determine municipal rates and taxes are enacted by national legislation and the administrative rules for their application are laid down by the central government. This relates particularly to the income tax, which has gradually become the principal source of municipal revenue. The rules governing taxes on property are also outside the control of the municipality. Finally, such services as education, public assistance and social insurance are all deeply rooted in national legislation.

In Sweden there is a strong tradition of local self-government which is jealously guarded. Some of the functions carried out by the city government of Stockholm are based on national legislation, which lays down precise provisions regarding each service. In these circumstances the municipality has little discretion in respect either of policy or of the scale of expenditure. On the other hand, in certain spheres of activity, such as public utility undertakings and the management of real property, the city is much freer and enjoys as much independence as a commercial company.

Stockholm is a county as well as a city: it corresponds to a county borough in England. For that reason it has—like other counties—its own lord lieutenant or governor (*överstathallare*) who is appointed by the central government. The approval of this high official must be obtained for certain decisions of the administrative organs. He can also disallow a decision of the council concerning the budget or the rate of tax on grounds of illegality, although he cannot substitute his own decision in its place. The general position is, however, that the central government exercises no direct power of a positive kind over the city administration and little general control.

The main reason for the independence which Stockholm enjoys lies in the financial strength of the city government. The city receives grants-in-aid, in respect of housing for example; and these involve various forms of state control. But the financial strength of the municipality is so great that the city, if forced to do so, could dispense with state grants and rely on its own resources. In consequence, the city is able to maintain a considerable measure of independence in its dealings with the national government.

In Britain, as in Sweden, there is a strong tradition of local self-government, and historically local government preceded central government in point of time. Nevertheless, local authorities are subject to a considerable amount of control by the central government and this has been increasing in recent years. Manchester, like all other English local authorities, is governed by a council whose constitution, organization, powers, procedure, officers and much else are laid down by general legislation. Manchester, like the much more ancient City Corporation of London, has in addition a royal charter conferring on the municipality certain powers and privileges. All the principal services, such as housing, education and public health, are the subject of Parliamentary statutes which lay down the national policy in these spheres, prescribe the powers and duties of local authorities, and also confer powers on ministers. Lastly, municipalities have the right—which the larger ones like the London County Council and Manchester City Council exercise frequently—of petitioning Parliament for a Private Bill: that is, a law applying only to their area, and giving them powers which are not generally accorded elsewhere.

The central executive also exercises a considerable degree of control over the local authorities of Manchester and London; but this is neither more or less than that which exists in the case of any other local authority in Britain—leaving aside special features which are to be found only in the capital, such as the administration of the police force by commissioners responsible to the Home Secretary.

In Manchester or London one finds no one corresponding to the prefect or governor of a continental city, nor does the central government have any voice in the appointment of the Lord Mayor or Chairman of the Council. The municipal budget does not have to be approved by a higher authority; and the central government cannot suspend or dissolve the council. There are no officials who act in a dual capacity as officers of the municipality and representatives of the central government. All these well-known forms of central control are absent.

The principal instrument of central control is the grant-in-aid paid by the national exchequer. This has been a factor of increasing importance in British municipal finance for many years, and in the aggregate the sum paid by the central government to local authorities is about as much as they now raise by local taxation. With this increase in central grants has come an increase in central control—a phenomenon which is in no way surprising. Central government departments have statutory powers of many different kinds affecting the activities of local authorities. Thus, Manchester City Council and the London County Council must submit their development plans for approval to the Ministry of Housing and Local Government and their school development plan to the Ministry of Education. The Ministry of Education makes regulations laying down the standards to be observed by local authorities in building new schools. Manchester and London must pay such salaries to school teachers and require such qualifications as the minister approves. The appointment and dismissal of the medical officer of health, of the chief police officer, and certain other chief officers employed by the local authority, must be approved by the appropriate ministry. The accounts of all the local authorities in London (except the ancient city corporation) and some of the accounts of Manchester City Council are audited by a central government auditor, who can disallow unlawful or excessive expenditure and recover the sums in question from the responsible persons. In some spheres of activity, ministers have 'default' powers: that is, if the local authority fails to carry out its duties they can intervene and arrange for the function to be carried out by another agency. Many individual acts of the local authority require the assent of a minister. But above all and beyond all, there looms the power of the minister to reduce or even to withhold entirely the grant-in-aid in the event of his not being satisfied that a particular service is being carried out with economy and efficiency. This applies both to the percentage grants for specific services, such as education and police, and to the general grants like the Exchequer Equalization Grant. Finally, there is the central government's control over local government loans.

The municipal authorities in London and Manchester are subject to the general requirements of this complex relationship between local and central government. It bears on them less heavily than on many smaller towns for several reasons. First, they are both highly efficient and progressive municipalities, leading the way for the country as a whole in many activities; they therefore do not have to be goaded or prodded into activity by the central government in order to achieve the national minimum standards of health, housing, education, etc.: it must be recognized that the enforcement of a national minimum is the object of many of the central controls mentioned above. Second, they are very wealthy cities, and the magnitude of their financial resources gives them a degree of independence not usually possessed by the poorer municipalities.

18. MUNICIPAL FINANCE

This brings us to the subject of municipal finance. In no sphere is there greater disparity among the cities comprised in this volume. The methods by which great cities obtain their revenue, and the amount and objects of their expenditures, differ so greatly that it would be futile to attempt any kind of comparison. A mere description of the phenomena would be tedious and unilluminating. We shall only attempt, therefore, to draw attention to a few outstanding points.

Real property is by far the commonest subject of local taxation, and property taxes are usually the chief source of municipal revenue. Nevertheless there are some striking exceptions. In Moscow, surplus or profit obtained from local industries yields about 60 per cent of the municipal revenue, the remainder coming from income tax and profits on municipal services. Copenhagen relies for most of its revenue on a municipal income tax. Thus, about 82 per cent of its revenue from taxation came from this source in 1949.[1] Zürich also draws most of its revenue from a surtax

---

[1] One of the interesting features of the Danish system is that a town may not increase by more than one-fifth the basic amount of the municipal income tax. If the expenditure provided for in the budget cannot be met unless the municipal income tax is increased by a larger amount, an extraordinary election, called a taxation-election, is held for the town council. After the election the new council decides whether the rate of income tax shall be raised or not. If the increased expenditure is caused by new legislation, the Minister of Home Affairs can exempt the municipality from holding a taxation-election.

levied by the commune on the cantonal taxes on income and capital, and the tax-surcharge levied by the commune is always larger than the cantonal tax. Calcutta imposes a small tax on professions and vocations which is 'a sort of miniature income tax.'

A graduated tax on income has many financial and social advantages over taxes levied on real property, whether assessed on annual or capital value; and this has led a number of people to advocate a municipal income tax. Whatever the theoretical merits of the proposal may be, it is quite impractical in countries such as Britain, where income tax levied by the central government for national purposes has already attained levels which are oppressive. It may not be an accident that both Denmark and Switzerland are countries whose national expenditure on defence is relatively low. It is worth noting that since 1929 Dutch municipalities have been prohibited by law from raising local income tax and surtax on government income tax. In place of this, a municipal fund was established fed by a special tax and distributed to local authorities according to their population.

Despite the predominance of taxes on real property, no country except Britain relies on them exclusively for municipal revenue derived from local sources. Elsewhere one finds a wide variety of local taxes. Bombay has a wheel tax; Amsterdam has municipal taxes on entertainments, dogs, and fire insurance; Chicago has taxes on entertainments for which admission charges are made, and also a tax on motor vehicles. Taxes on sales are a common method of raising municipal revenue. Thus, Los Angeles derives nearly 10 per cent of its total revenue from a municipal sales tax, Montreal has a sales tax which contributes nearly 15 per cent of the city's revenue, while Rio de Janeiro obtains nearly half of its total tax revenue from a levy on sales and commissions. New York City has a sales tax which produced about $131 millions in 1950–51 compared with less than $600 millions from taxes on real estate, and New York City also has a tax on businesses. The Ville de Paris levies a wide range of taxes. Some Paris municipal taxes are added to those collected by the state on, for example, furnished rooms, rents, and business premises. The municipality can at its discretion tax balconies, domestic servants, the consumption of gas, and many other articles or services. The city is obliged to impose taxes on entertainments, cafés, bars, stamp duties and, most important of all, on business transactions. Rome employs equally diverse methods of financing municipal services by means of taxes on rent receipts, industry, trade, the professions, domestic servants, dogs, pianofortes, billiard tables, coffee-making machines, advertising posters and billboards, stamps on official documents, and even a hearth tax. In Wellington rates on real property account for more than three-quarters of the revenue from taxation, but this is supplemented by a petrol tax, although it does not produce much.

Without attempting to evaluate the merits of all these diverse taxes, it is clearly an advantage for local authorities to be able to draw their revenue from several different sources. London and Manchester are at a disadvantage here, compared with Rome or Paris. Indeed, the rigid and exclusive emphasis which the British system lays on real property as a source of local taxes contrasts unfavourably with the wider basis of municipal taxation permitted to practically every other great city comprised in this book.

The inadequate and unsatisfactory basis of the rating system in Britain partly explains the much greater role which grants-in-aid from the central government play in municipal finance there compared with other countries. In 1950–51 government grants were equal to the sum raised by local authorities from local taxation, taking the country as a whole. The wealthier municipalities receive a smaller amount from grants-in-aid in proportion to their total revenue than the poorer areas. In 1949–50 the London County Council obtained 26 per cent of its total expenditure from government grants and 45 per cent from rates and unspent balances carried forward from previous years. In the same year, Manchester City Council obtained 28 per cent of its total revenue from government grants and 44 per cent from taxes on real property. So far as we are aware, in no other great city do subventions from central authorities reach so high a figure in relation to local tax revenue. In Bombay, government grants were only about 12 per cent of the total revenue in 1949–50; in Los Angeles, subventions and grants were 15 per cent in 1950; in Toronto, grants and subsidies were less than 7 per cent. But precise information is lacking on this point, and in the case of some cities with a highly centralized form of government it is not always easy to distinguish the central grants from the local taxes.

The financing of municipal services in the great city presents a problem of very great difficulty. This is as true in the wealthiest and most industrialized centres like New York, Chicago, or Paris as it is in much poorer cities such as Bombay or Calcutta. 'It is no exaggeration by now to say that New York City exists in a state of chronic bankruptcy,' writes Professor Rex Tugwell. 'It is not that municipal bills are not paid—although there have been crises when obligations were far greater than resources and further borrowing seemed impossible—but that, facing the fiscal problem, budget makers have had to reduce expenditure until the municipal services reached an almost impossibly low level. Streets have not been properly cleaned and repaired for decades; the school system is miserably maintained; water is chronically short; transportation on the municipally-owned lines is such that daily travel is an ordeal—and similar shortcomings affect every one of the three hundred odd services the city pretends to perform for its citizens.'

The reasons for this are complex. First, it is probable that the cost of providing some services increases disproportionately in the great metropolitan centre:[1] we have in mind the enormously expensive works required to provide the huge modern city with main drainage and sewage disposal, highways capable of carrying the continually growing mass of motor vehicles, water supplies which may have to be brought a distance of a hundred miles or so, housing and slum clearance, and so forth. Second, as the transport and traffic facilities improve, the number of workers coming into the city from outside each day increases, and these daily migrants must be provided with expensive services to the cost of which they contribute little or nothing. Third, the huge rise in the value of land increases enormously the cost of providing municipal works or services for which land is required, such as housing, highways, parks and open spaces, schools and playgrounds. Indeed, the cost of these services has risen so much in some of the largest metropolitan centres that no serious attempt is being made to provide them on an adequate scale. On Manhattan Island the motor traffic has almost seized up; the motor car has become the slowest and most tiring method of transport, because the streets are too narrow. If anyone should suppose this to be due to the fact that Manhattan is a narrow island, he will find a similar situation in central London, where almost no major highway improvements have taken place during the past half-century owing to the prohibitive cost. Paris and Moscow enjoy a much more favourable position in regard to main highways; but the housing shortage in those cities is deplorable.

[1] For statistics showing that the cost of urban government in the U.S.A. varies directly with the size of cities, not only in the aggregate but also function by function, see Albert Lepawsky: *Urban Government* (National Resources Committee, Washington, 1939), Vol. I, p. 33.

The Ontario Municipal Board, after an exhaustive enquiry into the local government of the metropolitan area of Toronto, concluded that in the larger municipalities with complex administrative problems costs tend to increase with the size of the municipality, chiefly because of the larger number of employees per unit of population. (*Report of Decisions and Recommendations* dated January 20, 1953, p. 29.)

It is for such reasons as these that central or provincial governments have been led in many countries to make grants-in-aid towards the cost of municipal services even in the largest and wealthiest cities. If, however, grants-in-aid become too large in relation to sources of taxation within the control of the local authorities they lead to central domination which endangers the independence, freedom and responsibility of the municipality.

The method of sharing taxes on an agreed basis between the local authority and the national or state government has certain advantages, since it does not necessarily lead to any great increase of central control. Among cities sharing taxes with higher authorities are Chicago and Los Angeles, which receive part of the state petrol tax from Illinois and California respectively; Montreal and Rome, which get a share of the amusement tax levied by the provincial or state government; and Toronto, which has a share in the revenue of the provincial liquor control board. Many taxes are shared in Buenos Aires between the national government and the municipality, but the absence of local self-government in Buenos Aires, and the financial maladministration from which the city suffers, confuses the situation.

Lastly, mention should be made of the equalization systems in operation in Denmark and Britain. In Denmark, an inter-municipal adjustment fund was established in 1937. Into this fund are paid moneys derived from special inter-municipal taxes levied on all ratepayers in all municipalities. Contributions are made from this fund to the cost of public assistance, hospitals, and education provided by the municipalities. In addition there is an equalization arrangement whereby municipalities with a low level of wealth and a high expenditure on public assistance receive contributions from those which are richer and with a lower expenditure on public assistance. In England, the Exchequer Equalization Grant, introduced by the Local Government Act, 1948, makes the government in effect a ratepayer in those local government areas where the rateable value per head is below the average for the whole country. In addition to these national schemes, there are special schemes applicable to the capital cities of the two countries. In the Danish capital the object is to equalize the proceeds from a number of rates and taxes in the three adjoining municipalities of Copenhagen, Frederiksberg, and Gentofte, which are separate local authorities but together comprise the capital. In London, the object of the equalization scheme in force in the administrative county is to reduce the disparities of taxable capacity between the twenty-eight metropolitan boroughs.

19. PLANNING THE GREAT CITY

Most of the world's great cities grew up without any serious attempt at planning in the modern sense of the term, although this does not mean that there was a complete absence of public regulation. It was obviously necessary to control in the social interest the width and building line of streets, the height and construction of buildings, the provision of open spaces, nuisances and noxious trades, the disposal of refuse, the drainage of water, and many other vital aspects of urban life; and such matters have been subject to municipal regulation for centuries. But it is a far cry from these elementary necessities to the positive control over growth and development which is inherent in modern town and country planning. Very few great cities bear the marks of deliberate planning of a creative kind: Paris, Washington, Stockholm and a few others are exceptional in this respect. The rectangular pattern of streets found in most American cities is a form of deliberate control over the layout and development of the city, though it is a long way behind modern ideas of city planning.

Often one finds in cities which are not planned in any general sense of the term, particular aspects which bear the planner's imprint. The park system of Chicago, and the parks and squares of London, are admirable examples of the planning of a single important feature.

The need for comprehensive planning is felt more strongly and more quickly in the great metropolitan city under modern conditions than elsewhere, for the dire consequences of unplanned development soon become painfully obvious in a vast industrial and commercial centre. It is not surprising to find, therefore, that most great cities have either adopted a master plan or are in process of doing so. The Stalin general plan for the reconstruction of Moscow was adopted by the Soviet Council of Ministers in 1935. It embodies a comprehensive and ambitious scheme for transforming the city. It prescribes a maximum population limit of five millions (which has been already exceeded, apparently) and extends the area from 70,000 to 150,000 acres. It provides for a green belt of parks and forests, for the reconstruction of streets and squares, and many other important features. Some of the main items, like the construction of the great Moscow–Volga canal, and the Saratov gas pipe line, have already been completed. It may be noted that the population density has recently been raised above the average of 33 per acre laid down in the 1935 plan. It has often been alleged that the high and increasing densities found in large cities where land and buildings are privately owned is due to profit-seeking capitalists trying to exploit a limited area of land to the utmost possible extent. The public ownership system in the Soviet Union seems to yield similar results in practice, though the reasons for it may be different. It is claimed that improved methods of construction make large high buildings cheaper to construct than smaller ones; and that the cost of transport is lowered by increasing the density. Neither of these contentions can be accepted without qualifications or evidence.

A master plan has been prepared for Bombay and approved by the corporation. This also recommends that the city's population be limited to three and three-quarter millions. The density is not to exceed 140 an acre in the old city and 50 in the suburbs. The Bombay plan embodies ambitious proposals for industry, docks, housing, parks, traffic and transport, water supply and the creation of satellite towns and community centres. This plan, like many others elsewhere, awaits the means for its realization.

Amsterdam drew up a master plan in 1934 to guide the development of the city until the year 2000. It calls for a new town hall and an opera house, neither of which have so far been built. It embodies a scheme for providing the city with a magnificent forest park, and this is now one of the glories of Amsterdam. The acute housing shortage is being met partly by the development of a housing estate at Slotermeer to accommodate 30,000 persons. This housing estate (now in process of construction) presents a dreary spectacle with blocks of flats in serried rows. It is far from approaching the garden city concept.

In Chicago the modern planning movement dates from the publication of the *Plan of Chicago* in 1909 by a group of businessmen and civic leaders. The plan aimed at improving the aesthetic appearance of the city, its housing, transport and traffic facilities, highways, parks and amenities. There was much idealism embodied in this early plan, but its authors were reluctant to impose restrictions on the use of private property.[1] Shortly after it was promulgated, the municipality appointed an official City Plan Commission to carry it into effect. Since then large sums of money have been raised to finance the capital expenditure required to carry out particular features of the plan. These include the acquisition of outlying forest preserves, street improvements (including two-level highways on Michigan Avenue and Wacker Drive), parks, airports, straightening the course of the river, and reclaiming the lakefront. Much has been done, but much more remains to be accomplished before

---

[1] For a severe criticism of the City Beautiful movement inaugurated at Chicago, see Lewis Mumford: *City Development* (London, 1946), p. 19.

Chicago, which still has vast areas of derelict dwellings and slum quarters, can be regarded as a well-planned city. The idea of city planning in at least its visual aspects has, however, to some extent entered the minds of the citizens. The real problem is the planning of Greater Chicago; and here the only unifying agency is an unofficial body known as the Chicago Regional Planning Association. This has not so far exerted any considerable influence on the planning and development of the metropolitan area.

In New York a remarkable report on Regional Planning for New York State was drawn up by Henry Wright in 1926. This was followed by a much more elaborate planning report covering a wider area directed by Thomas Adams, under the auspices of public spirited citizens and financed by the Russell Sage Foundation. The Regional Plan of New York and Environs was published in 1929–31 and in its day attracted great interest. The region comprised in the plan lies in three states, and includes hundreds of public authorities, public utility undertakers and railway companies. As in Chicago, there are no governmental organs possessing planning powers over the whole region, and under the conditions hitherto prevailing in the United States, no comprehensive plan can be carried out in a metropolitan region except through the voluntary co-operation of all the many authorities concerned—a result which is almost impossible to attain.

We can thus see that the New York region is not legally recognized as an entity; and, as Professor Tugwell explains, because it has no comprehensive government, it has no continuing and comprehensive planning. Whatever the merits or defects of the New York regional plan may be— and this is a matter on which opinions differ—the political, legal, and administrative obstacles to the fulfilment of that or any other plan of development covering the region are insuperable.

Impediments of a similar kind exist in nearly all countries where the metropolitan area presents a major planning problem. In Denmark, for example, a town planning law passed in 1938 requires every town or built-up area containing more than 1,000 inhabitants to prepare a town plan; but the act makes no provision for the comprehensive planning of several municipalities forming a single region, nor did it create a regional planning authority. To fill this gap in the official machinery, an unofficial committee (which some years earlier had considered the question of open spaces in Copenhagen) took on the task in 1945 of drawing up a plan for Greater Copenhagen. In 1948 it issued an outline plan for the metropolitan region. This plan envisages that future development should be connected with the railway lines which radiate from the centre of the city. This proposal would result in a formation resembling the human hand,[1] the palm consisting of Copenhagen and the five fingers representing the building development along the five radial railways. The spaces in between these spread out fingers would be mainly devoted to green areas.

The regional planning committee which was responsible for this work possessed no executive power. It nevertheless contained members representing the municipalities of the Copenhagen region and of many other authorities and institutions interested in planning. Unfortunately, its proposals are in conflict with the recommendations of the Metropolitan Committee, an official body set up to report on local government reorganization in the metropolitan area.[2]

The situation in Copenhagen resembles in essentials the position in New York, in that regional plans have been made by unofficial bodies but no public authorities exist to carry them out in practice.

Paris has had a regional plan since 1934–35. Greater Paris for this purpose comprises the departments of the Seine (which includes the Ville de Paris), Seine-et-Oise, Seine-et-Marne, and five cantons of the Oise. The individual communes included within this extensive region are required to prepare plans which conform with the regional master plan. The communal plans, and also the regional plan, have to be submitted for approval to an advisory body known as the *Comité d'amenagement de la région parisienne*, which contains senior civil servants, representatives of local authorities, members nominated by associations, and experts.

The most important proposal in the master plan for the region was the creation of a satellite town at Orly, to be built by private enterprise with the help of assistance from the City of Paris. This proposal has not been carried out for various reasons. The site proved to be unsuitable for large-scale building, private enterprise was not attracted by the proposition, and the financial situation of the Ville de Paris did not enable it to provide the funds required to pay for the highways, transport, and public utility services needed for the Orly development.

Compared with other cities, Mr. Chapman truly remarks, Paris is in

---

[1] See page 252.

[2] This body, which reported in 1948, considered three possible methods of reform and finally decided in favour of a compromise solution based partly on extending the boundaries of Copenhagen to include seven neighbouring municipalities, and partly on a weak type of federation which would provide a league of suburban municipalities with a metropolitan council to look after matters affecting the entire region. From ministerial statements it appears that these proposals are unlikely to be carried out.

many respects already a model of town planning. With so rich a heritage of beauty, spaciousness, dignity and splendour, one can readily understand the tendency of town planners in Paris to concentrate on aesthetic and conservation aims. Certainly planning in Greater Paris seems to adopt an unduly negative attitude towards development. French governments are not inclined to favour direct development by public authorities, but tend rather to assist private enterprise and encourage it to move in desirable directions. Housing development since 1945 in the Paris area seems to have been totally inadequate to the needs of the community by any standard; yet the action taken by public authorities to stimulate and foster the building of new dwellings is unimpressive. It is unlikely that planning will be effective and aggressive unless the public authorities initiate or participate actively in development, not only with respect to schools, roads, public utilities, etc., but particularly in regard to housing. Such direct action has been especially needed in France, where for many years the legal restrictions on rent have been so severe that they have discouraged new building by private enterprise for investment purposes or occupation, and also made it impossible for many existing blocks of flats and houses to be maintained in good condition by the owners.

The planning problem in Rome is rendered exceptionally difficult by the need to provide for the increased traffic needs of the city and to remedy the housing shortage while preserving and even improving the display of historical monuments and artistic buildings which adorn the Italian capital. Moreover, the overspill of the poorer classes from the overcrowded tenements into a ring of suburbs outside the main city has created a considerable danger, for in many of these outlying areas the general living conditions are primitive and the public services inadequate. Squalor, disease, and poverty are rampant in these suburbs, which are a potential menace to the whole metropolitan community. Several attempts have been made to produce a master plan, notably in 1931 and again in 1940-41; but so far the planning of Greater Rome—that is, the whole metropolitan area—has not been achieved, nor are there any organs available for the purpose.

Wellington is the smallest city described in this book, but it nonetheless has its regional planning problem. As early as 1929, New Zealand legislation permitted neighbouring local authorities, with the approval of the Town Planning Board, to set up a regional planning council to survey the resources of the area and to prepare a regional planning scheme. The Wellington Regional Planning Council was established under this Act, representing all the local authorities in the region. It is at present engaged in surveying the natural resources of the region. In due course it will produce a regional plan. This plan will not be legally binding, but it will serve as a guide to the constituent planning authorities in preparing their own schemes. There are weaknesses in the administration of this planning legislation, but Wellington is in advance of many larger cities in having a regional planning council, even though its powers are only advisory.

In Sydney also an attempt to deal with the planning of the metropolis was made by the Local Government (Town and Country Planning) Amendment Act, 1945. This constituted the Cumberland and County District and provided for a county council to be elected by the councils of the cities, municipalities, and shires of the area from among their own members. The county council has been given responsibility for preparing a master plan for an area exceeding 1,500 square miles. The master plan has been prepared in collaboration with the individual local authorities; and it deals with the comprehensive development of the metropolis. The plan has been submitted to Parliament by the appropriate minister.

In Toronto a city planning board was appointed in 1942. Its plan was approved by the city council in 1949, and by the Ontario Minister of Planning and Development in 1950. An attempt to plan the metropolitan area of Greater Toronto began with the creation of the Toronto and York Planning Board, composed of five representatives from the city and four from the county of York, which contains most of the outlying portions of Greater Toronto. The board was required to prepare a comprehensive development plan covering all the main aspects of regional planning—industry, commerce, agriculture, housing, highways, open spaces, a green belt, and so forth.

A further important development took place in 1953, when a law was passed by the Ontario legislature setting up a metropolitan government for Greater Toronto.[1] The metropolitan council has become the principal planning authority in the Greater Toronto area, but the provincial Minister of Planning and Development is required to define a larger area which will form a complete planning unit, to be known as the Metropolitan Toronto Planning Area. A planning board will be set up in this area, composed of representatives of the metropolitan municipality and other local authorities whose territory is comprised in the area. The powers of the planning board will include land uses in the broad sense, ways of communication, sanitation, green belts and parks, and public transportation.

[1] *Ante*, pp. 29 and 33.

Despite a good deal of town and country planning legislation from 1909 onwards, local authorities in Britain made only slight use of their powers before 1939. An outstanding achievement during the period between the two world wars was the development by Manchester of Wythenshawe, an estate of more than 5,000 acres, which was incorporated in the city in 1930. Wythenshawe has been planned and developed by Manchester City Council as a delightful garden suburb containing houses, shops, factories, schools, parks, cinemas, inns, and other amenities, to accommodate a population of about 90,000. It represents an original conception distinct alike from an ordinary municipal housing estate, an independent garden city like Letchworth, or a satellite town such as Welwyn. It was financed by Manchester out of local rates and is now nearing completion.

Regional plans have been prepared for the metropolitan area centred on Manchester, in addition to the separate plan for the city itself. The first one, published in 1945,[1] was the work of the Manchester and District Regional Planning Committee. This foreshadowed the more detailed proposals made by the South Lancashire and North Cheshire Advisory Planning Committee in 1947[2] for moving population and industry out of Manchester, while preserving a green belt and a sufficient amount of open space within the built-up area of the city.

The plan for the city envisages a radical redevelopment of Manchester on bold and progressive lines. The central area will be devoted to commerce and administration, the great hospitals, the university, and other cultural institutions. Residential areas will be designed to form neighbourhood units, consisting of small communities of 5,000-10,000 people, which will be combined to make larger districts of about 50,000 population. Each district will have its own schools, health centres, libraries, churches, cinemas, shopping centres, and other community institutions. The population density in the more crowded parts of the city will be drastically reduced by providing open spaces and individual houses with gardens rather than blocks of flats. In consequence, an overspill of at least 150,000 persons will have to be rehoused outside the city boundaries; and similar demands are being made by the neighbouring towns of Salford, Stockport, and Stretford. In the light of these demands the regional plan takes on a heightened significance as an instrument for

[1] Report on the tentative Planning Proposals made to the Manchester and District Regional Planning Committee by R. J. Nicholas (Jarrold, 1945).

[2] See the advisory plan prepared by R. J. Nicholas and M. J. Hellier for the South Lancashire and North Cheshire Advisory Planning Committee (Manchester, Richard Bates Ltd., 1947).

redistributing population, industry and housing on saner lines throughout the metropolitan area.

London has produced even more ambitious plans than Manchester. There are three related plans. First in point of time came the County of London plan, completed in 1943. It was accepted in principle by the London County Council and has formed the basis for the Council's subsequent planning activities. This plan provides for the decentralization of 500,000–600,000 persons from the County of London to outlying areas. It calls for a vast amount of rebuilding and redevelopment of all kinds—commercial, residential and industrial. Congested business and shopping areas in the central core are to be redesigned on new lines to fit them for commercial purposes in the present age. New ring and radial roads, electrification and improvement of the railway system (including the re-siting of main terminals) and the creation of 'precincts' dedicated to special purposes, such as government offices, the legal profession, the university, are leading features of the plan.

In 1944, Sir Patrick Abercrombie, one of the authors of the county plan, produced the Greater London Plan at the invitation of the Minister of Town and Country Planning. The Greater London area extends over 2,525 square miles, and affects a population of about 10 million persons, of whom about $6\frac{1}{2}$ millions live in Outer London and slightly less than $3\frac{1}{2}$ millions in the county.

We have already noted that the county plan provides for the decentralization of 500,000–600,000 persons. The Greater London Plan adds a further 415,000 to this figure, making a total of about a million inhabitants to be moved. About three quarters of them will be decentralized within the region—or just outside it—by means of new towns, the development of existing towns, and quasi-satellite housing estates. The remainder will, it is hoped, be dispersed outside the region.

The most striking proposal is the creation of eight new self-contained towns outside the green belt ring. These new towns are being planned as complete entities on sites designated by the Minister of Housing and Local Government. The responsibility for constructing them falls on development corporations established under the New Towns Act, 1946. The new towns are certain to be a notable contribution to modern planning practice. Each one is designed for a specified maximum population which varies between 25,000 and 80,000. A new town is intended to contain a balanced community most of whose members will both live and work there. It is therefore necessary to induce industrialists, businessmen and public authorities to establish factories, shops, offices, laboratories, etc., in the new towns. The new towns which are taking shape belong to the garden city type, and although they will be largely self-contained they will for certain purposes be satellites of the central city.

Severe restrictions are imposed on the entry of new industry into London or the development of existing factories there. The part of the Greater London plan dealing with industry is a masterly contribution. The problem is approached from the standpoint of decentralization, redistribution of industry, and its orderly development. The plan provides for the removal of a substantial amount of industry and commerce from central London, and recommends that new industrial development shall be guided to the new towns or to numerous existing small towns whose expansion is desired. New factories or mills will be prohibited elsewhere in the region unless a special case can be made out in each instance on economic or technical grounds.

The Greater London plan embodies very far reaching recommendations regarding communications and transportation. Roads, airfields, inland waterways, railways and markets are all treated comprehensively. Open spaces and amenities are also dealt with at length and in detail for the whole region. There are recommendations dealing with the preservation of the countryside, the protection of areas of special scenic beauty, amenities connected with the Thames and lesser rivers or waterways, the construction of parkways and footpaths, bridle and bicycle tracks, rest gardens, children's playgrounds, recreation and sports centres, town squares and town parks.

Finally, a plan for the ancient City of London, the small central core from which the great modern metropolis has sprung, was published in 1947. This aims at large improvements for this famous area which accommodates finance and trade, insurance and shipping, and other important economic interests.

So far as we are aware, the plans for London excel in magnitude, breadth of imagination, and boldness of conception, those for any other great city. The plan for Moscow, fine as it is, deals with a much smaller area and a population only half as great as that of Greater London. Indeed, the Moscow plan can hardly be said to cover a metropolitan region in any way comparable to that of Greater London. The New York Regional Plan covers a comparable population; but it has no official status, since it was produced by an organization of private citizens. Moreover, the New York Regional Plan never embarked on such adventurous proposals as the planning and building of eight or nine self-contained towns in the outlying parts of the metropolitan area, or the decentralization on a vast scale of population and industry from the central city.

But although the plans for Greater London were prepared by public authorities exercising statutory powers and spending public money, Greater London is in the same position as Greater New York in having no regional authority which is responsible for carrying out the plan as a whole. There is no Greater London municipal government, only the London County Council and a strange medley of local authorities of various kinds outside the county. Any co-ordinated action which takes place depends on the influence exerted by central government departments, and in particular the Ministry of Housing and Local Government. But there are strict limits to what central departments can do in those spheres of development where the initiative and discretionary power belong legally to the local authority or a public utility undertaking. In such circumstances, the ministry must rely on persuasion combined usually with the power of the purse.

So once again we are thrown back on the political and administrative problem of the great city: how to obtain a metropolitan government which will be able to plan and develop the metropolitan area as a whole, and to govern the metropolitan community as an entity. It is a principle of town and country planning administration that the whole area to be planned shall come under the jurisdiction of a single planning authority, in order that the needs of its different parts may be considered 'as with a single eye,' and the competing claims of sectional or local interests subordinated to the overriding good of the whole.

Town planning has made considerable headway in Stockholm and the city has a comprehensive development plan. The population of Greater Stockholm was 980,000 in 1950 and that of the city proper, 740,000. Planners estimate that the figure for Greater Stockholm will have risen by 400,000 within the next twenty years. The living space standards of the city are already extremely low, most of the housing improvements in recent decades having taken the form of better kitchen equipment, central heating, bathrooms and refrigerators. As recently as 1947 about half of all the dwellings consisted only of one room and kitchen. A much higher proportion of more commodious dwellings is now being built, and this, combined with the rehabilitation and laying out afresh of the overcrowded parts of the city, will result in an overspill of 140,000 persons to be accommodated outside Stockholm in the fairly near future. To these must be added the anticipated increase of population already mentioned, making a total of 540,000.

D

Building in Stockholm has developed in the direction of large blocks of flats containing very small individual apartments in which restricted families lead a cramped or overcrowded existence. According to the criterion usually accepted in Sweden, no less than 25 per cent of the entire population of the city, and 43 per cent of all the children in Stockholm under 15 years of age, were living in overcrowded dwellings in 1945.

In Sweden, as in England and the United States, there is an overwhelming preference for living in individual houses rather than in flats. Yet 90 per cent of the housing accommodation now being built in the northern capital consists of flats. There are signs of a slight change to a more rational housing policy based on meeting the known demand for one-family houses, but so far the public authorities have not committed themselves on this vital question. One highly favourable factor which should facilitate intelligent planning and housing in Stockholm is that a third of the land is publicly owned.

Despite the formidable overspill problem facing the Swedish capital, regional planning has not yet advanced to the stage of envisaging satellite towns similar to the new towns of Britain. Attempts are being made, however, to encourage business men and industrialists to locate their establishments and factories outside the city; and it is hoped that eventually half of all gainfully employed workers will both work and live in the outlying suburbs. The plan for Stockholm proper embodies many large-scale improvements in the way of arterial highways, a new underground railway in course of construction, and more open spaces.

20. THE PROBLEMS SUMMARIZED

In this introductory analysis we have surveyed broadly the many different kinds of problems which confront great cities of the world today. In the chapters dealing with individual cities contained in Part Two the contributors discuss in greater detail the incidence of these problems as they exist in the circumstances peculiar to each metropolis. The details vary from country to country and from city to city; and no metropolitan community has to cope with all the problems in an intense form at any one moment. But in general there is sufficient similarity between the difficulties which have arisen in many different countries to enable us to say that great cities all over the world are facing common problems which are mainly due to similar causes.

These problems may be classified under the following five headings:—

1. *Organization of areas and authorities*

Under this heading comes the question of evolving a constitutional framework for the great municipality which corresponds with the facts of population, territorial size, social and economic life in the mid-twentieth century. It involves the administrative integration of the whole metropolitan area for the large-scale services which require unified planning, co-ordination or administration. It also involves smaller, more compact units of local government to perform the functions which can best be administered by smaller municipal organs. The reform of metropolitan government thus demands both more centralization and more decentralization; in other words, both larger and smaller areas and authorities. It also requires the absorption of many *ad hoc* authorities which have been created on a widespread scale to provide one or more services in a special district. The fragmentation of metropolitan government has already proceeded too far in many cities. Even where an *ad hoc* authority can be justified on rational grounds, and not merely explained by the specious excuse that this method makes it possible to avoid much-needed reforms which are too difficult politically to carry through, the proper relationship of *ad hoc* bodies with the city government still requires careful thought. 'Divide and Misrule' seems to be the maxim which legislators have hitherto followed in devising the constitutional organization of many great cities.

2. *Popular interest and democratic participation*

The problem of evoking and holding the interest of the citizens in the affairs of the great city, and of securing their active participation in its government, is one of great importance. It has nowhere been solved, despite exaggerated claims put forward in regard to Moscow. Electoral apathy is usually more marked in the great city than in towns of smaller size; and popular interest in the government of the metropolis also compares unfavourably with that shown in government at the national level. Many of the defects of metropolitan government are directly traceable to public indifference or popular ignorance. The low level of popular interest in, and public understanding of, the affairs of the great city even among the more educated classes largely explain the difficulty of finding men and women of high mental and moral calibre who are willing to serve on the city council or its executive organs. They explain, too, the nepotism and corruption which occur in some great cities. They account for the ease with which self-interested politicians or officials obstruct or defeat attempts at improving the constitutional organization of the metropolis. Above all, it is only the low level of civic interest in city affairs which makes it possible for great cities like Paris, Rio de Janeiro, Buenos Aires, Bombay and Calcutta to be virtually governed by prefects, mayors or commissioners appointed by the central or provincial government, with only nominal or limited power, if any, given to the representative organs.

The public apathy towards the government of the great city is the result of many different factors. One of them is the divided interest and allegiance resulting from the separation of workplace and home in different municipal areas; another is the huge size of the very large metropolis. But whatever the causes, only a deliberate and sustained effort to educate metropolitan man in the politics and government of the vast metropolitan community of which he is a member will solve the problems arising under this head.

3. *Efficiency of the municipal services*

The product or end-result of city government in the great municipality is by no means satisfactory. Recurring themes in the studies which follow are the inadequacy of highways to carry modern traffic, the public transport problem, congested living quarters, the growth of slums, excessive density of population in the poorer parts of the city, shortage of open spaces, housing inadequacies, the problem of overspill, difficulties of carrying out major engineering works such as those connected with water supply or drainage. The great city of today lives by a miracle. It is not operating on a satisfactory standard in regard to schools, housing, open spaces and playing fields, traffic circulation, public transport, police protection and many other services. The causes are complex and diverse. Sometimes they spring from maldistribution of functions between the great municipality, *ad hoc* bodies, and national or provincial authorities. Sometimes they derive from defective organization within the metropolitan region, particularly in the allocation of functions among many small, poor and unco-ordinated authorities. Sometimes the chief causes are corruption or sheer incompetence on the part of the city government, which are in turn due to the public apathy or ignorance already referred to. An excessive bureaucratization of the city government may result in public needs being overlooked or sidetracked by stiff-necked officials. This can easily happen when the affected classes or groups are not articulate or do not possess effective methods of making their needs known to the authorities.

4. *Finance*

From many cities comes the complaint of inadequate financial resources to provide the municipal services of the great city. Wealthy New York City joins with impoverished Calcutta, Sydney with Bombay, in declaring their inability to find the money required to finance current municipal services and new developments. A chronic state of financial stringency appears to be a permanent condition of many metropolitan cities. This condition is one which afflicts both the wealthiest and the poorest cities. It is this which explains the fact that in many great cities highway development lags so far behind the proved needs of motor car traffic that it is scarcely an exaggeration to say that the central areas have almost seized up. In London the average rate of locomotion through the central highways is estimated to be slower than it was in the days of horsedrawn vehicles at the beginning of the century! Housing, education, water supply, the police, even street paving, are often less adequate in quantity and quality in the great metropolis than in the smaller towns, though some great cities have the highest achievements in every branch of municipal activity to their credit. The point I am making is that we cannot assume that mere size will by itself guarantee high standards of government and administration. Finance is all too often the Achilles heel of the metropolis.

5. *Planning the metropolitan region*

Many of the most difficult problems which confront the great city can only be solved by far reaching and imaginative planning. The elimination of overcrowding in the densest quarters; the provision of a reasonable amount of land devoted to parks, playing fields and other out-door amenities; the opening up of the suffocating alleyways in the older parts of ancient cities, like Stockholm or Copenhagen, or the narrow cañons which shut out light, sun and air in downtown New York or Chicago; the reclamation of slums and derelict residential, commercial or industrial areas; the avoidance and elimination of high population densities in the central areas in order to exploit the enormous land values which accrue in the great metropolis; the reduction and prevention of long, exhausting, and expensive journeys to work which waste the money, health, energy and time of those who live in the outlying suburbs of the great city: these and cognate aims to cure the maladies which are commonly found in the great metropolis can only be achieved by well considered, drastic and creative planning.

Such planning must secure control of the whole metropolitan region which is centred on the great city. It must comprise both town and country within its scope. It must be positive as well as negative: it must prescribe what shall be done and not merely what is forbidden. Authorities to carry out the plans must exist both at the regional and local levels, and they must have effective powers and adequate territorial jurisdiction. Much positive development must be entrusted to them where private enterprise is unwilling or unable to carry it out.

At this stage of the argument we reach a point where the question we are discussing is really the future of the great city. For although primarily this book is concerned with the government and the politics of great cities, it is also concerned with their planning. And in considering town planning one is bound to examine the social life of the vast communities which live in these huge urban aggregations.

21. THE METROPOLITAN REGION OF TOMORROW

The easy-going assumption that the bigger the city the better existence would be for everybody concerned, prevailed throughout the nineteenth century. It is no longer accepted in the leading centres of social and political thought. Sociologists and political scientists have in recent decades made some penetrating studies of great cities, beginning with Charles Booth's great pioneering study on *The Life and Labour of the People in London*. This was published in no less than eighteen volumes in 1903, and is a landmark of social investigation which had profound consequences both in the world of thought and the world of action. The most profound study of a general kind is Lewis Mumford's fine book *The Culture of Cities* (Secker and Warburg, 1938). A more journalistic and less balanced account is Robert Sinclair's *Metropolitan Man* (Allen and Unwin, 1937).

Mumford delivers a coruscating criticism of the modern metropolis. He considers it to be economically unsound, politically unstable, biologically degenerate, and socially unsatisfying.[1] Beneath the proud surface glitter of the great city and its artificially induced prestige and glamour he detects much that is meretricious and unreal. He sees the great city as too dependent on paper, on finance, insurance and advertising;[2] he sees the bulk of its inhabitants leading a drab and unhealthy existence too remote from the sights, sounds and smells of the countryside; he sees them wasting enormous quantities of time, energy, money and vitality on subway journeys and other debilitating forms of travel devoid of value

---

[1] *The Culture of Cities*, Chapter IV.   [2] *Ibid.*, p. 256.

in themselves.[1] 'The metropolitan world,' he contends, 'is a world where flesh and blood is less real than paper and ink and celluloid. It is a world where the great masses of people, unable to have direct contact with more satisfying means of living, take life vicariously, as readers, spectators, passive observers. . . .'[2] The physical form of the great metropolis he declares to be shapeless: to consist of mere size without significance. Even when one finds order, design and meaning in the central area, as in Paris, Madrid or Buenos Aires, the surrounding districts present a picture of a 'vast enveloping aimlessness.'[3]

There is much truth in Lewis Mumford's indictment, though it is not the whole truth about the matter. His work is profound, imaginative, and deeply-felt. That a revaluation is needed of our ideas about the great city as a habitat for man can scarcely be doubted.

In the nineteenth century, mere size was regarded as a desirable attribute of cities, and 'the bigger the better' was a maxim which embodied public opinion on the subject. There was little or no attempt to limit or control urban growth. Today, the overblown, dropsical city of elephantine proportions can no longer be regarded as desirable or even tolerable in its present condition.

The period of unqualified acceptance and unfettered growth of the metropolitan city is passing. We have already moved into an era in which the amorphous, shapeless, sprawling metropolis is being challenged by new ideas which will eventually transform it into something very different.

It is now clear that effective planning of a metropolis is impossible unless a limit is placed on its maximum size and population. The essential instrument for enforcing this limit is control over the location of industry. If new industry is permitted to enter the metropolitan area freely, and existing industries are allowed to expand indefinitely, no real control over either the ultimate size of the metropolis or its design is possible.

In the second place, the abolition of slums, the clearance of overcrowded tenements, the widening of narrow, airless streets, and the rehousing of the people at a reasonable standard, inevitably produces a big problem of urban dispersal and overspill. This can only be solved on rational lines by redistributing population between the overcrowded quarters of the central city and the suburbs, towns, villages and housing estates in the outlying parts of the metropolitan region.

Decongestion means decentralization; but it is senseless to move people out to the suburbs or semi-rural housing estates if they are obliged to work

[1] *The Culture of Cities*, Chapter IV, p. 243.   [2] *Ibid.*, p. 258.   [3] *Ibid.*, p. 234.

in the great city and to make costly, exhausting and time-consuming journeys to and from their work. It is far better also to move out the factories, warehouses, offices, and workplaces, or to encourage new industrial development in satellite towns or garden cities where the work-people can enjoy not only good housing conditions, but an agreeable environment and easy access to the countryside. By doing so, we can once more bring together the home, the workplace, and the playground or recreational centre: that essential triad of human life which has become scattered and disunited in the welter of the modern metropolis. We shall thereby restore family life to the primacy it once enjoyed but has since long lost in the turmoil of the great city. Finally, in offering easy access to the countryside to the citizens who move out from the congested central area to the outlying towns and settlements, we must insist on drawing a clear line of demarcation between town and country, in order to ensure that rural amenities are not destroyed by those who would enjoy them. The worst feature of many metropolitan areas today is that much of their environs consists of neither town nor country, but merely of land suffering from urban blight.

When we have done all this and much else which is in the minds of our most creative planners, our most imaginative social thinkers, and our leading practical reformers, we shall find that we have transformed the great metropolis. In place of an inchoate metropolitan community continually shuttling back and forth between suburb and central city, wasting its energy, time and money in nerve-shattering journeys to and from work, or living in overcrowded apartment houses in back streets, we shall gradually create an ordered, coherent, decentralized metropolitan region.

The central city will contain the institutions which need vast resources and the support of large numbers of people: the university, the opera house, the chief theatres; the great museums and art galleries; the principal reference libraries; the concert halls for symphony orchestras and famous musicians; the exhibitions of a specialized or highly cultivated character; the leading centres of scientific teaching and research; the superior courts of law; government offices of many kinds; the principal broadcasting and television studios; the headquarters of banks, insurance companies, and other financial institutions; the wholesale markets; the most important shopping centres for such goods as furniture, jewellery, motor cars, works of art, fashionable clothes, etc.; the headquarter offices of great commercial undertakings; the principal publishers of books, periodicals and newspapers; the more elaborate and highly developed hospitals for teaching and research; and much else which can best flourish in the great city and cannot easily be decentralized. The outlying towns or settlements of the garden city type will contain houses and flats, factories and offices, shops and laboratories, schools, cinemas, civic centres, lending libraries, and many other institutions to enable the local communities to lead a largely self-contained life for much of their time, but with recourse to the central city when occasion demands.

The metropolitan city has played a great part in the history of mankind. It can continue to play a great part in future, if we guide its destiny wisely and well. I have referred in these concluding pages to some of the severe criticisms which have recently been levelled at the metropolis. But there is much to love and to admire in the great city. It is the home of the highest achievements of man in art, literature and science; the source from which the forces of freedom and emancipation have sprung. It is the place where the spirit of humanism and of democracy have grown and flourished, where man's quest for knowledge and justice has been pursued most constantly, and truth revealed most faithfully and fearlessly.

Let us bear all this in mind, and remain the friends and lovers of the great city in all that we do to improve its government, its politics and planning, and thereby raise the quality of life of its citizens.

PART TWO

Amsterdam

J. P. WILDSCHUT

Born 1917. Educated: Gymnasium B and Municipal University of Amsterdam. LL.D.

Administrative planning expert, State Institute of Town and Country Planning. Formerly Deputy-Secretary to the Board of Burgomaster and Aldermen and assistant to the Town Clerk of the municipality of Amsterdam. Was for one and a half years scientific collaborator at the State Institute of War Documentation.

Lecturer in administrative law at the municipality and in philosophy at university tutorial classes. Secretary-editor of the philosophical section of the new Winkler Prins Encyclopaedia.

Amsterdam

IT is probable that at the beginning of the twelfth century Amsterdam was formed as a small settlement of fishermen and merchants where the river Amstel flows into the Y, a deep arm of the former Zuiderzee (now Yssellake). Since olden times the citizens of Amsterdam have been confronted with the problem of the all-surrounding water. The first inhabitants canalized the Amstel by building a dam about a quarter of a mile from its mouth. That is why the settlement was called Amstel-dam. At the same time they constructed a dam along the banks of the Amstel and furthermore along the Y and the Zuiderzee, thus protecting the hinterland against tidal waves.

The first inhabitants settled down on the dyke by the two banks of the river Amstel and round the dam. This dam is still called the 'Dam' and it still forms the centre of the city. There we find the Royal Palace built (1648–55) by the famous Dutch architect, Jacob van Campen; next to it the *Nieuwe Kerk* (New Church) where Dutch kings and queens have been inaugurated since 1814. Moreover this oldest part of Amsterdam is a trade centre; the commodities exchange, the stock exchange, large warehouses, and office buildings are established there. One of the dykes (*Nieuwendijk*, or New Dyke) is one of the busiest shopping centres of the city.

Because of fires in the fifteenth century, the oldest archives of Amsterdam have been destroyed. The first document in which the city was mentioned dates from 1275, which is rather late in comparison with other Dutch cities, e.g. Utrecht and Haarlem. That the waterland round Amsterdam must have been inhabited long before is proved by the fact that already in 900 the fishing rights of Wormer, a small village about twelve miles north of Amsterdam, were rented by the Bishop of Utrecht.

In the document of 1275, freedom from toll dues was granted to the inhabitants of Amsterdam by Floris the Fifth, Count of Holland, for the merchandise passing through his county. It is thus evident that in those times there were merchants as well as fishermen in Amsterdam, and that it was a centre of transit trade in the county. It is probable that trade was carried on with the hinterland (Amstelland) and the other side of the Y. There is, however, no historical evidence of this.

The first expansion of the city—which had already taken place at the beginning of the fourteenth century—was eastward. The *Oude Kerk* (Old Church) in this part of the city dates from the beginning of the thirteenth century; it is still the oldest church of the capital. Canals were dug parallel with the Amstel, partly as fortifications, partly as junction canals.

Shortly after this first expansion, another enlargement took place, this time westward. In the first half of the fifteenth century the city was surrounded by walls. Only a few fortified gates and relics of foundations are left, such as the *Schreierstoren*, the *Waag* (the Weigh House where Rembrandt painted his 'Anatomy Lesson'), the *Munttoren* (Mint Tower, nowadays a notorious traffic obstacle!) and the *Torensluis* (Tower Sluice). On the seaside, the city was protected by a palisade wall at about the same place where later the Central Station was built.

Evidently the city's geographical position was exceptionally favourable, for sea-going vessels could reach Hamburg and the Baltic via the inland seas, Zuiderzee and Waddenzee, thus not risking the dangers of the open North Sea and of piracy. Although Amsterdam has never been a member of the Hansa league, it carried on a lively trade with the Hanseatic cities, mostly in wood, beer, and skins. In the second half of the fourteenth century, 25 per cent of the Netherlands Baltic trade was conducted through Amsterdam.

Shipping and trading activity increased considerably when the Hollanders started their travels to the East Indies after the Act of Seclusion. Pre-eminent among them were the Amsterdammers who ousted the Portuguese and made repeated voyages of discovery. Amsterdam became the staple market for Indian products and within a century it was one of the richest and most prosperous cities of the continent.

Naturally, this caused a considerable expansion of the city. In the beginning of the seventeenth century, three semi-circular canals were projected around the old city. It was this plan that gave Amsterdam its typical form and remarkable beauty; foreigners soon named it 'the Venice of the North.' The extension proved to be so perfect that during two and a half centuries it met the demands of the inhabitants. Merchant houses were built along the canals, often serving both as living accommodation and warehouse space. The 'Golden Age' gave Amsterdam a beautiful belt of canals (called *grachten*) and a large number of distinguished monumental buildings.

After the rapid development of the previous century, the eighteenth century brought a period of consolidation of the wealth and trade obtained from the overseas settlements. But at the end of this century

I. AMSTERDAM
An aerial view of the old city.

[Photo: *Fototechnisch en Cartografisch Bedrijf K.L.M.*]

2. AMSTERDAM

An example of the beautiful seventeenth-century architecture which still abounds in the city.

and the beginning of the nineteenth century a severe blow was struck
to the economic development of Amsterdam by the Napoleonic wars.
It was only in the second half of the nineteenth century that Amsterdam
recovered from this decline and regained its previous wealth and acquired
new prosperity. Colonization of the East Indies began to give profits;
the 'culture system'[1] introduced there, benefited Amsterdam to a great
extent. King William I (1815–31) had a clear view of Amsterdam's
interests overseas and so took the initiative of establishing the *Nederlandsche Handel Maatschappij* (Netherlands Trading Company), which is
still the representative body on overseas banking affairs. The King also
initiated the first railway junction from Amsterdam to Germany.

There was, however, another problem. In the nineteenth century
sea-going vessels became too big to reach Amsterdam via the ancient
connecting route of the Zuiderzee-Y. For technical reasons it was not
deemed prudent to dig straight through North Holland. Again it was
William I who initiated the laying-out of the Noord-Hollands canal
which was dug from the much lower bog land to the north of North
Holland (1825). This canal, however, never gave satisfaction; sea-going
vessels had to pass sixteen bridges to reach Amsterdam.

It was not until 1876 that a canal was constructed straight through the
dunes on the coast of Holland (North Sea Canal). The canal has been
broadened and deepened repeatedly since; the work is still going on at
the moment. Amsterdam can be reached now by the largest sea-going
vessels, for the locks at Ymuiden are 45 feet deep and the canal itself has
a depth of about 40 feet. In 1950 a French ship, the *Pasteur* (30,447 B.R.T.),
arrived in the port of Amsterdam; she was the largest ship ever to reach
the capital's harbour.

Still there was no good waterway to the hinterland of Germany and
Central Europe; it was partially provided in 1896 when, with the Merwede
Canal, a connection was made with the Rhine. In 1952, however, a new
canal (Amsterdam-Rhine Canal) was opened to form an up-to-date
shipping communication with the national and international hinterland.
From the frontier, all Rhine vessels can reach Amsterdam along a route
of about 80 miles (formerly 105 miles).

In the second half of the nineteenth century, Amsterdam manifested
economic and cultural activity in several directions. Its diamond-cutting
industry became famous all over the world. The Amsterdam Bourse
brought a lively money trade by its international stock exchange prices.

[1] The 'culture system' was the system of regulating colonial plantations in the Dutch
East Indies so as to control the economic trade of these former colonies.

Even after abolition of the 'culture system' in the Netherlands East Indies, trade with the colonies flourished. Apart from that, Amsterdam once again became a centre for poets (the writers of the 80's), painters (G. H. Breitner, Jan Veth, etc.), and musicians (the Concertgebouw Orchestra under the direction of the late Professor Willem Mengelberg). The Athenaeum became the University (1877). In 1922 a faculty for economics and commerce was founded. Nowadays Amsterdam has two universities, a municipal one and a private Orthodox one.

The revival of the last century caused a new expansion of the city. The first phase does not bear witness to pre-conceived planning or any aesthetic feeling. Fortunately, soon after more responsible planning was done so that only a few quarters of the city now disfigure the beauty of Holland's capital.

THE SIZE AND STATUS OF AMSTERDAM

Amsterdam is the capital of the Netherlands without being the seat of the Government, which resides in The Hague. H.M. Queen Juliana lives in Soestdijk Palace in the province of Utrecht. This peculiar situation has an historical explanation. The Republic of the Seven Provinces—each with their own capital and with full autonomy—used to hold its meetings in The Hague. When the Dutch kingdom, after the Congress of Vienna, was definitely formed in 1815, a revival of provincial power was much feared, so Holland, by far the largest province, was divided into North and South Holland. (To say 'Holland' for 'the Netherlands,' in vogue abroad, is a misnomer like saying 'England' for 'Great Britain.')

Before the kingdom was formed, Amsterdam performed a very special and predominant function in the whole. Therefore it was decided to honour the most important city of the most important province with the title 'capital.' The representatives of the different provinces, however, having their governing apparatus in The Hague, remained there in the capacity of Government of the Netherlands. Nowadays, the Ministry of Foreign Affairs is established in the former building of the Amsterdam representatives in The Hague. The arms of Amsterdam can still be seen on its façade.

Amsterdam stretches over a surface of 17,112 acres, of which 2,479 acres is water, 5,945 acres is occupied by the city proper, and 11,167 acres by a rural area. An area of 243 acres is now being prepared for building purposes; while about 200 acres is set aside for recreation purposes.

On January 1, 1950, Amsterdam had a population of 835,834 (1951:

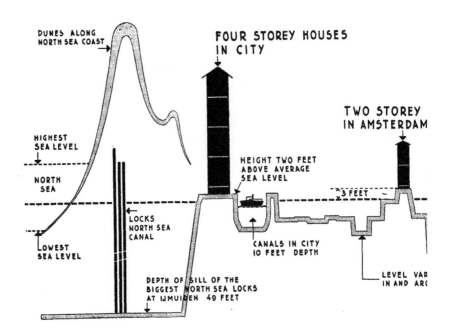

DUNES ALONG
NORTH SEA COAST

FOUR STOREY HOUSES
IN CITY

TWO STOREY
IN AMSTERDAM

HIGHEST
SEA LEVEL

NORTH
SEA

HEIGHT TWO FEET
ABOVE AVERAGE
SEA LEVEL

3 FEET

LOCKS
NORTH SEA
CANAL

LOWEST
SEA LEVEL

CANALS IN CITY
10 FEET DEPTH

DEPTH OF SILL OF THE
BIGGEST NORTH SEA LOCKS
AT IJMUIDEN 49 FEET

LEVEL VAR
IN AND AR(

VARIOUS LEVELS IN AND AROUND AMSTERDAM

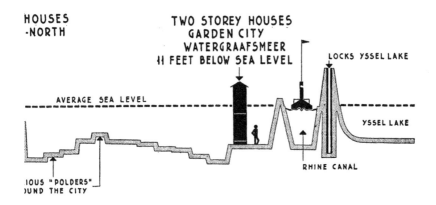

HOUSES
-NORTH

TWO STOREY HOUSES
GARDEN CITY
WATERGRAAFSMEER
11 FEET BELOW SEA LEVEL

LOCKS YSSEL LAKE

AVERAGE SEA LEVEL

YSSEL LAKE

RHINE CANAL

IOUS "POLDERS"
)UND THE CITY

846,000) or roughly 8 per cent of the total population of the Netherlands (10 million). The increase in population from 1930 to 1940 was 45,687 persons, from 1940 to 1950, 32,761 persons, in spite of the fact that more than 100,000 Jewish citizens were massacred during the German occupation.

Like other cities, Amsterdam has to cope with housing difficulties which means that houses that do not meet modern requirements have been put into use again. Dilapidated hovels in the old city are inhabited by whole families, so that proportionally the number of inhabitants of the city proper shows an increase in comparison with that on the outskirts.

Before the Second World War, Amsterdam's population remained fairly stationary; the increase in population was counter-balanced by the fact that many people working in Amsterdam preferred living outside the city.

Nowadays the situation has altogether changed; because of the housing shortage Dutchmen have often found themselves in a position resembling that of the Middle Ages, when a special licence of the municipality was necessary for living in a certain town.

THE PHYSICAL DEVELOPMENT OF THE CITY

In order to satisfy this ever increasing housing need, a new plan for expansion, accepted by the council in 1939, is now being carried out. The construction of the housing estate called '*Slotermeer*' will provide for the housing needs of 30,000 persons (10,000 dwellings). There are however, peculiar difficulties to be coped with for the nature of the soil makes it necessary to raise the ground level with sand. If the housing estate were to be brought up to the same level as other parts of the city, a layer of sand of about 12 feet would be necessary. Sand prices, however, have risen excessively (about 5s. per cubic metre); moreover, it is very hard to obtain the necessary quantities. It was therefore decided to raise the land level with a layer of only 6 feet (which means raising it above polder level instead of above basin level). Yet the expenses are not reduced correspondingly as artificial works such as locks, electric pumping-engines, etc., must still be installed to prevent inundation and to make navigation possible. Facing this page is a sketch of the various land and water levels in and around Amsterdam.

Just outside the municipal territory a new airport is being built with the co-operation of the Government. Pre-war Schiphol was totally unfit for use after the liberation as the Germans razed the airport and blew up all the runways during the victorious advance of the allied armies through France and Belgium. Under the new extension plan, Schiphol will be changed into a 'tangential' aerodrome (so-called as the runways are laid at a tangent), where aircraft of the heaviest type will be able to land and take off under all weather conditions and in every wind direction. It will be the largest aerodrome in Western Europe; it is expected to be ready in 1958.

The legal arrangements for this project provide that the municipality builds the airport jointly with the Government, on an area belonging to Haarlemmermeer municipality. The construction of the runways will be entirely on account of the Government, but the station building and the hangars only partially. The buildings will be rented to the K.L.M. (Royal Dutch Airlines) and other operators. The Fokker aircraft factory is also concerned with the new airport. Because of the need for a new aerodrome, reconstruction was started very soon after the liberation, without a legal basis. The form of management has not yet been settled, so it is still uncertain how expenses and financial investments of the new airport will be divided between the Government and the municipality of Amsterdam.

Finally, a brief remark should be made on the problems connected with the North Sea Canal. A garden city to the north of the river Y was constructed on a former dredge-depository at the beginning of the First World War. The population of this district amounts to about 60,000 now. Communication, however, between Amsterdam North—apart from the garden city, also the centre of industry—and the city centre, takes place by ferry-boat. It need not be emphasized that this is a serious disadvantage. A sub-municipality in this district offers only a small part of the service which the central town clerk's office can give; there is no hospital, so that ambulances have to use the ferry-boats, even in serious cases of illness, to take patients to one of the hospitals south of the Y. These inconveniences resulted in the decision that this northern district should not be subject to further extension. A canal around it has been projected.

For several decades, plans have been made for the construction of one or more tunnels under the river Y; until now, however, these plans have not been realized. At present, three tunnels under the North Sea Canal are projected: at Velsen (railway and motor tunnel), Hembrug (also a railway and motor tunnel), and Y. Only at Velsen has work been started. Discussions are taking place as to whether priority should be given to the Hembrug tunnel (in which case the last bridge over the

North Sea Canal would be taken away), or to the Y tunnel. The latter would consist of both a motor tunnel and a cyclist tunnel (most Dutchmen go to work on a bicycle). For shipping to Amsterdam the Hem tunnel is of first importance because of the obstacle formed by the railway bridge, but, on the other hand, the interest of the people living or working in Amsterdam North cannot be neglected. It must be considered, however, that if the Y tunnel were to be given priority, much extra traffic would be attracted as it would avoid the ferry-boats on the Hembrug route.

GOVERNMENT OF THE CITY

From a political point of view, the Netherlands is a decentralized unitary state. There is an historical explanation for this. In the Middle Ages the towns became separate authorities. Thereafter the area, nowadays called the Netherlands, formed a league of seven provinces, plus a few States-General countries, i.e. countries governed by the States-General of the provinces as 'colonies.' In this framework Amsterdam held an exceptional, if not an independent position. For it was mainly Amsterdam that provided money for conducting wars, organizing voyages of discovery, and equipping the mercantile marine.

After a brief period of centralized unity—firstly in imitation, later under compulsion, of Napoleon—a partial return to former conditions came in 1814. Fear of a revival of provincial power was stronger than that of municipal power. It is for this reason that the constitution of 1814 (repeatedly revised since, and for the last time in 1952) emphatically keeps provincial power under restraint. The constitution promised autonomy and self-government in large measure to towns and villages, now all called municipalities.

The composition, organization, and authority of the municipalities were from that date subject to legislation. An act was passed in 1851, the Municipal Corporations Act, which was drafted by the famous Dutch liberal statesman J. R. Thorbecke (1798–1872), Prime Minister and professor of constitutional law at Leiden University.

At the head of the municipality is the city council (*gemeenteraad*), whose members are elected directly by the inhabitants for a period of four years, according to a system of proportional representation. Every Dutchman is qualified to vote on condition that he has reached the age of twenty-one and has been living in the municipality for a period of at least twelve months.

The municipal council consists of 45 members (the maximum permitted under Dutch law for any city). The party membership of the council, resulting from the election held in 1948, is as follows: 14 Labour, 12 Communists, 9 Catholics, 4 Liberals, and 6 Orthodox Protestants. In the Dutch political framework, therefore, Amsterdam inclines to the left.

The executive organ is formed by a board consisting of burgomaster and aldermen. The aldermen (who are full-time paid executives) are elected by the councillors from their midst; their number also varies according to the population of the city. The number of aldermen in Amsterdam is now 7, having been increased from 6 in 1953. Party membership on the board comprises 4 Labour, 2 Catholics, and 1 Protestant. The Communists, who originally occupied two seats, have been excluded from the board by an amendment of the municipal law.[1] Although quite understandable, this modification has meant an infringement of the system of proportional representation for Dutch municipalities.

The burgomaster (*burgemeester*) is appointed by the Crown for a period of six years and he can be renominated for a further period of office. This implies that his function is not primarily a political one; he is rather an independent local officer. He represents the city on the one hand, and acts for the Government on the other. This double function hardly ever gives rise to difficulties, but juridically it often forms a puzzle. The burgomaster is chairman of the city council, and can be a member of it although this is not necessary. If he is a member he has a right to vote in the council's proceedings, but not otherwise. He is also chairman of the board of burgomaster and aldermen, in which he has a full right of voting, with a casting vote when necessary.

The burgomaster is head of the police. This applies to Amsterdam which has a municipal police force, as after the liberation most Dutch municipalities no longer provided their own police. The centralization of the police introduced during the German occupation has been only partially broken down by a return to pre-war conditions.

The town clerk assists the city council and the board of burgomaster and aldermen. He is appointed by the council and is responsible for putting into effect certain orders given by both authorities.

Finally, the municipal tax collector is responsible for the collection of municipal taxes and for financial affairs in general (municipal budget and accounts). He also is appointed by the council.

It is interesting to consider the above mentioned local authorities more

[1] Act of July 15, 1948, Section 87 (*a*). See *post*, p. 126. The law provides that the council is empowered to dismiss an alderman if he has ceased to possess the trust of the council. An alderman thus dismissed remains a member of the council.

closely, First, it is notable that when the Municipal Corporations Act was enacted, the political situation in the Netherlands differed considerably from what it is at the present time. The Municipal Corporations Act assumed that a number of citizens, elected by the people, would all be anxious to serve the interests of the municipality. In reality, however, the citizens who are qualified to vote, no longer vote for a person but for a party. It is the party that decides which members will be on the list of candidates, and in which order of preference. In the city council, the councillors are tied to the policy of the party, which means that in matters of any importance, the groups consult one another beforehand. When these matters are put to the vote the result is surprising only to outsiders.

Further, the council is important to the press, which gives lengthy accounts of the meetings, coloured according to their political conviction. Speeches often are made, therefore, more with the aim that the spoken words will appear in the party press than that colleagues of another opinion will be influenced.

In spite of the fact that the present practice has become far removed from the original intention, a new important aspect has been gained, namely publicity.

COMMITTEES OF THE COUNCIL

From a governing authority, the council has become a controlling one. Something of the governing function, however, is to be found in committees of assistance, formed out of the city council to assist the board of burgomaster and aldermen. Amsterdam has twenty of these committees. The most important are those concerned with finance and taxes, municipal services, sea and airport, art, education, public health, public works, housing, traffic, and accounts. The number of members varies from three to five. Proposals of the board of burgomaster and aldermen are dealt with by the committee of assistance relating to the matter in question, before they are submitted to the council. Thus members of the board of burgomaster and aldermen have the opportunity of going deeper into the subject, whereas this committee can probe the opinion of the council. Amendments from the committee of assistance are often embodied in proposals; it sometimes happens that the board of burgomaster and aldermen withdraws a proposal because the committee of assistance rejects it. This is not obligatory, however, for the committees of assistance only have an advisory task; the board of burgomaster and aldermen can make proposals to the council contrary to their advice. The advantage of this system is that the council can and does consider, when appointing members of the committees of assistance, both the political relations and the individual expertness of the councillors in a certain sphere. In passing we may note that, although in the council itself the communists are nearly always in opposition, the situation in the committees of assistance is much better, at least when matters under discussion are not directly concerned with party policy.

It is very difficult for most of the parties to interest their adherents in council membership. The council meets once a fortnight, every alternate Wednesday afternoon and evening and, if necessary, Thursday afternoon and evening; only in the month of August the council does not meet. The attendance allowance has been raised recently from fl.5 (10s.) to fl.10 (£1) per session. This applies also for the committees of assistance.

Finally the chairmen of the five political groups on the council form the so-called 'senior convention' which can be consulted in certain cases.

RELATIONS WITH HIGHER AUTHORITIES

The government of the municipality is carried out by the city council. This means that autonomy is given for matters only concerning local interests, which have not been regulated by a higher authority (province or central government). Thus the council has the right to make by-laws (e.g. general police by-laws, building by-laws, etc.). Moreover the municipality has the right to give effect to laws made by higher authorities. There is a tendency to restrict municipal autonomy in favour of government authority.

Decisions of the council concerning municipal property and other acts carrying legal significance, as well as the budget, are subject to the approval of provincial government authorities ('Deputed States'). The Crown has the right to suspend or nullify decisions of the municipality which it considers contrary to the law or the public interest. If the provincial government does not approve of a municipal resolution, the municipality can appeal to the Crown. Although, in general, an administrative jurisdiction does not exist, the law makes provisions for exceptional cases of the kind. Administrative-judicial affairs are dealt with in the last resort by the Council of State, in the section handling administrative disputes.

In general, co-operation of the 'Deputed States' with the city council leaves nothing to be desired. This may be due to the fact that in many cases Amsterdam has a larger and technically superior administrative machine than the province. The council itself can serve as an administrative appeal tribunal in case of summonses under the building by-law, Housing Act and Reconstruction Act.

BOARD OF BURGOMASTER AND ALDERMEN

The board of burgomaster and aldermen meets usually once a week in private. We have mentioned already that the board is assisted by the town clerk. The work of the executive is divided between the burgomaster and the seven aldermen who manage the following departments (in order of seniority):

Burgomaster:

> General affairs (police, fire brigade, archives, civil defence), national service, information and tourist office, Schiphol airport organisation and methods.

Aldermen:

1. Labour affairs, supplies, pensions, cattle market, abattoir, food inspection;
2. Education, arts (including museums and municipal theatre);
3. Finance, taxes, legal liability, insurance affairs, public utilities, street cleaning;
4. Public works and housing;
5. Social affairs, registrar's office, register of population, elections, statistics, publicity, cemeteries, municipal pawnshop;
6. Public health, hospitals, laundries, public baths and swimming pools.
7. Inspection of housing and lodging; public monuments, taxicabs, relations with private industry.

The town clerk's office is divided into sections which practically correspond with the above. At the head of the office is the town clerk; the heads of the departments are hierarchically his inferiors; in reality, however, they directly assist respectively the burgomaster or aldermen. Apart from these departments, the municipality has many services and works which, in principle, have a commercial basis. Proposals of their directors are handed in to the alderman through the head of the department under whose jurisdiction they come. It is the alderman who decides whether, and if so in what way, a proposal should be submitted to the board of burgomaster and aldermen.

Municipal services and works are often large; the directors enjoy considerable independence. For although all questions of any interest are formally the responsibility of the alderman concerned, and although the city council has to vote moneys for the construction of new works, the recommendations of the services cannot be put aside or altered without good reason.

Generally, the prevailing spirit in the board of burgomaster and aldermen is fraternal, in contrast to that in the city council. Three parties are represented on the board. On the whole they listen to each other's arguments, and try to find a solution which allows for different opinions. Only rarely is a matter put to the vote; in most cases it is discussed until a solution is found which is acceptable to everyone.

At each meeting, hundreds of items are put on the agenda. Most of them are dismissed, but many give rise to some form of debate. The discussions of the board are minuted; moreover, excerpts are made of the decisions, and the drafted resolutions are collected in the minute book of the burgomaster and aldermen. Copies of these excerpts are handed to the heads of the departments concerned.

THE BURGOMASTER

Amsterdam's burgomaster is appointed by the Crown for a term of six years on recommendation of the cabinet council, in consultation with the provincial governor of North Holland. Normally application has to be made to the Queen for the position of burgomaster, after which a visit is paid to the provincial governor concerned. Before the Second World War he had a very important voice in the matter, but after the liberation the influence of the political parties increased considerably so that mostly only those who belong to the party with the largest number of seats in the council come under consideration. Promotion to a bigger municipality was, and is, usual; just as in any other function, a burgo-master is promoted to a big town after serving in a smaller one. Such rules do not hold in this case, however, because Amsterdam's burgomaster is never chosen from among burgomasters of other cities, but is selected by preference from Amsterdam. The present burgomaster, Dr. Arnold J. d'Ailly, was, before his appointment in September 1946, one of the directors of the Netherlands Bank.

In most municipalities the burgomaster is deeply involved in the daily course of affairs; in Amsterdam this is impossible. Here, other qualities are required. To outsiders his function has a much more representative character; in some ways Amsterdam still forms its own community, reminiscent of its position in the Golden Age. To the insider, the burgo-master has the task of subduing the political passions that have always been fiercer in Amsterdam than elsewhere in the Netherlands. This requires not only statesmanship and tact, but also greater popularity with the people than is normally expected. For sometimes an unimportant event can cause a strike on a large scale; if this happens in Amsterdam, repercussions will be felt immediately in other parts of the country. On these grounds, the burgomaster often acts as a mediator, particularly in those cases that are normally outside his jurisdiction.

Let us now review his official duties. Although appointed by the Crown he is mainly a municipal official; with due responsibility to the council and the board of burgomaster and aldermen, but not to the Minister of Internal Affairs. Heading the municipality, he represents it generally and in law; this means that in legal proceedings against the municipality the burgomaster is summoned, and that the burgomaster appears on behalf of the municipality. On official occasions he acts on behalf of Amsterdam's citizens.

The burgomaster is chairman of the council. Unless he is also a councillor he has no right to vote but only an advisory voice. He conducts the meetings. Because the municipality has a certain measure of independence, but is not a completely autonomous body, the function of the burgomaster is also a double one, being subservient to governmental, as well as to municipal interest. The burgomaster can submit a resolution of the council to the Crown for nullification, if he does not approve of it; in fact, he does not make use of this right.

On the board of burgomaster and aldermen he has the right to vote. Moreover, he brings up matters for discussion and defends proposals from departments under his management.

Finally the burgomaster is charged by law with several explicit duties, such as the issuing of passports, birth certificates, etc. Besides these administrative functions the burgomaster holds several other posts, such as, for example, president of the Board of the Reparations Bank, commissioner of Hoogovens Ltd. (blast furnaces), and president of the Board of Radio Netherland World Broadcasting Corporation.

The burgomaster's salary is too small (fl.28,000, or £2,800) to meet all the responsibilities of the office. Thanks to a gift made to Amsterdam, its burgomaster has an official residence on one of the beautiful canals, and this helps to cover the large expenditure. Moreover, representation expenses are paid by the municipality.

THE TOWN CLERK

The town clerk has an advisory voice in the board of burgomaster and aldermen, but no voice in the council. He is appointed by the council on the recommendation of the board of burgomaster and aldermen. In smaller municipalities (under five thousand inhabitants) the burgomaster can also be the town clerk. His annual salary is determined by the provincial government authorities after hearing the views of the council. Approval of the Crown is required.

The town clerk's office in Amsterdam consists of twenty departments with a total of twelve hundred officials. While in smaller municipalities the town clerk puts into effect all kinds of resolutions, in Amsterdam his task is restricted to advising the board of burgomaster and aldermen in legal matters, and to supervising the work of the heads of departments. Proposals for the promotion of an official are handed in to the town clerk by the head of the department; if he agrees with the promotion, he presents it to the board of burgomaster and aldermen after obtaining the opinion of the alderman in control of labour affairs. Appointment and promotion of all municipal officials (also those working in municipal services and works) is effected by the board of burgomaster and aldermen. The employment and dismissal of workmen is effected by the head of the service or of the secretariat.

Furthermore, the town clerk's task has a representative character. Amsterdam is an attractive city for holding congresses, and authorities from abroad often visit the capital while staying in the Netherlands. The number of receptions given by the municipality increased considerably after the liberation and amounted to eighty in 1950, exclusive of the receptions which the burgomaster gave personally. As the town hall lacks adequate reception rooms, receptions are mainly held in the Municipal Museum, which has the added advantage that the guests can enjoy the pleasures of art at the same time.

THE TAX COLLECTOR

The last functionary mentioned in the Municipal Corporations Act is the municipal tax collector. He is a responsible official, and his function cannot be combined with that of burgomaster or town clerk. The council, the board of burgomaster and aldermen, and the tax collector co-operate in the field of municipal finance. The council manages finances, the most important part of which is the approval of the budget. The board of burgomaster and aldermen is charged with the control of municipal receipts and expenditure, except, for instance, in the case of purely monetary matters which are handled by the tax collector. Hence he is charged with the timely collection of taxes and all other municipal receipts, the making of payments from municipal funds, accounting, and rendering the annual accounts of municipal receipts and expenditure to the board of burgomaster and aldermen. The tax collector cannot influence the amount of the money that must be collected, and he can only make payments on warrants signed by the board of burgomaster and aldermen.

THE SCOPE OF MUNICIPAL FUNCTIONS

The main functions for which the city government is responsible include the administration of harbours and port installations; Schiphol airport; the operation of a municipal clearing bank; a municipal pawn-shop; the provision of a university library; the water supply; the supply of gas and electricity, and also the supply of steam for industrial purposes; local transport undertakings including tramways, motor buses and ferry-boats; public works; public cleansing and refuse collection; public health; public assistance; welfare services for old persons, orphans and widows; baths and wash-houses; the police force; fire brigades; housing construction and regulation of housing conditions; wholesale markets and an abattoir.

THE POLITICS OF THE CITY

We have already explained that the election of members of the municipal council takes place through the political parties, each of which presents a list of candidates. In smaller towns, political groups often unite on the basis of common municipal interest. Amsterdam has known such groups, but nowadays they are too insignificant to be able to gain a seat in the City Council.

As election takes place through the recognized political parties it is important to say something more about the general political situation. At the last election the Labour Party, the Catholic Party, the Communist Party, the Liberal Party and two Protestant parties—coming out with one list of candidates—gained seats in the City Council. There is co-operation between Labour and Catholics, just as there is in the government. These two parties together have a majority.[1]

Shortly after the liberation in 1945, for the first composition of the new municipal council—during the occupation it had been eliminated gradually—aldermen were chosen proportionally from representatives of each party. The board of burgomaster and aldermen consisted of 2 Labour, 1 Catholic, 2 Communist and 1 Protestant aldermen. In 1948 the Municipal Corporations Act was amended so that aldermen who no longer had the council's confidence could be deprived of their seat

[1] *Ante*, p. 118.

in the board (though not in the council). On these grounds, both Communist aldermen were removed from the board of burgomaster and aldermen in 1949; a Labour and Catholic member were appointed in their place.[1] The present composition of the board can no longer be considered as reflecting the proportional representation of Amsterdam's population; it is only representative of the 'national-minded' part of it.

POLITICAL PROGRAMMES

The political programme of the Netherlands Labour Party resembles that of the English Labour Party. It arose from a core of former socialists aiming at a so-called 'break-through' of the many political parties in the Netherlands (fifty-four parties in 1935, twenty parties in 1937). This break-through was only a partial success: the excessive number of parties was reduced to six big parties and two small dissident parties. The six big parties are represented in Amsterdam's City Council.

The Catholic Party is linked with the Catholic part of the population. Its political attitude can be described as social-religious. The protestant parties can be considered as conservatives, in a way resembling the English Puritans. The Liberal Party holds the well-known liberal opinions. In Amsterdam it represents mainly the interests of the port.

The Communist Party in the Netherlands has many supporters in Amsterdam, especially in the dock-workers' area. Their influence is also felt in the *Eenheids Vakcentrale* (United Trade Unions) set up as a combine of various trade unions which soon became a pseudo-communist organization. Communists in the city council are far from co-operating now.

The various political parties exercise great influence through their trade unions, which are consulted regularly about all matters concerning labour conditions, in a central committee for joint consultation. Partly on political grounds and partly for reasons of organization, the communist organs are eliminated in these consultations too. Therefore it hands in its motions directly to the council.

Political consultation between the bona fide organizations often have results which are successful from a political point of view, but objectionable all the same. There are examples illustrating this point: when a professor at the municipal university is nominated, the faculty's recommendation comes via the governing body of the university to the board of burgomaster and aldermen. The recommendation is generally sent on to the council, which has the right of appointment, and is submitted for

[1] *Ante*, p. 118.

approval by them to the Minister of Education. However, now and then appointments are due to political influences.

Both Labour and Catholics are zealous for useful and necessary social work in the various districts of Amsterdam; *De Eilanden* (the Isles) in particular, a neglected workmen's quarter, are given special attention. Social centres in other parts of the city are in preparation. A certain amount of activity on the part of the population itself is required before the municipality can support the proposed social centres.

MUNICIPAL REVENUES

Let us now consider the municipal revenues. These are obtained in large part from taxes procured from various services; from municipal services and works; from municipal property not intended for public service; and finally from government contributions in respect of work carried out by the municipality in accordance with government instructions, or general law.

Municipal taxes are of longer standing than taxes levied by the government. However interesting it might be to delve into historical differences it is quite impossible to do so in this chapter. It must suffice to say that town import, export and transit duties (once a leading source of income) are now prohibited.

Two important laws merit attention. The first is the Act of July 15, 1929. This repealed the right to raise local income tax and surtax on government income tax. At the same time a municipal fund was established (May 1, 1931) into which the government paid moneys derived from the so-called municipal fund tax. This fund was distributed among the 1,000 or so Dutch municipalities, according to the ratio of their inhabitants, on the basis of the Act dealing with the financial relations between government and municipality. This measure was an important improvement. Previously there were considerable differences in the amounts of tax paid in various municipalities. As a result, many of the better off people moved to districts where taxes were lower.

During the occupation this arrangement was dropped and supplanted by government funds. This government grant has been superseded by the Act of July 15, 1948, which made provisional payments for municipal finances. The purpose of this Act is to revive as far as possible the financial independence of the municipalities which had been lost. A considerably higher payment from the municipal fund will be made. The budgets must now, however, be balanced. This emergency Act covered the years 1948, 1949 and 1950. In order to cope with the problems which have since arisen, the government has decided to make a payment of fl.6 (approx. 12s.) per inhabitant (a payment of fl.6,50 would have been possible) from the moneys which are left in the municipal funds from the occupation. This means an amount of approximately five million guilders (£500,000) for Amsterdam.

Some of the taxes which the municipality raises independently, after approval by the Crown, are as follows: entertainment tax, dog tax, fire insurance tax, street tax, and surtax on certain government taxes.

Futhermore, harbour dues are imposed, the proceeds of which go to the Port of Amsterdam. An arrangement has been made with the municipality of Zaandam, which can also be reached by sea-going vessels via the Y, that ships making such a trip and calling at the Port of Amsterdam as well as that of Zaandam are only required to pay one lot of harbour dues. An extension of this regulation to other municipalities concerned on the Zaan is in process of preparation. The quay, the majority of the cranes and many of the sheds and storage depots in the harbour belong to the municipality. These are leased out for purposes of revenue. The question of commercial profit rarely arises. In this way, the municipality deliberately subsidises a number of large companies, for whom this regulation has many advantages.

For many years Amsterdam has drawn considerable profits from the large public utilities and services: e.g. electricity and transport undertakings. Modifications have been made in recent years as a result of the sharp rises in the price of raw materials and in wages. It is true that there have also been increases in various tariffs, but these have been only sufficient to enable the undertakings to pay their way. There are no longer any notable surpluses.

PURCHASE OF LAND

In connection with the large-scale expansion of the built-up areas of Amsterdam in the twentieth century (which is still proceeding), the municipality has bought or expropriated land as a counter-measure against speculation in property. After this land had been prepared for building (by raising with piles) it was let out on long lease. The conditions of tenure were last fixed by the council in 1933. When the lease expires, the buildings on the land revert to the municipality. In actual fact, however, the regulation is rather more elastic. It is possible to redeem the ground rent of the lease, and extension of the lease is also permitted. A difficulty arises, however, in connection with the latter. Suppose a lessee has rented a property for fifty years and after thirty years wishes to extend the lease on account of

3 . AMSTERDAM

A canal in the central area. Both the canal and the houses along its banks date from the seventeenth century.

4. BOMBAY
Bazaar Gate Street.

[*Photo: Fox Photos*

investments connected with the premises established on that piece of ground. Should the ground rent of the lease be payable by the new lessee immediately, or should it be assessed straight away and only become payable when the first lessee's contract expires? There are objections to both systems. In the first case, the lessee would pay more with prices at their present level than he should according to the terms of the old contract; in the second case, the ground rent of the lease would be assessed, to be paid in twenty years time. The municipality runs a considerable risk in this case. A solution to this problem has not yet been found.

The municipality owns considerable house property, bought for various purposes, such as the widening of streets, reorganization of housing in residential districts, etc. These municipal dwellings are controlled by the municipal housing service. The premises will continue to yield rent until such time as work is begun on widening the streets in question, and reorganizing the residential districts—and there is little chance of this for the time being owing, on the one hand, to the difficult financial situation, and on the other to the housing shortage.

The total revenue of Amsterdam for the year 1949 amounted to fl.92,000,000 (£9,200,000). Of this, an amount of fl.55,000,000 (£5,500,000) was obtained from government contributions under the Act of July 15, 1948; municipal taxes supplied fl.37,000,000 (£3,700,000).

Finally, it should be mentioned that a council resolution of February 23, 1949, instituted a loan fund which aimed at a more efficient municipal loan administration in Amsterdam. This resolution was retrospective to January 1, 1948.

THE BUDGET AND ACCOUNTS

Estimates of local revenue and expenditure have to be submitted by the board of burgomaster and aldermen to the council at least four months before the beginning of the year. Estimates are modelled on a budget prepared by the Crown. The Municipal Act, 1951, Section 240, lays down a number of compulsory expenditures which must be met.

The council prepares the budget at least two months before the year to which the estimates apply. The approval of the provincial government as the competent authority is necessary and their decision is made before the beginning of the year in question. Amsterdam is in the province of North Holland, for which a provincial council exists at Haarlem, the provincial capital.

Rules have been made for supplementation, amendment and partial

E

approval of the budget in the case of the 'Deputed States' refusing approval, by means of an appeal to the Crown.

The municipal budget is divided into two sections: one section for ordinary services, in which all expenditure and revenue is entered for a certain year (salaries, wages, maintenance; annually recurring revenues); and one section for capital services, in which only capital income and expenditure is entered.

In practice it is not always easy to determine what comes under capital expenditure and what under ordinary expenditure. It may happen that cars are purchased from the capital account, while new sewerage (destined for at least forty years' service) is met from ordinary revenue.

Each section has more or less its own system for defining this expenditure. Although this system, judged on its own merits, is usually justified, the method does not promote the smooth working of the whole of the municipal revenue and expenditure.

The municipal accounts are presented to the council by the board of burgomaster and aldermen within seven months from the end of the year to which they relate. The council provisionally assesses accounts other than the budget. The 'Deputed States' close the accounts before the end of that year and assess the amount of expenditure and receipts. This discharges the tax collector and the board of burgomaster and aldermen from their duties (the former for his control of moneys, the latter for their financial guidance). Appeal to the Crown may also be made here.

The above-mentioned Act of July 15, 1948, enabled a balanced budget to be made for Amsterdam. The years 1948, 1949 and 1950 even saw a surplus. Publication of the budget, as well as the rendering of accounts, is compulsory.

The revenue and expenditure of services and utilities is not included in the municipal budget and accounts. Only the estimated credit or debit balances are included. This is in accordance with the independent character of these services and utilities, whose budget and accounts are assessed separately. These are arranged on the same lines as those of the municipal budget and accounts. The prescribed formulae include, among other things, a division of capital and ordinary services. This method increases the independent nature of the different services and utilities.

In view of the position of the municipal tax collector, individual regulations on this point are included in the Municipal Act, owing to the fact that the tax collector cannot control the revenue and expenditure of the services and utilities.

Finally, a few more remarks on the financial problems with which Amsterdam has recently had to deal. The same fate is shared by nearly all the other authorities.

Although on the strength of the Act for emergency provision of municipal finances a balanced budget was obtained and a relatively liberal policy could be carried out, it is now apparent that the tide has turned. The position of the municipal treasury has become exceedingly precarious. It is still too early to draw conclusions. But it is to be feared that, apart from the austerity measures which will have to be inflicted on the whole nation, Amsterdam will be firmly tied by its purse-strings to government policy at the cost of its autonomy.

PLANNING THE CITY

The following pages will give a brief survey of the plans that have been projected for Amsterdam during the coming years.

Amsterdam continued its old tradition dating from the seventeenth century by laying down a comprehensive plan of expansion. In 1926 a first attempt was made to draw up a development plan for the whole municipal area. From this effort sprouted the 'Plan for Greater Amsterdam 1926.' The municipal council decided to establish a special section for town planning. The section was organized under scientific leadership, with economists and sociologists attached to it.

In 1934 the definite plan, or Master Plan, was published. This plan provides for the development of a well-serviced and at the same time a really beautiful city. The development will probably be finally completed in the year 2000, when the estimated population of the city will have reached a figure of about 1,000,000.

The plan provides for the building of a new town hall which is badly needed. The old one houses only part of the town clerk's office; the departments providing the various services are spread all over the city. In 1936 the board of burgomaster and aldermen recommended the appropriation of a piece of ground on the city's outskirts for the new town hall. Tenders were invited for this plan, but the matter was dropped until after the Second World War. Because of the depopulation and demolition of the Amsterdam ghetto a unique opportunity presented itself for building the new town hall in the centre of the city along the banks of the Amstel. The council, however, rejected this plan on grounds which are still not very clear now. So it was decided to retain the plot previously allocated, although it had become far too small to supply the present need. Consequently the building of a new town hall has been postponed indefinitely. The municipality plods along with the old one, with no proper reception rooms, with inadequate accommodation for the town hall personnel, with no possibility of bringing together the municipal services and undertakings. The public works department have finally received approval for the building of a new edifice in which, apart from this service, accommodation can be given to the municipal housing committee and building inspection. Marriages take place in such small rooms that the last splendour of this ceremony is fading away.

Since the liberation Amsterdam has had an opera company but no opera hall! Plans are being projected for the construction of such a building, with an annex, meeting and reception rooms. This building should be erected on the same ground as that intended for the new town hall in 1936. Unfortunately, this means that no opera building can be constructed until the town hall problem is solved.

THE HOUSING SHORTAGE

Expansion of the city is an urgent problem because 30,000 persons, at present lodged with other families, are pressing for better housing. The Slotermeer plan, already mentioned, must supply this need. The government contributes a certain percentage to the cost of building flats, but because of the priority that has to be given to military defence, house building is seriously in danger of falling between two stools. This means that Amsterdam will have to reckon with a housing shortage for the next twenty years at least. Naturally, resourceful people manage to carry on by living in emergency dwellings, summerhouses, caravans or house-boats. On grounds of aesthetic considerations, however, jerry-building is rigorously quashed. From the point of view of planning, this may appear to be justified, but too little account is taken of present wants. In a large-sized municipality like Amsterdam it is inevitable that certain cases just remain cases without being considered as individual people. A waiting list has been made for urgent cases, which included in July 1947 1,200 families who should be assisted on social or medical grounds. By April 1950 the waiting list had grown to 6,500 families, and by January 1954 it had reached 9,000. The housing shortage is still increasing; the number of houses approved by the government for Amsterdam amounted to 3,327 for 1951. No more than this number could be built. Up to 1951 the government allotted separate housing quotas to each of the three largest municipalities; Amsterdam, The Hague and Rotterdam; the other housing quotas are allotted to the provinces, so that the provincial government authorities attend to distribution among the municipalities. The aldermen concerned for housing affairs of these cities took turns in the Distribution Committee of Housing Quotas. This has now been changed, and the provincial authorities are also responsible for the housing quota in Amsterdam.

The housing shortage in Amsterdam is very serious. The number of family dwellings required in January 1954 was 26,000. Hardly any houses are condemned in the present circumstances. Especially in the slums, thousands of families are housed together in conditions which challenge every modern concept of hygiene. Moreover, in cases of sub-tenancy in particular, exorbitant rents are asked. The government hesitates to invest the municipalities with authority to deal with this matter, as it would counteract the tendency to centralization. Nevertheless, it would be very important if the municipality of Amsterdam especially—where the problem of rent profiteering is more urgent than elsewhere—were to be given such authority. In this matter the Burgomaster is a fervent advocate of municipal intervention. This is an example of the influence that the Burgomaster can have on matters that are not strictly in his domain.

FOREST PARK

There are other projects under consideration besides those for housing and street planning. In 1928 a plan was formed for the laying out of a forest park, which was becoming a necessity owing to the constantly increasing population living at an ever-increasing tempo. Forest Park, on which thousands of people laboured for years, and on which many are still working, has a total area of 2,212 acres, including the Pool and the sector by the banks of *Nieuwe Meer* (New Lake). This is roughly equivalent to the Bois de Boulogne, and three and a half times the size of Hyde Park and Kensington Gardens put together. Up to now the total area of all municipal parks and squares amounted only to 675 acres.

The boat-race course, with a length of 2,200 yards and a width of 213 feet, is unique in Europe.

WATER AND ELECTRICITY SUPPLIES

A new water-supply system is under consideration. Hitherto, water was obtained from the dunes. As rain-water was insufficient to supply Amsterdam's needs, deeper borings were made which revealed a fresh water sack, partly under the dunes and partly under the sea. This is evidently water that has been falling through the upper layer to deeper regions in the course of centuries. Millions of cubic metres are taken from this sack annually, resulting in a silting up of the dunes and a shortage of water. Plans are now being projected to draw water from the river Vecht. This water is led to the dunes, where it is filtered through the sand and pumped up again. It is hoped to obtain adequate supplies of water by this method, not only for Amsterdam but also for a large part of North Holland. Municipality and province are co-operating closely in this scheme.

The electricity supply of Amsterdam badly needs renewing, being coupled to the rest of the Netherlands and even to Western Europe. A new electric power-station is under construction north of the waterway known as the Y. The total cost of the work is estimated at £7,000,000. In the winter months electricity has to be used very economically during the peak-hours until such time as the new power-station is completed. When it is ready it will have a capacity to carry e.g. a load of 400,000 electric stoves burning simultaneously.

Finally, a brief remark about the removal of household refuse and sewage. The sewage of Amsterdam drains into the *Yssellake* (formerly *Zuiderzee*); refuse is burned as far as possible to generate electricity. Consideration is at present being given to the question of the extent to which this can be used for compost. Experiments have been made with favourable results but no definite steps can be taken yet.

A certain materialistic outlook in handling municipal affairs is unmistakable. It is remarkable that the planning of a new town hall and opera hall which has been in progress for several decades does not advance beyond its first stage. Realization seems farther off than ever before. On the other hand, works considered a technical necessity like the new power-station or the enlargement of the municipal waterworks, are carried out at short notice.

CONCLUSION

The subjects dealt with in this essay could not be other than incomplete; lack of space did not permit a detailed dissertation. Knowledge of Dutch constitutional law could not be assumed, so apart from Amsterdam's particular problems a bird's-eye view had to be given of general constitutional problems. It must be emphasized that for a deeper study it will be necessary to consult the handbooks mentioned in the bibliography. Unfortunately, however, juridical works are only published in Dutch as a general rule.

The essay only gives the opinion of the author; it can by no means be considered the point of view of the municipal authorities.

In general it can be said that the brilliant framework of Thorbecke's Municipal Corporations Act of 1851[1] still forms a very useful governing instrument. Much has changed since, but adaptation to circumstances could take place without too radical alterations of the law. The present situation, however, cannot be called satisfactory. Not because the system as such is no good, but because the interest of the inhabitants in municipal affairs can only be developed through the political parties.

Another defect is the lack of an administrative jurisdiction, i.e. an administrative court or tribunal. If an inhabitant thinks he has been treated unjustly and lodges a complaint, the decision in most cases rests with the same authority that has caused the injustice. Appeal is possible only in rare cases.

An excellent system of appeals has, however, been provided for officials who feel they have a grievance against the municipality. They can always lodge an appeal against decisions regarding them. This applies both to disciplinary and other measures. A start has been made with the regulation of appeals regarding housing claims, but jurisdiction in this complicated matter works slowly. The separation of powers principle, recognized in Dutch legislation, has been set aside in this case.

It must be expected that in future municipal autonomy—territorial decentralization—will lose ground in favour of functional decentralization. Naturally much depends on the development of the world's history in general, of Europe's destiny in particular. Now that considerable amounts must be spent for defence purposes, centralization of financial management is necessary, and with the loss of financial control by the municipality a loss of autonomy is inevitable. Trying to build a dam against its lessening influence, the municipality of Amsterdam seeks to interest the inhabitants in local affairs, but it is difficult to cultivate the degree of solidarity such as exists in smaller towns. Stimulation of the establishment of social centres, etc., has already been mentioned.

Economically Amsterdam recuperated after the liberation with remarkable swiftness. This is noteworthy because much has been lost which before the Second World War contributed to Amsterdam's wealth and power. The staple market of colonial goods suffered a severe blow in the loss of the former Netherlands East Indies. Foreign exchange and banking, before the war an important source of revenue, have been reduced to a minimum in consequence of foreign currency regulations.

[1] *Ante*, p. 117.

The German hinterland has been in large part lost. Amsterdam was a staple market even for goods destined for Austria and other Danube countries, which are now behind the Iron Curtain.

It is hard to predict anything regarding the development and possibilities of Amsterdam in the future. It is a fact that Amsterdam has to orientate itself once again in the international world. Its geographical situation in Europe is not altogether favourable. With respect even to its own country it has to cope with many difficulties. Many important industries such as the manufacture of electrical goods, artificial silk, textiles, etc., have been established on the sandy soil east and south of the country where factories can be built without ramming. It is therefore an urgent necessity that the municipality of Amsterdam continues its efforts in this field; up to now it has shown its keen interest in this matter and has given full co-operation.

Nevertheless, Amsterdam will always be more a trading and shipping city than an industrial city. It is self-evident that liberalization of international trade is an indispensable condition for further development. Unlike Paris for instance, Amsterdam cannot exist merely as the capital of the Netherlands. Certainly, however, Amsterdam can establish itself in the trading world, making use of its great experience in this field in the past. We have fallen on evil days before, but the typical business-like character of Amsterdam's population has always won through. From this 'historical' point of view the outlook for the future is propitious.

Bombay and Calcutta

M. VENKATARANGAIYA

M.A. (Triple First Class), University of Madras, 1907; awarded University Northwick Prize; Lecturer in Politics, Economics and History in colleges affiliated to the University of Madras, 1908–28; Principal, Venkatagiri Raja's College, Nellore (Madras), 1928–31; Head of Department of Politics, History and Economics, Andhra University, Waltair, 1931–44; Sir Pherozesha Mehta Professor of Politics and Civics, University of Bombay, 1944–52.

President, Indian Political Science Association, 1945–46; Member, Indian Historical Records Commission, 1932–42; Member, Planning Committee of the Telugu Encyclopaedia, Madras; Member, Madras Education Reorganization Committee, 1938; Member, Special Expert Committee on Village Panchyats Bill (Madras), 1942–43; Member, University Grants Committee, 1946–48; Member, Syndicate, Senate and Academic Council of Andhra University, Waltair, between 1926 and 1950; editor, *Educational India*.

Author of: *Beginnings of Local Taxation in the Madras Presidency* (1928); *Federalism in Government* (1935); *Development of Local Boards in the Madras Presidency* (1938); *The Fundamental Rights of Man* (1944); *The Case for a Constituent Assembly for India* (1945); *India's Draft Constitution* (1948); *Fair and Free Elections* (1950); Chapter on 'Local Finance' in *Modern Indian Economic Problems*, Vol. II (1938); *The General Election in Bombay*, 1952.

Bombay and Calcutta

THE GROWTH OF BOMBAY AND CALCUTTA

BOTH Bombay and Calcutta occupy a high place among the great cities of the world by virtue of their size, the numbers and density of their populations and their importance as political capitals and as centres of trade, industry, finance, education and culture.

There are many points of similarity between these cities. Both are quite modern, being about three hundred years old. They are in this respect unlike Benares and other famous cities of India which have a long history behind them.[1] They had their beginnings in the commercial contacts established between India and Europe as a result of the geographical discoveries of the fifteenth and the sixteenth centuries. It was mainly through the efforts of British economic and political enterprise that they grew.

Many of the great cities of the world owe their development to the operation of only one or two factors like commerce or industry, defence or strategic advantage, administrative importance or education.[2] In the case of Bombay and Calcutta all these factors have exercised their influence. As centres for the collection and distribution of the principal commodities of export and import they obtained a considerable degree of commercial importance. Both cities are busy centres of industry. Bombay is the home of the cotton and Calcutta of the jute industry of India. These industries expanded rapidly with the development of railways and other quick means of transport for which these cities became converging points. The establishment of other concerns like railway workshops and engineering factories, the growth of numerous banking and financial institutions, and other economic factors have led to a large and steady growth of the population of Bombay and Calcutta. These factors also account for the presence of numerous sections of people belonging to the lower middle class and the working class, and the emergence of numerous problems connected with planning, slum clearance, housing and transport.

Calcutta has from the outset been the capital of Bengal, which up to the partition of the country in 1947 was one of the richest and the biggest

[1] D. E. Wacha: *Rise and Growth of Bombay Municipal Government*, Chapter I.
[2] Griffith Taylor: *Urban Geography*, p. 416.

of the provinces. It was also the capital of the British Dominion in India up to 1912. This has brought into it people from almost every part of the country giving to its inhabitants a more or less all-India character.[1] There are certain Bengalee observers who attribute to this the lack of civic interest in the population of the city, a feature which has adversely affected the character of municipal administration from time to time. Bombay is the capital of another rich province and has also a population of an all-India character.

Both cities have been for nearly a century the seats of two of the best known universities of modern India, and as such they contain numerous colleges and other institutions of higher learning and research.

While these influences have been responsible for a fairly continuous increase in their populations, the Second World War and the events that followed it have brought about a sudden and unexpected rise in numbers.[2] The two cities became important supply bases during the war, and this naturally led to a large influx of people. The war was also responsible for a deterioration in the conditions of rural life in the surrounding areas and this brought—and continues to bring—more immigrants into the two cities.

Finally came the political partition of the country in 1947 and the flight of millions of Hindus from Pakistan into India. Among these more than a million have found a home in Calcutta and its neighbourhood, and between one hundred and two hundred thousand in Bombay. Between 1941 and 1946-47 the population of the two cities nearly doubled and this has placed the city governments under an unprecedented strain. In the case of Calcutta the strain was so great that it led to the complete breakdown of municipal administration. The government of the state had to supersede the municipal corporation and the government of the city was placed in the hands of a single administrator. The supersession continued until April 1952 and during that period there was no local self-government in the city.

The tables opposite illustrate the growth of the population of the two cities.[3]

[1] *Imperial Gazetteer of India*, Vol. IX—Calcutta.

[2] *The Calcutta Municipal Gazette Silver Jubilee Number*, p. 14. Memorandum submitted by the Corporation to the Bombay Municipal Finances (Shroff) Committee, paras. 2, 9 and 10.

[3] *Census of India*, 1931, Vol. IX, pp. 2, 6, 7 and 10; also S. M. Edwardes: *The Gazetteer of Bombay City and Island*, Chapters III and IV.

(Complete and correct figures are not available for Calcutta in all cases. The figures given here are calculated partly from the data in the Census Report for 1911 and partly

BOMBAY

Year	Area in acres	Population	Mean density per acre	Highest density per acre	Lowest density per acre
1872	11,930	644,405	54	–	–
1881	14,229	773,196	54	778	4
1891	14,080	821,764	58	699	5
1901	14,342	776,006	54	598	6
1911	14,576	979,445	64	602	7
1921	15,066	1,175,914	78	737	16
1931	15,480	1,161,383	75	727	18
1941	16,761	1,489,883	88	710	6.6
1949	16,761	2,553,735	151	–	–

CALCUTTA[1]

Year	Area in acres	Population	Mean density per acre	Highest density per acre	Lowest density per acre
1872	3,766	633,009	–	–	–
1881	3,766	579,966	–	–	–
1891	11,965	813,620	68	282	14·2
1901	11,965	861,501	72	255	15·9
1921	11,965	885,815	74	–	–
1931	19,253	1,123,790	56	218	14
1941	18,121	2,070,619	114	400	22
1949	18,121	2,993,443	165	454	9

CITY POPULATIONS AND CITY BOUNDARIES

What gives to a locality the character of a city is the concentration of a large number of people in a comparatively small area. Almost all the problems of city life arise out of this; and they have been accentuated by the late origin of the city planning movement and the slow pace at which it has been making progress. Overcrowding is invariably one of the more important of these problems and city governments have always been faced with it. The tables given above make it clear that the populations of Bombay and Calcutta have been increasing at a higher rate than their areas and this has naturally led to gross overcrowding. In Bombay the mean density rose from 54 per acre in 1901 to 151 in 1949, while in Calcutta the corresponding figures are 68 and 165. Mean density,

from the material found in the Report (1949) of the Corporation of Calcutta Investigation Commission which is the most comprehensive and authoritative study of the city government of Calcutta. Where correct figures are not available the columns are left blank.)

[1] *Census of India*, 1911. *Corporation of Calcutta Investigation Commission Report* (*C.C.I.C.R.*), Vol. I, para. 130.

however, is not a correct index of real congestion, as the population is never distributed evenly over all the wards of a city. A correct estimate of the degree of overcrowding should be based on a detailed examination of figures of density in the various wards into which a city is divided. Such an examination reveals that in Bombay there were in 1901 certain wards with a density as high as 598 per acre and that this figure has gone up to more than 710 by 1949. In Calcutta the highest density today is 454, while it was only 282 in 1901.

This overcrowding necessitates a study of the relation between the city's boundaries on one hand and its growing population on the other. In the case of Bombay and Calcutta three aspects of this relation deserve to be emphasized. (1) During recent decades there was a movement of the population from the more congested areas. This was facilitated by the extension of quick means of transport enabling persons to travel easily from their places of residence to places of work.[1] (2) A variant of this process is the tendency among some persons to reside in suburbs outside the city limits. This spillover of the population is observed in all countries. It became strong in Bombay in 1897 when the bubonic plague broke out and it has been going on ever since. This process has not given all the relief needed either in Bombay or Calcutta, owing to the fact that the services provided by suburban municipalities are not as good as those provided by the city corporations and it is therefore only as a last resort that people prefer to move to the suburbs. (3) The third aspect is the move for including the neighbouring suburbs within the city limits. This would enable better services to be provided in them and also make the suburban residents pay for the conveniences and comforts they enjoy in the city during the day-time when they go to work there—a payment from which they escape so long as they live beyond the administrative limits of the city.

On two occasions—in 1888 and again in 1923—certain suburban areas were thus included in Calcutta. There is now a movement for the further expansion of the city and the creation of Greater Calcutta. The Investigation Commission has, however, expressed itself against this move in view of the financial and administrative crisis through which the Corporation is passing at present although they recognize that such an expansion is needed for providing residential accommodation for the huge surplus population of the city, for quickening the pace of slum clearance and for shifting a number of noxious and offensive trades. It

[1] *Bombay Government Gazette*, February 18, 1950, Part V, pp. 35–8. *Calcutta Municipal Gazette*, September 17, 1949, pp. 498–500. *C.C.I.C. Report, op. cit.*, Vol. I, pp. 129–32.

is possible that when normal conditions are restored the five suburban municipalities will be included in the city. Its present area of 36 square miles would thereby be increased to 56 square miles.

An attempt to create a Greater Bombay by including within the city limits a number of neighbouring suburbs was made as early as 1918.[1] But it was only in 1950 that legislation for the purpose was enacted. As a result of this the city now extends to an area of 94 square miles and has therefore all the space required to put an end to overcrowding. An up-to-date master plan has already been prepared. It has recommended that the city's population should be limited to a maximum of $3\frac{3}{4}$ millions. It has also provided for so planning and zoning the city that the density in the suburban areas should not exceed fifty per acre and that in the old city it should be gradually reduced to one hundred and forty per acre.[2] If this plan is put into execution most of the problems arising out of the phenomenon of the overspilling of the population will be effectively solved. In Calcutta also a master plan is under preparation.

THE PROCESS OF DEMOCRATIZATION

The city governments that are now functioning in Bombay and Calcutta are for all practical purposes the creations of the British who ruled the country for a century and a half up to 1947. Though the first steps to organize municipal government in the two cities were taken in the days of the East India Company's rule, they were of a rudimentary character both in respect of the administrative organization and the range of local services. It was only after the government of the country was taken over directly by the British Crown in 1858 that real efforts were made to create municipal institutions. A municipal corporation was established in Calcutta in 1863 and in Bombay in 1865. At their start the corporations were composed only of justices of the peace, all nominated by the provincial government. They had no elected or democratic element in them. The first step in introducing such an element was taken in Bombay in 1872 and in Calcutta in 1878. But for a long time a nominated element continued to exist by the side of the elected element and it was only in 1938 that the Bombay Corporation became a wholly elected body, except for three officials who continued to be *ex officio* members until 1952. In Calcutta it was only in 1947 that nominations by government were completely abolished. This long interval between the

---

[1] N. V. Modak: *Note on Greater Bombay*, Introduction.
[2] N. V. Modak and Albert Mayer: *The Master Plan*, p. 11.

establishment of the Corporations and their democratization deserves to be noted.[1]

In the beginning the franchise was restricted to a small number of ratepayers. In Calcutta, for instance, their number was 5,044 out of a population of 612,000. It was only in 1947 that elections on the basis of adult suffrage were held for the first time in Bombay. In Calcutta there is no provision for adult suffrage even today although the Investigation Commission recommended it in the strongest terms. Under the Municipal Act which was in force up to 1948 when the Corporation was superseded, the number of voters was 69,264 out of a population of 2,070,619. It is expected that under the provisions of the amending Act of 1950 the number may go up to 600,000 out of the estimated population of three millions with permanent residence. Only those who are assessed to certain taxes, or occupy premises of a specified rental value or have received education up to the matriculation standard are entitled to vote.[2] We shall describe later the methods by which the corporations are elected.[3]

Although in this movement towards democratization there is a certain similarity between Bombay and Calcutta, there are also several important points of contrast. It is no exaggeration to say that on the whole they represent two different types of city government. This contrast is in the main the outcome of two or three features peculiar to Calcutta. This city was up to 1912 the capital of the British Empire in India and not, like Bombay, a mere provincial capital. Democratization of its municipal government was considered to be attended with greater risks. Moreover, Calcutta contained a larger European population than Bombay, and they were generally averse to any large-scale participation of the inhabitants of the city in municipal affairs. They shared with Sir Henry Maine, the great jurist, who was for some years a member of the Viceroy's Council, the view that it was surprising that 'the natives of India should be fit for municipal government at all.'[4] Another feature of Calcutta was that it was the capital of a province where the Muslims were in a majority and where political issues were very much influenced by their spokesmen. They were opposed to any real and rapid democratization of the city government of Calcutta for the reason that in the city itself the Hindus were in a majority and democratization meant in their view the passing

[1] *C.C.I.C. Report*, Vol. I, Chapter II. R. P. Masani: *Evolution of Local Self-Government in Bombay*, Chapters XXI, XXVII and XXXII.
[2] The Calcutta Municipal Amendment Act (1950), Section 6.
[3] See pages 148–149 below.   [4] *Proceedings of the Imperial Legislative Council*, 1868.

5. BOMBAY

The Town Hall. A mixture of Victorian Gothic and Muslim architecture.

[*Photo: Press Information Bureau, Government of India*

6. BOMBAY

A view of the bay. The buildings facing the sea consist mainly of luxury flats.

[*Photo: Gordon Pix*]

7. BOMBAY
Suburbia coming to town (Churchgate).

[Photo: Dorset, Bombay

8. BOMBAY
City End.

[Photo: *Dorset, Bombay*]

of power into the hands of the Hindus. The British authorities attached a great deal of weight to this view. It had its effect not only in slackening the pace of democratization but also in bringing into the Corporation the system of communal representation through separate communal electorates.

All this led to the evolution of the city government of Calcutta on lines somewhat different from that of Bombay. Democratization there was not only slower but less continuous. There were occasions—as under the Act of 1899—when the movement was a little away from the degree of democracy previously achieved. Besides minority representation, institutional representation attained a larger place in Calcutta as it helped the European community to secure a larger number of seats in the Corporation than they were entitled to on the basis of their numbers. It was mainly for this purpose that chambers of commerce, trade associations and other bodies dominated by the European community were given representation in the Corporation. Another feature of Calcutta was the system of plural voting at elections which was in operation from 1888 to 1923. In consequence of this, a person qualified to vote as owner or occupier of a house was given additional votes up to a maximum of ten in any ward according to the valuation of his properties.[1] A further point that deserves notice in this connection is that in Calcutta there was a larger degree of state control over the Corporation than in Bombay. There is no need to emphasize how all these characteristics are the result of political considerations peculiar to Calcutta.

THE BALANCE BETWEEN DEMOCRACY AND ADMINISTRATIVE EFFICIENCY

One of the controversial problems that arise in the organization of city governments is the kind of relationship that should exist between the legislative and policy-making function on one side and the policy-execution function on the other. It is more or less identical with the problem of reconciliation between the principle of democracy and the needs of administrative efficiency. Such a problem arose in the case of Bombay and Calcutta. Bombay was lucky in discovering a solution for it as early as 1865 and still more lucky in adhering to it all these years. In Calcutta, however, two solutions have been tried so far, one up to 1923 and another subsequent to it. Today a third solution has been recommended by the Investigation Commission.[2] In this respect also there is a noteworthy difference between the two cities in the evolution of their municipal institutions.

---

[1] *C.C.I.C. Report*, Vol. I, para. 59.   [2] *Ibid.*, Vol. II, Chapter II.

The Bombay system rests on three principles. In the first place it recognizes the distinction between the policy-making and the policy-execution functions. In the second place it entrusts the former function to a fairly large representative body and the latter function to a single individual. In the third place it makes the individual head of the executive more or less independent of the representative body although the two work not in isolation, but in co-operation. The head of the executive is styled the Municipal Commissioner. He is appointed by the government of the state for a renewable period of three years at a salary fixed by the Municipal Act itself.[1] The only limitation on his complete independence is that he has to be removed from office if the removal is demanded by a resolution of the Corporation voted for by not less than five-eighths out of its present strength of 124 members. He has a right to attend all meetings of the Corporation and its committees and take part in their proceedings without having the right to vote. This system secures democracy in policymaking and also ensures administrative efficiency.

The first two of the three principles of the Bombay system found a place in Calcutta, but difficulties arose in regard to the third one. Broadly speaking, until 1923 the executive dominated the legislature. The chairman (as the head of the executive was then styled) not only exercised the powers that legitimately belonged to the chief administrative officer but also controlled policy, by virtue of his position as the presiding officer of the Corporation Council and of the various standing and special committees.[2] As in Bombay he was appointed to his office by the provincial government. His position was more or less like that of the *Burgermeister* in German municipalities. Like him he had the full right to participate in the discussions of the council, full voting rights, and a casting vote wherever necessary. This was resented by the more popular section of the Corporation as undemocratic; and when under the Government of India Act of 1919, the portfolio of local government was transferred for the first time to a responsible minister in the provincial government, a new Municipal Corporation Act was passed in 1923 which completely reversed the relation between the legislative body and the head of the executive. This Act created an executive officer in place of the chairman. He was appointed to his office by the Corporation and not by the Provincial Government. Even in the administrative sphere he had only such powers as were delegated to him by the Corporation.[3]

This reform was based on the analogy of the English municipal system

---

[1] City of Bombay Municipal Act, Section 54.
[2] *C.C.I.C. Report*, Vol. I, para. 76.   [3] Calcutta Municipal Act, Section 12.

in which the elected council is supreme in the legislative as well as the administrative sphere. But in Calcutta this did not work well. It led to the complete breakdown of administration, and the undermining of discipline and efficiency all round as the Investigation Commission observed. This is one factor that led in 1948 to the supersession of the elected council and the appointment of an administrative officer by the Provincial Government to manage all the affairs of the Corporation. Democracy fell into a temporary eclipse until it was restored in 1952. In 1951 a Calcutta Municipal Act was passed, and amended in 1952 and 1953. Under the new legislation a municipal commissioner is established on similar lines to those existing in Bombay. He is appointed by the state government (and not by the corporation) on the recommendation of the State Public Service Commission for a period of five years at a time, on such terms and conditions as the state government may determine. The only control that the corporation has over him is its right to demand his removal from office by a resolution voted upon by more than half of its members. The entire executive power is vested in the municipal commissioner, and all municipal officers and servants are made subordinate to him.

THE STRUCTURE OF THE CITY GOVERNMENTS

It can be seen that until recently the structure of the city government of Bombay was fundamentally different from that of Calcutta, but the changes introduced recently in Calcutta have now assimilated the system there in most respects to the Bombay model. The difference which previously existed arose out of the fact that while the elected Corporation in Calcutta was given the status of the supreme governing body in the whole field of municipal administration, the elected Corporation in Bombay has not that status. Technically, the Bombay Corporation is only one of a number of municipal authorities. The other organs of city government which are included in the category of such authorities possess certain powers of their own which cannot in any way be modified or interfered with by the Corporation.[1] There are at present six of them. The Standing Committee and the Municipal Commissioner have been included among statutory authorities for a long time. More recently the Education Committee (previously known as the Schools Committee) and the Improvements Committee secured a place among such authorities. This principle of having a number of co-ordinate authorities has been extended to the Electricity supply and Transport undertaking which

[1] Bombay Municipal Act, Section 4.

is the major public utility enterprise now owned and worked by the Corporation. The affairs of this enterprise are managed by a separate committee and a general manager with a large amount of independent jurisdiction.

The Bombay Corporation consists of 124 members. All of them are elected on the basis of adult suffrage at a general election held once in four years in the several wards into which the city is divided. The wards are multi-member constituencies electing from two to five councillors each. The method of voting is cumulative and this enables a well-organized minority—communal or political—to secure some representation.[1] The wards are also so distributed that in some of them voters belonging to minorities like Muslims have sufficient numerical strength to enable them to return members of their own community to the Corporation. It may

Size and Composition of the Corporation
(a) BOMBAY

Act of (1)	Size of the Corporation (2)	Nominated members (3)	Elected members (4)	Elected by rate-payers (5)	Elected by institutions (6)	Elected by minorities (7)	Co-opted members (8)
1865	190–400	190–400	–	–	–	–	–
1872	64	16	48	32	16	–	–
1888	72	16	56	36	20	–	–
1922	106	16	90	76	4	–	10
1928	108	14	94	76	8	–	10
1931	112	18	94	86	8	–	–
1938	117	3	114	106	8	–	–
1950 (No.VII)	135	3	132	124	8	–	–
1950 (No. XLVIII)	124	–	124	124	–	–	–

be noticed incidentally that the present constitution of the Corporation is simpler and more democratic than the one under the earlier Acts. It finds no place for *ex officio* members. There is also no provision for the election of a certain number of councillors by institutions like the trade unions, the Chamber of Commerce, the Indian Merchants' Chamber, the Bombay Millowners' Association, etc. The case for institutional representation was never a strong one. Among those elected in the territorial constituencies there has always been a large percentage of businessmen capable of looking after the interests of trade. With adult suffrage and the rise of a strong and well-organized Socialist Party, the working classes have also been able to return councillors who could safeguard their

[1] Bombay Act (No. XLVIII), 1950, Sections 5, 11 and 28.

interests. The case, therefore, for institutional representation became increasingly weak and this led to its abolition in the latest Act.

Under the latest Act governing the constitution of the Municipality of Calcutta—the Act of 1951 as amended in 1953 and 1955—the Corporation consists of 81 councillors and five aldermen. Of the councillors, one is the Chairman of the City Improvement Trust who holds his membership *ex officio*. The rest are elected at a general election held once in four years in the 80 single-member wards into which the city is divided. These 81 councillors select five aldermen who also hold their office for four years.[1] Communal and institutional representation have now been abolished. There are some who doubt the utility of having aldermen. The Investigation Commission has, however, recommended their continuance and there is no likelihood of this being abandoned. There is no adult suffrage in Calcutta. Only those among the residents who are

Size and Composition of the Corporation
(b) CALCUTTA

Act of (1)	Size of the Corporation (2)	Nominated members (3)	Elected members (4)	Elected by rate-payers (5)	Elected by institutions (6)	Elected by minorities (7)	Co-opted members (8)
1876	72	24	48	–	–	–	–
1888	75	15	60	50	10	–	–
1899	50	15	35	25	10	–	–
1923	90	10	63	50	12	13	5
1939	98	8	85	47	14	24	5
1951	81	1	75	75	–	–	5
1953	86	1 (*ex officio*)	80	80	–	–	5

assessed to certain taxes or occupy premises of a certain rental value or have received education up to the matriculation standard have the municipal franchise. They number about six hundred thousand in a population of three millions. The Investigation Commission reported strongly in favour of adult suffrage and its introduction may be said to be an immediate need.

In both Bombay and Calcutta the Corporations are presided over by mayors annually elected by them. They are just like the mayors in English boroughs. They do not have any special powers. In addition to presiding over the meetings of the corporations, they discharge certain ceremonial functions as the first citizens of their respective cities. In Calcutta there is also an annually elected deputy mayor. Neither the mayors nor the

[1] *Corporation of Calcutta Year Book* 1954-55; the Calcutta Municipal (Amendment) Act, 1953, Sections 4 and 5; Act 1955, Section 3.

councillors receive a salary for their work. There is, however, a move now in Bombay for granting a sumptuary allowance to the mayor and a salary to the councillors in view of the increase in their work. Though the Act requires that the Corporation should meet at least once in a month, it actually meets twice a week at present and the sessions have shown a tendency to become unduly prolonged. The time limit imposed on the speeches of councillors has not found favour with all parties in the Corporation. Besides attending meetings of the Corporation, the councillors have to attend the weekly and fortnightly meetings of the several committees of which they are members. All this accounts for the move in favour of paying a salary to them.

THE COMMITTEE SYSTEM

In both the cities there is provision made for work being carried on through committees. In practice they play a larger part in Bombay than in Calcutta. Committees in Bombay are either statutory or non-statutory. Statutory committees are those which have to be compulsorily established by the Corporation. Among them are (1) the Standing Committee, (2) the Improvements Committee, (3) the Education Committee, and (4) the Electric Supply and Transport Committee. The first consists of seventeen members of whom the Chairman of the Education Committee is an *ex-officio* member. The second and third consist of sixteen members and the last of nine, one of whom being the Chairman of the Standing Committee. The members of Committees are appointed by the Corporation, one half retiring at the end of each year. Four of the members of the Education Committee have to be non-councillors with high academic qualifications. These are expected to provide expert advice. As the Electric Supply and Transport Committee has to manage a business concern, the Corporation is authorized to appoint to it persons with experience of administration or of electric supply, or with similar qualifications even though they happen to be non-councillors. At present there are two such members on it.

Each committee annually elects its chairman. The chairman and members of the Electric Supply and Transport Committee are paid fees for attending meetings. Committees are empowered to appoint sub-committees of their own and may delegate powers to them by a two-thirds vote.

The Standing Committee is one of the unique institutions in the city government of Bombay. The budget has to be scrutinized by it before it is considered and approved by the Corporation. It sanctions contracts and the schedule of staff to be employed. It frames service regulations, prescribes the form in which the accounts are to be kept, examines and checks the weekly receipts and expenditure and arranges for the proper investment of funds. It is a sort of general committee and serves both as a cabinet of the Corporation and as the main co-ordinating agency.

The Electric Supply and Transport Committee exercises a general control over this municipal undertaking. The Education Committee is in charge of primary education. Formerly, when it was known as the Schools Committee, it had wide discretion in the expenditure of funds and in appointing and removing teachers and other employees. Under the latest Act, it is deprived of many of its executive powers and its discretion has become limited. The Improvements Committee is in charge of schemes relating to improvement and development, slum clearance, provision of accommodation for poorer classes and the purchase and sale of land. It has statutory powers of sanctioning leases up to a certain limit and this secures to it a large amount of independence.

The non-statutory committees in Bombay are of two kinds. They are either special or consultative committees. At present there are four special committees. They are (1) the Works Committee, (2) the Medical Relief and Public Health Committee, (3) Markets and Gardens Committee, and (4) Law, Revenue, and General Purpose Committee. Each consists of twenty-four councillors appointed by the Corporation on the recommendation of a general selection committee. Each elects its own chairman and deputy chairman. The Corporation is also empowered to appoint consultative committees to enquire and report on any subject relating to the administration of the city. There are two such committees at present. It should be noted that only councillors of the Corporation are to be elected to all committees—special as well as consultative.[1]

No subject is ordinarily taken up for consideration by the Corporation unless it is first referred to one or other of these committees and its opinion obtained. Generally the Corporation accepts their recommendations, and where it is not possible to do so the procedure is not summarily to reject them but to refer them back to the committees for further consideration. This results in compromise and in smooth relations between the Corporation and its committees. As the composition of the committees reflects more or less the party complexion of the Corporation, the occasions for acute differences of opinion between them are rare. It is also open to the Corporation to delegate the final power of decision to committees

[1] Municipal Corporation Year Book gives briefly all the essential facts.

by a specific resolution carried by a vote of at least two-thirds of its members present at the meeting.

The committee system in Calcutta differs from that in Bombay in a few respects. (1) There was here until 1952 only one statutory committee—the Primary Education Committee. It had only nine members, of whom three were outside experts, and its powers were advisory in character. Under the legislation of 1951–53 there are now nine statutory standing committees and they deal with education, accounts, taxation and finance, health, town planning and improvement, works, buildings, public utilities and markets, and water supply. Each committee consists of ten members who are either councillors or aldermen. (2) Non-statutory committees are either special committees or other committees, the latter corresponding to the consultative committees of Bombay. Some persons other than councillors or aldermen may be appointed to all committees—statutory or non-statutory. The Corporation may delegate to statutory standing committees powers of final decision on specific matters. (3) Among the standing committees are the borough committees, to which there is no parallel in Bombay. Each of these committees consists of all the councillors elected from the several wards comprising a borough. Three other persons who are not councillors are elected to sit on a borough committee. The business of the committee is to look after the needs of the residents in the area and to exercise any powers delegated to it. It is with a view to bringing about a certain amount of decentralization that these committees have been instituted. It is regarded as a fair compromise between the advocates of extreme centralization and those who plead for the creation of a number of smaller municipal councils for different parts of the city on the analogy of the metropolitan boroughs of London.[1]

A serious defect in the committee system of Calcutta prior to the reforms of 1951, was the absence of a statutory finance committee. The Investigation Commission recommended the establishment of such a committee to watch receipts and expenditure, to approve the budget before it is presented to the Corporation and to control variations from the budget during the course of the year.[2] The unwillingness of the corporation to make use of standing committees and delegate powers to them was one of the causes of its administrative breakdown.

THE MUNICIPAL STAFF

To carry on the day-to-day work of administration the Corporation of Bombay employs a staff of about 28,000 persons of all kinds, while

---

[1] Sections 14, 15, 16, 26 and 98 as amended.   [2] *C.C.I.C. Report*, Vol. II, p. 191.

the corresponding figure in Calcutta is a little over 26,000.[1] About 21,000 in Bombay and 18,000 in Calcutta belong to the category of labourers. Next comes the clerical and subordinate staff of a little more than six thousand, and the balance consists of the superior staff and heads of departments. In Bombay the total cost of establishment is about two crores of rupees out of a total expenditure of eight crores under all heads, while in Calcutta the cost in 1949–50 was about two crores and sixty lakhs of rupees out of a total expenditure of a little over seven crores.

In regard to the organization of the staff there is an essential difference between the two cities. In Bombay all control over municipal employees is concentrated in the Commissioner. He prescribes their duties and exercises supervision over them and disposes of all questions relating to their service, pay, privileges and allowances in accordance with regulations framed for the purpose by the Standing Committee. It has already been pointed out that the Commissioner is independent of the Corporation in the discharge of his administrative duties and this has resulted in party and political influences being reduced to a minimum in the recruitment, tenure and discipline of the staff.

In Calcutta there was also a provision in the Municipal Act under which the Chief Executive Officer was given control over all servants and officers of the Corporation. But the nature of that control was not defined by law and it was left to the Corporation to determine the powers to be delegated to him for the purpose. As the Investigation Commission has observed, 'the powers of delegation varied from time to time according to the whims and caprices of the councillors and reduced the Chief Executive Officer to a position of complete subservience, so much so that even in matters in which the Chief Executive Officer had statutory powers, individual councillors had their way.'[2] The fact of his being appointed by the Corporation and his salary being fixed by it made it impossible for him to assert himself against those whims and caprices. This naturally resulted in laxity and indiscipline, in inefficiency and corruption.

Under the new reforms introduced in 1951 the evils of jobbery and nepotism have been checked by providing that the Corporation shall appoint officers in superior grades only on the recommendation of the State Public Service Commission or of an independent Municipal Service Commission. The Municipal Commissioner appoints officers in the lower

---

[1] *Corporation Establishment Schedule*, 1950–51, Appendix XV. *C.C.I.C. Report*, Vol. II, p. 151. [2] *Ibid.*, Vol. II, p. 151.

grades in accordance with the rules and regulations laid down by the Municipal Service Commission.

Some of the officials in the superior grade are statutory and, therefore, must be appointed. In their cases, the state government prescribes their qualifications, salaries, and the conditions of their tenure of office. Among such officials are the city engineer, the executive health officer, the hydraulic engineer in Bombay and the chief executive officer, the chief engineer, the chief accountant and the health officer in Calcutta. In some of these cases the appointment and removal of individual officers are subject to the approval of state governments.

In respect of other officials, the Bombay system is on the whole conducive to economy and efficiency. It is so for the following reasons: (1) the whole establishment schedule is prepared by one officer—the Municipal Commissioner—and sanctioned by one committee—the Standing Committee. The result is the avoidance of overlapping and the determination of staff requirements on an overall instead of a piecemeal basis. The figures already given about the expenditure on establishment in the two corporations bear witness to this superiority of the Bombay system. (2) While appointments of officers on a salary of Rs.500 and below are made by the Commissioner and appointments carrying a higher salary by the Corporation, the Act and the rules made under it require that all vacancies should be advertised, that the Commissioner should make a preliminary scrutiny of the applications, and that his recommendations should be considered by a committee before the matter goes to the Corporation. (3) Recruitment to clerical and other subordinate services is made by selection on the basis of the highest percentage of marks obtained by the candidates at their matriculation or degree examinations.[1] The minimum educational qualification for recruitment is the passing of the matriculation examination or some other examination equivalent to it. The Commissioner has to choose his staff from a list so prepared. Some concessions are granted to applicants belonging to backward communities. Graduates are given a higher starting salary. Employees are in general encouraged to improve their qualifications, as advance increments are given to those who pass the examinations held by the Local Self Government Institute. The clerical staff which constitutes the main ingredient of the subordinate staff is recruited through the Employment Exchange. The work of selecting the staff drawing a minimum salary of Rs.300 and above is entrusted to the State Public Service Commission. The staff is, therefore, free from all

[1] Rules framed by the Corporation.

political and partisan influences. (4) The salaries, allowances, rules regarding promotions, retirement, etc., are more or less similar to those in state public services.

All this stood in contrast to the Calcutta system till about 1949. Under that system the power to appoint, to punish and to remove the members of the staff was vested in the Corporation itself and was ordinarily exercised through certain committees. Except in regard to officers drawing Rs.750 and above and employees drawing Rs.30 and below the entire patronage was in the hands of these committees. Appointments were filled in the period before supersession without regard to merit and sometimes even in disregard of the resolutions of the Corporation laying down minimum qualifications. The Investigation Commission had consequently observed thus: 'Abuse of patronage and practical immunity from punishment explain the malpractices we have found almost everywhere. No notice was taken of neglect of duty generally speaking and the attitude towards misconduct of every description in the inferior staff was one of great tolerance.'[1] Fortunately as a result of the Act of 1951 and of the creation of a Municipal Service Commission and the concentration of all administrative powers in the hands of the Commissioner most of these evils have now disappeared and the conditions in the Calcutta Corporation are in respect of economy and efficiency similar to those in Bombay.

The staff in both the corporations is organized into a number of departments and most of the departments are subdivided into sections and branches. There are 25 departments in Bombay and about the same number in Calcutta. Departments like Public Health, Waterworks, Engineering, and Accounts have qualified technical persons at their heads. In points of detail there are certain differences between the way in which the departments are grouped in Bombay and in Calcutta. For instance assessment and collection are grouped together under one officer in Bombay, but they work as separate departments in Calcutta. Moreover, the work of different departments is better co-ordinated in Bombay partly because the work of all of them comes up in a way for review by the Standing Committee. There are also periodical meetings of the heads of departments. Moreover, the number of special committees is only four[2] and each of them is necessarily in charge of several departments.

The machinery for co-ordination is not so good in Calcutta. Even departments like assessment, collections, and buildings which have to work in close co-operation have been made to work in isolation. In

[1] *C.C.I.C. Report*, Vol. II, p. 152. [2] *Ante*, p. 150.

neither city has the development of departments proceeded on any planned basis. In Calcutta this has resulted in the setting up of departments, like the Motor Vehicles Department, which do not serve any useful purpose.[1] In Bombay there has been no searching investigation into the subject of departmental organization as has been done in Calcutta recently, but a view was expressed by the (Shroff) Municipal Finances Committee appointed by the State Government in 1948 that 'The Municipality should introduce mechanical methods of accountancy and that it should secure a better layout of office arrangements in several offices.'[2] It recommended that an expert firm should be consulted regarding the organization of some of the departments. In pursuance of this recommendation an examination of the organization of departments is being carried on by a firm of consultants.

THE FUNCTIONS OF CITY GOVERNMENTS

The municipal governments thus constituted and organized are entrusted with a variety of functions for promoting the health, convenience, safety and education of the local inhabitants. Some of the functions are obligatory and some others optional.[3] Among the obligatory functions in Bombay are (1) those relating to public health like the construction, maintenance and cleansing of drains, the removal of refuse and rubbish, the reclamation of unhealthy localities, the prevention and checking of the spread of epidemic diseases, the provision of water supply, the establishment and maintenance of hospitals, public markets and slaughter-houses, and the regulation of offensive trades; (2) the construction and maintenance of roads, lighting of streets, the removal of dangerous buildings and obstructions, and the maintenance of a fire brigade; (3) the establishment, maintenance and aiding of schools for primary education; and (4) the improvement of the city through proper utilization and management of land and building sites. Among the optional functions are (1) the establishment and maintenance of libraries, museums, art galleries and the laying out of parks and open spaces; (2) the provision of transport; and (3) the supply of electricity. The Calcutta Act does not divide functions into these two clear-cut categories, but it is the spirit of the Act on the whole.

There are one or two other features of these functions which deserve to be noted. One is that neither police nor public assistance in the shape of the relief of the poor is included among them. In this respect the Corporations of Bombay and Calcutta differ from similar bodies in

[1] *C.C.I.C. Report*, Vol. II, pp. 146–85. [2] *Report, op. cit.*, p. 33. [3] Sections 61, 62 and 63.

England. Police is entirely a state function in India, and public assistance has not yet been recognized as a government responsibility. The other feature is that there are practically no *ad hoc* authorities to carry on any of the municipal functions except the City Improvement Trust in Calcutta. The experiment of having a similar trust was tried for some years in Bombay but finally abandoned. Each Corporation is therefore a compendious authority engaged in directly providing almost all the services which the state governments want to provide for the local inhabitants.

THE FINANCES OF CITY GOVERNMENTS

The sources from which the two corporations derive their income for meeting the cost of the services they provide are (*a*) taxation, (*b*) fees, (*c*) grants from the state, (*d*) trading enterprises and (*e*) loans.

Unlike the English local authorities which depend on only one tax —the rate on the rental value of fixed property—the Corporations of Bombay and Calcutta have a variety of taxes at their disposal. The tax on the rateable value of fixed property is common to both the corporations and this brings them a substantial portion of their tax revenue. The other taxes in Bombay are the wheel tax and the town duties on commodities imported into the city for purposes of consumption. In Calcutta there is a tax on animals and carriages corresponding to the Bombay wheel tax but the yield from it is small. Calcutta has no town duties but it has a tax on professions and callings—a sort of local miniature income tax. Fees are collected in both corporations for the sale of water mainly for non-domestic consumption, from markets and slaughterhouses and from the issue of licences.

Calcutta has not so far undertaken any trading enterprise. In Bombay the Corporation owns and manages an electricity supply and transport undertaking. This contributes at present a sum of 40 lakhs of rupees to the general revenues, which forms less than five per cent of those revenues. The Corporation is not hopeful of getting any substantial income from other utility services like this. In a recent memorandum it observed: 'Taking over existing services after paying compensation cannot be recommended as a general panacea in the present circumstances because the amount of compensation to be paid may come to such a large sum as would cut vitally into profits or make profits non-existent. Even regarding the starting of municipal utility services newly, progress should be tempered with extreme caution. Training of requisite technical and supervisory personnel will have to proceed apace.'[1] Even the presence

[1] Page 6, Replies to the Questionnaire issued by the Government of India Local Finances Enquiry Committee, 1948–50.

of a well-organized socialist party in the Corporation, which is the case now, is not likely to produce any modification in this policy of extreme caution.

The grants from state governments do not in normal years constitute more than 1 per cent of the total revenues of the two corporations. In this respect the governments have been less liberal towards them than towards rural local authorities and the municipalities in towns. In view, however, of the special difficulties created by the Second World War and the unprecedented growth of populations, the state governments agreed to make certain special contributions of a non-recurring nature. Bombay for instance got Rs.30 lakhs in 1947–48 and Rs.50 lakhs in 1948–49.

There is, however, a strong case for an increase in government grants. The rise in the general level of prices which lies at the root of the financial strain to which the corporations have become subject has been brought about by the policies of the state. In addition to this some of the grants, like those paid as compensation for the abolition of tolls or in connection with Motor Vehicle Taxation, were fixed years ago and need revision in the light of changed conditions.[1] The exemptions claimed by the central government from liability to pay property tax on its buildings and the wheel tax on defence vehicles has brought much loss of legitimate income. Rent Control Acts passed by state legislatures have prevented the corporations from assessing the rateable value of fixed property at a high figure. All these factors make it incumbent on state governments to increase their grants to the corporations and, as the Investigation Commission has pointed out, there is urgent need for both the central and state governments coming specially to the rescue of Calcutta.[2]

For meeting their capital expenditure the two corporations are permitted to raise loans up to a specified limit. In Bombay the total of loans must not exceed twice the rateable value of all the assessable property in the city. In Calcutta the condition is that the payments to be made annually towards interest and sinking fund charges shall not exceed 10 per cent of the annual rateable value of the properties. These borrowing powers are considered to be adequate and there is no demand for increasing the present limits. The outstanding loans in Bombay now amount to about Rs.28 crores. The corresponding figure in Calcutta is about Rs.10 crores. The borrowing capacity of the Calcutta Corporation was exhausted several years ago and since then few new loans have been raised. This is one of the aspects of the administrative breakdown with which it

---

[1] Shroff Committee Report, Appendix IV. Also Replies, *op. cit.*, para. 25.
[2] *C.C.I.C. Report*, Vol. II, p. 187.

was faced, a breakdown which stood in the way of reassessing the rateable value of properties correctly.

The following tables give an idea of the finances of the two corporations:

(a) *Receipts of the Bombay Corporation* (1954–55)[1]

	Lakhs of rupees
Property Taxes	488
Charges for Water	126
Town Duties	75
Wheel Tax	41
Markets and slaughter-houses	37
Sale of surplus lands	4
Government Grants	5
Miscellaneous	143
Total	919

(b) *Receipts of the Calcutta Corporation* (1954–55)[2]

	Lakhs of rupees
Consolidated rate	412
Charges for Water	18
Taxes on Professions	33
Markets and slaughter-houses	26
Government Grants	80
Miscellaneous	66
Total	635

(c) *Expenditure of the Bombay Corporation* (1954–55)[3]

	Lakhs of rupees
Primary Education	99
Conservancy	114
Maintenance of Roads and Lighting ..	50
Medical Relief and Education	90
Public Health	30
Debt Charges	215
Improvement Trust	40
General Establishment	25
Dearness Allowance	116
Other items	104
Total	883

[1] Municipal Budget Estimates, 1954–55.   [2] *Year Book*, 1954–55.
[3] Municipal Budget Estimates, 1954–55.

(d) *Expenditure of the Calcutta Corporation* (1954–55)[1]

	Lakhs of rupees
Primary Education	28
Debt Charges	72
Contribution to Improvement Trust ..	37
Establishment	61
Dearness Allowance	99
Medical Relief	23
Lighting of Roads	24
Conservancy, Roads and Miscellaneous Items	339
Total	683

A few observations may now be made on the general financial situation of the two corporations. Taxation contributes about 80 per cent of the receipts in Bombay and about 60 per cent in Calcutta. Of the taxes the most lucrative is the consolidated rate or the property tax levied on the annual rental value of fixed property. Its yield would have been greater in Bombay, but for the Rent Control Acts regulating rents of premises. In Calcutta it is administrative inefficiency, corruption and negligence that have been responsible for its low yield during the last few years. It is in the main to this source that the corporations have to look for the augmentation of their tax resources.

Expenditure on the staff employed by the corporations is high in both cities. It is due partly to the rise in the cost of living and partly to increasing self-assertion on the part of the employees in the labour and clerical grades. In Calcutta the employment of staff in excess of requirements has also been an important factor in this respect.[2]

Expenditure on essential services like water supply, drainage, medical relief and roads is low compared with the standards which obtain in the municipalities of Western Europe. The sudden growth in population and the increasing cost of labour and of materials have brought about a still further lowering of the standards today. Any improvement in this connection depends on a substantial increase in the financial resources of the corporations. Much of this increase must come from government grants which now contribute very little towards the income of the corporations. The Bombay Corporation has made out a case for its being given a share of the Betting Tax, Entertainment Tax and a number of other taxes now levied by the state government.[3] The Investigation

---

[1] *Year Book*, 1954–55.   [2] *C.C.I.C. Report*, Vol. II, pp. 181–2.   [3] Replies, *op. cit.*

9. CALCUTTA

The Masjid Nakhuda Mosque.

10. CALCUTTA

The West Bengal Secretariat, or State Government Offices, on Dalhousie Square. This is the central section of the city for governmental and commercial activity.

[Photo: *Press Information Bureau, Government of India*]

11. CALCUTTA

The High Court of Justice, a good example of Victorian Gothic architecture.

[*Photo: Press Information Bureau, Government of India*

12. CALCUTTA
The General Post Office.

[Photo: Press Information Bureau, Government of India

Commission recommended lump sum grants for giving the needed relief to the Corporation of Calcutta. But these are matters of detail. All are agreed, however, that government grants should have a larger place in the finances of the two corporations in future.

In the federal system of India local government comes under the jurisdiction of the states and not of the central government. It is therefore the states that exercise control over the corporations of Bombay and Calcutta. The control is legislative, judicial, and administrative as is the case in the English municipal system. The corporations can exercise only those powers that have been specifically conferred on them by the municipal acts passed by the state legislatures. Any citizen can question the validity of proceedings in excess of such powers through the ordinary courts of justice. The doctrine of *ultra vires* is as much in force here as in England.

Besides this legislative and judicial control there is a large amount of control exercised by the administrative departments of the states. This, however, is not a recent phenomenon. It is as old as the earliest municipal acts passed by the British after 1858. In this respect, as in several others, there is considerable difference between the Corporation of Bombay and that of Calcutta. On the whole, the control exercised by the state over the Bombay Corporation is much less than that exercised over Calcutta. There is real municipal self-government in Bombay but not in Calcutta.

One explanation of this difference is that in Bombay the Commissioner who is the head of the executive is a nominee of the state government and he is generally chosen from among the experienced members of the All-India Civil Service. It was felt that he could be relied on to serve as a check on any undesirable activities of the Corporation. The position of the executive head in Calcutta was quite different after the Municipal Act of 1923 was passed. He ceased to be the nominee of the state government and became a servant of the Corporation. Other devices therefore had to be specially adopted for exercising whatever control the state wanted to exercise, and provisions to that effect were introduced into the Act of 1923.

Certain kinds of control are common to both the corporations. Their by-laws and the loans they raise are subject to the approval of state governments. The appointment and removal of some of their superior officials are subject to confirmation by these governments. The powers

F

of the state governments to step in when the corporations are in default in the discharge of their duties are similar in both cases. But in several other matters the control to which the Calcutta Corporation is subject is greater than in the case of Bombay.

One of the extraordinary provisions in the Calcutta Municipal Act authorizes the state governments, 'to annul any proceedings of the Corporation which they consider not to be in conformity with law, or with the rules or by-laws in force,' and it is also authorized to do *all* things necessary to secure such conformity. This is a power of too sweeping a character and when it is put into effect it results in the nullification of all self-government. It is also open to the state government to depute any officer, 'to make an inspection or examination of any department, office, service, work or thing under the control of the Corporation,' and to take all action necessary on the report of such an officer.[1] In addition to this the accounts of the Calcutta Corporation are examined and audited by auditors appointed by the state government, and these auditors are authorized to exercise powers of disallowance and surcharge. In Bombay there is no such external audit. A municipal chief auditor appointed by the Corporation audits all accounts. This has been working satisfactorily as is shown by the fact that the power which the state government possesses to conduct when necessary a special audit through its own officers has never been so far exercised. Besides this, the state government has to approve all projects undertaken by the Calcutta Corporation when their cost exceeds five lakhs of rupees. Most projects fall into this category.[2] The rule-making power of the state is also greater in the case of Calcutta[3] than of Bombay. All this illustrates how drastic is the control of the state over the Calcutta Corporation.

It is, however, one thing to have such provisions in law and another to put them into effect. From this point of view it has to be said that the state government of Bengal was as negligent in the exercise of its powers of control as the Corporation was in the discharge of its obligations. If the government and the auditors appointed by it had exercised even a tithe of the powers conferred by the municipal act on them, they could have prevented the administrative collapse which overtook the Calcutta Corporation by 1948. All the same it should be noted that the spirit underlying the system of municipal government in Bombay is different from that in Calcutta.

[1] Calcutta Municipal Act, Sections 42–7.
[2] *Ibid.*, Section 109.
[3] *Ibid.*, Sections 531–6.

A GLANCE INTO THE FUTURE

Municipal government has been in existence in Bombay and Calcutta for nearly ninety years. It was set up with a view to provide services of a local character for promoting the health, safety, convenience, comfort and education of the inhabitants. In the course of these years the corporations organized for the purpose have achieved much. The filtered water supply, the system of drainage, the gas and electric lights, the parks and open spaces, the metalled and asphalt roads, are only a few among their achievements. Rates of mortality have come down. Epidemics have become less virulent. Streets have a cleaner appearance. But what has been achieved is very little when contrasted with what has still to be achieved even if what is aimed at is a minimum standard of health, comfort, and convenience. This disparity between need and its fulfilment would have called for comment even in normal times. But the conditions created by the Second World War have made the present situation abnormal in character. Both in Bombay and Calcutta populations have grown enormously within a very short period of time. In addition, there is an increasing political consciousness among classes of people who in the past were given to a certain kind of fatalism, but who now insist on being provided with at least a minimum of the essentials of life. This is a new factor in the situation, and it calls not merely for a provision of all municipal services on a larger scale, but their provision as speedily as possible. People are not in a mood to wait. This is the key to the emerging problems of the day.

It is now recognized that in the past difficulties in solving problems of municipal government were due to a failure to take a comprehensive view of all of them and appreciate their inter-connection. This is why there is so much attention paid to planning at present. In Bombay a master plan has been prepared and it has received the approval of the Corporation. In Calcutta such a plan is in course of preparation. The Bombay plan has laid down targets for population, industry, ports and docks, housing, parks and playgrounds, traffic and transport, water supply, and everything else needed to make the city an ideal place to live in. It has provided for zoning, for the creation of satellite towns and community centres. It has given an estimate of the cost of the plan, the stages through which it should be executed, the nature of the legislation to be enacted and the administrative boards to be set up.[1] The great

---

[1] *The Master Plan, op. cit.*, Section XIII.

problem today, therefore, is to find ways and means for giving effect to this plan.

The state government should immediately enact the necessary legislation and create a planning commission. Provision should be made for raising loans to the extent of nearly 65 crores during the next ten years. Finance is the crux of the whole question. What is needed in Bombay is not any change in the constitutional or administrative structure of the Corporation, but the provision of ample financial resources to carry out the schemes contemplated in the master plan.

In Calcutta, however, the situation was quite different until the passing of the Municipal Act of 1951. Serious problems of both an administrative and a political nature confronted the city. The constitutional structure had to be remodelled and a strong executive created. Administrative corruption had to be rooted out and nepotism and jobbery abolished. There was need for more grants from Government and for the strengthening of the city's financial resources—both contingent upon the establishment of a strong executive and the improvement of administration. Politically, the task was to secure democratic participation while yet ensuring efficient administration. Fortunately, the reforms inaugurated by the Act of 1951 have gone a long way in the direction of providing for very many of these needs. What now remains is for large groups of persons to learn to use the franchise. Able and enlightened representatives have to be persuaded to serve in the councils of the Corporation. The techniques and methods of influencing municipal policy must be taught to electors who have so far had little experience in the use of such tools. The future depends on the speed with which progress is made in all these directions.

Buenos Aires

RAFAEL BIELSA

Born 1889. Educated Universidad Nacional de Buenos Aires. Attorney and Doctor in Jurisprudence.

Professor of Administrative Law, University of Buenos Aires, 1924–52; Professor of University Nacional del Litoral (since its foundation, 1920–47); Director of the Instituto de Derecho Publico anexo (since its foundation, 1937–47); formerly Professor of Constitutional Law and Finance and Commercial Law in Escuela Superior Nacional de Comercio, and of Preceptive Literature, 1920–23.

Dean of the Faculty of Economics and Political Science, 1927–29; 1930–32; 1936–46; General Secretary of Municipalidad de Rosario; Under-Secretary of Justice and Public Instruction of Argentina Republic, 1932–33; Member of Academia Nacional de Derecho de la Universidad de Buenos Aires since July 1936; Member of Institut International de Droit Public, Paris, since 1929; Member of Academia de Ciencias Morales y Juridicas since 1939; Member of Institut International des Sciences Administratives, Brussels (Vice-President since 1933); Member of Academia de Legislacion y Jurisprudencia de Madrid since 1936; Member of Institut Royal des Sciences Administratives de Roumanie, 1937; Honorary Member of the Ordem dos Advogados Brasileiros since 1939.

Appointed Delegate by the Universidad Nacional del Litoral to the Third International Congress of Administrative Sciences at Paris, 1927; appointed by the National Government of the Argentine Republic as delegate to the Fifth Congress at Vienna, 1933.

Author of: *Derecho Administrativo* (4 editions, 1921, 1929, 1938 and 1947); *Ciencia de la Administracion* (1937); *Compendio de Derecho publico: Constitucional, Administrativo, Fiscal* (1952); *La Abogacia* (2 editions, 1934, 1945); *La Proteccion constitucional y el recurso extraordinario* (edited by Faculty of Law of Univ. Nac. de Buenos Aires), Primer Premio Nacional en Ciencias Juridicas y Sociales, 1938; *Estudios de Derecho Publico* (2nd ed. 1950–52). Contributor to *La Prensa*, Buenos Aires (1931–50).

Buenos Aires

THE population of the city of Buenos Aires was 3,000,371 in 1947, and the area of the federal district comprises 307 square kilometres. This is considered disproportionate in a country with 16,000,000 inhabitants and an area of 2,950,000 square kilometres, since it means that the population of the capital city amounts to nearly one-fifth of the nation's total population.

The remarkable growth and importance of Buenos Aires are due to the following causes. First, Buenos Aires is the political capital of Argentina, and administration has been excessively centralized in that city. This has led to an increase in the number of administrative bodies. Second, Buenos Aires is the country's main port. Indeed, it is of greater importance than all the other Argentine ports put together, from both the commercial and the tourist point of view. Third, the centres of education in all branches of learning are in Buenos Aires. The real cultural leadership is not necessarily to be found in that city, for it often shows itself more clearly in the provinces. Nevertheless students are drawn to the capital city in far greater numbers than they are to other educational centres. Fourth, the largest industries and the most important business and banking houses are in Buenos Aires, although industry is gradually moving out towards the suburbs.

The population has risen continuously although the rate of increase has fluctuated at times. In 1797 the city of Buenos Aires had a population of 40,000; in 1810, the year of the war of independence, it had 45,000 inhabitants. In 1863 it had a population of 120,000; in 1869, 187,126; in 1895, 663,198; in 1904, 950,890; in 1919, 1,231,698; in 1914, 1,575,814; in 1936, 2,415,142; and in 1947, 3,000,371. Thus in forty years the population was trebled.

In the first half of the nineteenth century, the average annual increase was under 20 per thousand. From 1852 until 1904 the population was considerably augmented by European immigration. Between 1863 and 1869 (in which year the first general census of the nation was taken) the yearly rate of increase was 78 per thousand; from 1887 to 1895, 52·4 per thousand; and from 1904 to 1909, 51 per thousand.[1] The same rate was maintained for the period 1909 to 1914. From 1914 to 1947 the rate of increase varied yearly between 19·1 and 19·6 per thousand.

During the period 1863–95, which followed the reconstruction of the nation, the federalization of the territory of the city of Buenos Aires was completed, and in 1880 the city was declared a federal district and the seat of the national government.

Although the rate of economic and demographic expansion was not greatly altered in 1914, the First World War brought profound changes to the country. For instance, European immigration, which was already on the wane, came now almost to a standstill. Moreover, imports decreased considerably owing to the many restrictions which the war imposed on international trade.

CHIEF STAGES OF MUNICIPAL GOVERNMENT

Although municipal government has been in existence in the city of Buenos Aires since the Constitution was established in 1853,[2] we can take as our starting-point law 1,260 of November 1, 1882, which was in force until the coming of the *de facto* government of 1943. (During this period the law was suspended on several occasions—sometimes *de jure*, at other times *de facto*, such as during the first *de facto* government, from September 1930 to February 1932.) We give below a brief account of the various phases:—

1. Law 1260 granted to municipal ratepayers[3] the right to elect city councillors. The franchise was restricted by means of a property qualification. This law was suspended in 1889 by law 2675. The city council

---

[1] These figures are taken from the 1938 Report of the National Bureau of Research, Statistics and Census. The average population density of the city of Buenos Aires (within its strict legal boundaries) is 15,230 inhabitants per square kilometre. Greater Buenos Aires, which includes suburbs and the city's whole zone of influence, has a total population of 4,644,000.

[2] On May 6, 1853, the Constitutional Congress itself sanctioned the organic law for the municipality of the federal capital, in accordance with the democratic principle of popular election. It has been said that this law is the 'authentic interpretation' of the constitution.
Gonzalez Calderon: 'Regimen municipal de la Capital Federal,' in *Rev. Estudiantes de Derecho* (1915), No. 53, p. 163.

[3] Under this law the ratepayer was the inhabitant who paid taxes or rates (ratepayer *de jure*), the minimum requirement being ten pesos for Argentine citizens and fifty pesos for foreigners. Suffrage was also granted to members of the liberal professions (physicians, barristers, engineers, dentists, surveyors, etc.), with six months of residence prior to registration in the case of citizens, and two years of residence in the case of foreigners.

was replaced by a citizens' committee appointed by the national executive. In 1890, law 2760 re-established the city council; but in 1901, law 4029 again set up a citizens' committee, or executive organ, composed of members designated by the President of Argentina.

2. In 1907 the national executive proposed the restoration of an elective local government in accordance with the provisions of law 1260, and that project took shape as law 5098, which was in force until March 22, 1915.

3. On March 22, 1915, the President of the nation, by means of a decree-law, dissolved the city council and appointed a committee of citizens similar to that created in 1901. On August 4, 1917, Congress enacted law 10,240, which re-established the representative character of the city council, but with two important innovations: first, there was to be universal suffrage, in the sense that all Argentine citizens domiciled in the federal capital, including the foreign-born, would qualify as electors; and second, the system of proportional representation was introduced.

This new electoral régime was clearly an advance towards local democracy, but the efficacy of the system was not taken into consideration. The new régime affected only representation in the city council, where decisions are made, and did not apply to the mayor, who is responsible for administration and for executing those decisions. The mayor is appointed by the national executive, that is the President of the republic.

Experience has shown that in Argentina this representative form of local government has been unfavourable to the growth of a true democratic system, whether considered from the financial or political viewpoint, because no account has been taken of the efficiency of the councillors.[1] In fact, the representatives—or electors—of the big national parties have brought to the sphere of local government the same directives and practices which have been typical of those parties in the national and provincial political spheres. The socialist councillors have enjoyed a reputation for efficiency and for seeking to promote financial integrity, but so far they have never been in power and their opportunities for effective action or influence have therefore been restricted.

In other countries, the good work of political parties in the sphere of local government has enabled those same parties to triumph later on in the national elections;[2] but it will be noticed that in the Argentine Republic the process has been reversed.

---

[1] R. Bielsa: *Principios de Regimen Municipal* (Buenos Aires, 1940), p. 31.

[2] In England, for example, the movement has taken place from the municipal to the national sphere. The Fabian Society of fifty years ago obtained some representation in

F*

4. Period of the *de facto* government, from June 4, 1943, to June 4, 1946. The national executive had dissolved the Buenos Aires City Council in 1942, and intervened in the city's affairs by appointing an *ad hoc* governing body, a citizens' committee. The *de facto* government, set up by a military rising, issued a decree, dated June 25, 1943, dissolving the citizens' committee and conferring upon the mayor the powers of the city council, with the following exceptions: (*a*) modification of tax ordinances and municipal budget ordinances; (*b*) raising of loans; (*c*) approval of the city's accounts, investment of funds, and similar matters. All these powers were now vested solely in the President of the republic.

This régime is still in force despite the fact that the constitution was formally restored on June 4, 1946. It was stated, at that time, that the municipal government would be re-established, but up to the present day the mayor is the sole municipal authority, though naturally his powers are limited. Representative municipal government has ceased to exist.

The present centralized form of local government is reprehensible. There is no discussion whatsoever. The budget has been enormously increased in order to carry out an electoral programme at the ratepayers' expense. Taxes are extremely high and are a burden on the least privileged classes, besides causing inflation. The régime is an expression of anti-democratic authoritarianism; it is a mistake to allow this state of affairs to continue to exist. It is necessary to establish control by representatives of popular forces not corrupted by political opportunism.

THE MUNICIPAL GOVERNMENT OF BUENOS AIRES

Buenos Aires has a special form of municipal government. Each of the fourteen provinces of Argentina has its own municipal type of government in accordance with laws enacted by the provincial legislature, for local government is a matter of provincial determination under the constitution. The city of Buenos Aires, however, is a federal district, and is regulated by legislation passed by the national congress.

The national character of the city of Buenos Aires is explained by the fact that it is the capital of the nation and the seat of the federal legislature, executive, and judiciary. It is inconceivable that the national capital should be situated in the territory of a province, for the provincial

English local government; since 1945 their successors, the Labour Party, brought to the national government the socialist force which gave birth to that Society. See W. W. Crouch: 'Local Government under the British Labour Government,' in *The Journal of Politics*, Vol. 12, May 1950, pp. 233–59.

government would then have political authority over the national capital. Therefore, the city of Buenos Aires remains within the orbit of national authority. The seat of federal government lies in federal territory, under the direct authority of the federal government[1] and it is not the 'guest of a provincial government.'

The Argentine constitution provides that the President of the republic shall be the direct and immediate head of the city. In that capacity, the chief executive is in direct control of the security police (whose functions we describe on a later page) and thereby ensures the maintenance of public order and the internal safety of the state.

Nevertheless, the role of the President as supreme political authority in the national capital is quite distinct from that of the municipal administration, which is primarily concerned with matters of purely local interest, such as public cleansing, street lighting, public health, entertainments, building regulation, and so forth. In consequence, the municipal administration of the city should be quite separate from, and untrammelled by, the national executive.

The municipal system of the city of Buenos Aires is similar to that of the provincial cities. It differs only because the city is established on federal territory. In this municipal system the political, commercial, and cultural factors specially affecting the capital have not been taken into consideration.

Under the law of 1882 (which is not applied at the present time), the political government of the federal capital resides in two authorities: (1) Congress, which is the local as well as the federal legislature (Art. 68, Section 23 of the national constitution) and (2) the President of Argentina, who is the immediate and local head of the capital (Art. 83, Section 3 of the national constitution). The President also has administrative authority within the capital in those functions which are exclusively his own, as chief of the armed forces and of the security police, which is partly militarized. This police force has also to co-operate with the courts, assists in the investigation of crimes, and performs other auxiliary functions.

---

[1] In 1881, when the whole question of the municipal law of Buenos Aires was being discussed, it was pointed out in the national congress that the United States of America had created a federal capital—Washington—because when the congress used to meet at Philadelphia it was often influenced by crowds which exerted undue pressure on its members and on their proceedings, and the local government did not suitably protect the assembly. This served as a lesson to all Americans.

Vide R. Bielsa: *El problema de la descentralizacion administrativa* (Buenos Aires, 1935), p. 56 *et seq.*

The municipal government differs from the political government in that it is essentially administrative. It deals with all matters relating to the public services (such as transport, light and power, telephones, etc.) and the use of public property. This municipal administration, traditionally carried out by local councils elected by the townspeople, was abolished in 1943.

Municipal administration of popular origin is proper to every city, and not only is it compatible with the political form of government, but it also exists in the letter and spirit of the constitution. In the constitution of 1853 it was deemed to be impliedly authorized by a constitutional requirement relative to the election of the President of the nation: the constitution laid it down that one of the electoral lists should be sent to the president of the municipality of the federal capital and another one to the president of the senate (Art. 81).

However, this traditional municipal system of government has in fact been abolished, perhaps by a strange or arbitrary interpretation of the text of the new constitution. Under the latter, the election of the President of the nation is direct; therefore, the president of the municipality no longer has the function of being the recipient of a list of electors. Furthermore, the new constitution has added a clause to the previous text. The constitution of 1853 stated that 'the President of the nation' is the 'immediate and local head of the capital of the nation.' The recently amended constitution adds: 'he (the President of the nation) may delegate those functions in the manner to be established by administrative regulations.'

This constitutional amendment in no way implies the legal suspension of the deliberative body representing the citizens. Moreover, the President's power to delegate his local functions is not incompatible with the existence of a deliberative council.

The abolition of an elected council is undoubtedly a step away from democracy. In the capital of Argentina the people have been deprived of the right to participate in the administration of their own affairs in matters which directly affect them, such as security, public health, morality, comfort and conveniences, and, above all, rates and taxes. Municipal autonomy has in fact been eliminated, for there is no self-government where the city cannot grant its own organic charter without government intervention.[1]

Under the provisions of the law of 1882, which was in force except

---

[1] R. Bielsa: *Derecho administrativo*, 4ª ed. V. III, pags 196 y sigtes. *Principios de Regimen Municipal*, 2ª ed. (Buenos Aires, 1940), pags 45 y seqtes.

for brief periods of suspension until the year 1943, the municipality established and authorized the concession of public services, and exercised all administrative powers concerning the regulation of public property, buildings, public health (except sewage disposal and the supply of drinking water), markets, factories, theatres, etc. It also exercised the power to tax.

The municipal authority was divided into two branches: (*a*) the municipal council, which decided administrative and fiscal policy; and (*b*) the mayor, who was empowered to execute such decisions. The law formerly in force called for thirty members of the municipal council, which allowed the proportional representation of the more important parties. The following persons could be councillors: (1) those citizens having the qualifications needed to become a national deputy (25 years of age or over) and two years' previous and uninterrupted residence in the federal capital; (2) foreigners of 25 years of age or over who also fulfilled the requirements of a voter and who had four years' previous and uninterrupted residence in the district. The city councillors held office for four years. Until 1933 councillors were unpaid, but in that year salaries were authorized.

The present trend towards inordinately high salaries is so blatant and obvious that it is justifiably criticized, and more so because of the fact that the party with overwhelming majorities in most of the city councils pays lip-service to democracy, and its members call themselves *descamisados* or shirtless ones (something like the *sans-culottes* of the French Revolution). This tendency manifests itself most acutely in the national congress and provincial legislatures, whose members' salaries have been tripled. It has been considered that the system which set up an elective council empowered to sanction taxation, organize public services, grant concessions for such services, regulate building, supervise public health, morality, etc., provided a kind of local autonomy. Strictly speaking, however, local autonomy was non-existent, inasmuch as the municipality could not issue its own organic charter and the mayor was not elected, but was named by the President of Argentina.

PERSONNEL AND FINANCE

The municipal personnel are classified in the following categories: (*a*) administrative or clerical staff; (*b*) technical staff, which implies special training in some subject; (*c*) professional staff, which comprises those who have a university degree or diploma granted by a secondary school;

(*d*) special services staff; (*e*) manual workers, who are divided into skilled and unskilled; (*f*) unclassified.

The municipal administration of the city of Buenos Aires is directed by the mayor,[1] who is the head of the city government. It is also called the executive department as distinct from the deliberative department, which was the former deliberative council.

The executive department has four secretariats. These are the big administrative divisions. In fact they are larger than some of our provincial ministries, both in number of officials or employees and the scale of expenditure. There is also a municipal court which decides violations of municipal ordinances.

The secretariats comprise: (*a*) finance and administration; (*b*) public works; (*c*) public health and supplies; (*d*) culture and municipal police.

The staff employed by these secretariats and the mayor's office number nearly 32,000. They are distributed as follows:

	Administrative or Clerical	Technical	Professional	Workmen or unclassified	Total
Mayor's executive office	359	19	69	—	1,597
Finance and Administration ..	1,801	299	242	879	3,221
Public Works	685	646	208	2,493	4,032
Public Health and Supply ..	1,704	3,520	170	4,975	20,578
Entertainment and Police	613	87	8	1,637	2,345

The salaries and wages of these employees amounted (in 1947) to more than 200 million pesos. The salaries of other personnel brings the city's wage bill to nearly 239 million pesos; excluding independent bodies with their own budgets such as the Department of Municipal Works, public utility undertakings operated by concessionaires, nationally-provided services such as sewage disposal and water supply, or separate organs like the Colon Opera House, which is heavily subsidized by the municipality, or the municipal broadcasting station.

The Colon Opera House is a revealing example of municipal extravagance. It employs no fewer than 677 persons, including a chorus of 100, a corps de ballet of 91, an orchestra numbering 98, 11 teachers, 131 skilled workers and 42 administrative employees. The staff costs the city 4·6 million pesos a year. The total cost of the opera house is 11·7 million pesos a year, which is as much as the entire expenditure of Rosario, the second largest city in Argentina, with a population of 600,000 before the

---

[1] Often called the 'Lord Mayor' in imitation of his opposite number in the City of London.

revolution of 1943. The city of Buenos Aires had to meet a net loss on the opera house of 9·1 million pesos.

The total expenditure of the municipality is about 462 million pesos. This includes 35·7 million pesos for servicing the municipal debt and 86·6 million pesos by way of grants or subsidies to independent or 'decentralized' bodies.

The mayor has a salary of 4,000 pesos per month. The mayor's male secretary is paid 2,000 pesos. The heads of the four secretariats each have a salary of 3,200 pesos per month. The Director of Public Festivals—a creation of the new government—receives 2,400 pesos per month, i.e. he has the same salary as that of national ministers before the 1943 revolution.

In our opinion the municipal budget could be reduced to one-third its present total. In the recent past, under constitutional government, the municipality was better administered on a budget of less than a quarter of the current one. We make this statement in all seriousness, after taking into account the depreciation of the currency.

The Department of Public Festivals employs, in addition to the director, a technical secretary at 1,600 pesos, an administrator at 1,400 pesos, and 200 persons with salaries varying between 300 and 800 pesos per month. This department is concerned with the celebration of four or five holidays in the course of a year, which remind observers of the *candombes* or street celebrations held by the coloured slaves and freedmen in colonial days. The real purpose of these celebrations is to make propaganda—in very poor taste but nonetheless most expensive—for the party in power.

For further evidence of maladministration and waste of public funds we may point to the Boxing Commission, with 14 employees. The administrator's salary is 1,000 pesos, and several of the other employees receive from 500 to 700 pesos. The cost of this commission is 48,000 pesos (which is about what a good school, of which this city is in dire need, would cost). A children's theatre has an administrator at a salary of 1,300 pesos per month, and 13 employees at from 375 to 700 pesos.

The main sources of regular fiscal income are: (*a*) charges for lighting, street sweeping and cleaning, estimated at 87,000,000 pesos a year: (*b*) licences for vehicles, 30,136,000 pesos a year.

The so-called 'special' resources are those derived from national taxes. The proceeds of these are shared with the municipality, which receives the following amounts from the various taxes: (*a*) from income tax, 50,000,000 pesos; (*b*) from sales tax, 9,900,000 pesos; (*c*) from excess profits tax, 12,000,000 pesos; (*d*) from capital gains tax, 8,600,000 pesos; (*e*) from land tax (i.e. real estate taxes), 39,200,000 pesos; (*f*) racecourse taxes, 8,500,000 pesos; total amount: 128,200,000 pesos. Any deficit in the municipal budget is met from the funds of the national government.

The contributions from public utility services are as follows: (*a*) electric light companies, 15,500,000 pesos; (*b*) gas undertaking, 1,900,000 pesos; total contribution; 17,400,000 pesos.

The public debt amounts to 532,029,977 pesos; of this, 36,554,079 pesos has been redeemed at the time of writing. Public debt services amount to 32,891,871 pesos (interest, 22,599,582 pesos; amortization, 9,292,289 pesos; various other items, 1,000,000 pesos).

POLICE POWERS, FEDERAL AND MUNICIPAL

In Argentina jurisprudence, the concept of 'police' is not as inclusive as in the American usage, nor is it as limited as in that of continental Europe. In the Argentine system the European concept prevails in principle, but in practice the influence of North American jurisprudence is to be noticed in our courts.

In the municipal sphere the police has certain characteristics defined by its limitations, which are of two kinds: (*a*) those which result from the centralization of the security police, which is always subject to the national or provincial government, depending on the status of the municipality; (*b*) those resulting from the basic municipal law which is in force in the city.

In the city of Buenos Aires the security police is federal and is under the authority of the President as chief executive of the nation. This police force, which is partly militarized, also has duties auxiliary to the judicial courts, especially in criminal matters; for example, it assists in the investigation of crime and in the apprehending of criminals. It also maintains order in public places, such as streets, squares, and places of entertainment. The security police control gambling, which is against the law in the city of Buenos Aires. This police force also contains a firemen's section for fighting fires and other disasters.

The municipal police supervises the security of buildings, public health, street cleaning, and food hygiene. It also controls matters connected with public morality such as prostitution and decorum on the streets, in theatres, and in other places of entertainment.

The main function of the security police should be to keep order and protect the citizens; not to indulge in spying activities of a political nature.

13. CALCUTTA

A pagoda in the Eden Gardens.

14. BUENOS AIRES
The Plaza Mayo.

It should be under the orders of the national executive, but should also carry out the duties of an auxiliary body to the municipal authorities; the fulfilment of its obligations in this respect should at no time be hampered or weakened by the political authorities. If the security police co-operate with the courts in the investigation of crime and in the apprehension and custody of lawbreakers, the judicial authorities' orders must not suffer alterations or limitations imposed by the executive.

The security police cannot be a municipal police, owing to the nature of its functions (public security allied to the national interest, i.e. the security of the state and the diplomatic representatives of foreign powers). If it were a municipal organization, its powers would cease at the city limits, and this would reduce its general effectiveness in the prosecution of criminals.

The municipal police carries out duties of inspection, proof, and enforcement, but it does not proceed *vi militari*, as only the security police can order the use of force.[1] The municipal authority must request the assistance of the national police in order to take coercive measures. The running of public utility services (tramways, omnibuses, light and power, telephones) is also supervised by the municipal police, who must ensure the continuity, regularity, safety, and efficiency of these services. The operation of public utility undertakings is entrusted to companies to whom concessions have been granted.

But the questions most frequently at issue are those dealing with territorial jurisdiction. A municipal public service is usually organized and administered within the municipal area. When the city expands in population and building beyond its own territorial boundaries, those services are in practice extended to the suburbs of the city. In such cases they become inter-city services and, in the particular case of Buenos Aires, inter-provincial services, for the federal district becomes the equivalent of a national territory—which it in fact is—largely surrounded by territory forming part of the province of Buenos Aires. Furthermore, public services such as railways and telephone systems are the chief means of communication between cities and provinces (sometimes between several provinces).

It should be understood that when it becomes important to extend a public service to a point outside the city the fact that it also benefits the surrounding towns does not detract from the municipal character of the public service in question. In such cases, however, it is always necessary

---

[1] R. Bielsa: *Ciencia de la Administración* (Rosario, 1937), Cap II, pags 55–108; *Derecho Administrativo*, ed. cit. V. 1, pags 156 y sigtes.

to obtain the authorization of the national government, or permission from the relevant communal authority or provincial government.

PLANNING AND DEVELOPMENT

The original outline and general planning of the city of Buenos Aires was the same as that of the other cities founded by the Spaniards in the New World, and these characteristics remained unchanged during the colonial period. Its area is divided into uniform squares, especially in the centre, except for some irregular portions due to the manner in which the land was distributed amongst the founders. Within certain zones the streets are of uniform width (twenty metres is the average).

In the early days building was for the most part of one storey only, as can still be seen in the outskirts of the city. Later on, it was more profitable to build houses of several stories for renting purposes. The letting of houses has been in certain periods one of the most profitable of business undertakings, and it is only fair to recognize that the incentive of such excellent profits was one of the principal factors in the general development of building in the city, the growth of which has been much greater than that of most European cities. This urban growth has taken place without any planning to regulate it. The reason for the large buildings erected many stories high is not the limited area of the city but rather the increasing value of the land, which in many instances was the result of excessive commercial speculation.

As the city grew in size and population, successive local governments opened new avenues or widened the main streets to facilitate the increase in traffic and also for aesthetic reasons. The widening of streets has at times given rise to legal restrictions such as the obligation to refrain from building on that portion of property reserved for the future widening of public thoroughfares. Where a restriction of this kind is imposed, property owners have been compelled to leave an empty space next to the street following future building lines. Needless to say, the sterilized portion of property is expropriated and the owner receives due compensation for his land, although expropriation proceedings are sometimes not instituted until many years after the municipal authorities have decided that a street is to be widened, and imposed the necessary restriction on the landowners. The subsequent street plans do not follow a basic, symmetrical, or unified plan for the city. They have rather been made piecemeal in order to facilitate urban traffic.

The city authorities have also tried to satisfy, at the same time, some rules of architectural design and urban health, but there has been no attempt to regulate the building of private houses. The municipality has, of course, always exercised its police powers in controlling building construction from the standpoint of safety, health, and aesthetic requirements.

In 1942 a bill was presented by which land occupied by the Rural Society would be transferred to the municipality of Buenos Aires and the Zoological Gardens moved to a new site. The main object of the plan was to use the area thus liberated to make a large park surrounded by buildings whose height and architectural style would resemble those of Hyde Park in London, Central Park in New York, and the Parc Monceau in Paris. This would effect a notable urban improvement in a large sector of the city and would better living conditions. The bill called for a system to finance the project which did not entail any disbursement of municipal funds, but did provide an incentive to the investment of private capital, whilst assuring work for many labourers. However, this bill was not passed.

On December 26, 1947, the mayor of Buenos Aires issued a decree which created a committee to study and formulate a plan for Buenos Aires under the Secretariat of Public Works and Town Planning. This body was to function for a period of three years. It was composed of one executive counsellor, four permanent counsellors, technical advisers, and representatives of the national and provincial authorities. Its principal functions were to be the preparation of a plan for the city of Buenos Aires; the planning of urban public works; advising on works; and the negotiation of agreements with the government of the province of Buenos Aires and with the neighbouring municipalities for the preparation of a plan for Greater Buenos Aires.

The planning committee's duties were to put the proposed plans into execution, to carry out technical studies, and to train specialized personnel. In its brief life-time of two years it did quite a lot of good work in town planning and drafted laws which revealed a careful study of the legal and financial aspects of the question. It also formulated rules for a building code dealing with the most convenient size for the city (density modulus), and dealt with such matters as transport and open spaces for common use and recreation. Housing would be regulated in its architectural, urban, legal, and financial aspects. The committee proposed the immediate construction of vertical building units as an experiment and an incentive to put the plan into effect on a large scale. The plan called for the unification of building sites and the abolition of squares. (A square is a built-up, quadrangular space, measuring 100 to 130 metres on each side, and opening on to the street. It is the result of the original laying-out of the city, and is now considered most inconvenient.)

Vertical housing, as the term indicates, consists of a building covering a large area, many stories high, and with a number of flats on each floor. These flats may be purchased individually, as the recent horizontal property law allows such a division to be made, although formerly it was not permitted by the Civil Code. Thus, small properties may be bought in a building with hundreds of flats; the unit is built on a green space, assuring for its future occupants sufficient fresh air, light and sunshine. The plan includes the establishment of residential areas, business districts, government offices, etc. However, a few months ago this plan was scrapped, and the work it proposed to do will probably not be put into effect.

Less than two years after its creation, the planning committee was dissolved, and its duties were assigned to the Secretariat of Public Works.

HOUSING AND TRANSPORT

The most urgent municipal problems requiring solution in the city of Buenos Aires are housing and transport.

The housing problem has been made worse by the *de facto* government's policy of rent restriction which was introduced in 1943. The obvious reason for this step was to win the tenant's goodwill, that is, the goodwill of the majority of the population. The results were disastrous for new tenants and for the thousands of persons who can find no place to live.

We mention the situation in which new tenants find themselves because, despite official maximum rents, those who have to rent a new place to live in must pay outrageous sums in order to secure a lease. This leads to manifestly unfair situations where one tenant is paying 100 pesos in rent, whilst his next-door neighbour has to pay 500 pesos for exactly similar premises.

The rationing of building materials and their allocation to unnecessary public works has made the situation worse than ever. And, as if that were not enough, the increase in municipal rates discourages would-be builders.

The housing shortage requires a rational solution based on careful planning for a regular supply of comprehensive and economically-constructed houses, which could also be let at reasonable prices. Those were the aims of the 'Plan for Buenos Aires' to which we have already referred.

The transport problem is also a very serious one. The city's considerable area, and the growing need of residential houses in the outlying districts or in the suburbs, call for increases in the number and capacity of the means of transport as well as the renewal of the rolling stock in use at present. The problem is extremely difficult to solve because of the lack of foreign exchange with which to purchase the new means of conveyance so badly needed by the city's millions. Underground transport has helped to ameliorate conditions, but it has been found necessary to increase fares. There is also a shortage of taxi-cabs, which causes considerable inconvenience to visitors from the provinces and tourists from abroad.

The municipal problem is complex. Its solution must be gradual: it should start with what the population most needs and go on from there to its comfort and convenience. Housing should receive first priority, transport facilities should come next, and then recreation and amusement for the physical health and spiritual welfare of the people, which are always affected by life in a big city.

DEMOCRACY AND EFFICIENCY IN LOCAL GOVERNMENT

Efficiency is not taken into account in designating the authorities which make up the two branches of local government: that is, the councillors who compose the deliberative organ and the mayor who forms the executive organ. The political parties name their candidates, and the people cast their votes for them, without knowing anything about their efficiency. Not even in the exclusively municipal parties is efficiency considered, though it is often enough cited in their propaganda. Possibly the sole exception is the Socialist Party, where the principle of division of labour and specialization is applied in local government as well as in parliament.[1]

As for the mayor, the national executive nearly always names a person with a certain political and party prestige, but not a 'municipalist.' The basis for a man's appointment to the mayoralty is that he is a 'good administrator.' The technical experts are the ones who fill executive and other posts in the administrative office, beginning with heads of departments, directors, section chiefs, advisers.

In a government elected by universal suffrage, the party spirit usually prevails over a selective criterion of efficiency. This is logical and to a certain extent democratic. Furthermore, the social and financial con-

[1] R. Bielsa: *Principios de Regimen Municipal*, ed. cit., Preface; *Reflexiones sobre sistemas políticos* (Buenos Aires, 1944), pags 188 y sigtes. *El estadista y su pueblo* (Buenos Aires, 1945). p. 71 *et seq.*

sideration of local problems presents similar characteristics to those of a national or provincial order. The professional politician is always preferred for his popularity and experience in the art of achieving political or economic aims.

In practice, democracy has had no influence on the progress of municipal government, and least of all has it made any improvement in efficiency or administrative morality.[1] There are many examples to prove this; but a few of them, such as public service concessions, are symptomatic.

In 1907, the municipality of the city of Buenos Aires granted a concession for the supply of electricity to the inhabitants and the provision of electric light to public and government buildings. The clauses setting forth the obligation and duties of the concessionaire, as well as the rights retained by the municipality, denote a degree of foresight and a well-developed sense of responsibility quite different from that shown by the municipality in 1936, when it granted an unjustified and expensive extension of that concession to the powerful concern which succeeded the one which obtained the original concession in 1907.

This extension became the object of an investigation which lasted from 1943 to 1945, but things were left as they were. The deliberative council of the year 1907 was not constituted as democratically as the 1936 council, inasmuch as its members were not elected under the system of universal suffrage but by the ratepayers only. And yet it must be admitted that the 1907 council acted as a wise administrator of the public interest, at least in the matter of the electric power concession, whereas the democratically-elected council of 1936 granted an extension of that concession which was inimical to the general interest. For that reason, criticism of its action in this instance was practically unanimous.

This fact and others of a similar nature which were the direct cause of the President's intervention in the year 1942, are an obvious argument against the system of representation of the old parties. However, the failure must not be attributed to the democratic system itself, but rather to the political parties which became involved in such irregular acts, and thereby discredited democracy. The reasons are well known. The

[1] This short article descriptive of municipal government is not the proper place to make recommendations on political sociology, but the fact we point out is of frequent occurrence in practically all countries in America, and in some European countries as well. The decisive electoral factor may be a local political leader without any sense of responsibility, whose main object is to enrich himself at the expense of the public by making use of the opportunities which his position affords him. Though councillors receive a stated salary, they are not always satisfied with it, and are prone to engage in unsavoury business deals in order to improve their private finances.

Argentine political parties are in the habit of choosing as candidates for office persons lacking in intellectual or moral integrity, and with no professional competence. More often than not they are habitués of local political clubs or district leaders[1] who are popular in their respective districts, or else they are persons with some influence in the party's central committee.

In actual practice these people have made of the deliberative council a miniature parliament, instead of a corporate organ for active administration. Unsound parliamentary practices and obstructionism, which have to a certain extent discredited the parliamentary system, have been reproduced in the city councils to the detriment of municipal administration. Moreover, a complete lack of responsibility and loyalty to the public interest has been revealed in systems of public service concessions and in the squandering of municipal funds. Some of these actions have brought the political parties into disrepute, because it is generally understood that the actions of local representatives are governed by directives issued by their party's central committees. The socialist councillors have had a salutary effect in exercising their functions of opposition, control and criticism of the irregular measures which we have mentioned, but in those instances where parliamentary absenteeism is admissible in order to prevent a quorum, they have unfortunately not applied this, the only effective remedy. The sole result of this political strategy is to discredit the rival parties—which they richly deserve—but it does not prevent the passing of measures which are harmful to the interests of the citizens and which deplete the municipal treasury, to the detriment of the honest ratepayer.

When speaking of municipal democracy—which is the best way to establish public services in the common interest, provided their administration is governed by an efficient system—it is important to bear in mind that its success or failure depends on the civic education of party representatives and on a severe control over these representatives. This control should be exerted by the party itself. This feature has always been characteristic of the Socialist Party, the principal reason being that it is an opposition party.

Socialist councillors, who are under strict party discipline, are efficient council members. But the system of proportional representation does not give them a majority; that is, they do not govern, though their greatest strength lies in the federal capital, which has at times given them majority representation of that district in parliament. The *criollo* (old-

[1] R. Bielsa: *El cacique en la función pública* (Buenos Aires, 1928).

established Argentine parties) are the Radical and the Conservative parties, or others resulting from the combination of the latter with elements of other parties, sometimes including former Radicals.[1] (It was that type of combination which made up the *Concordancia* from 1932 to 1940.) That type of so-called reform bore no real result in increased efficiency of government.

Under the system in force before 1917, council members were persons of integrity and responsibility, such as industrialists, businessmen and professional men (lawyers, doctors, etc.), who did not belong to any party. Municipal parties are formed on platforms of exclusively local or municipal interest. That municipal system—which we have criticized for reasons of principle—was called 'government of one's own.' Laboulaye's principle 'Point de représentation, point d'impôt' was interpreted conversely: 'Without taxation there is no representation.' In consequence, those who did not pay direct taxes were not included in the electoral list. Inasmuch as taxes are passed on to the consumer, the user, the tenant, etc., they are really paid indirectly. For this reason we believe that system was unfair.

Popular suffrage, established in 1917, brought democracy to municipal government. This modification had no effect, however, on efficiency in administration. Efficiency is a question not of suffrage but of planned and basic regulations and of the civic education of those in office.

Some parties make a point of formulating a set of principles of efficiency in government, but the important thing is to put such principles into practice. The tendency to convert deliberative councils into miniature municipal parliaments, instead of their being corporate bodies able to

[1] The principal parties in opposition to the present Government are the radicals the, socialists, conservatives and progressive democrats. The Radical Party is a popular democratic party, conservative in economic, social and religious matters. It was in power from 1916 until 1930. The Socialist Party has always been strong in the Argentine capital, where it has taken the place formerly occupied by the Radicals. The Socialist Party resembles the British Labour movement and is the principal opponent of the present government. It contains men of great ability and rectitude, and is at present united and morally strong. This has not always been the case, for in 1931 there was a split which resulted in the formation of the Independent Socialist Party, which damaged both the Socialist Party and the Radicals. The Conservative Party is republican and liberal. It held power for more than fifty years until 1916; and in 1932 it returned to power in a coalition which held power until the Government was deposed by a military *coup* in 1943. The Conservatives oppose the present régime and sympathize with the Socialists; and although they are accused of representing oligarchic interests they have not been opposed to legislation to protect the working class or to promote social progress. The Progressive Democratic Party was formerly powerful, especially in 1916. It fought the 1932 elections in alliance with the Socialists, but was defeated by a Conservative coalition. The Progressive Democrats are now more or less similar to the British Liberals.

deal with administrative and financial matters in the interest of the community, is the chief reason for their failure. Only too often have the proceedings of the council been used as a means of party propaganda.

The municipal problem is one of civic education,[1] of the republican spirit, of respect for the law and for the interests of the community of citizens. Excessive taxation reaches its highest point under régimes without popular representation, such as the government of Buenos Aires, where fiscal greed is probably without parallel in any other city in the world. Not even the worst governments in past Argentine history dared go so far. As a result of this mistaken fiscal policy the cost of living has reached dizzy heights, especially for the working classes who, dazzled by apparent benefits, propaganda, and demagogy, are not aware that they have lost not only their freedom but their well-being as well.

CONCLUSIONS

The double character of Buenos Aires, first as the political capital of the nation, and therefore situated in a federal district; and second, as a large city which is a residential, business and cultural centre, justifies a special legal system compatible with the political government of the capital by the national executive. That is the role of the municipality as an administrative body charged with promoting specific local interests of the whole community of citizens, with the right to govern its own administration in respect of safety, public health, morality, comfort, and entertainment or recreation.

We have seen the results, and are aware of the advantages and disadvantages, of municipal decentralization; that is, of administration effected by the democratic city councils, which owing to their deficient technical and moral capacity have not always been faithful to the best interests of the public. We have pointed out these failings in articles which have appeared in the public press, and in books on the subject, and have made known our criticism in public lectures. Despite all this we maintain that the representation of the citizens in the democratic form of a city council is necessary for good government; when such a council exists there is always discussion and criticism of proposed measures, and this frequently prevents the passing of arbitrary ordinances or misguided resolutions.

To avoid the consequences of the disloyalty or incapacity of the councillors we have advised that effective control be established over all fundamental actions taken by the council, and over all measures of financial

[1] R. Bielsa: *Reflexiones sobre sistemas políticos*, ed. cit., pag 188.

importance or of general public interest. Such control does exist generally
in provincial affairs. Monopolies and concessions with privileges or
exemptions must be approved by the provincial legislature.

In the capital of the nation the body exercising legislative control is the
national congress. Unfortunately, however, its power to approve or
disapprove actions taken by the municipality does not include public
service concessions, but only the issue of public loans.

The problem of administering a great city presents special characteristics, depending on the city's form of government, and on the degree
of political education and respect for law of the voters and elected
representatives. It is a problem of civic and political culture. But control
should always exist, especially over all matters of a financial nature.

Public control in the form of a referendum is advisable only in the
case of public service concessions or important public works. Such popular
control cannot be applied to decisions affecting rates or taxes, because
any ordinance establishing new rates or increasing those already in
existence would run the risk of being voted down.

The impeachment of public officials cannot be applied where the body
of voters is not conscientious or accustomed to think matters out, because
the party spirit, demagogic action, etc., would usually give the officials
of the majority party a bill of indemnity. It is preferable to have recourse
to the regular procedure for dismissal from public office, or for jurisdictional proceedings: accusation, defence, evidence, and a reasoned
verdict. Experience shows that the mechanism of the recall of officials,
and even the referendum, do not function as successfully in large cities
as they do in smaller ones, where it is easier for the voters to have direct
knowledge of the facts and personalities involved.

To make local democracy work efficiently, a charter or statute for
councillors should be instituted, establishing standards of professional and
moral fitness or capacity. Disqualification for incompetence, and the
incompatibility of various functions with the duties of a city councillor,
should also be established. The list of candidates should be published
sufficiently in advance to permit the deletion of one or more of the
names proposed by a kind of *actio popularis*. The reasons should be based
directly on the candidate's antecedents, and should be the responsibility
of the accuser.

In principle, no one should be entitled to vote who cannot prove that
he is a municipal ratepayer, either directly or indirectly. If indirectly, his
claim to belong to this category should be based on strong evidence.

As Buenos Aires is a federal district, the national congress is its local legislature, and the national executive its immediate and local head. The administrative powers of the national executive are those which concern national interests, especially those relative to internal and external security. Like every other political capital, Buenos Aires is the seat of the authorities which constitute and exercise the federal power, and is at the same time the seat of the diplomatic corps. These ambassadors and ministers are protected by a principle of immunity of jurisdiction, and this also concerns the protection which must be afforded them in accordance with international law. The maintenance of public order, i.e. personal safety and the basic attributes of authority, pertains to the national executive, which cannot renounce this duty.

As a highly populated urban centre on whose several millions of inhabitants circumstances have imposed a community of interest as regards cleanliness and sanitation, transport and housing, public morality and entertainment, Buenos Aires should have a body whose main business it would be to look after these common interests. This entity, the city council, should be representative of the various civic, economic and cultural forces of the city. It should be made up of representatives of the political parties in proportion to their importance, as it was up to the year 1943. Such a type of local government fulfils the requirements of a popular democracy.

The authority charged with putting into effect decisions affecting the aforementioned interests of the community is the mayor, who may be designated by the national executive or elected by the people, or, better still, by the city council.

The national executive is responsible at all times for local security through its principal local representative, i.e. the Commissioner of Police.

The people of Buenos Aires do not at present have any communally elected administrative body. This means that the historic, democratic principle of free communities pursuing their own specific interests as distinct from those of the nation as a whole, is not operative in the most important city of the country and the second largest Latin city in the world.

A logical corollary of administrative efficiency is that public officials and municipal employees should have their qualifications, tenure of office, and conditions of service determined by statute or regulation. Officials and municipal employees should only be dismissed or transferred for just cause, determined by regular administrative proceedings in which the accused is given the right to defend himself against the charges brought against him.

Efficiency is impossible without a stable régime, and without capable public servants whose number is not excessive, and who receive adequate pay for their services. One of the failings often imputed to demagogic régimes is the marked increase in the number of persons who hold public jobs, many of whom do no work at all and live off the public treasury. The inordinate growth of bureaucracy increases the burden on the taxpayer and saps the economic and moral strength of the people as a whole.

An administrative control of accounts should be set up. Such control should be the concern of a court or tribunal of accounts created by law and designated by the national executive, with the approval of the senate.

Public utility services of local interest, such as telephones, tramways, omnibus lines, electric light and power, should be under municipal ownership and administration even if they have connections with other nearby towns outside the capital. All these services are at present run by the central government. The municipal character of any public service should be determined by its main purpose and organic unity.

Practical experience has shown that sometimes the vote-catching policy of the national or provincial government in the direct operation of a public service is even more detrimental to the public interest than the system of concessions—a system of which in principle we do not approve. The reason for this is not that the state cannot provide services on terms which are advantageous to the people who use them. It is rather that demagogic governments without a civic conscience increase the number of people employed in those public services, pay them high salaries and relax discipline in order to attract a large number of votes from these same public employees. Thus are the services made more expensive, since their good functioning for the public is impaired and a further contribution to monetary inflation is made.

For this reason it is convenient that the representatives of the citizens should themselves organize and administer the public services which they use and pay for. It is also advisable to institute the referendum for deciding questions of policy (except for price-setting, which should be done by committees of experts). Moreover, despite what we have said above, it is desirable to institute the recall or similar legal machinery for the dismissal of the administrators of such public services, in cases of negligence, dishonesty or incompetence.

Chicago

ROBERT A. WALKER

Educated University of Chicago; A.B. 1934; Ph.D. 1940. Edward Hillman Fellow in Political Economy, University of Chicago, 1935–36; University Fellow in Political Science, University of Chicago, 1936–37; Pre-Doctoral Field Fellow, Social Science Research Council, 1937–38.

Assistant in Political Science, University of Chicago, 1937; Research Assistant to Professor C. E. Merriam, University of Chicago, 1938; Director, Institute of Citizenship and Professor of Political Science, Kansas State College, 1945–48; since 1949 Associate Professor of Political Science, Stanford University.

National Resources Planning Board, 1939; U.S. Department of Agriculture, Office of Budget and Finance, 1940–45; U.S. Department of State, Foreign Service Institute, 1948–49.

Post War Planning Committee, U.S. Department of Agriculture, 1941–45; American Society for Public Administration, San Francisco Bay Area Chapter, President, 1953–54.

Author of *The Planning Function in Urban Government* (rev. ed., 1950, The University of Chicago Press); Editor, *America's Manpower Crisis* (Chicago: Public Administration Service, 1952); Co-author, *How California is Governed* (New York: Dryden Press, 1953); and numerous articles in professional journals.

Chicago

WHEN Chicago became an incorporated unit of government in 1833, it was a remote frontier trading post of 350 persons. Prophetically, only thirteen of these were sufficiently interested in the town's affairs to attend a mid-summer meeting to consider the question of incorporation.[1] Five commissioners were elected shortly thereafter and on August 12, 1833, Chicago began its legal existence.

The new village was described by an English visitor, who arrived just one month later, as a 'chaos of mud, rubbish, and confusion.'[2] Chaos it undoubtedly was, as houses and stores were hastily put together to accommodate the steady stream of pioneers from the east. New arrivals huddled impatiently in covered wagons as builders struggled to keep pace. Within two years the population had increased tenfold. By 1850 it was 29,963; by 1870 it totalled 298,977. America was moving westward into a rich and fertile empire. Geography decreed that most of them would go by way of Chicago, that they would be supplied through Chicago, and that their produce would find its way back east after being processed in Chicago.

One hundred years after the thirteen incorporators had met, Chicago was a city of 3,376,438 people jammed into an area of 206·7 square miles. Well over another million lived within the metropolitan area. To understand this unparalleled growth, it is necessary to look at the city's geographic and economic setting. Like most great cities, Chicago is the beneficiary of impressive geographic advantages. Lake Michigan thrusts deeply into the mid-continental area, acting at once as a barrier to a more northerly land passage and as a convenient and economical means of transport. Without the Great Lakes, there would be no Chicago. Transshipment has been the basis of its economy from the beginning to the present day.[3] Even now, no railroad goes through Chicago. All stop or begin there.[4] To be sure, the coming of the railroads greatly reduced

[1] For a definitive treatment of the early history of Chicago see Bessie Louise Pierce: *A History of Chicago:* Vol. I. *The Beginning of a City, 1673–1848* (New York and London: Alfred A. Knopf, 1937).   [2] *Ibid.,* p. 49.

[3] See J. Paul Goode: *The Geographic Background of Chicago* (Chicago: The University of Chicago Press, 1926), p. 5.

[4] Only since the Second World War has it been possible for trans-continental passengers to go through Chicago without changing trains. Through cars are now shuttled from one railroad to another, although the delay is still substantial.

the relative importance of water shipment, but Chicago had already established itself as the principal outlet to the new West. The availability of water transport also gave Chicago shippers a continuing advantage over their competitors to the south, particularly in St. Louis and Kansas City, as the railroads indulged in 'deals' of various kinds to keep freight on the rails and off the boats.[1] Lake transportation remains, however, far more than merely a potential advantage to the city. The great iron ore deposits of Minnesota and Michigan lie within sixty miles of Lake Superior and Lake Michigan. This heavy and bulky cargo can be handled at unbelievably low rates by giant lake steamers. Shipped to the Chicago area and combined with coal from the rich fields to the south and east, it has made the southern tip of Lake Michigan the centre of a great iron and steel industry.

Almost as important to Chicago as the Great Lakes has been its proximity to the rich agricultural soils of the central plains. Rain in this area is well distributed throughout the year, and agricultural production is consistently good.[2] The value of farm crops produced in the United States has been heavily centred in the mid-western and the eastern Great Plains states, with practically all of the tremendous corn crop produced within a 500 mile radius of Chicago. Corn is fed to hogs. The hogs, together with cattle and sheep from the western plains, move to market through Chicago, providing the basis of Chicago's famed meat-packing industry. Wheat and other grains likewise follow the rails into Chicago, where the Grain Pit in the Board of Trade establishes the price and moves the grain to market.

Chicago's phenomenal growth, then, is explained largely by its location in the heart of a region of great mineral wealth, of fertile well-watered soils, and of the most extensive railroad network in the world. Today it is the hub, likewise, of the nation's airline transportation system and of a motor transport system which provides daily service to 30,000 communities.[3] Railways, highways, airlines, and waterways, when plotted on a map, spread out from Chicago like the spokes of a wheel. Even if one limits the perimeter of the wheel to a 500 mile radius from Chicago it will embrace:[4]

[1] Goode: *op. cit.*, p. 63.

[2] The semi-arid Great Plains are usually considered to begin with the 100th meridian, about 500 miles to the west of Chicago.

[3] See Gerald William Breese: *The Daytime Population of the Central Business District of Chicago* (Chicago: The University of Chicago Press, 1949), p. 23.

[4] From Breese, and sources there cited.

15. BUENOS AIRES

An aerial view which reveals the influence of the United States skyscraper in Latin America. The Cavanagh building on the left is the highest in the city and consists of luxury flats. In the foreground is the Plaza San Martin and the well-to-do residential district. In the background is the business quarter, which contains many skyscrapers.

[Photo: B.O.A.C.

16. BUENOS AIRES

Avenue 9 de Julio, said to be the widest street in the world. The skyscraper in the distance houses the Ministry of Public Works.

[Photo: B.O.A.C.

17. CHICAGO

A view of the lake front looking across the centre of the city known as the Loop. The double-deck Wacker Drive can be seen along the side of the Chicago River as it turns left towards the lake. [*Photo: Chicago Plan Commission*]

18. CHICAGO

The building development in the centre of the photograph is a public housing project known as Dearborn Homes.

[Photo: Chicago Plan Commission

(*a*) 36 per cent of the nation's population,
(*b*) 36 per cent of the nation's wholesale establishments,
(*c*) 38 per cent of the nation's retail stores,
(*d*) 39 per cent of the nation's manufacturing concerns,
(*e*) 40 per cent of the nation's farm output (dollar value of product),
(*f*) Seventeen of the nation's thirty-three major industrial areas.

It is little wonder that the young city prospered, for it was at once merchant, processor, and financier to a large and rich area.

THE PEOPLE

From every walk of life and nearly every nation of the world people were drawn to the booming, bustling, young metropolis of the West. Of the 3,376,438 people in the city proper, one hundred years after its founding, some 842,057 were foreign-born immigrants. Another 1,332,373 were the children of foreign-born or of mixed parentage.[1] Many sections of the city were in effect foreign quarters. At least forty languages were spoken. Time and the curtailment of immigration is, of course, steadily reducing the distinctiveness of nationality groupings. By 1940 the number of foreign-born had fallen to 672,705.[2] The negro population, on the other hand, has been steadily increasing. In 1910, there were 44,103; in 1940 there were 227,731. This group, unlike the immigrant whites, has been largely confined to segregated sections of the city through restrictive covenants and collusion among real estate owners and dealers. With the added influx of negroes during the Second World War, population pressure on these segregated areas has become intense and there has been slow but steady penetration of the borderline areas. Thus the assimilation of the negro is a mounting rather than a declining problem. Heterogeneity of population, with the resulting conflicts and tensions, has been a continuing source of governmental problems in Chicago, and the end is not yet in sight.[3]

[1] From tabulations in Louis Wirth and Eleanor H. Bernert (eds.): *Local Community Fact Book of Chicago* (Chicago: The University of Chicago Press, 1949). [2] *Ibid.*

[3] It is probable that no major city of the world has studied its nationality and minority problems more intensively than has Chicago, due principally to the continuing work of the Local Community Research Committee (later Social Science Research Committee) of the University of Chicago. This project was begun in 1924 and has given rise to a distinguished series of publications. For a review of the work of this Committee to 1929 see T. V. Smith and Leonard D. White (eds.): *Chicago: An Experiment in Social Science Research* (Chicago: The University of Chicago Press, 1929). The survey was continued in *A Decade of Social Science Research* (Chicago: The University of Chicago Press, 1940). See especially the series of publications in the 'University of Chicago Studies in Sociology.'

G

Despite its mixture of national and racial origins, Chicago has been predominantly a city of 'white collar,' skilled, and professional workers. The following breakdown is illuminating:[1]

Major Occupation Groups by Sex

Major Occupation Groups	Total		Male		Female	
	No.	Per cent	No.	Per cent	No.	Per cent
Total	1,352,218	100·0	942,365	100·0	409,853	100·0
Professional workers ..	85,165	6·3	49,529	5·3	35,636	8·6
Semi-professional workers	18,790	1·4	14,267	1·5	4,523	1·1
Proprietors, managers and officials	110,665	8·2	96,864	10·3	13,801	3·4
Clerical, sales and kindred workers	371,497	27·5	203,275	21·6	168,222	41·0
Craftsmen, foremen and kindred workers ..	191,399	14·1	185,472	19·7	5,927	1·4
Operatives and kindred workers	297,027	22·0	208,361	22·1	88,666	21·7
Domestic service workers	35,448	2·6	1,774	0·2	33,674	8·2
Service workers except domestic	147,190	10·9	95,705	10·1	51,485	12·7
Labourers	88,344	6·5	82,873	8·8	5,471	1·3
Occupation not reported ..	6,693	0·5	4,245	0·4	2,448	0·6

The occupational distribution of the working force suggests a number of other facts about the population which are of some interest. Incomes are steady and relatively high. Even in 1939, when the effects of the depression were still being felt, 44·4 per cent of all persons employed (including women) had incomes over $1,000 per year. Today the figures are substantially higher. In 1940, 99 per cent of the dwelling units had running water; 71·9 per cent had central heating; 64·1 per cent had mechanical refrigeration; 96·1 per cent had radios. Chicago has its extremes of wealth and poverty, but it is above all else a city of the middle class.

THE METROPOLITAN REGION

The population characteristics just cited were for the incorporated city, the population of which was 3,620,962 in 1950. Equally important for the government of Chicago is the distribution of population in the metropolitan area. For many years it has been 'the thing to do' for the upper-income groups to live in the suburbs, especially the string of North

[1] Table D on 'City of Chicago,' Wirth and Bernert, *op. cit.*

Shore towns along the lake and the group of residential cities to the west. To the south, running around the curve of Lake Michigan to Michigan City, Indiana, is a series of industrial centres with a large proportion of factory workers. Of the 4,920,861 persons in the urbanized metropolitan area, over 25 per cent live outside the city limits.[1] Most of them are distributed among 115 suburban municipalities. The city boundaries have remained virtually unchanged since 1889, with the result that congestion and obsolescence are holding the population of the central city static while the suburbs continue to grow. Resistance to annexation has been vigorous and effective in the Chicago area, as it has in most American urban centres,[2] although nowhere has the resulting confusion of governmental jurisdictions been more dramatically demonstrated. The urbanized Chicago metropolitan area today embraces a total of approximately 821 governmental units.[3] In addition to the municipalities already mentioned,

[1] The figure given is for the metropolitan area as defined by the United States Bureau of the Census for the census of 1950. In this definition, which was first introduced in 1950, the metropolitan area includes the total area of adjacent counties even though they may be only partially urbanized. Thus the official population of the Chicago metropolitan area in 1950 on the new basis was 5,495,364. In 1940 this area had a population of 4,825,527, which was 326,427 more than the population of the metropolitan district as then defined for census purposes. A new term, 'urbanized area' corresponds more closely with the previously used 'metropolitan district.' It includes each city of 50,000 or more persons, plus the more densely populated adjacent area. In 1950, the urbanized area of Chicago had a population of 4,920,816. None of these terms corresponds with a political unit, but the metropolitan area even as now broadly defined tends to be a more or less integrated area with common economic, social, and, often, administrative interests. See 1950 *Census, Vol. I, Number of Inhabitants*, pp. xxvii, xxxiii. The Chicago metropolitan region has been defined in somewhat different terms for unofficial research purposes. The Social Science Research Committee of the University of Chicago and the Chicago Regional Planning Association have used a fifteen-county region for their studies, including three counties in Wisconsin and three in Indiana. A study by Charles E. Merriam, Spencer D. Parratt, and Albert Lepawsky in 1933, *The Government of the Metropolitan Region of Chicago* (Chicago: The University of Chicago Press) defined the region as the area enclosed by drawing a circle at a 50-mile radius from the Chicago Loop.

[2] For a review of American experience with this problem, see the Supplementary Report of the Urbanism Committee to the National Resources Committee, *Urban Government* (Washington: Government Printing Office, 1939), pp. 27–35.

[3] *The Municipal Year Book, 1950* (Chicago: The International City Managers' Association, 1949), p. 21. The study by Merriam, Parratt and Lepawsky in 1933 revealed a total of 1,642 governmental units in the region described by drawing a circle at a 50-mile radius from the Chicago Loop. There has been some slight consolidation of governmental units since, but the difference between this figure and the total of 821 mentioned above is principally due to the fact that the Merriam study used a more inclusive area than does the official U.S. Census definition. The figures used here are based on the 1940 census definition of 'metropolitan area,' which corresponds closely with 'urbanized area' as defined for the 1950 census. For the earlier study see *The Government of the Metropolitan Region of Chicago*.

these include 5 counties, 42 townships, 593 school districts, and 66 special districts.[1]

The consequences of this dispersion of population and profusion of governments have been serious for the city of Chicago. It has lost politically important elements to the suburbs. The higher income, home-owning, and well-educated groups in the residential communities to the north and west have concentrated their efforts at civic improvement in those communities. Many of them are, therefore, distinguished for competent government, excellent schools, fine recreational facilities, and other advantages. This same leadership has been lost to the city, and observers have long been agreed that this has had a bearing on civic apathy and machine politics within Chicago. Lost to the city also has been the potential tax income from the industrial properties to the south. The fact that many of them are located in another state (Indiana) means that any change in this regard is highly improbable.

Another consequence of the multiplicity of governmental units is the costly duplication of services, with resulting variations in standards among different jurisdictions. Police administration, for example, is divided among the city police department, the park police, the police departments of the suburban municipalities, county sheriffs, and state highway patrols. Since local regulations and standards of enforcement differ, the result is harassment of the motorist and easier refuge for the criminal. School districts differ widely in income and facilities. Health, welfare, and security services are likely to operate on varying standards, especially where state lines are involved. And, embracing all of these, effective metropolitan planning is virtually impossible. A multiplicity of municipal planning agencies, state and federal departments, and local governmental bodies are all involved. They, of course, lack any effective governmental jurisdiction to bring them together. Highway planning has been a relatively happy exception, for here the co-operative effort of city, state, and federal agencies has brought about the construction of an excellent system of highways in the Chicago area. Planning to deal with the problems resulting from outward migration of the population, on the other hand, has been all but non-existent.

---

[1] The term 'special districts' includes a variety of governmental units, organized apart from municipal, county, or other local governments. They are usually set up under state law to provide only one public service, such as fire control, parks, drainage, water supply, water pollution control, irrigation, soil conservation, and so forth. Districts of this kind follow no common pattern of organization, function, or financial power. The Chicago Sanitary District and Chicago Park District are the most prominent examples in the Chicago area.

The pulsating heart of the Chicago region is the 'Loop.'[1] Into this central area of only 475 acres, considerably less than one square mile, come between 800,000 and one million persons per day.[2] The central business district has been prevented from expanding by the lakefront, the Chicago river, and the railway terminals, which together completely surround it. As a result, the skyscraper appeared early in Chicago.

A study made of persons coming into the central business district on a typical day in May 1940, showed that out of 603,013 persons of known origin, 113,790 came from outside the city limits.[3] Of this number, 30,729 came by automobile; 39,737 by elevated train; and 43,324 by railroad. The number of commuters travelling by railroad has continued to increase, but the proportion has not kept pace with the automobile and modern highway. By filling in the lakefront, Chicago has built a magnificent parkway running south from the Loop. An equally fine freeway has been constructed along the lake to the north. Neither one of these carries suburban motorists beyond the city limits, but they provide excellent highways for what would otherwise be the most congested part of the trip. Highways to the west, south-west, and north-west are less well developed, although parkways in the outer sections combine with good traffic control and a system of one-way traffic on major streets to make commuting by automobile entirely feasible.

Despite the competitive threat of the automobile, the railroads have been slow to modernize equipment. Fast electrified service is available to the north, south, and due west of the Loop. Obsolete steam-driven equipment continues, however, to serve many outlying areas. The terminal problem has been a continuing subject of agitation, and since commuters use the same terminals as long-distance travellers, it is tied in with the often-studied, never-solved question of terminal consolidation. The elevated lines are notorious for noisy, obsolete equipment, much of which dates from before the turn of the century. Chicago had no subway facilities until a decade ago, when the federal government financed the construction of a short section under the loop as a public works project for unemployment relief. This is used as the downtown terminus for some elevated trains. The surface lines provide slow but extensive coverage of the city, using a combination of ancient and modern street cars with new but overcrowded buses. Interurban buses do not as yet carry a significant share of the commuter traffic.

[1] So named from the elevated tracks which circle the four sides of the central business district, around which elevated trains loop to discharge and receive passengers.
[2] Breese: *op. cit.*, p. 104.　　　　　　　　[3] *Ibid.*, pp. 66–7.

The outward movement of population, combined with the freedom from public transit afforded by the automobile, has resulted in a marked decentralization of retail business in recent years. This is not entirely a suburban movement. One study made in 1936 showed that twenty major shopping districts within the city averaged 100,000 passengers entering and leaving each day.[1] The long-range effects of this trend on the central business districts are not yet clear, but to counteract it the downtown business interests in all metropolitan centres are waging a constant battle against congestion.

The continuing growth of outlying metropolitan areas at the expense of the central city is, of course, a problem common to virtually all of the ninety-six metropolitan districts in the United States. Preliminary figures from the 1950 census show this trend to be continuing unabated.[2] Many of these areas overlap state lines, all are beset by rivalries between the central city and the suburban municipalities. In most cases a continuing jealousy between urban and rural areas, with state legislatures dominated by the latter, has prevented sympathetic state assistance in dealing with metropolitan problems. This has been conspicuously the case in Illinois, where the state legislature has ignored since 1900 a constitutional requirement for redistricting in accordance with population, because to do so would greatly enhance Chicago's influence.

In brief, metropolitan government is one of the major unsolved problems in America. Our great cities—the economic, social, and cultural centres of a revolutionary technological civilization—are governed under a set of governmental patterns concocted to meet the needs of one-horse farming and stagecoach transportation. The problem was subjected to a thoroughgoing review by the National Resources Committee in 1937, but its far-sighted recommendations for federal and state legislation to facilitate co-operation and consolidation among urban governmental units have been all but ignored.[3] The political climate everywhere outside the central cities themselves is bitterly antagonistic to any development which would strengthen their position. Thus, there is little prospect of any early improvement in the present unsatisfactory situation.

[1] Study by Malcolm J. Proudfoot, 'The Major Outlying Business Districts of Chicago' (unpublished Ph.D. dissertation, University of Chicago, 1936). Cited by Breese, p. 25.

[2] See American Society of Planning Officials, *Newsletter* (Chicago: American Society of Planning Officials, 1313 East 60th Street), July 1950.

[3] Report of the Urbanism Committee to the National Resources Committee, *Our Cities: Their Role in the National Economy* (Washington: Government Printing Office, 1937), p. 80.

THE GOVERNMENT OF THE CITY

Chicago has a long record of direct interference in its affairs from the state legislature. Legally, all American cities are creatures of state government. Thus they have only those powers specifically delegated by state enactments. The growth of large metropolitan centres during the past century, with a host of governmental problems setting them apart from smaller municipalities, gave rise to the municipal home-rule movement, with its demand for a high degree of self-governing powers delegated to the cities.[1] Opposing the movement has been the fact of rural domination of state legislatures, with corresponding ignorance of urban problems and suspicion of the rising power of the cities. This battle has been waged in Illinois from the time that Chicago first became clearly the dominant urban centre of the state.

The first city charter of 1837 was soon obsolete. But not until after the constitution of 1870 was adopted did the city obtain any relief. Meanwhile it had been repeatedly subjected to the political raids of a corrupt state legislature. The classic example was an Act of 1865 granting private traction companies a 99-year franchise over the streets of Chicago, despite the vigorous objection of Chicago's representatives. The constitution of 1870 placed a check on legislative raids, and provided for liberal grants of power to all cities of the state. The Cities and Villages Act of 1872, adopted by Chicago in 1875, was generous for its period and despite repeated amendments remains the basis of the city's powers. It was not adequate to the needs of a rapidly growing metropolis, however, and the battle for more extensive self-governing authority was soon resumed. A constitutional amendment of 1904 authorized the legislature to set up a general plan of local government for Chicago, subject to referendum approval of the city. The legislature proved unwilling to propose an overall scheme which was acceptable to the city, with the result that only a few specific governmental changes have been brought about under this amendment.[2] Later efforts at constitutional amendments to grant home rule to Chicago likewise have run on the rocks of conflict between the city and down-state rural areas. Thus the city voters rejected a compromise

---

[1] One of the first major cities to be granted a substantial degree of home rule was St. Louis, Missouri, in 1875. For a discussion of the struggle for home rule in Chicago see Albert Lepawsky: *Home Rule for Metropolitan Chicago* (Chicago: University of Chicago Press, 1935); Merriam, Parratt, and Lepawsky, *op. cit.*; and Charles E. Merriam: *Chicago: A More Intimate View of Urban Politics* (New York: The Macmillan Company, 1929).

[2] Principally consolidations of the municipal court and park systems. See Lepawsky, *op. cit.*, p. 118.

proposal coming out of a stormy constitutional convention of 1919–20, by which the city was to be permanently limited in its representation in the lower house of the legislature in return for a grant of constitutional home rule.[1] Meanwhile the city remains grossly under-represented in the state legislature through the failure to reapportion seats noted above.

Some extension of municipal powers has taken place through laws classifying cities according to size, with Chicago the only city in the largest size category. Nonetheless the city has been seriously hamstrung in its taxing, licensing, and inspection activities by restrictive and duplicate state laws.[2] Special legislation must repeatedly be sought at the state capital to clarify the city's powers, and only constant vigilance by representatives of the city government at legislative sessions prevents the passage of many laws which would create further difficulties in city administration.[3] As far as municipal home rule for Chicago is concerned, the situation is essentially one of stalemate. The governmental powers of the city have, however, been gradually broadened by specific enactments and today city officials exercise substantial authority over a wide range of public activities. They enjoy particularly wide latitude over the organization of the city's administrative services.

Principal officers of the city are the mayor and a council of fifty aldermen. The mayor is elected by the city at large for a four-year term, and can be re-elected for successive terms of office indefinitely.[4] He receives a salary of $18,000 per year. The aldermen are elected from their respective wards for four-year terms, on a non-partisan ballot,[5] and are

---

[1] Merriam, *op. cit.*, p. 14.

[2] The city has suffered repeated tax crises, due largely to state limitations on its taxing powers. During the severe crisis of the early nineteen-thirties a careful study by a Citizens' Advisory Committee and the Chicago Recovery Administration revealed that statutory or constitutional amendment was required for each one of their detailed recommendations for financial relief. The state has taken over the licensing function, with corresponding income from fees, in a number of fields, although often without relieving the city of the necessity for continuing its own inspection services. Rural domination of the state legislature was reflected in a nineteenth-century law prohibiting milk supply inspection and control by municipal authorities, a law which Chicago was forced to ignore to protect the health of its citizens. See Lepawsky, *op. cit.*, especially Chapters 7–9.

[3] *Ibid.*, pp. 144–8.

[4] The term 'at large' in American political usage means that the success or failure of the candidates is determined by totalling the votes for the entire election area involved. Smaller electoral units within the area, which may be used for other purposes, are ignored. Thus the mayor of Chicago is elected by all qualified voters of the city; members of the council each represent only one ward and stand for election in that ward alone.

[5] A 'non-partisan ballot' is a ballot on which no party designation is shown for the candidates. Thus the ballot for a Chicago municipal election does not indicate the party affiliation of candidates for the council. The purpose of this type of ballot is to minimize

paid $5,000 per year. In the hands of a strong leader the mayor's office is likely to be the dominant factor in city government; otherwise the balance of power shifts to the council. Both have broad authority and are, in effect, co-ordinate centres of power. The council can create new departments and determine their powers, while the mayor appoints department heads (with council consent) and is directly responsible for administration. The annual budget is prepared under the mayor's direction and he can veto specific items in the appropriation bill,[1] but the council with the guidance of its powerful finance committee must appropriate the funds. The mayor can exert strong policy leadership through his legislative programme, but new ordinances can come only through council action.[2]

The potential strength of the council has been somewhat dissipated by the calibre of men elected to it. In general, the post of alderman carries prestige only in those wards where the party machines have been strongest. It is usually held in low repute in the better neighbourhoods. Certainly the low income attached to it has done nothing to attract persons of standing in the community. The almost single exception is the fifth ward, which houses the University of Chicago and where efforts at civic reform have been continuous over a period of years. More representative of the general situation has been the consistent return of such renowned characters as 'Bathhouse John' Caughlin and 'Hinky Dink' Kenna, whose earthy mannerisms and dubious election practices in no way interfered with long and influential careers in the city council.[3] As a matter of fact, it is not easy for an honest person, without private income, to make ends meet on the $5,000 salary an alderman receives. The city pays in addition only the salary of one secretary and minor incidental expenses. It is almost essential, moreover, that in addition to his downtown office he maintains a second office in his own ward, where his constituents can reach him. If he happens not to be the party ward committeeman, as he often is, this puts a substantial burden on his income. So acute has this problem become for one of the present independent

party influence and maximize consideration of the merits of the individual candidates in local elections. In practice, party endorsement or opposition remains an important factor in election results.

[1] The item veto is unusual in American government, and an especially strong instrument of policy. See Dennis J. Fleming, *The Relation of the Mayor to Budgetary Control in Chicago* (unpublished M.A. dissertation, Department of Political Science, The University of Chicago, 1949).

[2] See the analysis of this relationship between mayor and council in Merriam, *op. cit.*, pp. 222–9.

[3] *Ibid.*, and Warren H. Pierce: 'Chicago: Unfinished Anomaly,' in Robert S. Allen (ed.): *Our Fair City* (New York: The Vanguard Press Inc., 1947), p. 173.

G*

and progressive aldermen that friends recently took up a voluntary collection to help him meet the costs of his ward office. Public service on this basis requires the highest sense of social obligation on the part of persons who could easily command higher incomes and greater prestige without the calumnies of political life. It is not surprising, therefore, that the council has been dominated by machine politics, except for brief interludes of reform.

The power of the mayor in relation to the council has been repeatedly enhanced by his being, at the same time, the party leader or 'boss.' Such a mayor as Edward J. Kelly was able, during his long term, to maintain a high degree of discipline over 'organization' members in the council.[1] In this, as in other respects, the official structure of government in Chicago cannot be understood apart from organization politics. Of the latter, more will be said later.

Despite the broad powers exercised by the mayor and council, the overall structure of Chicago's government is complex and confused from the standpoint of the average voter. Figure I opposite shows the governmental units, each calling for the election of important officers, to which a resident of the city is subject.[2] It is notorious that the voter is called upon to elect many more officials than he can be reasonably expected to choose with real discrimination. This requirement finds him struggling periodically with unmanageably long ballots in small dimly-lit voting booths.

The principal operating departments and bureaus reporting directly to the mayor, with their basic appropriations for 1950 are:[3]

Department of Police	$28,266,169·00
Fire Department	13,241,235·00
Department of Buildings	1,054,544·00
Department for the Inspection of Steam Boilers	232,136·00
Department of Weights and Measures	203,928·00

[1] 'Organization' members of the council, in American usage, are those who customarily vote and otherwise function as instructed by the leaders of the political parties. These leaders, together with those who work actively in party affairs, are usually thought of as comprising the party 'organization.' Those who merely indicate party affiliation when registering to vote, or who vote with a particular party, are not ordinarily regarded as part of the party organization. One party may completely dominate urban politics for a considerable period of time, as has been true in Chicago, and thus come to be known as '*the* organization' in local parlance.

[2] This table is the latest available from official sources, but is dated 1941 and is somewhat out of date. The principal omissions are the Chicago Housing Authority and the Land Clearance Commission. These are discussed below.

[3] Adapted from *The Annual Appropriation Ordinance of the City of Chicago for the Year 1950*(Chicago: Ludwig D. Schreiber, City Clerk, 1950).

THE SIX LOCAL GOVERNMENTS
INCLUDING ELECTE

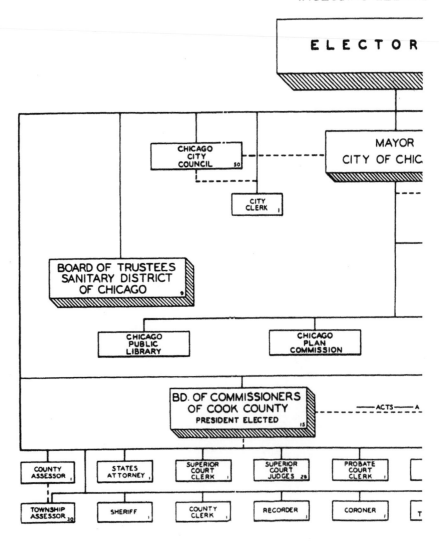

ELECTOR

CHICAGO
CITY
COUNCIL   50

MAYOR
CITY OF CHIC

CITY
CLERK

BOARD OF TRUSTEES
SANITARY DISTRICT
OF CHICAGO

CHICAGO
PUBLIC
LIBRARY

CHICAGO
PLAN
COMMISSION

BD. OF COMMISSIONERS
OF COOK COUNTY
PRESIDENT ELECTED   15

———ACTS———A

| COUNTY ASSESSOR | STATES ATTORNEY | SUPERIOR COURT CLERK | SUPERIOR COURT JUDGES 28 | PROBATE COURT CLERK | |
| TOWNSHIP ASSESSOR 30 | SHERIFF | COUNTY CLERK | RECORDER | CORONER | T |

OF THE CHICAGO AREA
D OFFICES

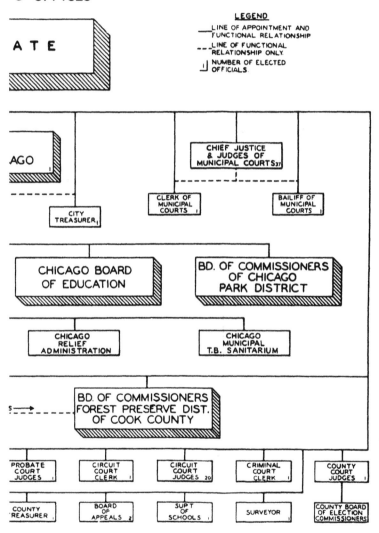

A T E

AGO |

CHIEF JUSTICE & JUDGES OF MUNICIPAL COURTS 37

CLERK OF MUNICIPAL COURTS |

BAILIFF OF MUNICIPAL COURTS |

CITY TREASURER |

CHICAGO BOARD OF EDUCATION

BD. OF COMMISSIONERS OF CHICAGO PARK DISTRICT

CHICAGO RELIEF ADMINISTRATION

CHICAGO MUNICIPAL T.B. SANITARIUM

BD. OF COMMISSIONERS FOREST PRESERVE DIST. OF COOK COUNTY

PROBATE COURT JUDGES |

CIRCUIT COURT CLERK |

CIRCUIT COURT JUDGES 20

CRIMINAL COURT CLERK |

COUNTY COURT JUDGES |

COUNTY TREASURER |

BOARD OF APPEALS 2

SUP'T OF SCHOOLS |

SURVEYOR

COUNTY BOARD OF ELECTION COMMISSIONERS

Department of Streets and Electricity:—
Commissioner's Office	$26,060·00
Bureau of Streets	257,813·00
Bureau of Sanitation	9,454,276·00
Bureau of Electricity	3,520,865·00

Department of Public Works:—
Commissioner's Office	88,034·00
Bureau of Maps and Plats[1]	76,500·00
Bureau of Architecture and Building Maintenance ..	2,466,910·00
Bureau of Engineering	327,950·00
Bureau of Rivers and Harbors	1,667,835·00
Bureau of Sewers	2,140,309·00
Bureau of Parks and Recreation	1,510,751·00
Bureau of Aviation	715,035·00
Department of Subways and Superhighways	1,250,000·00
Chicago Welfare Administration	5,035,100·00
Board of Health	3,622,522·00
Municipal Tuberculosis Sanitarium	6,560,000·00
Chicago Public Library	5,080,000·00
House of Correction..	1,107,047·00

These appropriations do not include the municipal water supply system, operated under a special Water Fund of $20,698,855. The primary staff agencies, providing centralized services to the operating departments and general supervisory assistance to the mayor, are:

Department of Finance (General)	$804,750·00
City Comptroller's Office	695,576·00
City Treasurer	141,217·00
City Collector	419,248·00
Department of Law	812,742·00
Department of Purchases, Contracts, and Supplies ..	811,568·00
Civil Service Commission	209,790·00
Chicago Plan Commission	146,571·00

The partial use of boards and commissions at the head of both operating and staff agencies greatly complicates the administrative authority of the mayor. Even more important, from the standpoint of maintaining administrative integration and centralizing responsibility for public policy, is the generous use of independent authorities: the Chicago Transit Authority, Chicago Housing Authority, Land Clearance Commission, Chicago Park District, and the Chicago Board of Education. Although

---

[1] A 'plat' in American planning and governmental usage is a map or plan of a defined section of land. It is most often used in referring to proposed land subdivisions. A plat must normally meet certain legal requirements as to size of lots, street layout, etc., and be filed with the proper local authority before public facilities or services will be provided in a newly developed area.

each of these carries on important governmental functions, they possess separate legal powers, considerable financial independence, and methods of appointment which insure a high degree of independence from other city officials.[1]

The Chicago Transit Authority is in effect a government corporation, financing itself through borrowing authority and operating revenues.[2] The Chicago Housing Authority has responsibility for public housing and related slum clearance activities. It was created under the state Housing Authorities Act of 1934, and derives both its powers and finances from a combination of state, city, and federal sources. The Land Clearance Commission was created under the Blighted Areas Redevelopment Act of 1947, and like the Housing Authority it is a distinct municipal corporation eligible to receive state, federal, and local funds. Its function is to acquire and clear slum land and blighted areas in accordance with a redevelopment plan. This is in part a duplication of the powers of the Housing Authority,[3] but the primary purpose of a redevelopment programme is to stimulate private investment in lands acquired and cleared under public authorities.[4] The Chicago Park District and the Board of Education, like the elected board of trustees of the Chicago Sanitary District, have tax support independent of city revenues. All three of these groups have an interesting history of relations with the city government, but space will not permit its elaboration here. They have each been under the cloud of political manipulation at various periods. Currently, the situation appears considerably improved. The Chicago Park District, in particular, is establishing something of a reputation for efficient conduct of its functions, including a sizable police department.[5]

---

[1] Members of these boards are in most cases appointed by the mayor with the consent of the council. In the case of the Chicago Transit Authority, however, three of the seven members are appointed by the governor of the state. The Chicago Housing Authority appointments are subject to approval by the State Housing Authority.

[2] For a comparative study of the Chicago Transit Authority and the London Passenger Transport Board, see George Adolph Tychsen: *A Comparative Study of Municipal Transport Corporations* (unpublished Ph.D. dissertation, Department of Political Science, The University of Chicago, 1948).

[3] On the history and governmental relationships of the Chicago Housing Authority, see Lloyd J. Mendelson: *The Chicago Housing Authority: An Administrative Study* (unpublished Ph.D. dissertation, Department of Political Science, The University of Chicago, 1948).

[4] The term 'urban redevelopment' has come into widespread usage only since 1940. For a discussion of the movement see Robert A. Walker: *The Planning Function in Urban Government* (Chicago: The University of Chicago Press, 2nd ed., 1950), pp. 349–54.

[5] See A. Lepawsky: 'Chicago: Metropolis in the Making,' *National Municipal Review*, XXX, No. 4 (April 1941), pp. 211–16.

It will be clear that the structure of authority within the city of Chicago is highly complicated. This is due, in part, to an underlying American distrust of public officials. The proliferation of both elected offices and independent boards reflects this state of mind. Practically all new governmental functions are first placed under board or commission direction in American municipalities, only later and gradually taking their place as administrative departments of general purpose authorities.[1] In Chicago an additional factor of importance has been the state constitutional limitation on the indebtedness permitted municipal corporations. This limitation, 5 per cent on the assessed valuation of property within the jurisdiction, has made it necessary to maintain separate corporate bodies in order to gain essential revenues.[2] Any consolidation of governmental authorities will have to await solution of this problem.

The range of governmental activities in Chicago is tremendous, and is not wholly represented by the agencies mentioned thus far. It is informative simply to list some of the representative items from the Annual Appropriation Ordinance:

Advance sewer planning	Local transportation
Airports	Milk Inspection
Ambulance service	Motion-picture censorship
Boiler inspection	Plumbing inspection
Bridge operation	Planning
Hospitals	Police
Housing	Poor relief
Recreation	Rat control
Welfare administration	Slum clearance
Construction	Smoke abatement
Waterworks system	Subways
Crime detection laboratory	Superhighways
Dead animal removal	Swimming pools
Electrical inspection	Venereal disease control
Fire Department	Ventilation inspection
Infant welfare	Libraries
Committee on Labour-Management Relations	Water works

---

[1] This development in American municipal government, particularly as applied to planning, is discussed more fully by the writer in *The Planning Function in Urban Government*, pp. 134–65.

[2] The State Constitution of 1870 was made extraordinarily difficult to amend because of the requirement that any amendment must be approved by a majority of all voters voting at that election, not just a majority of those voting on the amendment itself. Most amendments proposed in the past have received a majority of favourable votes from those voting on the issue, but not of all voters casting ballots in that election. As a result the Constitution has not been amended since 1908.

The scope of governmental functions has been steadily increasing, for, as in the case of federal and state governments, demands for cutting the cost of government have been powerless against the public support of new services. As a result, the budget of Chicago exceeds that of most states, including until recently the state of Illinois. Appropriations for all purposes in 1950 amounted to $268,979,884. The separate budgets of the Park District, School Board, and Cook County add well over another $100,000,000 to local government expenditures in and near the city. This still would not include substantial federal and state funds expended in Chicago for housing, slum clearance, highways, and other purposes.[1]

The money for carrying on municipal services in Chicago comes principally from real and personal property taxes. This dependence has led to repeated tax crises, particularly in times of economic depression when collections are difficult. Personal property assessments have been notoriously unsatisfactory. Unexplained differences in the tax levied on persons of comparable means, lax enforcement, and political influence have been the order of the day. But the tax is a significant source of revenue and is still used. Chicago levies a motor vehicle tax and receives from the state a portion of the state gasoline tax. It collects a 3 per cent entertainment admissions tax which yields some two million dollars a year. It does not levy a separate city gasoline tax, sales tax, utility tax, gross receipts business tax, income tax, tobacco tax, or alcoholic beverages tax, all of which are authorized and collected in varying combinations by many other large cities of the United States.[2] In a somewhat different category, it collects water fees which are used to maintain the water supply and distribution system. A summary of the sources from which appropriations were made for the year 1950 is shown in Table I.[3]

The history of public finance in Chicago has been extremely complicated. State legislation and court decisions have created a complex pattern of tax and debt limits, contracting limitations, and budgetary requirements which cannot be explored here.[4] Reference should be made, however, to the frequent use of tax anticipation warrants to tide the

---

[1] No reliable estimate of total governmental expenditures and collections in the city of Chicago is available, as far as the writer is aware. Such an estimate would involve innumerable federal and state, as well as local, governmental agencies. Its preparation would be a major research undertaking.

[2] Data on non-property tax revenues are from the *Municipal Year Book*, 1950, Table XII, pp. 198–202. The estimated revenue from the admissions tax is from the 1949 *Year Book*.

[3] From the *Annual Appropriation Ordinance of the City of Chicago for* 1950.

[4] A detailed study of the background of Chicago's public finances will be found in Lepawsky: *Home Rule for Metropolitan Chicago*, pp. 88–113.

TABLE I

SUMMARY OF RESOURCES FROM WHICH APPROPRIATIONS ARE MADE

Fund No.	Funds	Gross Tax Levy	Other Revenue	Total Revenue	Surplus and Other [1]	Total Appropriable
1.	Corporate Purposes Fund	$41,000,000·00	$41,042,500·00	$82,042,500·00	$3,387,823·60	$85,430,323·60
2.	Water Fund		19,565,000·00	19,565,000·00	1,135,000·00	20,700,000·00
3.	Water Works Certificates Funds				5,700,000·00	5,700,000·00
4.	Vehicle Tax Fund		10,145,000·00	10,145,000·00	1,421,000·00	11,566,000·00
5.	Bond Funds				117,059,090·47	117,059,090·47
6.	Penalties and interest on Special Assessments		31,000·00	31,000·00	99,200·00	130,200·00
7.	Motor Fuel Tax Fund		8,150,000·00	8,150,000·00	19,450,000·00	27,600,000·00
8.	City Traction Fund				1,675,000·00	1,675,000·00
8A.	City Transit Fund		137,362·00	137,362·00	1,451,880·67	1,589,242·67
8B.	Transit Revenue Bonds				2,000,000·00	2,000,000·00
9.	Judgment Tax Fund	1,096,774·97		1,096,774·97	257,225·03	1,354,000·00
10.	Bond Redemption and Interest Fund	17,615,817·00		17,615,817·00		17,615,817·00
11.	City Relief Fund	4,000,000·00		4,000,000·00	1,035,613·66	5,035,613·66
12.	Library Fund	4,000,000·00	20,000·00	4,020,000·00	1,060,503·88	5,080,503·88
13.	Library—Building and Sites	250,000·00		250,000·00	312,489·41	562,489·41
14.	Municipal Tuberculosis Sanitarium Fund	4,500,000·00	1,421,814·00	5,921,814·00		
15–21.	Pension Funds	18,641,500·00		18,641,500·00		18,641,500·00
	Totals	$91,104,091·97	$80,512,676·00	$171,616,767·97	$156,683,039·99	$328,299,807·96
	Deduct: Reimbursement between funds	4,545,000·00	4,545,000·00	4,545,000·00		4,545,000·00
	Net totals	$91,104,091·97	$75,967,676·00	$167,071,767·97	$156,683,039·99	$323,754,807·96

[1] Surplus and other includes bonds authorized but unsold and proceeds from proposed sale of Water Works System Certificates of Indebtedness and Transit Revenue Bonds.

city over periods of financial shortage. These warrants are a form of script issued to banks in exchange for cash, to employees in payment for work, and to merchants in payment for goods delivered. The security is taxes to be collected in the future. Often they circulate at less than face value, and they tend to perpetuate the fiscal crisis by committing future revenue in advance. Nonetheless, they have been repeatedly used by the city of Chicago. That the practice is an unhappy one, from the fiscal standpoint, is little questioned, but it has been employed with some state legislative encouragement since as far back as 1837.[1] The only defence of the practice is that it has enabled the city to compensate its employees and to meet its other obligations in some manner when cash was not available.

To supply its numerous public services, Chicago employs some 29,890 persons, of whom 23,345 are permanent full-time employees. All employment is under civil service regulations, with a few minor exceptions. The latter would include labourers, clerks and secretaries of elected officials, and election clerks. The system is administered by a Civil Service Commission of three, appointed by the mayor. Employees work a thirty-eight hour week, with only a skeleton force on Saturdays, and receive compensatory time off for overtime work. They receive ten days annual paid vacation and all are covered by some kind of retirement plan. A variety of employee organizations are active, including the American Association of State, County, and Municipal Employees; the United Public Workers of America; and the International Association of Fire Fighters. There is little doubt that 'politics' has been rife in the Chicago public service, but the service now appears to be reasonably secure for the great majority of employees and the conditions of work are not unattractive. Salaries tend to be lower than federal levels for comparable work, but as in other municipalities the trend has been upward in recent years.[2]

CHICAGO POLITICS

Chicago's politics have been bawdy, corrupt, and unashamed. At times the machine politicians have overstrained the public's wide tolerance and been thrown out for a while, but they have always returned. At times,

---

[1] *Home Rule for Metropolitan Chicago*, pp. 103–6.

[2] Data on personnel do not include school employees, and are from *The Municipal Year Book*, 1950, pp. 134–47. The best study of the public service in Chicago, although it is now out of date, is Leonard D. White: *Conditions of Municipal Employment in Chicago: A Study in Morale* (submitted to the City Council of the City of Chicago, June 10, 1925). The effects of political influence on morale are well analysed on pp. 62–5.

too, the newspapers have been on the side of good government, but more often they have promoted the private prejudices and ambitions of their owners—notoriously so in the case of Colonel McCormick's *Tribune* and Hearst's *Herald-American*. The indifference of these papers to truth and their ready use of personal attack have undoubtedly contributed to the dearth of civic and political leadership in the city. In any case, Chicagoans have been hard to arouse to civic action. It has been, instead, a city of strong personal political followings and of amazing loyalty to party organizations.

For a generation the name of Carter Harrison, the elder and the younger, loomed large in Chicago politics. Between them they captured the city hall ten times, despite the machine influence and ample financing of rival leaders. Later the names of Dunne, Dever, Deneen, Brennan, Lorimer, Lundin, and 'Big Bill' Thompson were rallying points for rival party factions. The story of the varying fortunes of these groups has been well told by Charles E. Merriam in his intimate report on Chicago politics.[1] The career of William Hale Thompson is worthy of special note here, for he enjoys the dubious distinction of carrying Chicago politics to their lowest depths.

Running on a platform of 'America First' and the promise to 'punch King George in the snoot,' Thompson gained the mayoralty for two terms beginning in 1915 and again for one term in 1927. He was essentially a showman and had little stomach for administrative responsibilities. He is reported to have delegated at one time many of his functions to Fred Lundin, himself a political power, as chairman of a patronage committee. The machine and his political cohorts had a field day exacting tribute from public business. An example of the Thompson technique was the exorbitant fees paid real estate experts in connection with the extensive public improvements incident to carrying out the Plan of Chicago.[2] Charges of corruption in this connection were first aired by one of the aldermen in 1920. They were followed up by a reporter for the *Chicago Tribune* and a taxpayers' suit for recovery was instituted on June 24, 1921. Mayor Thompson, Public Works Commissioner Michael J. Faherty, and City Comptroller George F. Harding were named defendants. After seven years of litigation, the Circuit Court ordered the repayment of $1,732,279.23 paid out of public funds for so-called experts fees. In concluding its opinion the Court said:

[1] See his *Chicago: A More Intimate View of Urban Politics*, especially Chapter VI.
[2] The Plan and work toward its accomplishment are discussed below.

The Court's findings as to the existence and accomplishment of the conspiracy, for the two-fold purpose of financing the activities of the Thompson organization and for the private benefit of members of the conspiracy, lead to the conclusion that all the payments made . . . aggregating $1,732,279·23 were utterly illegal and void, and that restitution must be made for this amount, under a decree of this Court, finding the defendents jointly liable therefor.[1]

Mayor Thompson was by this time already in his third term. This decision and other factors in Chicago politics sounded his political death knell.

One of the principal new factors was the rising tide of gangsterism. Stimulated by the national prohibition law, and abetted by political 'protection' payments of unfathomed dimensions, the underworld was becoming increasingly bold. But even if the citizens of Chicago enriched the underworld with a steady demand for illicit liquor, gambling, and women, they had no stomach for the blatant displays of lawlessness which were destined to make the name 'Chicago' synonymous with gang-warfare the world over. The tempo mounted as rival gang leaders—'Big Jim' Colosimo, Johnny Torrio, and finally Al Capone—fought for domination. And as it mounted, the public was gradually stirred out of its indifference. Finally it was brought to shocked attention by the St. Valentine's Day massacre in February 1928, when seven men were machine-gunned to death in a garage. The people demanded action.

Another decisive factor in cutting the ground out from under the Thompson machine for good was the economic collapse of 1929 and 1930. Unemployment mounted, and the demand for a 'new deal,' upon which Franklin D. Roosevelt was to capitalize so effectively, was already being felt when the mayoralty election of 1931 came due. In that year the city was bankrupt. This writer was witness to rows of homeless, jobless men sleeping on newspapers on the lower level of Michigan Avenue; to men and women abjectly poking in garbage cans behind Loop restaurants for something to eat. The 'do nothing' policy of the national Republican administration had already embittered the urban population. The resounding defeat of the city administration and the return of Anton J. Cermak's Democratic ticket were but the prelude of the national Democratic sweep of 1932.

But Cermak was by no means a political accident, nor did he mark the end of machine politics. He and his associates, Patrick A. Nash and Edward J. Kelly, had been carefully building an organization for a number of years. By carrying the city in this election, and controlling the appoint-

[1] The opinion appears in full in the *Chicago Tribune*, June 21, 1928.

ment of Cermak's successor as President of the County Board of Supervisors, the Democratic organization won virtually complete control of
both the city and county offices. When Mayor Cermak was killed two years
later by an assassin's bullet meant for President Roosevelt, Pat Nash and
his organization, through its control of the city council, chose Edward
J. Kelly to succeed him. Thus was begun another personal rule of major
proportions, for Kelly remained in office from 1933 to 1947, the longest
continuous period ever served by any Chicago mayor.

Mayor Kelly, like Cermak and Thompson before him, was essentially
a machine politician. He kept a firm hand on the helm, however, and
never permitted the excesses of the Thompson era. The Chicago of his
day would not have tolerated them. His attitude was expressed in his oft-
quoted statement, 'You've gotta be a boss!', but he remained sensitive to
the demands of public opinion and was in many ways a liberal. He actively
supported the repeated nominations of Franklin D. Roosevelt for
President, he backed public housing in Chicago, and he supported the
transfer of transit facilities from private to public ownership through the
creation of the Chicago Transit Authority. If he failed to eliminate
organized crime and gambling, he at least kept them within bounds. His
political astuteness failed him, however, on his appointments to the Board
of Education. Here his selections raised a religious issue and this, coupled
with widely-supported charges of incompetence in school administration,
created a rising tide of criticism which finally led directly to his withdrawal from the mayoralty campaign of 1947.[1]

The first overt suggestion that the centre of political power might be
shifting away from Kelly was the selection of Jake Arvey as Chairman
of the Cook County Central Democratic Committee in 1946. Arvey
had been active in organization politics before the Second World War, but
reputedly grew tired of the machine's lack of ethics and the superficiality
of its programme. He had been greatly influenced by the social philosophy
of Franklin D. Roosevelt. When he re-entered politics after the war,
he indicated that his primary interest was in the social programme of
the Democratic party. He saw that Kelly's vote-getting power was
waning; that Chicago was growing restive under prolonged machine
domination and the mounting dissatisfaction over the school situation.
In the face of possible defeat, therefore, he persuaded Kelly to withdraw

[1] Sympathetic interpretations of the Kelly era are found in R. Madison: 'Letter from
Chicago,' *New Republic*, CXII (April 23, 1945), pp. 549–51; and V. Rubin: 'You've Gotta
Be a Boss,' *Collier's*, CXVI (August 25, 1945), p. 20. Edward J. Kelly remained a political
force in the city until his death in October 1950.

in favour of Martin J. Kennelly, a successful businessman with little political experience but with a reputation for personal integrity. At the same time the chairman of the school board announced that he would resign at the end of Kelly's term of office. Thus, Arvey adroitly pulled the rug from under the mounting opposition cry for reform and Kennelly won easily over a weak Republican opponent.[1]

The people of Chicago voted Mayor Kennelly into office as a reform candidate, even though he ran with organization support. His performance as mayor, however, has left something to be desired. He has been reluctant to accept the responsibilities for leadership inherent in the office, despite early efforts to strengthen the civil service system and to establish stronger controls over public purchasing. His lack of forcefulness on matters of public policy has allowed free rein to sectional and special-interest politics. In a recent battle over public housing between the city council and the Chicago Housing Authority, for example, individual councilmen have been unashamed in their efforts to appease local real estate interests by keeping public housing out of their wards. Kennelly's belated and rather ineffectual involvement in the issue has done much to impair the public housing programme and hence to alienate his liberal support. He has shown little ability, either, to cope with the problem of organized crime in the city. Thus, the future of the immediate 'reform' movement in Chicago is highly uncertain.

On the other hand the determination of the voters to repudiate machine politics in its more blatant forms was again demonstrated in the elections of November 7, 1950. Despite his good record, Jake Arvey appears to have made a major error in his backing of a somewhat unsavoury candidate for county sheriff at this election. This nomination, which was opposed by other liberals in the party, contributed greatly to a sweeping victory for Republican candidates for this and other important county offices. These reverses for the party organization which has so long controlled Chicago politics are difficult to evaluate, since the Democratic Party suffered a nationwide setback at this election. The situation is further complicated by the resignation of Jake Arvey as Chairman of the County Democratic Committee, announced the day following the election. The one thing that appears fairly certain is that the political machine can no

---

[1] For a discussion of the rise of Jake Arvey and Martin J. Kennelly in Chicago politics, see A. Hepner: 'Call Me Jake,' *New Republic*, CXVI (March 24, 1947), pp. 20–3; M. P. Akers: 'Chicago Dumps Kelly,' *Nation*, CLXIV (January 4, 1947), pp. 7–9; S. Frankel and H. Alexander: 'Arvey of Illinois: New Style Political Boss,' *Collier's*, CXXIV (July 23, 1949), pp. 9–11; Charles B. Cleveland: 'Look What He's Doing to Chicago,' *Saturday Evening Post* (July 3, 1948), p. 15.

longer deliver great blocks of votes without reference to the merits of the candidates.

GOVERNMENTAL EFFICIENCY V. DEMOCRACY

Chicago government cannot be understood apart from party organization. The continuing strength of the party 'machine' in Chicago, as in other large cities, has been firmly rooted in the fulfilment of certain functions for which the people have felt a need, a need not met for one reason or another through regular governmental channels. This has been most evident among those groups suffering from a high degree of economic, social, or psychological insecurity. This element has loomed large in Chicago. To these people, disadvantaged or uncertain in their relation to the rest of society, the party has been a democratizing influence, in the sense of bringing government to a level at which they could understand and deal with it.

The party has been, first, a channel to needed assistance. Party representatives close to home, the ward and precinct committeemen, have been service institutions in time of need—persons to whom the lowliest citizen out of a job, in trouble with the law, about to be evicted from his home, or worrying about his hungry family, could go and get a sympathetic hearing. More often than not, he left with some relief for his problem. The 'machine,' able to turn out solid blocks of votes at election time, has been built to a large extent on the resulting gratitude. Thus, at the roots it has been rather philanthropic, sensitive to the problems of the poor, the immigrant, and the uneducated; securing for them services which were either not available or which they did not know how to get through established governmental channels.[1] Many of these services were taken over by relief and welfare agencies after the middle nineteen-thirties. Rising incomes since then have made them decreasingly important. But the party as 'broker' for governmental services has undoubtedly been one of its important sources of strength.

Equally important has been its function as a source of individual recognition, reassurance, and influence. Chicago grew rapidly, which meant that from rural America and foreign countries came persons new to city life and new to American culture. The foreign immigrants particularly encountered severe problems of social adjustment, but they were often serious for the native groups as well. Family beliefs, customs,

---

[1] This aspect of the party activity in Chicago has been best described by Harold F. Gosnell: *Machine Politics: Chicago Model* (Chicago: The University of Chicago Press, 1937), especially pp. 76–81.

language, and colour were likely to be ridiculed. The newcomer was all too likely to be mystified, then hurt, then embittered as he endured sneering references to 'wop,' 'kike,' 'dago,' 'nigger,' 'hick,' and similar street terms for minority or nationality groups. Terms of this kind are, of course, most common among the least educated and lowest economic level of the indigenous urban population, where they serve the psychological function of reassuring the user of his self-importance. They have, however, a disastrous effect upon the not-yet-assimilated groups to which they are applied. The terms and the attitudes they reflect often create deep-lying resentments and aggression against the unfriendly environment. They may lead to a conscious effort to abandon family customs and ideas, resulting in frequent intra-family conflicts and, more seriously, to conscious or unconscious feelings of doubt, guilt, uncertainty, self-censure and inferiority, as both actions and thoughts conflict with early training. The result of psychological stress of this kind is a wide variety of neurotic symptoms and traits. The victim is almost certain to want others to show him affection and to reasssure him that he is really all right. He is likely to become over-assertive, to show off, or to seek power and prestige in an effort to gain social approval. Above all, he wants to 'belong,' to be part of a sympathetic group.

This process of adjustment has great political significance, for in Chicago one of the most accessible forms of group identification has been the political party. No matter what a man's origin, the party welcomed him. The candidates, the committeemen, the precinct captain all treated him as if he were important. They welcomed him to meetings, lent a sympathetic ear to his problems and helped him in time of need. In brief, the party helped to span the gap between his origins or his ignorance, on the one hand, and an unfamiliar or unfriendly environment on the other. In addition, diligent work for the party might well lead to a position of importance in the community, either in a public job or through new sources of income. Here, then, was reassurance, prestige, even power. The party filled a deep-lying psychological need.

It should perhaps be noted in this connection that another obvious outlet for the aggressions of the maladjusted was to revolt against society, to turn 'tough guy' and flaunt one's disdain of a hostile environment. It is no accident that Chicago's criminal underworld recruited generously from immigrant neighbourhoods, that 'foreign' names like Colosimo, Torrio, and Capone were conspicuous among gangland leaders. Crime was an obvious, if dangerous, route to wealth, status, and self-assertion.

Needless to say, maintenance of the political machine often meant gross waste and inefficiency in government. Persons appointed to public office because of service to the party were often incompetent, dishonest, or both. Government employees were sometimes taxed a percentage of their salaries for the party war chest. Since a politically ridden civil service was not attractive to able and honest persons, the city could not recruit the best talents for its work. Gambling and other forms of crime were likely to be condoned so that loyal party workers or important sources of money could 'get theirs.' At the other end of the social scale, utility, real estate, and industrial interests have been willing to contribute to the party to get the special favours from government which only party domination made possible.[1] Thus, the political machine has meant incompetence, favouritism, distortion of the public interest, and contempt for the public service. This is the 'undemocratic,' the invidious, aspect of machine politics. In the last analysis, of course, the paternalistic and philanthropic side of the party exists only to gain support, so that those in public office may retain their positions and so that those wishing to influence governmental policy may continue to do so. It must be recognized, nevertheless, that in the process the party had meant something of real importance to the least privileged portions of the urban population.

The future of the political machine is uncertain, as recent events in Chicago indicate. Rising incomes, better education, reduction of the immigrant population, and the widening sphere of governmental services all mean a steady reduction in the kinds of ignorance and acute need which the party has exploited. The same forces are hastening the social and psychological amalgamation of the population. The one big exception is the negro, whose colour so clearly identifies him. But on the whole the kinds of persons upon whom the party machine has relied in the past are declining in numbers. From another standpoint, policy and principle are looming ever larger in party affairs. The Kelly-Nash machine in Chicago maintained itself in no small part by identifying itself with the national social programme of Franklin D. Roosevelt. Literacy, the radio, and now television are constantly expanding sources of information and exposure to political leaders. Thus, the uninformed party-follower, voting meekly as he is told, is bound to decline in numbers.

Dangerous as it is to generalize, it may well be that as governmental policy in the United States has itself become more 'democratic,' in the sense of meeting an increasing share of the needs of the people (through

---

[1] The classic statement of the role of the 'good citizen' and respectable elements in society in maintaining political machines is found in Lincoln Steffens: *Autobiography* (New York: Harcourt, Brace and Co., 1931), *passim*.

relief, welfare, education, social security, and related programmes), it has effectively undermined the political machine. To this extent, therefore, it has become increasingly difficult for any 'boss' or small group of party leaders to deliver party and governmental support to the forces of special privilege or corruption without self-jeopardy. If this be true, we may look forward to an era in which the party will have to adhere to principle and be sensitive indeed to changing political opinion if it is to remain in power. This is strongly suggested by what has happened in Chicago. The revolution may be far from complete, but the future promise for a more effective, efficient, and responsive democracy is bright.

PLANNING THE CITY OF CHICAGO

The 'chaos of mud, rubbish and confusion' which our English visitor noted in 1833 is today a city of magnificent lakefront drives, great parks, delightful forest preserves, two-level streets in the Loop, and fine highways. These are the product of one of the most ambitious and successfully executed city planning programmes to be found in America. But behind the lakefront and between the parks stretch mile upon mile of dirty, congested, and deteriorated slum areas. Chicago is a great contradiction. Within it lies the best and the worst of city building.

If ever a city grew without plan, it was Chicago. Its rise from a village of 350 people to a metropolis of four and a half millions within one hundred years is an epic of modern industrialization. But the men who built it were preoccupied with material things. Human values and the future alike were left to take care of themselves. Thus, its people were packed into endless rows of badly-lighted, unimaginative apartments which needed only the depredations of time to become today's congested and noxious slums. Chicago, like London two hundred years earlier, disregarded the opportunity presented by the great fire of 1871 to rebuild according to plan. The only result was the outlawing of wood construction in the central district, and eventually in the entire city. Thousands of people had been left homeless. The city was growing apace. Haste was of the essence. Buildings went up in record time, and if such matters as light, air, and good planning were ignored no one seemed to mind.

Only two things relieved the general lack of interest in the future among Chicago's early builders. One was the rectangular street system, with relatively generous street widths, to which the city rigidly adhered from the beginning. Thus, it was spared one of the worst evils of uncontrolled subdividing—unco-ordinated streets with their multitude of jogs and dead ends.[1] The second factor was the development of the park system. The beginnings of this system date from 1869, when the voters approved a two million dollar bond issue for parks on the south side. A truly comprehensive park plan emerged only gradually, and partially at the insistence of aldermen who wanted neighbourhood playgrounds provided in their wards. But Chicago's modern park and playground system, together with the outer belt of parks created by the forest preserve purchases, make it one of the leading cities of the country in park facilities.

Chicago's parks and its lakefront have done much to ameliorate the social effects of overcrowding and congestion. Nonetheless, by the late-nineteenth century the city embodied some of the worst consequences of haphazard speculative growth. A large proportion of the immigrant labourers and the transient population were housed in run-down slum areas. Here poverty combined with crowded quarters, filth, and social disorganization to foster delinquency, vice, adolescent mobs, organized crime, and political demoralization. This was the side of Chicago that was much publicized. It gained a reputation for wickedness which all but concealed from the eyes of a shocked nation the more constructive movement for civic reform that was taking shape just before the turn of the century. This movement began roughly with the plans for the Chicago World's Fair of 1893. It culminated in the preparation and execution of the Plan of Chicago.

The World's Fair, which was designed to present to the world a Chicago of civic pride and aesthetic accomplishment, was not the only evidence of an awakening civic consciousness. One group of reformers was making a frontal attack on the city's social problems, particularly those originating in the slums.[2] Closely allied to this agitation for social betterment was the movement for political reform which developed in this same period. The Civic Federation and the Municipal Voters' League, organized in 1894 and 1896 respectively, took the lead in attacking political corruption. A model civil service system was instituted in 1895, one of the first in a major American city, and the Municipal Voters' League went on to capture control of the city council in remarkably short

[1] A 'jog' in this sense is a break or angle in the line of a street. It is usually the product of poor planning in laying out successive subdivisions of land as a city grows. Thus, streets which should be continuous do not meet properly at intersections and are likely to become traffic hazards.

[2] This paralleled a similar movement in New York City in the same era. In Chicago it was marked by the publication in 1895 of the *Hull House Maps and Papers* (New York: Thomas Y. Crowell and Co., 1895) edited by Jane Addams, and a report on *Tenement Conditions in Chicago* (Chicago: City Homes Association, 1901).

time.[1] Hence when Lincoln Steffens came to Chicago to write the 'sensationally wicked story' he wanted to climax his series of 'muckraking' articles, which did so much to awaken the civic conscience of America in this period: he wrote instead a story of reform.[2] The reform was incomplete, but it was real. And from this spirit of unrest was drawn the energy which produced the Plan of Chicago and carried it to a degree of completion that remains unequalled in the annals of American city planning.

The *Plan of Chicago* was published in 1909 by the Commercial Club of Chicago. Its architect was Daniel H. Burnham, who had played a major part in designing the 1893 World's Fair. The men who instigated the Plan were likewise those who had managed the Fair. Prominent in the early work toward the Plan were Frederic A. Delano and Charles D. Norton, both of whom were later leaders in the preparation of the *Regional Plan of New York and Its Environs*.[3] Another of the leaders was Charles H. Wacker, whose long and vigorous guidance was to see the Plan of Chicago well toward realization. There was a strong element of civic advertising in the origin of the Plan, but in the outcome Burnham and those who worked with him were making a tremendous advance in the practice of urban planning in the United States.

The spirit in which the Plan was conceived is indicated by the first paragraph, which states that:

men are becoming convinced that the formless growth of the city is neither economical nor satisfactory; and that overcrowding and congestion of traffic paralyze the vital functions of the city. The complicated problems which the great city develops are now seen not to be beyond the control of aroused public sentiment; and practical men of affairs are turning their attention to working out the means whereby the city may be made an efficient instrument for providing all its people with the best possible conditions of living.[4]

This spirit was at least a generation in advance of contemporary efforts at city beautification. The principal parts of the plan dealt with rapid

[1] Merriam, *op. cit.*, p. 21.

[2] On the circumstances surrounding his work in Chicago, see his *Autobiography*, p. 423. His report on Chicago politics appears in his *The Shame of the Cities* (New York: McClure, Phillips and Co., 1904).

[3] The New York Regional Plan, completed in 1929, remains the most extensive and ambitious regional planning effort in the United States. Mr. Delano was to continue his planning career as Chairman of the National Capital Park and Planning Commission, the planning agency for Washington, D.C., and later as Chairman of the National Resources Board, National Resources Committee, and National Resources Planning Board.

[4] *Plan of Chicago*, p. 1.

transit and suburban growth, a comprehensive park system, transportation and terminals, streets and subdivision control, and problems of the central city. The last covered a number of subjects, including the civic centre, but one of its main themes was congestion. Noting that the slum is 'a menace to the moral and physical health of the community,' the report demands better sanitary regulations and goes on to say:

> Chicago has not yet reached the point where it will be necessary for the municipality to provide at its own expense, as does the city of London, for the rehousing of persons forced out of congested quarters; but unless the matter shall be taken in hand at once, such a course will be required in common justice to men and women so degraded by long life in the slums that thay have lost all power of caring for themselves.[1]

When it is considered that this Plan was the product of a group enjoying great economic privilege and political influence at the time, this language is remarkable.

Despite the social vision shown by its authors, controls over the use of private property played little part in the Plan itself. It was essentially a plan for improving public facilities. This aspect of the Plan was pressed forward with unremitting energy. Immediately upon presentation of the plan to the city, the mayor and council appointed an official City Plan Commission to oversee its execution. At the outset it consisted of 328 members, an unwieldy number that permitted bringing it together only occasionally. The work was done by an executive committee of twenty-seven, with Charles H. Wacker as chairman. The men who had fostered the plan from the beginning were, therefore, in complete control of the commission and at once launched a promotional campaign of the first magnitude. No significant group was neglected. The first publication, *Chicago's Greatest Issue—An Official Plan*, was distributed to every property owner and to every renter paying over $25 per month rent. This was followed by Wacker's *Manual of the Plan of Chicago*, a grade-school text which was adopted by the Board of Education in 1912 as part of the eighth-grade curriculum and used actively for over a decade. Civic organizations could make use of a public lecture service which provided over four hundred lectures in the first seven years of the commission's work. A feature motion picture, *A Tale of One City*, was prepared and shown widely in the city's theatres. Even the ministers were not overlooked. To them went a document entitled *Seed Thoughts for Sermons*, described as 'a compilation of the humanitarian and social arguments that had been advanced from time to time in the various publications of

---

[1] *Plan of Chicago*, pp. 108 f.

the Plan Commission.' Appeals to business men, to public spirited citizens, and to every influential segment of the population sounded an insistent call to action, prodding the citizens to support the Plan of Chicago.[1]

The Plan was indeed well supported. Between 1912 and 1931 the people of Chicago voted favourably upon some Chicago Plan Bond Issues in every year except two. The total value of these issues was $233,985,000, while another $57,596,000 was raised by special assessments against the properties immediately benefited. Thus, the total expenditures for Chicago Plan projects in the twenty year period approximated $291,581,000.[2] This included substantial amounts for the acquisition of outlying forest preserves, street improvements (including the two-level development of Michigan Avenue and Wacker Drive), city parks, airports, river straightening, and bridge construction. Meanwhile, the lakefront was being filled in by the use of city waste, as recommended in the Plan, and the State of Illinois was persuaded to finance the parkway construction on the land thus reclaimed. As noted earlier, actual construction of certain city projects was tainted with the corrupt practices of Mayor Thompson. The work was, nonetheless, pushed forward with impressive results. The vigorous leadership of the Plan Commission, plus the willingness of the citizens to bond and tax themselves, carried the Plan to a remarkable degree of accomplishment.

Neither the vigour of the Commission nor the willingness to pay survived the depression years of the nineteen-thirties. Chicago planning drifted into a period of somnolence from which it was aroused in 1940 only after energetic resuscitative efforts. This time the drive stemmed primarily from groups and individuals interested in the social problems of the city, particularly in more direct application of planning to such matters as slum clearance, public housing, and rehabilitation of the city's blighted areas. Unfortunately, the Chicago Plan Commission had up to this time largely ignored this field of planning, despite the notice taken of it in the original Plan. It had likewise failed to utilize the vast body of information about the city, its population, and its problem areas which was being assembled by private research organizations. Most prominent among these was the social science research group at the University of Chicago, whose work on the social, economic, and political problems of the city has already been mentioned. It is not surprising, therefore, that this group played a leading part in arousing public and official interest

---

[1] The background of the Plan of Chicago, and steps in its implementation, are described in greater detail in Walker, *op. cit.*, Chapter VIII.   [2] *Ibid.*, pp. 216–19.

in the need for a revitalized Plan Commission. Support came from many sides. The Metropolitan Housing Council and the Chicago Housing Authority were interested in the housing problem. Downtown business men and local improvement associations were concerned with the creeping deterioration of business properties in the Loop and outlying business areas, as well as with such matters as subway extension, super-highway proposals, and other aspects of transit which affected their interests. Finally, in 1940 a new planning ordinance was passed, adding planning for the rehabilitation of blighted areas to the specific duties of the Commission and creating a wholly new Plan Commission.

Since 1940 the Commission has been more active and it has had the funds to employ a technically trained staff.[1] It has issued a number of interesting and useful reports. These include a report on land use in Chicago,[2] a study of the slum clearance problem,[3] a study of industrial development,[4] and an analysis of population trends.[5] Shortly after the Second World War, however, the Commission brought down upon itself a flood of criticism from the very groups that had been most active in securing its reorganization in 1940, occasioned by its release of a 'Preliminary Master Plan' and a report on *Housing Goals for Chicago*.[6] The principal criticism of the Preliminary Master Plan was that it was not really a plan, but a device for preserving existing land uses and boundaries containing certain population groups. Professor Louis Wirth of the University of Chicago, long a leader in the social science research group and former chairman of the Illinois Post-War Planning Committee, took the leadership in the ensuing criticism. His sharp critique was issued by the Metropolitan Housing and Planning Council and became the basis of formal protests to the chairman of the Plan Commission and the mayor. It is difficult to appraise the merits of the controversy, but circumstances strongly suggest that conservative business and political influences within the Commission itself preferred to avoid the highly controversial economic, racial, and social issues incident to an aggressive attack on land use planning and slum rehabilitation. This is confirmed in part by the emphasis on the role of private enterprise, *vis-à-vis* public housing projects, in the report on *Housing Goals for Chicago*. It is consistent also with the statement of the Commission's executive director, when he

---

[1] The appropriation of the Chicago Plan Commission for 1950 was $146,571.00.
[2] *Residential Chicago* and *Land Use in Chicago*, 1942.
[3] *Rebuilding Old Chicago*, 1941.
[4] *Industrial and Commercial Background for Planning Chicago*, 1942.
[5] *Population Facts for Planning Chicago*, 1942.   [6] Issued in 1946.

told the writer, n June 1949, that the Commission was staying pretty well out of the 'housing picture' except as it was asked to do specific jobs for the housing agencies.

Planning in Chicago is at present finding much of its impetus in these housing agencies—the Chicago Housing Authority and the Land Clearance Commission—and in specific redevelopment projects on the south side of the city. In the latter case, the Illinois Institute of Technology and Michael Reese Hospital have taken the lead in planning the redevelopment of the badly deteriorated area surrounding them; securing the assistance of the Housing Authority and the Land Clearance Commission in acquiring, clearing, and rebuilding a large sector of slum land. Nevertheless, the federal impetus to public housing, urban redevelopment, and the related planning which stems from the National Housing Act of 1949 makes neglect of this area by the principal city planning agency rather serious. Strong central leadership in planning would seem to be crucial to the co-ordinated and effective operation of the several different agencies involved.

In closing this discussion of Chicago planning, a word should be said about metropolitan planning. Many of the municipalities in the metropolitan area have their own separate planning agencies, but the only unifying agency among them has been the Chicago Regional Planning Association. This is a private organization, first organized in 1925 for the purpose of co-ordinating planning efforts for the region as a whole. It has stimulated and participated in a number of region-wide planning projects, but these have been confined to special problem areas. Thus, it helped secure agreement among construction and planning agencies to a tentative highway plan some years ago. The major effort of the Association has gone into assisting outlying communities with their immediate planning problems on a consulting fee basis. It has not been a strong integrating force for planning land use and population distribution in the metropolitan area as a whole.

EMERGING PROBLEMS IN CHICAGO

Chicago has yet to solve most of the problems which have been touched upon in the foregoing pages. Some of them are perhaps dwindling in importance, as in the heterogeneity of its nationality groups. Some are likely to become more serious with the passage of time. One such crucial problem is the growing negro population and its peaceful assimilation into residential areas from which it is now excluded. At present the excess population in the so-called 'Black Belt' is reliably estimated at 75,000 to 100,000 persons.[1] Tensions between white and negro are chronic. In 1919 a race riot of major proportions took place,[2] and while nothing similar has occurred in Chicago since, repeated local incidents and a serious outbreak in Detroit in 1943 point to an important continuing problem. Immediately after the Detroit riot, Mayor Kelly appointed a Committee on Race Relations. Through its efforts, and careful handling by police officials,[3] minor difficulties have been kept within bounds. The problem will be solved, however, only by an aggressive programme of slum clearance, rehousing, and public education.

This suggests a second major problem that is by no means confined to the negro sectors—that of reclaiming the deteriorated areas of the city. From these come a high proportion of the city's delinquency and crime. Costs of police and fire protection are extremely high, while tax receipts are likely to be low. Much of the property in slum areas is held by owners hoping to convert it profitably to commercial or industrial use, or simply holding it to continue to collect rentals which are often generous for the delapidated and overcrowded facilities offered. Consequently, only a slum clearance and rehabilitation programme under public authorities can be expected to deal adequately with the problem. Public powers are now reasonably adequate, under the national and state housing and redevelopment acts, and agencies exist which have the responsibility for utilizing them. Strong political opposition continues, however, and, as has been noted, vigorous central leadership in planning and carrying out a slum clearance and rehousing programme is currently lacking. The most optimistic note for the future, in Chicago as in most major cities, is the federal government's programme of financial aid to state and local governments in public housing and the redevelopment of publicly-acquired land through controlled private financing.

Critical governmental problems in Chicago have long been observed by political scientists, but they remain unsolved. They include adequate representation of the city in the state legislature, by reapportionment of districts as required by the state constitution; a greater degree of home rule for the city, including more adequate taxing authority; the con-

---

[1] Mayor's Commission on Human Relations, *Report for 1945*, p. 29.

[2] See the report of the Chicago Commission on Race Relations, *The Negro in Chicago: A Study of Race Relations and a Race Riot* (Chicago: The University of Chicago Press, 1922), and Carl Sandburg: *The Chicago Race Riots* (New York: Harcourt, Brace, and Howe, 1919).

[3] An excellent manual is used by the Chicago Park District in its police training programme. It was prepared by a sociologist, Joseph D. Lohman of the University of Chicago. See *The Police and Minority Problems* (Chicago: Chicago Park District, 1947).

solidation of governmental units and a greater degree of administrative integration within the city, so that responsibility may be centred in a manageable number of elected officials; and some form of overall governmental authority for the metropolitan area as a whole. The last is perhaps the most difficult of all, and furthest from solution. Like most other metropolitan areas in the United States, rivalries and jealousies among the governmental units surrounding Chicago appear an impenetrable barrier to the rationalization of the governmental structure for the area. The best hope for the future may lie in some form of federated authority, with responsibility only for functions of urgent common concern. The immediate prospect, however, is for continued confusion.

An urban problem of the future, and one which has prompted considerable discussion in the past few years, is whether or not large cities should be dispersed as a national defence measure. The National Security Resources Board and the President's Air Policy Commission both issued reports in 1948 calling attention to the extreme vulnerability of large urban centres to atomic bombs in wartime.[1] There is great difference of opinion over the merits of dispersal.[2] It is not at the moment being taken very seriously by local authorities or the population at large. The only tangible efforts in this direction are at the federal level, including some attention to geographic factors in allocating national defence production contracts and a proposal to decentralize government agencies over a larger area in the Washington, D.C. vicinity. The discussion, and even the remote possibility of a decentralization programme, points, nonetheless, to the complete inadequacy of existing metropolitan planning and governmental machinery to do anything about it. There is at present no local authority, in the Chicago area or elsewhere, with the jurisdiction to act. State and local defence planning currently is limited to programmes of training covering what to do after a bomb has struck.

In the long run, of course, Chicago's most critical governmental problem is maintaining a high level of civic interest and individual political responsibility. This is a highly complex matter, involving the newspapers, the schools, civic organizations, and other channels of public education. The newspapers have been more a deterrent than an aid.

[1] See report of the Air Policy Commission, *Survival in the Air Age* (Washington: Government Printing Office, 1948), and National Security Resources Board, *National Security Factors in Industrial Location* (Washington: Government Printing Office, 1948).

[2] For the arguments for and against dispersal see Tracy B. Augur, 'Decentralization Can't Wait,' and Charles E. Merriam, 'Problems of Reorganizing Our Great Cities,' in *Planning, 1948: Proceedings of the National Planning Conference* (Chicago: American Society of Planning Officials, 1313 East 60th Street, 1948).

19. COPENHAGEN

A meeting of the Town Council. The chairman sits in the centre under the canopy. On his right hand is the chief mayor and on his left
the clerk. The members of the Executive Council are seated on either side of the dais.   [*Photo: B. Schleifer, Copenhagen*]

20. COPENHAGEN

A meeting of the Executive Council. The chief mayor, who presides, is seated in the centre of the right-hand side of the table. With him are the other mayors and aldermen.

Proto. B *Schleifer, Copenhagen*

On the other hand, the frequent rejection of candidates favoured by the *Chicago Tribune* and the Hearst paper over the past twenty years suggests that radio and now television may be making the political role of the newspaper far less decisive than formerly. The schools have scarcely been a brilliant success in political education, if past levels of civic consciousness are any criteria. Again, this may improve with less political manipulation in the school system. Chicago's civic organizations have been distinguished for remarkable bursts of energy at periodic intervals, as we have seen; but they have not heretofore seemed to be capable of sustained effort on behalf of good government. At the moment the situation looks favourable. The people are supporting an administration dedicated to honesty and efficiency; the economic, social, and psychological supports of the political machines show signs of advanced decay. But the fact remains that all future improvements in the government of Chicago are contingent upon an enlightened citizenry, able and willing to take an active part in the solution of public problems. It is to be hoped that the maturing city will have a sense of responsibility which the rapidly growing adolescent of an earlier day so conspicuously lacked.

Copenhagen

AXEL HOLM

Born 1897. Higher School Certificate, 1916; *candidatus politices*, University of Copenhagen, 1922.

Deputy Director of the Statistical Office of Copenhagen. Instructor to the Public Service School of Copenhagen. Visiting examiner in National Science at Denmark's Technical High School; Statistical Adviser to the District Planning Committee for Copenhagen and its Environs.

Author of: *A Textbook on Local Government of Copenhagen* (1938 and 1954); *Economy of the Municipality of Copenhagen, 1840–1940* (1941); *Parliamentary Elections through 100 years* (1949); also several publications of the Statistical Office, and newspaper articles.

Copenhagen

LIKE most other great cities, Copenhagen long ago burst the formal, legal boundaries set for the municipality of Copenhagen. Moreover, Copenhagen has not merely spread beyond its formal boundaries into the neighbouring municipalities but has, because of incorporations effected in the early part of this century, encircled a totally independent municipality, Frederiksberg. These two municipalities together with the directly bordering municipality, Gentofte, have during the last thirty years been treated in all official, administrative, and statistical publications as a corporate body known as the capital of Denmark.[1]

The total population of this capital numbered on January 1, 1952, 970,100 inhabitants, of whom 764,100 lived in Copenhagen, 117,900 in Frederiksberg, and 88,100 in Gentofte. The three metropolitan municipalities together cover an area of 10,876 hectares, of which Copenhagen has 7,468 hectares, Frederiksberg 870 hectares, and Gentofte 2,538 hectares.

The fact that the administration of the capital is split up into three independent municipalities has naturally produced several difficult problems, especially as regards taxation. These problems have recently been combined with the question of a rational settlement of the administrative, economic, and traffic problems of the whole metropolitan area. This area comprises, in addition to the metropolitan municipalities already mentioned, a considerable number of suburban municipalities.[2]

Up to the beginning of this century the greater part of the suburban municipalities were more or less rural districts, whose inhabitants were chiefly engaged in farming and gardening, but the development of modern traffic and increasing industrialization have gradually made the structure of these suburbs purely, or at any rate chiefly, urban. At the same time

[1] The capital of Denmark is in the strictest legal sense only the municipality of Copenhagen, but in the practical administration and in all statistical publications 'the capital' means the three municipalities of Copenhagen, Frederiksberg and Gentofte, which are also called the metropolitan municipalities (*Hovedstadskommunerne*). These three municipalities co-operate in several ways, but have no common administrative organ.

[2] The term suburban municipality is only a convenient town-planning and statistical expression, but has as yet acquired no definite legal sense, and this also applies to the terms metropolitan area and metropolitan community, which comprise: Copenhagen, Frederiksberg and Gentofte besides nineteen suburban municipalities (see p. 230).

those very factors have induced an ever increasing number of the population engaged in industrial and business pursuits in Copenhagen, or attached to the central administration or the cultural institutions of the capital, to take up their abodes in these suburbs. Added to this is the fact that the trade activities found in suburban municipalities are often branches of those in the capital.

On these criteria the Statistical Office of Copenhagen in 1930 defined the metropolitan community as including, besides the three municipalities of Copenhagen, Frederiksberg, and Gentofte, those of Taarnby, Hvidovre, Rødovre, Glostrup, Brøndbyøster-vester, Herlev, Gladsaxe, Lyngby-Taarbaek, Søllerød, Hørsholm and Birkerød.

Since 1930 the metropolitan area has, however, been further extended. This is chiefly owing to the fact that the municipalities of Copenhagen and Frederiksberg have gradually been almost completely developed, so that the new buildings necessary to absorb the excess of births over deaths, to provide for the influx of new citizens, to meet the housing requirements caused by the conversion of old residential houses into business premises or the pulling down of old, unhealthy residential blocks, and finally to cater for the changed age distribution and habits of the population, have to be placed in the suburban municipalities. In addition the electrification of the railways connecting the capital and the suburbs, and the extension of the motor coach service, have rendered possible a considerable distance between residence and working place.

After a renewed investigation by the Statistical Office of Copenhagen in 1950, eight more suburban municipalities will have to be included in the metropolitan area, so that the entire area (as will appear from the map facing page 256) comprises twenty-two municipalities in all, namely, the three metropolitan municipalities, the eleven suburban municipalities mentioned before, and also those of Farum, Vaerløse, Ballerup-Maaløv, Herstederne, Høje-Taastrup, Vallensbaek, and St. Magleby and Dragør in the Island of Amager.

The total metropolitan area would thus be 55,000 hectares, of which 11,000 are covered by the three metropolitan municipalities and 44,000 by the nineteen suburban municipalities. Only 40 per cent of the area of the latter municipalities is now farm-land, compared with 80 per cent in 1907.

The total population of the twenty-two municipalities on January 1, 1952, numbered 1·2 million inhabitants, viz. 970,000 in the capital and 260,000 in the suburban municipalities. To illustrate the development of these municipalities we may note that the population in 1901 numbered only about 40,000; in 1930 it had risen to about 105,000, and during the last twenty years it was further increased by 155,000 people.

This great increase in population has resulted in the populations of several of the suburban municipalities exceeding those of most Danish provincial towns in number. Thus, Lyngby-Taarbaek had on January 1, 1952, 48,500, Gladsaxe 42,400, Taarnby 25,500, Hvidovre 25,600, Rødovre 19,700, and Søllerød 17,000 inhabitants.

That the inhabitants of the suburban municipalities, as mentioned before, are drawing their incomes largely from Copenhagen is revealed by the fact that of the total income earned by the suburban population in 1948—about 700 million kr.—40 per cent originated from independent trading or employment in the municipality of Copenhagen.

THE IMPORTANCE OF COPENHAGEN TO DENMARK

A few figures will suffice to illustrate the importance of Copenhagen to the entire Kingdom of Denmark. As mentioned above, 970,100 persons, or 23 per cent of the total population of the whole country (4·3 million), were in 1952 resident in the capital; and of the total urban population 48 per cent lived in Copenhagen. To illustrate the concentration of trade in the capital it will be appropriate to mention that 35 per cent of the total manpower employed in trade and industry is found there. Moreover, the big industry is chiefly centred here, for the capital absorbs 52 per cent of the total manpower of the whole country employed in firms of at least 100 employees. Again, 54 per cent of the total manpower in wholesale trade was employed in Copenhagen; and no less than 83 per cent of those engaged in the biggest concerns, which employ more than 100 persons.

Another sign of this concentration is that three-fourths of the share capital of the total number of joint stock companies belong to companies situated in the capital; 61 per cent of the imports to Denmark went via Copenhagen, while 32 per cent of the exports of the country left via Copenhagen; and 70 per cent of the total mercantile shipping was registered in Copenhagen.

The importance of Copenhagen as a port is shown by the fact that in 1951 it was a port of call for no less than 24,300 ships aggregating 9·0 million tons. Of these, 9,400 (aggregating 5·4 million tons) were ships from foreign ports.

As regards aviation, Copenhagen also occupies a dominating position in northern Europe. In 1951, the passenger traffic using the airport amounted to 392,400 persons.

But it is not merely in trade and industry that Copenhagen predominates. The capital may also justly be named the cultural centre of the country, because art and science are primarily associated with the oldest and greatest university of the country, the Technical High School, the Veterinary and Agricultural High School, and many other scientific institutions, museums, and art galleries, which are located in Copenhagen. Finally, as the capital of the country, Copenhagen is the residence of the King, the seat of the central government, Parliament, and the superior law courts.

THE CONSTITUTION OF COPENHAGEN

The government of the city of Copenhagen is founded on an act of March 4, 1857, with amendments of March 18, 1938, in conjunction with various other acts, the most important of which is the electoral law of 1924, with later amendments, and the city statutes of 1920, with amendments of 1938 and later. The fundamental constitution of Copenhagen is embodied in the Danish constitutional Law of June 5, 1953, which says that the right of municipalities to manage their own local affairs under supervision of the state shall be established by law. In our time, this fundamental principle of local self-government, which was originally adopted in Danish administration by the first democratic constitution of 1849, has in reality been considerably curtailed through the many encroachments on local administration and finance effected by the central government.

According to the constitution now in force, the municipal corporation of Copenhagen consists of an executive council (*magistrat*) and a town council (*borgerrepraesentation*). The town council, whose members are elected directly by the electors, constitutes the approving authority in municipal affairs, whereas the executive council, i.e. the mayors and aldermen, who are elected by the town council, constitute the executive or administrative power. Compared with central government institutions, the town council corresponds to parliament, and the mayors and the aldermen to the ministry.

THE TOWN COUNCIL

The town council consists of fifty-five members, all of whom are elected for four years. The last election took place in March 1950.

Every citizen, man or woman, aged 23 years or over, who has lived in the town since February 1st in the year of the election, and has paid rates for a certain period, is entitled to vote. Aliens are disqualified from voting, and so are persons who have received certain kinds of relief.

Except persons who have been punished for a criminal offence everybody who has a right to vote is eligible for election, i.e. may be nominated as a candidate for election to a municipal council, the only condition being that he must be nominated by from five to fifteen electors.

The election takes place every fourth year in the first half of March. After their names have been checked on the electoral register, the electors at some thirty polling-stations pass into a secret room and there place a cross on a slip of paper against the list they prefer. Elections are conducted on party lines, each list representing a political party, e.g. List A = Labour, List C = Conservatives, and so on. It is also possible for an elector to write the name of the candidate to whom he wishes to give his personal vote. Few electors, however, avail themselves of this right. The number of candidates to be allotted to each list is calculated according to the d'Hondt system of proportional representation.[1]

During the last fifty years municipal elections have reflected fairly well the elections to Parliament (House of Commons), for the same political parties gained a majority at both elections.

At the election to the town council in March 1950, 453,700 persons had a right to vote. Seventy-nine per cent of these voted, namely 84 per cent of the men and 76 per cent of the women.

---

[1] d'Hondt's system of proportional representation may be explained as follows: Suppose that the following number of votes have been registered in a municipality where 9 councillors are to be elected from three lists:

List A	List B	List C
5,050	3,460	1,490

These figures are then divided by 1, 2, 3 and so on until the 9 greatest quotients have been calculated:

Division by	List A		List B		List C	
1	5,050	*(1)*	3,460	*(2)*	1,490	*(6)*
2	2,525	*(3)*	1,730	*(4)*	745	
3	1,683 $\frac{1}{3}$	*(5)*	1,153 $\frac{1}{3}$	*(8)*	496 $\frac{2}{3}$	
4	1,262 $\frac{2}{4}$	*(7)*	865		372 $\frac{2}{4}$	
5	1,010	*(9)*	692		298	
6	841 $\frac{4}{6}$		576 $\frac{4}{6}$		248 $\frac{2}{6}$	

The figures in italics after the quotients indicate the succession of candidates elected from the three parties. It will be seen that party A returns 5 councillors, party B 3, and party C only 1. If the quotients are equal, there will be a drawing of lots.

H*

The number of votes received by the various lists and the number of members returned were as follows:—

List	Number of votes	Per cent	Candidates elected
A—Labour	172,355	48·0	28
B—Radicals (left wing of Liberals) ..	20,050	5·6	3
C—Conservatives	75,914	21·1	12
E—Danmarks retsforbund (Georgists)	42,223	11·8	6
K—Communists	38,984	10·9	6
Other Lists	9,282	2·6	–
Total	358,808	100·0	55

Thus Labour has a narrow majority in the council, and apart from the period 1946–50, when the party obtained only twenty-seven seats, it has been in a majority since 1917.

Fifteen of the fifty-five town councillors are women. The persons elected are largely recruited from the trade unions and political organizations. In many cases participation in local politics serves as a nursery for parliament.

Theoretically, every citizen is bound to accept membership of the town council, which is unpaid. Members receive an annual remuneration of 800 kroner, but that is merely an allowance for the extra meals and travelling involved.

As to the jurisdiction of the town council, the Act of 1938 enacts that the town council has the approving power in municipal affairs, and is entitled to debate and decide upon any municipal matter. In consequence of this power, the town council alone can grant money, which in turn means that no municipal expenditure can be defrayed without the consent of the town council, nor can the real property and money constituting the capital of the town be touched or any loan negotiated without the approval of the town council.

As this body elects every eighth year the mayors and aldermen (see below), the centre of gravity in the administration of Copenhagen lies not only nominally but actually in the town council.

The standing orders of the town council are embodied in the rules of procedure passed by the council itself, according to which the members every year elect a chairman from among themselves, also a first and a second deputy chairman. The principal task of the chairman is to preside at the meetings of the town council, which are held at the Town Hall of

Copenhagen, generally on every second or third Thursday at 6.30 p.m., except in July and August.

The matters treated by the town council may be either motions from the mayors and aldermen or motions, questions and petitions from the members of the council or from persons and institutions outside the council. Motions from outside the council have to be introduced through a member; he passes them on to the chairman, who is bound to include them in the agenda as soon as practicable. By far the greater number of the matters dealt with by the town council are, however, motions by the mayors and aldermen. During the year from April 1, 1952 to March 31, 1953, the council dealt with 985 matters, 894 of which initiated from the mayors and aldermen; 76 from the members, the standing committees, or the municipal auditors; 11 from other institutions or authorities; and only 4 were petitions from outsiders.

If a question discussed by the town council is not initiated by the mayors and aldermen, it must, before any final decision is taken, be submitted to the mayors and aldermen, who are allowed a certain time to express their views.

Motions from the mayors and aldermen are introduced by the mayor to whose department the matter relates. All motions involving expenditure must be read twice, at intervals of at least three days; other matters require only one reading, but often motions of supply as well as other motions are referred to one of the committees appointed by the council. These include the ten permanent or standing committees, which are appointed every year for the chief municipal services, such as lighting, hospitals, education, tramways, etc., the important budget committee, and select committees appointed for special purposes. The members are elected by proportional representation. The object of the committees may be partly to work out the details and implications of a motion, and partly to make suggestions on certain matters.

A fundamental difference between the government of Copenhagen and that of other Danish municipalities, as well as many foreign ones, is that the committees appointed by the elected council do not directly participate in the administration, which is entrusted to the mayors and aldermen alone. The committees deal only with matters of fact, which are referred to them for examination and report to the town council; and on the basis of this report the motion is read a second time. The question is then opened by the spokesman of the committee, and the spokesmen of possible minorities propound their views.

When the matter has been debated, it is put to the vote. The issue is decided by a simple majority of those present and voting. The presence of at least 50 per cent of the members constitutes a quorum. An absolute majority of the whole assembly (which means at least twenty-eight members in favour) is required for the election of the chairman and other individual elections.

The mayors and aldermen are entitled to attend the meetings of the town council and address the assembly; and as the law requires the presence of the respective mayor or alderman at the readings of all motions of supply in order to offer information, the meetings are practically always attended by them. The first meeting in each month is specially reserved for the answering of questions, the time for the asking and answering of questions being limited to ten minutes. The importance of these 'question-hours,' which were introduced in 1948, has not, however, proved to be very great.

The meetings of the town council are held in public. Certain matters of a confidential nature, such as appointments to office and transactions in real property are, however, dealt with at meetings from which the public is excluded. The deliberations are taken down in shorthand and published in print. The publication named 'Deliberations of the Town Council of Copenhagen' dates back to 1840.

THE BUDGET

The most important matter dealt with by the town council in the course of the year is the municipal budget, and the additional grants made in connection with it. The budget is divided into two main parts: the working budget and the capital budget.

The working budget is an estimate of the operating revenue and working expenses of all the municipal institutions during the coming financial year, which runs from April 1st to March 31st, and of the rates that will have to be collected to balance the budget. The various institutions begin working on the budget in the month of August. Estimates are sent by the heads of the various institutions to the mayor of the respective department. When the respective mayors have drawn up the estimates submitted by their departments, the proposals are sent to the chief mayor before the end of September. Any amendments desired by him are discussed with the respective mayors. On the basis of the figures representing expenditure and revenue the chief mayor draws up his motion concerning the rates to be collected during the coming financial year, and after that the whole budget is presented to all the mayors and aldermen. After it has been considered by the latter, and possibly criticized by minorities, the budget is presented by the chief mayor to the town council for approval. This has to be done before the end of November. After the opening speech by the chief mayor, which takes the form of a general survey of the whole economic situation, the budget debate begins, when not only the financial status of the municipality is discussed by the members, but all sorts of local political problems, and the mayors have to answer numerous questions.

When the budget has been through the first reading, usually in the first part of December, it is sent to the budget committee, which at present consists of fifteen members, to be examined in detail.

Here it is gone through very carefully and finally reported upon, after which the second reading takes place. The budget must have been finished by February 15th, though the Minister for Home Affairs may extend the time limit until the end of March.

Nominally the passing of the budget does not require the confirmation of a superior authority. But this does not really mean that the town council has a free hand concerning the nature or the extent of the expenditure and revenue contained in the budget. This aspect will be considered later.[1]

The budget determines the scope within which the activities of the administration (i.e. the mayors and aldermen) are confined during the financial year. Should questions arise in the course of the year involving additional expenditure not included in the budget, a further grant will be necessary. This, as mentioned before, has to be read twice by the town council, possibly with an intervening committee stage. Small amounts of less than 3,000 kr. may, however, be granted by a standing committee appointed by the town council. At the end of each financial year a report and accounts are published.

The audit of the municipal accounts is done partly by a special directorate of auditors, which is responsible for the correctness of the municipal accounts, and partly by the municipal auditors, who number five persons elected for a period of four years by the town council, one of whom is elected chairman. These auditors have to see that no expenditure is incurred which has not been granted, and that the dispositions of the administration conform with the decisions taken by the corporation, and the existing laws and regulations. When the municipal auditors have finished their report it is sent to a decision committee, which in its turn reports to the town council, by which the municipal accounts must be finally approved.

[1] *Post*, pp. 243–5.

Unlike the working budget, the municipal capital budget cannot be made for a certain period, but results from the special grants made for new plant and extensions of existing plant and activities according as they are required. But every year a so-called working plan is made showing the amounts available for capital expenditure during the coming year.

Whereas the working budget, as already mentioned, does not require the approval of higher authorities, all purchases, sales and mortgages of municipal property must be sanctioned by the Home Office. Likewise the negotiations of loans to a larger amount than can be repaid out of the annual revenue. This last provision is, of course, of great importance to all municipal activity which is largely financed through loans, such as the building and extension of gas and electricity works, streets, and schools.

THE EXECUTIVE COUNCIL

The executive council (*magistraten*) is composed of one chief mayor (*overborgmester*), five mayors (*borgmestre*) and five aldermen (*raadmaend*), who are all elected by proportional representation every eight years by the town council. Everyone who has a right to vote for the lower house of Parliament is eligible. As no one can be at the same time a mayor or an alderman and a member of the town council, a town councillor has to retire from the council if he is elected a mayor or an alderman. Whereas a mayor was formerly chiefly an official he is now more of a politician, and mayors as well as aldermen are mostly recruited from among the town councillors, who are elected on political lines. Of the present mayors, who were elected in 1946, four (among them the chief mayor) were elected by the Labour Party, one by the Conservatives, and one by the Communists. Of the five aldermen, two were elected by the Labour Party, one by the Conservatives, one by the Radicals, and one by the Communists.

The mayors receive salaries and are entitled to pensions on retirement. The chief mayor receives a basic salary of 25,000 kr., together with certain additional payments dependent on the price-index, which brings his present total salary up to 43,600 kr. He also receives an entertainment allowance of 4,800 kr. The basic salaries of the mayors amount to 22,000 kr. which with similar additions are at present raised to 39,000 kr.

The aldermen are not entitled to pensions, but receive a basic fee of 6,600 kr. with additional payments, so that each alderman at present receives 11,600 kr.

As mentioned before, the mayors and aldermen constitute the executive power in municipal affairs. They act partly as a body through collective decisions made by all of them, and partly through departments, each headed by the chief mayor or a mayor.

Matters concerning the municipality in general, such as the statutes and the annual budget, or which affect several branches of the administration, or are of vital importance to the administration and economy of the town, are treated by the executive council. As chairman the chief mayor may cause any matter to be brought before it, just as each of the mayors can submit matters belonging to his department.

In pursuance of the city ordinances matters concerning the appointment and discharge of officials and employees are decided by the executive council, with a few exceptions. The highest municipal officials, such as directors of the departments, are nominated by the mayors and aldermen but appointed by the town council; moreover, the town council, or a committee appointed by it, is in certain cases called upon to decide questions relating to the pensioning of an official.

The executive council normally meets every Monday at 10 a.m. at the town hall, except in July and August. The meetings are not public, like those of the town council, and even officials cannot attend except the Lord Lieutenant, who attends in his capacity of state superintendent. Matters are decided by a majority. Resolutions are drawn up in the name of the executive council and signed by those present, unless the matter belongs to a single department.

As chairman of the executive council the chief mayor is invested with certain special powers. If, for instance, he finds that a decision taken by a mayor, or by an alderman acting on his own initiative, exceeds his authority, or is in conflict with the law, or is detrimental to the municipality, or neglects a duty incumbent on the municipality, it is his duty to protest, and the decision is suspended. If the matter is not settled by arrangement between the chief mayor and the mayor or alderman in question, the former may cause it to be submitted to the executive council. If he considers it necessary, he can insist upon the matter being finally settled by the town council.

To make sure that the chief mayor (and consequently all the mayors and aldermen) keeps in touch with the work of the town council, he has a statutory right to be informed, not later than the night before each meeting of the town council, as to the business to be debated.

Finally, the chief mayor is entitled temporarily to transfer the business of a mayor or alderman to one of his colleagues. Similarly, pending the decision of the town council, he can appoint one of the mayors or aldermen to be acting chief mayor during his absence.

The most important thing about the chief mayor is that he is the head of the municipal treasury. In consequence, every motion of supply has to be submitted to him to receive his comments before it is sent to the executive council, as well as before it is brought before the town council. In this way he is enabled at his leisure to decide whether he considers the contemplated financial measure justified or desirable, having regard to the financial position of the municipality as a whole.

As previously explained, the chief mayor or a mayor is at the head of each department. The election itself determines who is to be chief mayor and the city ordinances contain special rules for the distribution of departments among the mayors. Thus the elected mayors and aldermen, at the first meeting of the executive council, choose in accordance with their seniority on the executive council the departments they would like to direct, so that one mayor and one alderman are allotted to each department. If any of the elected mayors or aldermen has at any time served in the same capacity, a request to remain in the same department will, however, be complied with, if the chief mayor has been notified before the meeting takes place.

Each mayor has charge of all matters referred to his department, and acts independently and on his own responsibility. No appeal from his decision can be made to the executive council, but if anyone has cause for complaint an appeal lies to the central government provided that the matter does not belong to the Law Courts. Nominally the mayors in Copenhagen thus have great administrative powers, and unlike the mayors of other towns, they are not practically tied by committees appointed by the elected council.[1] However, the above-mentioned powers of the chief mayor must be borne in mind, and likewise that all the money for the administration is granted by the town council.

The city ordinances define how the municipal tasks and business are assigned to the six departments.

To the chief mayor's department belong in the first place the institutions dealing with municipal accounts, audits, wages, pensions, and rates.

To the first department are assigned (among other things) certain legal functions, education, libraries, statistics, the register of population, and the funeral service.

To the second department are assigned (among other things) hospitals, lunatic asylums, and housing.

[1] *Ante*, p. 235.

21. COPENHAGEN

The Town Hall

22. COPENHAGEN

Amalienborg Palace, the winter residence of the King. Behind it is the Marble Church.

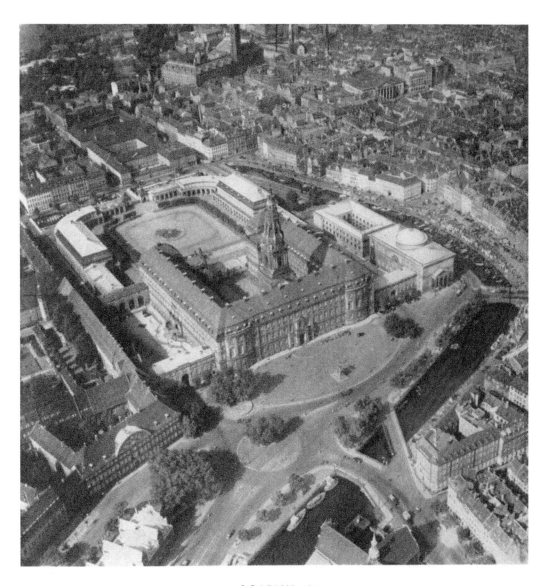

23. COPENHAGEN

Christiansborg, the seat of Parliament, the Government, and the Supreme Court of Appeal.

24. LONDON

Big Ben, the famous clock, seen from Parliament Square. The clock tower forms part of the Palace of Westminster, the building in which Parliament sits. The statue is of Lord Palmerston.

[*Photo: H. D. Keilor*

To the third department are assigned public assistance, social insurance, and the child guidance service.

To the fourth department are assigned the technical institutions such as the town engineer's office, the town architect's office, the surveyor's office, and the fire brigade.

Finally the fifth department is in charge of the municipal trading services, such as the supplies of gas, electricity and water, tramways, etc.

As previously mentioned, the executive council includes five paid but unpensioned aldermen, who are supposed to act as intermediaries between the mayors' naturally more official outlook and the councillors' more political point of view. In the executive council the aldermen are on equal terms with the mayors, and their votes are of equal weight. Within their respective departments the aldermen may be entrusted with certain administrative tasks, but in reality this does not amount to much. On the other hand, the aldermen are largely employed in various committees as representatives of the executive council.

Certain institutions are directed by special authorities, committees or boards. Local education is thus administratively and educationally under the supervision of an education committee consisting of the mayor of the first department as chairman, the alderman of the same department, four members elected by the town council and four members elected by and from among the boards of managers attached to each school. These school boards of managers in their turn consist of two members elected by the town council and three elected by the parents who have children attending the school in question. The idea of the boards is to obtain information about the working of the school, and they are entitled to deal with all questions concerning the school and to make recommendations to the education committee.

The education committee appoints the teachers, draws up the syllabus and curriculum in accordance with the legislative prescriptions, whereas the corporation decides all questions respecting the economy or finance of education.

Health services are in the hands of a public health committee consisting of the police director (a government official) as chairman, the mayor of the second department, the medical officer, and two members elected by the town council.

The municipal libraries also have an administration of their own, presided over by the mayor of the first department. This committee, which plans and superintends the working of the libraries, is invested with rather limited powers concerning the appointment and dismissal of the staff.

Of institutions in Copenhagen which are not administered by the corporation, in contradistinction to what is the case in certain other towns, we shall only mention the Port of Copenhagen. This is a self-supporting institution administered by a board consisting of the Lord Lieutenant of Copenhagen[1] and sixteen members, of whom only four are elected by the corporation of Copenhagen, while the others are elected by the central government, Parliament, and the trade organizations.

RELATIONS BETWEEN THE STATE AND THE MUNICIPALITY

An account of the constitution of Copenhagen would be incomplete without some mention of the relations between the state and the municipality. The influence of the central government on municipal administration is steadily increasing.

The formal relations between the state and the municipality appear already from the words of our constitutional law, which declares that the right of municipalities *under the supervision of the state* to govern their own affairs shall be established *by law*. This implies that the very rules according to which the municipalities are governed are embodied in acts of Parliament, and consequently municipal statutes are to be approved by the Home Office. As the statutes contain a list showing the number of public servants and their wages, the latter provision involves among other things that the creation, amendment, and abolition of posts within the local government service have to be approved by the Home Office. In Copenhagen, however, this provision has been limited to superior officials only.

The right of the Home Office to sanction the raising of loans in connection with capital expenditure, and the purchase, sale or mortgaging of municipal property has already been mentioned.[2]

The state has also to supervise the municipalities. In Copenhagen this supervision is performed by a lord lieutenant (*overpraesident*), whose office dates back to the charter of 1661, when all the mayors of the town were chosen by the King, and the first mayor got the title of '*praesident*,' later on '*overpraesident*.' After the constitution of the town had been rendered more democratic in 1840 and 1857 the lord lieutenant, however, survived as president of the executive council, whose members were by

[1] *Ante*, p. 239.　　　　　　　　　　　　　　[2] *Ante*, p. 238.

now all elected by the town council. In 1938 the lord lieutenant left the corporation, and is now only the superintendent of the state. In this capacity he has a right to demand information from, and to negotiate with, the corporation and to attend the meetings of both the executive council and the town council. If the lord lieutenant finds that a decision taken by the corporation exceeds the authority conferred upon it or is otherwise contrary to law, or tends to prevent the performance of a duty incumbent on the municipality, he should without delay report to the Home Office, who after hearing the corporation decides whether the decision shall be cancelled. It goes without saying that the Home Office may also take action without the intervention of the lord lieutenant.

The local administration is in practice largely dependent on the central administration. To show this in detail would be beyond the scope of this work, so we must confine ourselves to the main points. In the first place it should be emphasized that the corporation has no real authority as regards taxation. The provisions according to which municipal rates and taxes are levied, have all been made by central legislation—the King and Parliament—and the administrative rules for their execution have been largely fixed by the central government.

THE MUNICIPAL INCOME TAX

Thus the most important of the taxation laws, the act dealing with income tax in Copenhagen, fixes a certain progressive scale to be applied to rateable incomes. The percentages contained in this scale may be reduced by the corporation, but not increased by more than one-fifth.

According to the law the municipal expenditure of Copenhagen which cannot be defrayed through other municipal rates and duties or out of the municipal revenue (e.g. surplus of the municipal works) is covered by a local income tax.

This income tax, which was introduced in 1861 (i.e. about forty years before income tax was introduced in the central taxing system) has gradually become the dominating source of municipal income. As will appear from the table on page 247 the total revenue of the municipality in 1951–52 amounted to kr. 358 millions, of which rates and taxes represented kr. 276 millions, and the local income tax 230 millions or four-fifths of the total amount of taxes.

The local income tax is levied on individuals, associations, joint stock companies, and other companies with limited liability.

All the inhabitants of the municipality who have an annual income of at least 800 kr. are liable to pay local income tax. The tax is paid according to a progressive scale fixed in 1943, at the following rates:

On taxable incomes
Between   1,000–1,950 kr.:   4 kr. on 1,000 kr. and 7·6 per cent on the remaining sum
Between   2,000–3,950 kr.:   80 kr. on 2,000 kr. and 8 per cent on the remaining sum
Between   4,000–9,950 kr.: 240 kr. on 4,000 kr. and 9 per cent on the remaining sum

. . . .
. . . .
. . . .

Between 400,000–999,950 kr.: 72,000 on 400,000 kr. and 23 per cent on the remaining
sum
On 1,000,000 kr. or above: 210,000 kr. on 1,000,000 kr. and 25 per cent on the portion
that exceeds 1,000,000 kr.

Before the taxable income is assessed certain allowances are made. Thus for one child 600 kr. is allowed; for two children 1,300 kr. and for each successive child, 800 kr. For married couples, whose joint earnings in Denmark constitute a taxable unit, certain allowances are made for the wife's income. Special allowances are made for old age and infirmity pensioners.

The amount resulting from the application of the above scale to the taxable incomes is called the basic amount. If the budget cannot be approved unless the income tax is increased by more than one-fifth, an extraordinary election called a 'taxation-election,' must be held for the town council. After the election the new town council decides whether the tax already levied shall remain in force or be reduced. It will be understood that this provision means a serious check upon any inclination on the part of the corporation to impose an income tax that exceeds the limit mentioned. There is, however, this exception to the rule: if the increased income tax is caused by new acts of Parliament, the Minister for Home Affairs may exempt the corporation from holding a taxation-election for the town council.

The company tax, levied on commercial companies and associations, is not progressive, but amounts to 5 per cent on the net surplus.

The rules governing the municipal property tax are also fixed by national legislation, and include among other things certain per mille rates not to be exceeded by taxation.

As to the expenditure, it should be pointed out that in Denmark a so-called inter-municipal adjustment fund was established in 1937, from which contributions are made to the cost of public assistance, education, and hospitals provided by the various municipalities. The means necessary to meet this expenditure are derived from special inter-municipal taxes collected from the ratepayers of all municipalities. It will be understood that this system means a considerable economic infiltration between the state and the municipalities as well as between the municipalities themselves. In addition there is a sort of inter-municipal clearing system, so that municipalities with a large expenditure on public assistance and a low level of wealth receive contributions from municipalities with a low expenditure and a high level of wealth. It should be borne in mind, however, that recently a taxation committee proposed that these arrangements and clearing systems should be abolished and replaced by a redistribution of the tasks and burdens of the state and the local administration, and a brand new taxation system introduced. Several years will probably elapse before these fundamental changes can be carried through.

Of municipal institutions principally rooted in central legislation and central administration, education should be mentioned, together with public assistance and social insurance, for the whole educational system and structure has been largely determined by the state. The corporation is supreme, however, as regards the construction and equipment of school buildings.

The regulations of the funeral service and the local libraries must be approved by the Ministries of Church and Education respectively. The register of population was established under an act of Parliament and is administered according to regulations made by the Home Office. Likewise the health service is administered largely in accordance with regulations made or approved by the central government. Also the technical departments, which are in charge of sites, streets, roads, sewers, etc., must be administered on the basis of special acts concerning streets and roads, building laws, town planning, etc.

These central controls do not amount to an abolition, but only a limitation, of local self-government. In a municipality of the size of Copenhagen the elected corporation will naturally find ample scope for self-government and consequently for the realization of the local political views of the majority in the corporation.

SOME DIFFERENCES BETWEEN COPENHAGEN AND OTHER DANISH TOWNS

The units of local government in Denmark may be divided into two main groups (1) *købstadskommuner* and (2) *landkommuner*. A *købstadskommune* is about the same thing as an English borough or county borough, the status being conferred on them by special charter. A borough is administered by a town (borough) council, which governs the local affairs of the borough independently under the supervision of the Home Office. Denmark has eighty boroughs or county boroughs proper and six other urban municipalities which are governed on similar lines.

The rural units, known as *landkommuner*, are divided into two groups, viz.:

(*a*) About 7,300 parish-units roughly corresponding to the English rural districts and parishes. They are administered by parish councils, which attend to all local business, except hospitals and high roads, which are administered by

(*b*) The counties (*amtskommunerne*), governed by *amtsraadne*, corresponding to the English county council. The county councils, moreover, supervise the parish councils within the area of the county, e.g. their budgets must be approved by the county council. The total number of county councils in Denmark is twenty-five.

The most important difference in the administration of Copenhagen and the other towns is that, whereas the approving authority in the former is separated from the executive, the councils of the latter, though elected on the same lines as the town council of Copenhagen, are in possession of the executive as well as the approving power. This will at once appear from the fact that, whereas the mayors and aldermen in Copenhagen cannot at the same time be members of the town council, the opposite is the case in other towns. Secondly, the administration is in other towns largely carried out by committees appointed by the town council, whereas the committees appointed by the town council of Copenhagen do not participate in the administration of the capital. This difference actually means that in Copenhagen the administration is entrusted to the 'efficient' element (i.e. the chief officials) in a higher degree than elsewhere, though the mayors are responsible to the town council, whereas in the smaller towns, at least, the town council—the 'democratic' element—is in closer touch with the daily administration.

Another difference that may be mentioned is that whereas Copenhagen has a special state superintendent, the lord lieutenant, the other towns are under the direct supervision of the Home Office. Finally, different rules apply to the appointment of auditors.

MUNICIPAL FINANCE

Before we begin a brief survey of municipal administration and its financial basis in terms of revenue and expenditure, it is desirable to give a broad survey of the net revenue and net expenditure of Copenhagen for the year 1951–52.

Net Revenue	1,000 kr.	Net Expenditure	1,000 kr.
1. Taxes and rates ..	276,191	I. Central administration ..	38,439
2. Contributions from Frederiksberg and Gentofte as per intermunicipal clearing ..	20,569	II. Social insurance, assistance, and relief	84,289
3. Surplus on municipal services, utilities, and real property	15,940	III. Hospitals and Mental Welfare	60,384
Of this:		IV. Public Health	2,875
Gasworks	4,711	V. Education	35,472
Electricity works	11,557	VI. Public lighting, scavenging and watering of streets and roads ..	10,180
Heating works	1,069		
Water supply	3,209	VII. Roads and sewage service	16,181
Public markets	129	VIII. Fire brigade	8,849
Tramways	—2,835	IX. Interest on debt	49,561
Funeral Service	—898	X. Contributions to the pension account	35,434
Land and real property ..	—314		
Other municipal activities	—688	XI. Other expenditure ..	12,985
4. Interest	44,801	Surplus	2,857
5. Other income	5		
Total	357,506	Total	357,506

The most important rates and taxes collected in 1951–52 by the municipality of Copenhagen for its own expenditure were:

	Mill. kr.
Income tax	229,5
Property tax	34,0
Taxes and duties on consumption, etc...	12,7
Total	276,2

The object of the equalization arrangements between Copenhagen and its two neighbouring municipalities, all of which constitute one social and economic community, is to equalize the varying taxable capacity of the inhabitants of the three municipalities, seeing that the income level of the citizens of Frederiksberg and particularly Gentofte is somewhat higher than that of the citizens of Copenhagen. Particulars of this clearing fund are given in the appendix to this chapter.

As a supplement to this very brief survey of what the administration of the municipality of Copenhagen costs we give the following table showing the personnel employed in the most important municipal institutions.

Number of Persons Employed, 1951–52

	Permanent Officers	Wage Earners	Other Employees	Total
Taxing Department	682	—	440	1,122
Statistical Office and Register of Population	178	—	184	362
Libraries	100	35	179	314
Funeral Service	162	294	105	561
Education	3,110	1,041	713	4,864
Public Markets	139	171	15	325
Hospitals and Mental Welfare ..	2,180	2,272	2,689	7,141
Public Health	120	63	209	392
Directory of Municipal Residential Properties of Copenhagen ..	55	4	1,295	1,354
Administration of the Town Hall	356	600	53	1,009
Social Insurance and Public Assistance	1,412	730	1,166	3,308
Fire Brigade	554	82	56	692
Administration of Streets, Roads and Sewers	404	875	90	1,369
Town Architect's and Engineer's Offices, etc.	298	—	70	368
Electricity and Gas Supply ..	1,027	1,516	262	2,805
Water Supply	221	231	32	484
Tramways	3,669	1,700	60	5,429
Department for Handicraft ..	35	412	31	478
Other Institutions	389	417	253	1,059
Total	15,091	10,443	7,902	33,436
Of these women	4,955	4,527	5,812	15,294

According to the budget for 1951–52 the municipality of Copenhagen consequently employs a total number of 33,400 persons, of whom 15,100 are permanent officers, 10,400 wage-earners (e.g. the charwomen employed in hospitals and schools), and 7,900 belong to the group 'other employees,' i.e. periodically employed office workers, apprentices, etc.

THE RELATION BETWEEN COPENHAGEN AND THE NEIGHBOURING MUNICIPALITIES

In the foregoing we have dealt with the municipality of Copenhagen as a constitutional and administrative unit alone, but, as already mentioned, the formal boundaries of the municipality were burst long ago by the economic and political development of the city.

The fact that the twenty-two municipalities constitute one economic community naturally involves a considerable interest in the administration of public affairs on the part of the inhabitants, not merely in the municipality where they live, but also in the one where they work. They are interested in reliable and fast means of communication, roads, bus-lines, railways, and tramways; and the location and efficiency of gas, water, electricity, and heat works become matters of importance to a circle of consumers extending beyond the formal boundaries of the individual municipalities. The same thing applies to hospitals, schools, etc. A real co-ordination of all these matters of mutual interest is beset with considerable difficulties, not only because the metropolitan area is split up into twenty-two different independent municipal units, but also because the nineteen municipalities outside the capital are parish councils, which together with three other parishes are under the supervision of the county council of Copenhagen, which also administers the hospital service of these municipalities, along with the most important main roads.

Besides the mutual interests mentioned, which are bound to result in the demand for a rational administration of the metropolitan area, attention should be turned to another important problem, which is not a question of mutual interest to all the inhabitants of the metropolitan area, but rather a question of guarding the legitimate economic interests of some municipalities against others within the area. The distribution of vocational and income classes, and consequently the social structure of the population, vary considerably in the different municipalities. Some municipalities are inhabited chiefly by the more well-to-do business people and superior civil or public servants and employees, others by members of the working classes. The public expenditure per head being about the same in the municipalities—somewhat higher in the 'poor' municipalities than in the 'rich' ones—a rather considerable discrepancy arises as regards municipal taxation. Endeavours have been made to remedy this discrepancy by the imposition of a so-called trade tax on persons receiving an income exceeding a certain amount from a municipality outside the one where they are domiciled. This tax is added to the revenue of the municipality in which the income is earned, but is refunded to the ratepayer partly or entirely by the municipality in which he resides. Between Copenhagen on one hand and Frederiksberg and Gentofte on the other the trade taxation has, however, been replaced by a clearing system, as mentioned above.

For the purpose of enquiring into these problems the Home Office in 1939 appointed a committee, called the Metropolitan Committee, to consider what amendments of the existing municipal conditions the economic, technical, and popular developments in the metropolitan area render desirable.

During the German occupation of Denmark from 1940–45 the work of the committee was of course considerably hampered, because public administration was engrossed in many other tasks. But when the occupation came to an end, the work was resumed in earnest, and in January 1948 the committee made a report.

This report began by pointing out that the municipal boundaries fixed centuries ago under widely different administrative and economic conditions, do not at all suit modern developments as regards the built-up area, trade, commerce, and traffic, which often utterly disregard these boundaries. In the metropolitan area the problem is exceptionally acute owing to an almost explosive development.

The report then considers the various possibilities of reform, specially dwelling upon the arguments pro and con three separate solutions: (1) An arrangement for the purpose of adjusting the economic discrepancy between the municipalities already mentioned; (2) the establishment of a sort of municipal league so that certain business of an intermunicipal nature could be assigned to a major municipality, whereas the more local business would be left to the primary municipalities of the league, i.e. a system somewhat corresponding to the relations between the county of London and the metropolitan boroughs, and (3) the incorporation in Copenhagen of a number of municipalities and the amalgamation of others.

It would require too much space to deal with the various arguments for and against the several possibilities and combinations. We shall therefore only say that the committee did not succeed in making a unanimous recommendation, and the majority obtained in favour of a certain solution could only be secured by the sacrifice of principles all round. The majority recommendation suggested that the seven municipalities of Gladsaxe, Herlev, Rødovre, Hvidovre, Taarnby, St. Magleby and Dragør, together with parts of Glostrup and the municipalities of Brøndbyøster-Brøndbyvester should be incorporated in the municipality of Copenhagen, and that the other municipalities in the area should be amalgamated into four new municipalities, which were to be raised to the status of *købstadskommuner* (county boroughs). These last municipalities together with Gentofte were to form a league of suburban municipalities with a special council. The other municipalities within the county of Copenhagen were to be assigned to the county of Roskilde, and the county council of Copenhagen was to be abolished. A metropolitan council would in that case look after the affairs of Copenhagen, Frederiksberg, and the suburban municipalities.

The majority of the committee made their recommendation because they thought the reforms it proposed could be carried through politically, but on this point they were disappointed. Within the suburban municipalities in question, and principally within those to be included in Copenhagen, very strong opposition against the recommendation has arisen. This opposition is primarily based on the belief that in the big municipality that would come into being if the municipalities in question were incorporated in Copenhagen, the inhabitants of the latter would lose a very considerable part of their influence on their own local affairs.

Despite many reminders from the corporation of Copenhagen as well as from other expert quarters, who consider it most unfortunate that the capital should still be split up into twenty or twenty-five municipal units, the government have not yet introduced any Bill concerning a reorganization of the metropolitan area.[1]

THE PLANNING OF GREATER COPENHAGEN

The fact that neither the previous government nor the present one have promoted or intend to implement the recommendation of the Metropolitan Committee must be ascribed to the disagreement between the corporations concerned as to the proper solution, and maybe also to the fact that the metropolitan problem has been approached on other than political lines, for the recommendation was dictated by rational, town planning considerations.

So far no general town planning authority has existed for a majority of municipalities or a part of the country. In 1938 a town planning act was indeed passed, which made the drawing up of a town plan compulsory in every town or built-up area containing more than 1,000 inhabitants. This act, however, had the serious drawback that it contained no directions for the comprehensive planning of several municipalities forming a single region, just as a responsible authority for such regional planning is totally missing.

While considerable planning work has been done in Copenhagen and

---

[1] By an act of May 1952 the said suburban municipalities have obtained a more independent position than before, somewhat like that of county boroughs. But as yet no institution to watch the interests of the entire metropolitan area has been created.

environs of late years this is primarily due to a regional planning committee appointed on private initiative. This committee was originally appointed in January 1928 at the request of the Danish Town Planning Laboratory. But after the committee in 1936 had finished a work which at that time was rather urgent and important, concerning the green areas of Copenhagen, i.e. woodland, parks, areas for open-air recreation, etc.,

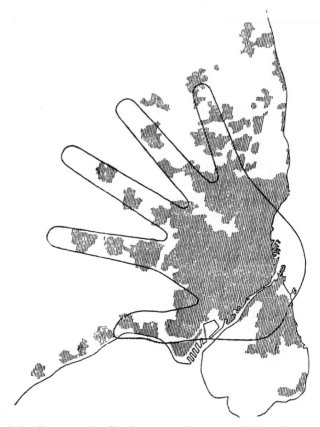

The proposed development plan for the metropolitan area of Copenhagen. The spaces between the 'fingers' are intended to be used for green areas.

the committee came to a standstill. As the work of the Metropolitan Committee dragged on, however, the demand for a general plan for the Copenhagen region grew more and more urgent, especially after the town planning act of 1938 had required the municipalities of the area to draw up separate town plans. In March 1945 the regional planning committee then resumed its work. The membership of the committee consisted of representatives of the municipalities of the Copenhagen area and representatives of a great number of authorities and institutions interested in planning. It was, however, an unofficial body without executive power.

In January 1948 the regional planning committee issued an outline plan for Greater Copenhagen. This is a rather comprehensive work treating all the problems connected with the planning of a future Greater Copenhagen. One of the points of departure is, of course, the number of inhabitants, and here the figures are based on a population prognosis, made by the Statistical Office of Copenhagen, which on certain assumptions estimates that in 1965 the metropolitan area will contain 1·4 million inhabitants, and at a somewhat longer sight the figure is estimated at 1·6 to 1·7 million. The plan proceeds to show which areas are most suitable for residential building, specified according to housing, and indicates the localities for the different kinds of industrial and commercial development. It deals with traffic routes, recreation grounds, etc.

The plan proposes that the future development of the metropolitan area should be connected with the radial railway lines. These lines are the electrified north and north-west railways, and the partly constructed and partly projected electric railways to the west. The structure of the future metropolitan community can therefore, as shown in the outline plan, be illustrated by a hand, the palm forming the municipality of Copenhagen while the five spreading fingers indicate the building along the five radial railways. The spaces between the fingers are intended to be largely laid out as green areas. This division of the metropolitan area into traffic districts greatly favours a similar division of Greater Copenhagen into administrative areas roughly corresponding to these main traffic districts. But such a reorganization is completely opposed to the report of the Metropolitan Committee, with its recommendation of incorporations and the creation of the great municipality of Copenhagen and several big *kobstaeder* (county boroughs).

On one point only, though a very essential one—the future siting of new development—there seems to be a chance of realizing ideas of the sort suggested by the regional planning committee, because an act was passed in 1949 on the regulation of urban building.

According to this act town development plans may be drawn up for periods of not less than fifteen years for areas further defined, and of course also for the metropolitan area, prescribing which areas may be used for urban building (inner zones), which may be laid out as garden links for later development (intermediary zones), and which are to be kept free of urban building (outer zones). In the autumn of 1949 a town development committee was appointed for the Copenhagen region to draw up these plans. This committee finished its work in 1951, and the plans now only need the sanction of the ministry.

At the present moment nothing can be said for certain about the final constitutional and administrative structure of the metropolitan area, but everyone who is interested in the sound development of the capital must hope that it will not be too long before rational solutions are adopted for the housing, traffic and economic problems of the metropolis as well as for the administration of its numerous relations with the official authorities. Briefly, we need from every point of view the best possible organization of the government of the Danish capital.

APPENDIX
THE EQUALIZATION SYSTEM IN COPENHAGEN

The object of the equalization between Copenhagen and Frederiksberg and between Copenhagen and Gentofte is to make the total net proceeds from a number of rates and taxes per inhabitant in each of the municipalities equal to the corresponding proceeds per inhabitant in Copenhagen plus Frederiksberg and Copenhagen plus Gentofte respectively.

First an equalization is made between Copenhagen and Frederiksberg, and the following taxes are included in the calculation: personal income tax, company tax and property rates.

To make it possible to ascertain the varying tax-paying capacity in the municipalities on the basis of the tax revenue as it appears in the municipal accounts, it is necessary that the taxes and rates should be levied according to a uniform system. However, this is not so, because the aforementioned scale for the taxation of personal incomes is differently applied in the three municipalities, and the same per mille is not used in the taxation of property. When equalizing it is therefore necessary to recalculate the actual tax revenue.

The calculation of the amount of income tax to be included in the equalization presents the greatest difficulty.

To arrive at this amount two calculations are made every year partly of an expenditure figure and partly of a tax figure. By the *expenditure figure* is understood the average expenditure per inhabitant in both municipalities together on the central administration, social welfare, hospitals, mental care, public health, and education. The greater the expenditure figure, the higher the level on which the income tax will be assessed. By the *tax figure* is understood the average amount of tax revenue calculated per inhabitant provided that the income tax is paid according to the scale of rates without increase or reduction, and after the deduction of certain special tax regulations, that the same percentage of company tax is collected, and finally that the property rates are levied according to the per mille of Copenhagen. The tax figure affects the equalization calculation conversely seeing that the income tax becomes proportionally lower when the tax figure rises.

The calculation of the amount of taxes subject to equalization is thus effected in the following way:

First the basic amount of the income tax (i.e. the amount that the tax would yield according to the unamended application of the legal scale[1]) by a percentage representing the difference between the expenditure-figure and the tax figure, after this difference has been reduced according to the relation between the retail price index in 1935 and that of July within the respective financial year.

To the reduced amount of income tax evolving from the above process are added partly the amount of tax on joint stock companies and partly the rates on real property calculated in such a way that the per mille of Frederiksberg are recalculated in conformity with those of Copenhagen. The total amount of taxes is calculated partly for each municipality separately and partly for both municipalities jointly, and after that the three average amounts per inhabitant are calculated. The municipality that exceeds the joint average has to pay to the other one so large an amount that after the equalization it comes down to the joint average. The equalization contribution is then made out by multiplying this difference by the number of inhabitants of the contributing municipality.

The following statement from 1952–53 will explain the calculation of the equalization:

Expenditure Figure

Total expenses (1,000 kr.)	233,673
Per capita (Copenhagen plus Frederiksberg) (kr.)	265

Tax Figure

	Copenhagen	Frederiksberg	Copenhagen + Frederiksberg
Basic amount of income tax (1,000 kr.) ..	169,702	39,143	208,845
Special tax regulation	7,659	−432	7,227
Company tax and property rates (1,000 kr.) ..	46,851	6,378	53,229
Total (1,000 kr.)	224,212	45,089	269,301
Per capita (kr.)			306

[1] *Ante*, p. 244.

The difference between the tax figure 306 and the expenditure figure 265 is 41, but as the retail price index in July 1952 was 220 when that of 1935 is put at 100, the 41 will have to be reduced by $\frac{100}{220}$, which is equal to 19. This means that in the amount of income tax on basis of which the calculation is made, the aforementioned amounts are reduced by 19 per cent, and consequently the taxes will be:

	Copenhagen	Frederiksberg	Copenhagen + Frederiksberg
Income tax (1,000 kr.)	145,118	31,273	176,391
Company and property rates (1,000 kr.) ..	46,851	6,378	53,229
Total (1,000 kr.)	191,969	37,651	229,620
Per capita (kr.)	251,99	320,12	261,10

Accordingly Frederiksberg has to pay *per capita* to Copenhagen 320,12 less 261,10 = 59,02. The number of inhabitants of Frederiksberg being 117,615 the total contribution to Copenhagen 1952–53 will be 6,941,637 kr.

After the contribution from Frederiksberg has been made out, the equalization between Copenhagen and Gentofte is calculated in all essentials according to the same rules, except that whereas the contribution received from Frederiksberg is included in the tax revenue of Copenhagen, Gentofte for various reasons is granted a deduction. It should be noticed that it is merely an equalization of income that is effected between the three municipalities, and not of expenditure, as originally proposed in 1937.

Copenhagen

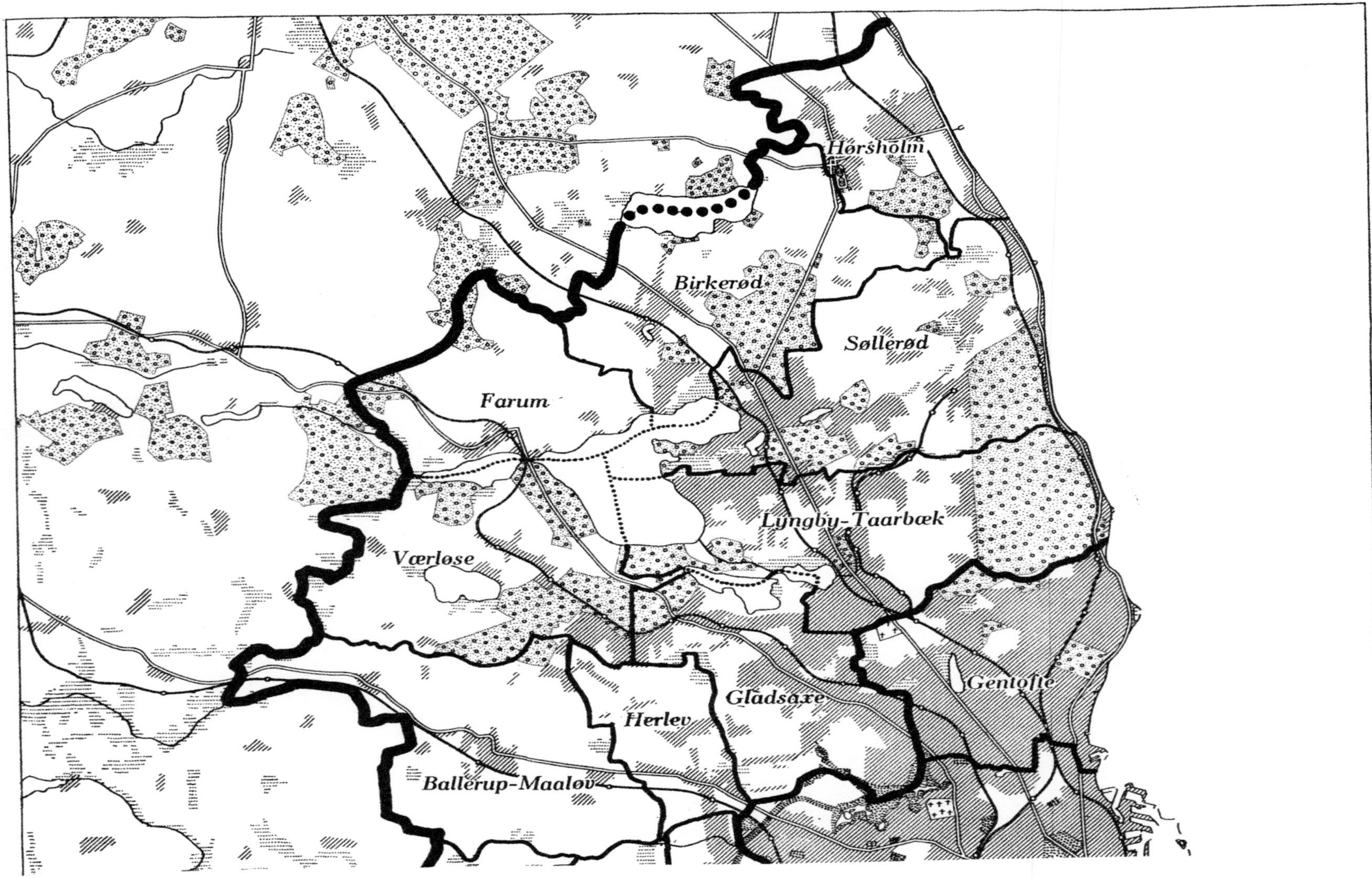

Hørsholm

Søllerød

Birkerød

Lyngby–Taarbæk

Gentofte

Gladsaxe

Herlev

Farum

Værløse

Ballerup–Maaløv

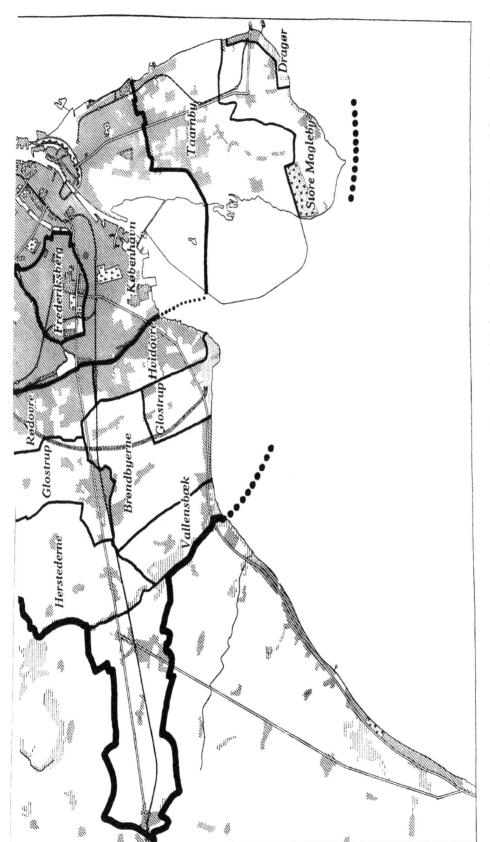

THE METROPOLITAN COMMUNITY OF COPENHAGEN. The municipality of Copenhagen (København) is the capital only in the strict legal sense, and for most administrative and statistical purposes the capital is taken to mean the three municipalities of Copenhagen, Frederiksberg, and Gentofte. As the above map shows, the urban population (represented by shaded portions) has spread out over the nineteen suburban areas surrounding the capital.

London

WILLIAM A. ROBSON

WILLIAM A. ROBSON

Educated University of London (London School of Economics and Political Science). B.Sc.(Econ.), 1922; Ph.D. 1924; LL.M. 1928. Barrister-at-law. Lecturer at the London School of Economics, 1926–33; Reader in Administrative Law, 1933–46; Professor of Public Administration in the University of London since 1947. Visiting Professor, University of Chicago, 1933; University of North Carolina, 1951; University of Patra and other Indian Universities, 1953; Principal, Mines Department, 1940–42; Ministry of Fuel and Power, 1942–43; Assistant Secretary, Air Ministry, 1943–45; Ministry of Civil Aviation, 1945–46.

President, International Political Science Association, since 1952; Executive Council, Institute of Public Administration; Member of Council, Town and Country Planning Association; Departmental Committee on Administration of Greater London Plan; Local Government Examinations Board; Executive Committee, British Political Studies Association.

London

LONDON owes much of its size and power to its commanding position
astride the Thames. Mackinder remarked that London Bridge is the pith
and cause of London.[1] The river is navigable for sea-going vessels as far
as London, but not much higher. Hence London became a point of
trans-shipment from water to land, and from river to sea. Downstream,
the river would in past centuries have been impossible to bridge owing
to its great width, and dangerous to ford on account of the heavy tides
and broad marshes. The estuary is well placed for ships voyaging to and
from the continent. London cannot be attacked or captured from the
sea alone, although it could be defended almost entirely by seapower
in the days before aviation. There are no physical obstacles to development
until one reaches the Chiltern Hills in Hertfordshire or the North Downs
in Surrey and Kent.

THE GROWTH AND SIGNIFICANCE OF LONDON

These are the elementary geographical factors which led to the establishment of an early settlement more than 2,000 years ago,[2] and its
subsequent survival and expansion. The growth of the modern metropolis
from the Londinium of Roman Britain is the result of many social,
political, and economic causes which cannot be mentioned here. The
history of London is a large subject on which many books have been
written. We may observe that the growth of London has been a continuous process for many centuries; and that its leading position in the
national, international, imperial and commonwealth affairs of the British
people has never been challenged.

London is the political capital of Britain, the focus of the Commonwealth, the centre of the British Empire, and a metropolis of world
importance. It is the place where the sovereign resides, where Parliament
assembles, and the Cabinet meets. It is the headquarters of all the great
government departments and the centre of the judicial system. It is the
financial and commercial hub of Britain, the home of famous institutions

[1] Sir Halford Mackinder: *Britain and the British Seas*, p. 204.

[2] Tacitus tells us that Londinium, although not distinguished by the title of colony,
was nevertheless a busy centre, notable for its crowd of merchants and stores. *Annals*,
Book XIV, Ch. xxxiii.

like the Bank of England, Lloyd's, or the Baltic Exchange. It is also a great manufacturing city. The port of London, with its thirty-six miles of quays, has the largest system of docks in the world. It carries on a larger entrepôt trade than any other port, in addition to supplying the London region with commodities and goods of all kinds.

The cultural resources of London are immense. The great national collections housed there include the British Museum, the Record Office, the Victoria and Albert Museum, the National Gallery, and the Tate Gallery. The metropolis is the home of the Royal Society, the Royal College of Physicians, the Royal College of Surgeons, the Inns of Court, and many other learned bodies. Most of the leading hospitals at which teaching and research are carried out are located in London. The University of London is the largest university in Britain and the principal centre for post-graduate study and research, to which students come from all parts of the world. London is the principal domain in Britain of dramatic and musical enterprise, of the opera and ballet, of film production and broadcasting. It is the directing centre for book, newspaper and periodical publishing. Its arenas provide the setting for many great sporting and athletic contests of national or international interest.

The combined influence of all these factors, together with the great size and wealth of its population, have given London a unique position not only in Britain but among the great cities of the world.

THE SIZE OF LONDON

In considering the size of London we are confronted with the problem of defining the area to be taken for the purpose. This is a serious but very common difficulty in dealing with great cities. The population of London has spread far beyond the present administrative boundaries, which were laid down for the metropolis nearly a century ago. In consequence the facts of social life no longer correspond even remotely with the area formally allotted to London government. London is not a city in the ordinary sense of the word, but a region. There is, however, no general municipal authority, no regional council, which administers the metropolitan region or represents the eight or nine million persons who live in it.

The size of the population and territory of London can best be described in terms of its several parts. In the heart of the metropolis is the ancient City of London, an area measuring approximately one square mile. It was for long a place in which people both lived and worked, but nowadays it has become almost entirely a financial and business quarter. Hundreds of thousands of men and women earn their living in the City, but scarcely more than 5,000 persons reside there. For nearly a thousand years the ancient city has been governed by the Corporation of the City of London. For various reasons the City Corporation failed to expand its boundaries or to develop its municipal organization. It thus remains an obstinate relic of mediaeval structure sticking out like a rock in the sea.[1]

The largest and most important area in the metropolis is the County of London. The County of London and the City of London are together known as the administrative county of London, and this is the area in which the London County Council carries out most of its work. The administrative county comprises nearly 117 square miles. It measures $16\frac{1}{4}$ miles in extreme length from east to west, and $11\frac{3}{4}$ miles in extreme breadth from north to south. This area was originally designed for the now defunct Metropolitan Board of Works in 1855 and has remained practically unchanged. It was handed over to the London County Council in 1888. Even a century ago this area did not include the whole of London, and since then the proportion outside its boundaries has progressively increased.

In 1939 the administrative county contained 4,013,400 persons. The largest population ever recorded was 4,536,267 in 1901, with a slight decline to 4,521,685 in 1911, to 4,484,523 in 1921, and to 4,397,003 in 1931.[2] On the outbreak of the Second World War a vast scheme of evacuation was put into operation. Hundreds of thousands of children, young mothers, university students, and other special classes were moved to safer areas. Many thousands of officials, both civil and military, were also transferred to less vulnerable places. In consequence of these and other results of war, such as casualties and conscription, the population of the administrative county sank to 3,084,100 in 1940 and 2,320,000 in 1941.[3] After the war people flocked back to London, and the population of the administrative county had risen by 1951 to 3,348,336. It is unlikely that the population of this area will regain its former size, for several reasons. The most important is that a policy of decentralizing industry and population from the central core to the outer areas of Greater London is part of the town planning policy adopted for London.

Greater London is usually used to denote the area of the Metropolitan

[1] W. O. Hart: *Introduction to the Law of Local Government and Administration* (5th edition), p. 242.

[2] Statistical Abstract for London, 1939–48, Vol. XXXI, published by the London County Council, 1950. p. 3. The figure for 1939 is the Registrar General's estimate.

[3] *Ibid.*, p. 5.

Police District, within which the metropolitan police force operates, together with the City of London, which has its own separate police force. It covers 721·6 square miles and contained in 1951 a population of 8,346,137, or one-fifth of the total population of England and Wales. The Metropolitan Police District was created as long ago as 1839, and its boundaries remained unchanged until 1947, when some minor adjustments were made.[1] By subtracting the population of the county from that of Greater London, we see that some 4,997,801 persons live in outer London: that is, the area between the boundaries of the administrative county and the Metropolitan Police District.

There are numerous regional areas for special purposes. Thus, the Metropolitan Water Board supplies water in an area comprising 575·8 square miles containing a population of about 7¼ millions. The London Electricity Board, which distributes electricity under the recent nationalization scheme, supplies an area of 253 square miles containing a population of 6,273,000. The London Passenger Transport system, a vast complex of underground and surface railways, motor-buses, trolley-buses and motor-coaches, originally placed under unified management and public ownership in 1933, and now forming a distinct part of the nationalized transport industry, operates in an area of 1,975 square miles serving a population of 9,700,000. There are other regional areas for the port of London; for the gas boards which supply London north and south of the Thames; for the hospital boards that form part of the national health service; for traffic regulation and the licensing of passenger road services; and for town and country planning purposes.[2]

Even if we confine ourselves to Greater London and leave aside the larger areas mentioned above, we are dealing with a population bigger than any one of fourteen European states. Greater London contains about one-fifth of the population of Great Britain, and a much larger share of its wealth, political leadership, commercial management, administrative power, and cultural activity.

THE CHARACTER OF THE METROPOLIS

The metropolis presents a continuous built-up territory exceeding 400 square miles. It stretches from Enfield in the north to Croydon in

[1] These adjustments resulted in 43·8 square miles of territory, situated mainly in Surrey and Kent, being added to the district.

[2] For details, see Herbert Morrison: *How London is Governed* (2nd edition, 1949), Chapter I; W. A. Robson: *The Government and Misgovernment of London* (2nd edition, 1948), pp. 163-74.

the south, and from Hounslow in the west to Ilford in the east. Beyond this are numerous suburban settlements and dormitory towns situated along the railway lines and main roads. The central core of London developed largely through the spread and agglomeration of separate villages and urban settlements. A similar process has taken place in recent decades on a larger scale in outer London where the distances are much greater. The process has now been restrained or stopped by town and country planning control. Many disadvantages have resulted from this type of sporadic growth, and much valuable land has been misused or spoilt. One good feature is that many districts have preserved a distinctive neighbourhood life and community sense of their own. It has been truly said that 'although London today may seem like one huge town, it is in fact a collection of inter-related communities which still retain many of their individual characteristics. This community structure is more marked in London than in any other national capital.'[1]

London is a great industrial city. The County of London contains between 30,000–40,000 factories and workshops and there are many others in outer London. Most of the factories within the county are on a small scale, though there are some large firms engaged in manufacturing food-stuffs, for example. The larger firms naturally prefer sites outside the built-up core of the county, and tend to go to outer London.

A feature of London is the concentration of particular industries, trades and professions in certain districts. In or near Fleet Street one finds nearly all the newspaper offices and newspaper printing presses. Shoreditch is a centre of the furniture making industry, Bermondsey of canned food, biscuit and jam manufacture, Deptford of ship repairing. The Temple, Gray's Inn and Lincoln's Inn form the legal quarter. Harley Street is the medical centre; Hatton Garden the street in which the dealers in precious stones congregate. In the period between the two world wars a vast industrial expansion took place in London. Many new factories were set up in the metropolis, and many firms migrated there from provincial towns and the areas where unemployment was most severe, in order to be near the nation's biggest market or to avoid high local taxes. Industrial development in London is now controlled by the local planning authorities and the Board of Trade. Industrialists cannot build or extend their factories in the metropolis unless they satisfy certain conditions laid down by the Board of Trade and obtain a certificate from that department.

The east end of London is occupied by factories, workshops, and the

[1] *The Youngest County* (published by the London County Council, 1951), pp. 20–1.

vast paraphernalia of docks, wharves, quays and warehouses associated with the port. It also houses an immense number of working class families in unattractive surroundings. Immediately south of the Thames, in places like Southwark, Battersea, Balham, Brixton, Peckham, Camberwell, Lewisham, are vast areas of lower middle class housing. There are similar areas north of the river in such places as Finsbury, St. Pancras, Islington, Hackney and Stoke Newington. Further south, as one approaches the lovely grass-covered rolling hills of the North Downs, the character of the buildings changes and the standard of housing improves. Here are the homes of better-off Londoners who have escaped from the noise and hustle of the town. A similar trend occurs to the north and north-west of outer London. The most favoured residential districts in the central area are to be found in Chelsea, situated near an attractive stretch of the river, and in Westminster, Marylebone, Kensington, all within easy reach of the magnificent parks which are the finest feature of inner London. Hyde Park, Kensington Gardens, St. James's Park, Green Park and Regent's Park are 'royal' parks, which means they are situated on Crown land, are managed by the Ministry of Works, and are paid for out of national taxes.

Fortunately for the material, mental and moral health of Londoners, poverty and wealth are not segregated as completely as the preceding remarks might suggest. Overcrowded tenements and slums sometimes exist within a stone's throw of modern blocks of luxury flats; one can often find a lovely eighteenth-century Georgian house in an old-world garden standing amid dreary rows of monotonous nineteenth-century dwellings built for low-paid artisans and clerks. Age cannot wither nor custom stale the infinite variety of London. It is this which charms visitors from other lands and evokes the love of its citizens.

THE CITY CORPORATION

The Corporation of the City of London is the oldest local authority in Britain, and the only one which has escaped reform by modern legislation. The corporation has a constitution which does not accord with modern democratic principles. It enjoys great privileges, powers and property; and it has been permitted to retain these not because of its value or importance as a local government unit, but on account of its political influence, symbolic traditions, and the splendour of its ceremonial entertainments and hospitality.

The origin of the City Corporation is lost in the mists of antiquity. Its earliest charter, granted in 1070 by William the Conqueror, confirmed existing privileges and did not purport to confer new ones. The corporation consists of three separate chambers known as the Court of Common Hall, the Court of Common Council, and the Court of Aldermen.

The Court of Common Hall is an annual assembly of the lord mayor, aldermen and liverymen of the so-called city companies or guilds. The basis of representation in medieval times was linked with the crafts and burgesses. This explains the association of the Court of Common Hall with the city companies, and the absence of any relation between the City Corporation and the population of the city or that of the metropolis outside the narrow boundary. The city companies, however, long ago ceased to be guilds regulating particular trades, and have become voluntary organizations composed of business and professional men following diverse occupations, associated for social and philanthropic purposes. The liverymen of these companies or guilds, numbering less than 10,000, are alone entitled to attend Common Hall, apart from the lord mayor and aldermen. The Court of Common Hall chooses two aldermen who have filled the office of sheriff; and from these the aldermen select the lord mayor. Common Hall also elects the sheriffs, the chamberlain or treasurer, the bridgemaster, who looks after the cross-river bridges maintained by the City, and the city auditors. The lord mayor of London is thus the nominee of a tiny oligarchy of archaic guildsmen,[1] and his official status is confined to a small obsolete fragment of London government.

The Court of Common Council is composed of the lord mayor, 25 aldermen and 206 common councilmen elected annually at the wardmotes by the local government electors of the ward. Common Council is an excessively large body to control the affairs of a local authority administering one square mile of territory and proposals have been made to reduce its size by a quarter. It is the principal governing organ of the City Corporation. It exercises both legislative and administrative functions. It appoints the judicial officers of the City's courts, the town clerk, remembrancer, solicitor and other chief officers. It determines administrative and financial policy.

The third chamber is composed of twenty-five aldermen presided over by the lord mayor. The aldermen are elected separately by the wards. They hold office for life, subject to approval by their own court, but are otherwise irremovable. They appoint certain specified officers, and in

---

[1] W. A. Robson: *The Government and Misgovernment of London* (2nd edition), pp. 28-31. Herbert Morrison: *How London is Governed*, pp. 11-17.

I*

former centuries much of the administrative power (now exercised by Common Council) was in the hands of the aldermen. They still possess extensive financial power, for they can authorize payments out of the city cash, a large fund derived from property owned by the City Corporation. This property, acquired by gift, grant, purchase or bequest, is under the sole control of the corporation.

Both common councilmen and aldermen are required to have a property qualification to be eligible for office. In practice, they are chosen from men of substance; and this again reinforces the oligarchical character of the City's government.

The City Corporation is responsible for maintaining four cross-river bridges.[1] It owns and operates the great wholesale food markets which supply the bulk of the fish, meat, poultry, game, eggs, cattle, sheep and horses consumed by the entire metropolis.[2] It owns and maintains several thousand acres of parks and open spaces situated outside its own area. In providing these amenities we see the most beneficent side of the City's activities, for without its interest and money, splendid woods like Epping Forest and Burnham Beeches which provide healthy enjoyment for Londoners, would have been lost for ever.

The City Corporation has its own highly efficient police force. It is the sanitary authority for the port of London. It cleans, lights and repairs its streets, and is responsible for highway improvements. It provides two excellent schools for boys, and one for girls; and also the Guildhall School of Music. The City Corporation is not the local education authority for its area, and these schools are maintained from bequests or grants from the city cash. The City Corporation exercises a number of minor regulatory functions concerning such matters as weights and measures, shop hours, the storage of petroleum and explosives, etc.[3]

THE SYSTEM OF LOCAL GOVERNMENT

All the large towns in England and Wales[4] except London are county boroughs, of which there are eighty-three at the present time. A county borough is governed by a county borough council, which exercises all local government powers within its area. The county borough is thus a unified or single-tier form of local government.

---

[1] Tower Bridge, London Bridge, Southwark Bridge, and Blackfriars Bridge.
[2] W. A. Robson: *The Government and Misgovernment of London* (2nd edition), Chapter X, pp. 243-7.　　　　　　　　　　[3] *Ibid.*, p. 33.　Herbert Morrison: *op cit.*, pp. 22-4.
[4] Scotland has its own form of local government which differs in certain respects from that obtaining in England and Wales.

The remainder of England and Wales is divided for local government purposes into sixty-two administrative counties. Each administrative county has two tiers, and in rural areas three tiers, of local authorities. The major authority in an administrative county is the county council, a directly elected body which either provides services requiring large-scale organization or carries out planning and co-ordination throughout the county.

The minor authorities in a county are called generically county district councils. They are of three kinds: non-county borough councils, urban district councils, and rural district councils. The distinction between these three types of district council is one of form rather than of substance. One important difference is that in rural districts we find a third tier of authorities, namely, parish meetings and parish councils. These exist or can be created in almost every village or hamlet.

County councils were established by the Local Government Act, 1888, which also conferred county borough status on those towns considered too large or important to be subjected to the jurisdiction of the county councils. In general, the county borough form was deemed suitable for large, wealthy, independent, proud, energetic towns with a strong sense of community like Manchester, Birmingham, Leeds, Bristol, Sheffield or Bradford, while county government was regarded as appropriate for the rest of the country. The counties have existed in Britain since time immemorial, and have served all kinds of social, political, military and administrative purposes. There was, therefore, a powerful sentiment of county 'patriotism' or loyalty to which the county council system was attached, although the administrative counties for local government purposes often differ considerably from the historic shires.

It was anomalous in a sense to make London a county for local government purposes in 1888, and to give this newly-created county a county council. Yet the decision was clearly right in principle, for it envisaged a two-tier system of local government for London, and this is undoubtedly desirable for a community numbering millions. It would have been a fundamental mistake to have made London a county borough possessing only a single authority responsible for all local government services, large and small, throughout the whole area. This would have been a grossly over-centralized solution of the problem of London government. The error which was made in 1888 was to perpetuate, under the name of the county of London, the totally inadequate area which had served the Metropolitan Board of Works since 1855. The metropolis has suffered continuously from the failure of Parliament to create a major organ of government corresponding even remotely to the social, physical, political, administrative and financial facts of the region.

THE LONDON COUNTY COUNCIL

The London County Council is the principal organ of local government in London. It is also the largest and most important local authority in Britain.

The London County Council consists of 150 members, of whom 129 are councillors and 21 aldermen. The councillors are directly elected on a popular vote by 43 electoral divisions, each of which returns three members. A general election is held every three years. The aldermen are elected by the councillors and hold office for six years, half their number retiring every three years. There are thus 10 or 11 aldermanic vacancies to be filled by the new council after a general election. The aldermen may be chosen either from among existing councillors (in which case by-elections are held to replace the councillors) or from among outside persons possessing the necessary qualifications. The practice is to elect persons from outside the council to be aldermen, though defeated candidates are often chosen. Aldermen do not form a special body but sit as ordinary members of the council and possess the same powers as councillors.

The council elects each year a chairman, who is the ceremonial head of the council. He presides at council meetings and represents the council at many important functions both inside and outside County Hall. He may be chosen from inside or outside the council.[1] The usual practice is to appoint a member of the council nominated by the majority party. The council also elects a vice-chairman and a deputy-chairman, the former being nominated by the majority party, while the deputy-chairman is nominated by the opposition party. The vice-chairman and deputy-chairman must be chosen from among members of the council. The chairman of the council, although a party nominee and nearly always a supporter of the party in power, carefully avoids political partisanship during his term of office.[2]

The chairman does not possess any powers of an exceptional nature, but when the council is in recess he or one of his deputies can decide any urgent matter on behalf of the council, provided he acts on the recommendation of the appropriate committee or its chairman.

The London County Council, like some of the other large local

[1] If he is chosen from outside the size of the London County Council is increased from 150 to 151 members.   [2] Morrison, *op cit.*, pp. 38–9.

authorities in Britain, has a leader of the council, and a leader of the opposition. These are political appointments held by the leaders of the majority and minority parties on the council. The functions of a leader are to keep his party together, to preside at party meetings, to keep an eye on the activities of the council as a whole, to see that his party supporters on the various committees are consistent in their attitudes, and (in the case of a leader of the council) to advise chairmen of committees on difficult points.

The London County Council meets once a fortnight in the afternoon. Its meetings usually last only a few hours and are open to the public. A novel feature not commonly found in English local government, consists of questions about the council's work. Any member can, by giving previous notice, ask a question, and it will be answered by the chairman of the appropriate committee. This is an imitation of the celebrated parliamentary question by means of which ministers can be interpellated in the House of Commons about their duties.

The reason why the council meets infrequently[1] and for such short sessions is because the main work of the London County Council is done by committees. This relieves the council of nearly all the detailed work and its business is concerned mainly with discussing, approving or rejecting the reports of committees.

The distinction made in many countries between the legislative and the executive organs of city government does not exist in Britain. The council is the supreme body which passes regulations or by-laws (i.e. legislative ordinances), votes the budget, and determines administrative policy within the powers conferred by law.

COMMITTEES OF THE COUNCIL

There are fifteen standing committees of the council,[2] as follows:

Children's	General Purposes	Restaurants and Catering
Education	Health	Rivers and Drainage
Establishment	Housing	Supplies
Finance	Parks	Town Planning
Fire Brigade	Public Control	Welfare

[1] The London County Council does not meet infrequently in comparison with other county councils, which usually meet once a month or even once every three months. Nor are its meetings unduly short compared with other local authorities, who also rely on the committee system for day to day administrative decisions.

[2] The London Government Act, 1939, which consolidated the legal framework relating to local government within the administrative county of London, authorizes the London County Council (or a borough council) to appoint a committee for any general or special

In addition, *ad hoc* committees are appointed from time to time to deal with matters requiring special attention.[1] There are many sub-committees attached to the committees.

Some committees deal with matters concerning all branches of the council's work. The establishment committee, for example, deals with staff questions, while the supplies committee is responsible for central purchasing of goods and services. Other committees are concerned with particular services provided for the public or with particular functions.

Five of the committees—those dealing with children, housing, welfare, education and health—consist, not only of aldermen and elected councillors, but also of a minority of co-opted persons: that is, members appointed from outside the council. The council has legal power to appoint co-opted members to all committees except the finance committee, provided that at least two-thirds of a committee's members are councillors or aldermen. But not much use has been made of the principle of co-option at County Hall. Salaried officers of the London County Council are never appointed to a committee.

Nearly every member of the council serves on two committees and usually on one or more sub-committees. Each committee is constituted so as to reflect the relative strength of the political parties on the council. The majority party nominates the chairman of each committee, who occupies a position of great importance. He answers for the work of the committee at meetings of the council. He is frequently required to decide matters of detail which will later come before the committee. He can

purpose which the council considers would be better regulated and managed by means of a committee. The council fixes the number of members of each committee, their terms of reference, and their tenure of office. Many other Acts of Parliament dealing with particular services require local authorities in London and elsewhere to appoint committees for those services. For example, the Education Act, 1944, requires the London County Council to appoint an education committee. The establishment of such committees in no way derogates from the authority of the council, with whom responsibility ultimately rests.

[1] The term *ad hoc* committee is not clearly defined. The standing committees meet regularly for the conduct of business, but in addition the council appoints certain special committees for specific purposes which meet only when required, which may be at long intervals. There is, for example, a special committee to hear staff appeals where questions of discipline are involved. Another instance is the committee on the welfare of old people, appointed in 1951. The purpose of this committee was to secure the fullest co-ordination between the standing committees concerned with welfare, health, housing, and restaurants and catering in regard to the general welfare of old people, and similarly to co-ordinate the work of the London County Council and the hospital boards. The committee met regularly until October 1952, but it will not meet again unless its advice is particularly required by a standing committee. Many of the sub-committees and sections set up by standing committees are *ad hoc*.

control in large measure the agenda of its meetings, though in practice this is usually left in the hands of the clerk of the council, who provides each committee with a committee clerk. The efficiency and harmony of a committee's proceedings will depend to a considerable extent on the chairman's tact and ability.

Every committee of the London County Council is advised by the chief officer of the department or departments responsible for conducting the service with which they are concerned. A clear distinction is made between the sphere of a committee on the one hand, and that of the paid officials of the council on the other. Policy is for the former; execution of policy for the latter. Thus, the health committee is advised by the medical officer, who is in charge of the public health department of the council. The education committee is advised by the education officer, the fire brigade committee by the chief officer of the London Fire Brigade, and so forth. In addition to these 'service' aspects, there are 'functional' aspects of the business coming before committees on which advice will be often sought from other departments. Thus, the architect will advise the education committee on the design of school buildings, and the medical officer will advise the restaurants and catering committee on matters of hygiene in regard to food. Each chief officer, with the exception of the clerk to the council and the comptroller, is both the head of a service and also responsible for a function which enters into other services.

The chief officers—and, indeed, the whole official staff—maintain a strictly non-political attitude towards the work of the council. All members of the council, whatever their political party, can rely with confidence on the complete loyalty and obedience of the official staff to the council and its declared policies. Every party knows that if by obtaining a majority it can determine policy on the council and on its committees, it will obtain the genuine co-operation of the staff in implementing that policy. When it is remembered that the council is composed of politicians who are usually not specialists in the matters for which the council is responsible, the importance of maintaining a satisfactory relationship between the elected and bureaucratic elements is obvious.

Members of the council are unpaid, although since 1948 small allowances can be claimed to meet out-of-pocket expenses and loss of earnings incurred in the course of official duties. The chairman of the council receives no salary or expenses allowance, but he is provided with the use of a car and his official entertainments are organized and paid for by the council. The work of a councillor or alderman is very heavy, and may absorb as much as two days a week. The demands made on the time and energy of a committee chairman, the chairman of the council, or the leader of the council are extremely arduous. It is a remarkable fact that so much honest and devoted work should be done without remuneration, and it is a sign of the public spirit of those who do it. For one thing is certain: there is no graft or corruption or nepotism at County Hall to compensate the councillor by illicit means.

The heavy demand for unpaid service made on members of the council has certain disadvantages, for the number of persons who can afford to serve on the London County Council is limited. In consequence there are too many married women, trade union officers, and company directors on the council but insufficient men and women engaged in industry, scientific or university work, or the professions other than law.

THE WORK OF THE LONDON COUNTY COUNCIL

We can now consider the functions performed by the London County Council. The first in order of cost is education. The London County Council is the local education authority for the administrative county, and provides primary, secondary and technical education on a vast scale. The University of London is entirely independent of the council, though the London County Council grants money to the University.

Education in London is a big job. There are about 950 primary schools and some 350 secondary schools employing more than 15,000 teachers. A large programme of expansion in secondary education is in hand. In addition, education is provided for 300,000 older students at nearly 200 polytechnics, technical colleges, evening institutes, and day continuation schools. London affords a system of technical, vocational, and cultural education for day and evening students of all ages unrivalled in range, comprehensiveness and quality.

Fire fighting and fire protection are the responsibility of the London County Council. The London Fire Brigade is a large and highly efficient body whose heroic exploits during the blitz over London in 1940–41 won for it well-earned fame in many lands.[1]

Main drainage is another basic service provided by the London County Council. Owing to the difficulty and expense of constructing and operating a drainage and sewage disposal system many local authorities outside the administrative county have asked the London County Council to provide this service for them. In this way London's main drainage system has

---

[1] The fire brigades were taken over by the central government in 1941 but handed back to local authorities after the war.

25. LONDON

Trafalgar Square. Behind Nelson's Column is the National Gallery. On the right is the church of St. Martin-in-the-Fields.

[*Photo: H. D. Keilor*]

26. LONDON

The Tower of London. On the left is the headquarters building of the Port of London Authority. The launch is a patrol of the river police, a branch of the Metropolitan Police force.

[*Photo: H. D. Keilor*]

27. LONDON

Tower Bridge, with the Tower of London in the foreground. Across the river is the commercial and industrial district of Bermondsey.

[*Photo: Fox Photos*

28. LOS ANGELES

The metropolitan area shown on a relief map. The light portions are areas forming part of the City of Los Angeles. The dark portions comprise unincorporated county areas, and satellite towns or suburbs. The black lines show major highways, either already completed or projected.

[Photo: Los Angeles City Planning Commission

been extended to cover an area of nearly 180 square miles, containing a population of 4·6 millions.[1]

The London County Council is the major housing authority within the administrative county, though there are minor authorities possessing concurrent powers. The London County Council's housing activities are on a vast scale.[2] By the outbreak of the Second World War the council had provided nearly 100,000 dwellings of all types ranging from blocks of working class flats to individual houses with gardens. It had demolished 17,000 slum dwellings.

During the war new building was entirely stopped and London's housing suffered terrible damage from enemy air raids. In consequence, by 1945 the housing problem had become one of extreme urgency and difficulty. To deal with it the London County Council has undertaken ambitious schemes with great energy. From the end of the war in 1945 to April 1952 they had constructed 67,196 new dwellings; 15,220 of them were temporary houses, of which 7,355 have been handed over to metropolitan borough councils for management. Of the rest, 10,847 are rebuildings, reinstatements, or repair of dwellings rendered uninhabitable during the war and 41,129 are new houses and flats.[3] In addition, nearly 80,000 dwellings damaged during the war have been repaired.

More than half of these dwellings are situated outside the administrative county, providing homes for more than 80,000 persons on housing estates like those of Debden, Headstone Lane, Oxhey, St. Paul's Cray and Mastham. Some of these out-county estates are 20 or 25 miles from the centre of London, though many are much nearer. Post-war plans for developing distant estates include efforts to attract industry to the site or its neighbourhood and to develop a sense of community and a local self-sufficiency, as in the 'New Towns' scheme. These achievements by no means exhaust the L.C.C.'s housing programme, for sites have been approved (and many of them acquired) for many more dwellings. In March 1952, its new housing projects totalled 51,000 dwellings, to be

---

[1] W. A. Robson: *The Government and Misgovernment of London* (2nd edition), pp. 231–7. Local sewers are provided by the minor authorities. There are 3,000 miles of such sewers in the main drainage area referred to above. *The Youngest County*, pp. 62–73.

[2] On March 31, 1951, the number of dwellings belonging to the London County Council was 130,723 containing 449,974 rooms accommodating 556,781 persons. Rather less than half of these dwellings (63,454) were situated in the London County Council's own area, and the remainder (67,269) in outer London. *London Housing Statistics*, 1950–51, published by the London County Council, Table 13, pp. 52–3.

[3] If we add these 41,129 new houses and flats constructed between 1945 and 1952 to the 100,000 dwellings completed by 1939, the London County Council has provided accommodation for about 500,000 persons in about 140,000 dwellings.

constructed on 7,375 acres of land, (including some already partly developed) and building had been begun or contracts had been placed for 18,500 of these dwellings.[1]

The London County Council has been compelled to carry out so many of its housing schemes outside its own boundaries owing to the lack of land suitable for new building within the administrative county. At present 13 estates, comprising 5,500 acres of the new sites mentioned above, are located outside the county.

The largest housing estate in the world, at Becontree and Dagenham in Essex, was constructed by the London County Council several miles outside its boundary. This covers 2,775 acres, or more than four square miles of land, on which 25,000 dwellings to accommodate 115,000 persons have been built at a cost of £15 millions. The estate, like many others built or about to be built by the London County Council, is really a town containing churches, schools, shops, public houses, cinemas, clinics, libraries, and other institutions.[2] These out-county housing estates create serious financial and administrative problems not only for the London County Council but also for the local authorities of the area in which they are situated, for the latter have the duty of providing municipal services of all kinds for these immigrants from inner London, whose presence is often resented by the local inhabitants. The London County Council, on the other hand, is compelled to expend much money and effort on housing activities which deplete its own area of population and rateable value. This demonstrates in a vivid manner the need for a regional authority.

The London County Council is responsible not only for housing, but also for town planning. More will be said about this aspect of its work later. It provides parks, open spaces, and recreation grounds. Here, too, the land available for this purpose in the county of London is quite inadequate. The London County Council has undertaken to contribute substantially[3] towards the cost of acquiring a green belt circling London at a distance of about fifteen miles from Charing Cross. This, again, is outside its boundaries; and the acquisition and management of the green belt land is carried out by outlying local authorities.

The London County Council is not generally responsible for the construction, maintenance and improvement of streets; but it is the authority for highway improvements of more than local importance. The council

---

[1] *The Youngest County*, pp. 165–73.

[2] W. A. Robson: *The Government and Misgovernment of London* (2nd edition), pp. 430–8.

[3] Up to a maximum of £2 millions.

has a poor record in this sphere and its achievements are conspicuous by their absence. There has been an almost total absence of any main road development in central London during the past forty years and the traffic problem has become insoluble without large-scale improvements involving huge expenditure. The council provides cross-river bridges outside the City area, and has replaced most of the older structures with fine modern bridges, the most recent improvement being the rebuilding of Waterloo Bridge which was completed in 1945.

Prior to 1948 the council was the largest hospital authority in the world, but its hospitals were then transferred to the national health service. It continues to perform a number of public health functions. Thus, it administers maternity and child welfare centres; it provides a midwifery service for women who are confined at home; it administers day nurseries for young children whose mothers go out to work; it supplies home nursing and home helps for cases of domiciliary sickness; it employs health visitors to advise households how to prevent disease from spreading, and on similar matters. The council has an extensive school medical service. It maintains the London ambulance service, with a fleet of 350 vehicles.

The London County Council is the welfare authority. It cares for deprived or handicapped children. It looks after old folk, blind persons, and cripples in need of help, pregnant women without homes of their own, and others in special need. The London County Council deals with these groups with considerable skill and great humanity. The council is highly conscious of its role as an agent of the welfare state. It maintains more than seventy residential establishments for persons needing care and attention.

The London County Council has for some years been authorized[1] to provide entertainments of various kinds and to finance them out of local taxes. The programme of open-air entertainments presented by the parks department in 1951[2] shows the way in which the London County Council is contributing to the opportunities for enjoyment in London. It includes orchestras, bands, concert parties, ballets, puppet shows, athletic com-

---

[1] The council first obtained powers by a private bill entitled the London County Council (General Powers) Act, 1935. These were extended by a similar measure in 1947. The following year a general statute (the Local Government Act, 1948) conferred powers to provide entertainments on all local authorities, so long as the net deficit does not exceed the product of a sixpenny rate. The Act of 1948 applied in London only to the metropolitan borough councils. The London County Council continues to operate under its own private Acts and is not subject to the rate limitation.

[2] *Open Air Entertainments*. Season 1951. Published by London County Council Parks Department.

petitions, games tournaments, performances for children, open-air theatres, dances, cinema shows, fairs, and other events. These cultural and recreational activities are doing much to enrich and brighten the lives of Londoners. In 1951 the London County Council built the Royal Festival Hall, a magnificent auditorium which has become the chief centre for concerts in the metropolis. The hall was designed by the council's own architect.

These are the main services carried out by the London County Council. It is also responsible for a number of regulatory functions to protect the public against various dangers.

THE STAFF OF THE LONDON COUNTY COUNCIL

The staff of the council numbers over 60,000. It is divided into three main groups: (1) 9,000 administrative and clerical officers; (2) 23,000 professional and technical officers, including the teaching, medical and nursing staff; and (3) 28,500 tradesmen, manual workers, and other operatives.

The following is a classification by departments or functions:

Headquarters administration	11,000
Teachers	17,500
Management and maintenance of housing estates	6,000
Public health work	5,700
Restaurants and catering service	5,000
Fire brigade operating staff	2,500
Parks service	1,500
Main drainage work	1,000
Welfare work	2,400

This accounts for about five-sixths of the total staff.

The administrative and clerical grades are divided into a general clerical class and a major establishment. Recruitment to both classes was formerly by open competitive examinations in academic subjects, but this method is now used only for the major establishment. It was abandoned for the general clerical class a few years ago in order to recruit staff from among the older groups of ex-service men who were not prepared to take a literary competitive examination for low grade posts not specially attractive to them.

Promotion from the general clerical class is mainly by competitive examination to the major establishment, by merit to the higher clerical class, by selection without examination to a small number of major establishment positions, or by selection for training to a wide range of specialized appointments. Many of the highest administrative positions in the service are filled from the senior ranks of the major establishment, in which promotion is by merit rather than seniority. There are fifteen chief officers in charge of departments who receive salaries ranging from £2,200 to £4,700 a year. All appointments are normally of a permanent and pensionable character though they can be terminated by either side at any time.[1] There is no 'spoils' system, and political considerations play a part only to the extent that in appointing chief officers councillors doubtless take account of candidates' known attitudes to the policies they would be required to carry out.

THE METROPOLITAN BOROUGH COUNCILS

So far we have been concerned with the major organ of London government, apart from a brief account of the ancient City. We may turn now to the minor authorities within the county. The lower tier is composed of 28 metropolitan borough councils. Each of these is a separately elected council, having a membership varying from 35 to 70, and possessing a mayor, aldermen, and councillors. The metropolitan borough councils are elected for three years.

The metropolitan boroughs vary greatly in area, population, and rateable value. Holborn has the smallest population (24,806) and Wandsworth the largest (330,328). Holborn occupies only 406 acres of land while Wandsworth extends to 9,107 acres. There are similar disparities of wealth and taxable capacity.

The status and structure of the metropolitan boroughs can only be understood in the light of London's political history. From the moment when the London County Council was first set up in 1889 it transformed the atmosphere of London government and attracted to its ranks many of the most able and vigorous reformist politicians of the day. Men like Sidney Webb, Will Crooks, and Ramsay MacDonald became eager to play an active part in London government. After generations of frustration and restraint, a great accumulation of pent-up energy and aspiration was released. The London County Council was from its birth a 'progressive' or radical body because municipal life in the metropolis had been starved and frustrated for decades and there was an immense amount of creative work which urgently needed doing.

---

[1] Cf. *A career in the service of London*; and *Handbook for New Entrants* published by the London County Council for further details.

The drive, vigour and radical aspirations displayed at County Hall filled Lord Salisbury, the Conservative prime minister, with deep distrust and hostility.[1] He was, moreover, incapable of conceiving the metropolis as a whole. To him it was a mere agglomeration of small and unrelated communities which happened to have been placed in juxtaposition with one another almost by accident.

These influences inspired the London Government Act of 1899 which set up the metropolitan boroughs.[2] This measure magnified the importance and the independence of the minor authorities in every possible way. It gave each metropolitan borough council a mayor and aldermen with robes of office, gilt chains, a mace, and all the other insignia likely to enhance the feeling of their separate civic consciousness. Far from attempting to bring them into an organic partnership with the London County Council, the object was rather to engender jealousy and friction, and to foster the parochial spirit.

The plan succeeded only too well, and conflict and dissension between the London County Council and the metropolitan borough councils has often detracted from the good government of London during the past fifty years. Mr. Herbert Morrison, who served for many years both as a member of the Hackney Borough Council and as leader of the London County Council, remarks that owing to friction 'difficulties were not only created in London government, but in Parliament itself on local government questions affecting London. Until recent times, the metropolitan borough councils tended to have more influence or frightening-powers with London members of Parliament than the London County Council, and this, together with Parliament's fear or jealousy of the Council, used to lead in most cases to County Hall getting the worst of the battle.'[3] He claims that relations between the major and minor authorities improved after the Labour Party gained control of the London County Council in 1934. This may be due to the fact that the Labour Party also secured control of a majority of metropolitan borough councils.[4]

The main functions of the metropolitan borough councils are the cleansing, lighting, paving and improvement of streets; the collection and disposal of refuse; the provision of public libraries and museums; public

[1] Lord Salisbury served as prime minister in three governments which were in power 1885–86, 1886–92, and 1895–1902. See also W. A. Robson: *The Government and Misgovernment of London* (2nd edition), pp. 84–92.

[2] The legal basis of the metropolitan borough councils is now contained in the London Government Act, 1939.

[3] *How London is Governed* (2nd edition), p. 90. County Hall is the headquarters of the London County Council. [4] See below pages 290–1.

baths, swimming pools, gymnasia, and washhouses; burial grounds, crematoria, and mortuaries; the registration of births, deaths and marriages; sanitary supervision of factories and shops; food and drugs inspection; slum clearance and the provision of housing;[1] small parks and public gardens; local drainage; the regulation of offensive trades, and various other matters.

The efficiency and extent to which all these powers are exercised by each of the metropolitan borough councils naturally varies according to its resources and ability.

A body known as the Metropolitan Boroughs Standing Joint Committee represents the common interests of all the metropolitan borough councils. It negotiates with government departments and the London County Council, and acts as a powerful defence organ for the metropolitan boroughs.

It also advises and assists the constituent councils in the administration of their services. In pursuance of this aim, it has secured the appointment of a metropolitan boroughs library committee, to co-ordinate the work of the public libraries run by member councils and to facilitate co-operation between them; and this has greatly improved the library service. A more ambitious example is the formation of a committee to promote improved organization and methods in the work of the metropolitan borough councils. This committee undertakes reviews of matters of general interest to all the constituent councils, and conducts investigations of particular problems affecting an individual council, in order to advise on methods of securing maximum efficiency in the executive machinery.

LOCAL AUTHORITIES IN OUTER LONDON

Outer London is divided for local government purposes among a mass of local authorities of all kinds, none of them in any way related to the special needs of the metropolis. Apart from the *ad hoc* bodies to be described later, London government virtually ends at the boundary of the county of London, although most of the population of the metropolis lives beyond it. Outside that boundary we find the ordinary system of local government which obtains throughout England and Wales.[2]

The county of London is surrounded by the 'Home Counties.' These comprise Middlesex, Essex, Hertfordshire, Kent, Surrey, Buckingham-

---

[1] The metropolitan borough councils possess concurrent powers with the London County Council in regard to housing and slum clearance. [2] See *ante* pages 266–8.

shire and Berkshire. Each of them has a county council and a number of county district councils.

Let us consider so-called Outer London (i.e. the Metropolitan Police District), omitting the administrative county of London which we have already examined in detail. The boundary of Greater London cuts right across the counties of Kent, Essex, Surrey and Hertfordshire: only Middlesex is entirely inside it. Within the broad band of territory lying between the boundaries of the administrative county of London and Greater London, local government is in the hands of 3 county borough councils,[1] 36 non-county borough councils, 26 urban district councils, 3 rural district councils, 3 parish councils, and the county councils mentioned above. In addition there are a number of joint boards and committees formed between neighbouring local authorities for particular purposes, such as cemeteries, drainage and sewerage, and fire brigades.[2] This version of Greater London cannot be regarded as comprising the whole of the metropolis, and is taken only for illustrative purposes.

There are thus at least seven types of local authority in outer London, possessing different powers, and displaying an immense diversity of size, population and wealth. There is no regional organ responsible for guiding, co-ordinating, planning, controlling or administering Greater London. The 8 or 9 million people living in the metropolitan region have no representative organ responsible for the good government and development of the region as a whole. Chaos, conflict, and confusion are bound to exist in these circumstances.

AD HOC AUTHORITIES

A number of functions have been entirely removed from the jurisdiction of the general local authorities in London and entrusted to *ad hoc* bodies.

An early example of this was the establishment in 1829 of a unified metropolitan police force within the Metropolitan Police District. Ten years later the force was placed under the control of a commissioner and an assistant commissioner, acting under the direction of the Home Secretary. This system of central government control has continued with minor changes to the present day.

[1] Croydon, East Ham and West Ham.

[2] For details see: *Statistical Abstract for London*, 1939–48 (published by the London County Council); Part vi, Statistics relating to local areas in Greater London, pp. 80–1; Herbert Morrison: *op cit.*, pp. 96–105; W. A. Robson: *The Government and Misgovernment of London* (2nd edition), pp. 370–80.

Elsewhere in Britain the police forces are under the control of the county borough councils, and, in the counties, of standing joint committees composed of equal numbers of county councillors and justices of the peace for the county. In the City of London, the ancient corporation maintains its own police force headed by a commissioner appointed by the Crown on the recommendation of the Home Secretary. The placing of the metropolitan police under central government control was due to the absence of any organized system of government in the large area rightly considered necessary for police administration even a century ago, rather than to a desire on the part of the central government to have the preservation of law and order within the capital in its own hands.[1]

The water supply of London was in the hands of a number of commercial companies until the beginning of the twentieth century. Great efforts were made to persuade Parliament to empower the London County Council to take over the companies' functions and undertakings. It was decided, however, to organize the water supply on a larger scale than the county, and this led to the setting up of the Metropolitan Water Board,[2] an indirectly elected body composed of sixty-six representatives chosen (in varying proportions) by the London County Council, the county councils of the Home Counties, the City corporation, the metropolitan borough councils, and other local authorities in outer London.

The port of London and the river Thames are also administered by two special bodies. The Port of London Authority is a statutory public corporation created in 1908 by Lloyd George, then President of the Board of Trade. Previously the London docks were owned and operated by several private companies, which had exhausted both their resources and their credit, with detrimental effects on the port facilities.

The Port of London Authority is governed by a board of 28 members serving for three years, of whom 17 are elected by the immediate users of the port, such as wharfingers, payers of rates, and river craft owners. The other members are appointed by the Admiralty, Minister of Transport, the London County Council, and other public authorities. The authority controls the tidal portion of the river and its estuary, a distance of 69 miles. It owns and operates the largest system of docks in the world, equipped with an enormous mass of wharves, warehouses, and cold

---

[1] Sir John Moylan: *Scotland Yard*, p. 63-5.

[2] Created by the Metropolitan Water Act, 1902. See W. A. Robson: *The Government and Misgovernment of London* (2nd edition), pp. 101-20. For the size of the area served by the Metropolitan Water Board, see *ante*, p. 262.

storage facilities.[1] It is responsible for dredging the river bed, improving and constructing whatever facilities are required, regulating navigation, and licensing river craft.

The upper reaches of the river come under the care of the Thames Conservancy, which represents government departments, local authorities, the Metropolitan Water Board and the Port of London Authority.

Until 1933 public transport in London was owned and operated by commercial companies except for the tramways, which were mostly owned by local authorities. The London Passenger Transport Act, 1933, set up a public corporation known as the London Passenger Transport Board and transferred to it the entire mass of transport undertakings in the London region, except the main line railways.[2] This board was amalgamated with the nationalized transport system in 1947.

Gas and electricity are also supplied by public corporations administering nationalized industries. Electricity is distributed in the metropolis by the London Electricity Board, a body appointed by the Minister of Fuel and Power. It operates entirely within Greater London, though its area of supply is not coterminous with it. Three other area electricity boards supply small portions of Greater London. Prior to nationalization in 1947, electrical energy was distributed by 16 metropolitan borough councils and 14 joint stock companies. Co-ordination was supposed to be carried out by the London and Home Counties Joint Electricity Authority, but this feeble body was never effective.[3]

London's gas supply was entirely in the hands of commercial companies prior to 1948, when the industry was nationalized. It is now entrusted to two public corporations operating on the north and south banks of the Thames respectively.

RELATIONS OF LOCAL AND CENTRAL GOVERNMENT

We have seen that the metropolitan police is under the control of the Home Secretary, and that ministers appoint some or all the members

---

[1] Lincoln Gordon: 'The Port of London Authority' in *Public Enterprise*, edited by W. A. Robson; and see also Lincoln Gordon: *The Public Corporation in Great Britain*; W. A. Robson: *The Government and Misgovernment of London* (2nd edition), pp. 131-8. A substantial amount of wharf and warehouse accommodation is in the hands of commercial companies.

[2] See Ernest Davies: 'The London Passenger Transport Board' in *Public Enterprise*, edited by W. A. Robson; W. A. Robson: *The Government and Misgovernment of London* (2nd edition), pp. 150-3; Herbert Morrison: *How London is Governed*, pp. 128-30; *Socialisation and Transport, passim*.

[3] W. A. Robson: *Problems of Nationalized Industry* for an account of the nationalized industries; *The Government and Misgovernment of London* (2nd edition), pp. 238-42.

of the *ad hoc* bodies which are responsible for the port of London, the Thames conservancy, the London passenger transport system, gas and electricity supply. Apart from these special bodies, the relations between local and central government are almost the same in London as elsewhere in Britain. The central government does not possess any larger degree of control over local government in the capital than it does over that of a provincial town. There is in London no officer corresponding to the Prefect of the Seine in Paris, no centrally-appointed governor with power to override the elected councils.

The relations between central and local government are too complicated to be described here in detail.[1]

There is not one department of the central government which is exclusively concerned with local government, but several. The Ministry of Education is the central authority for education; the Home Office for police, fire brigades, the care of deprived children, the regulation of shops and other local government functions; the Ministry of Transport for highways, bridges, transport and traffic regulation; the Ministry of Health for matters concerning the national health service; the Ministry of Housing and Local Government for housing, parks and open spaces, town and country planning, drainage and sewerage, and many other matters. This last-mentioned department is the department which possesses general responsibility for local government.

There is not a single, uniform relationship existing between these departments on the one hand and local authorities on the other. Each department had evolved its own characteristic pattern of relations more or less suited to the services with which it is concerned. Nevertheless, certain common features can be detected in the patterns of central and local relations.

Local authorities are required to formulate long-term plans for developing important local government services, and to submit them to the appropriate central department. Thus, the London County Council has had to submit to the appropriate department a school development plan; a town plan for the county of London; programmes of development for highways, health, welfare and other services. The minister concerned can approve, amend, reject, or refer back for further consideration a development plan or scheme submitted to him. When it is approved it becomes binding on the local authority until such time as it is amended by a similar procedure.

Grants-in-aid figure largely in the relations between central and local government. Subventions from national funds form a substantial propor-

[1] See D. N. Chester: *Central and Local Government, passim.*

tion of local authorities' revenue. The central government has, through these subventions, acquired in varying degree the power to inspect, audit, influence, and supervise the activities of local authorities.

Grants-in-aid are of several kinds. Those for education, health, fire brigades, and town planning are based on a percentage of approved local expenditure on each of the services, or on a percentage of prescribed items of expenditure. Housing grants are based on the number and type of housing units provided by the local authority. An Exchequer equalization grant assists areas where the rateable value per head is lower than the average for the country as a whole, but as the county of London does not receive anything from this source, it need not be described here.

There are several other methods of central supervision or influence. Ministers are often empowered by statute to make regulations affecting local government. They must approve proposals for capital expenditure which involve long term loans. They often act as administrative tribunals, or appoint such tribunals, to determine appeals in matters affecting local authorities. Sometimes a minister can act in default of a recalcitrant local authority which has failed to carry out its duties.

Let us glance at the relations which subsist in a major service such as education. The London County Council must obtain approval of the Ministry of Education for its school development plan. Teachers' salaries are determined nationally by a joint negotiating council, which fixes what are known as the Burnham scales; and when the Minister makes an order the London County Council (like all other local education authorities) is obliged to pay salaries in accordance with this scale. The Minister of Education issues regulations determining the standards of building and accommodation to be provided in new schools, and the London County Council must comply with them. The grant regulations of the ministry lay down the conditions to be observed by the London County Council regarding such matters as staffing, the size of classes, and school meals, in order to qualify for its education grant. The ministry appoints inspectors of schools, who occasionally inspect the council's schools, give advice, obtain information, and see that the ministry's requirements are fulfilled. On the other hand, the scope and content of education is entirely free except that the ministry specifies certain subjects which must be taught. The curriculum is determined by the school in conjunction with the London County Council. The central department has no voice either in appointing, promoting or dismissing teachers, choosing the school books, or determining the methods of instruction.

Central control over local government has been increasing in Britain in recent years, and this tendency is still at work. It nevertheless remains true to say that the local authorities continue to enjoy a substantial measure of freedom and independence, though less than they formerly did. Their budget is not subject to approval; nor can the council be suspended or dissolved by the central government. In London the elected local authorities possess far more autonomy and responsibility than in most capital cities.

FINANCE

The following table summarizes for the year 1951–52 the gross expenditure of the principal local authorities in London and shows the main sources of their income. It also gives the outstanding indebtedness.

TABLE I

Gross Expenditure of Principal Local Authorities, 1951–52

Authorities	Total Gross Expenditure	Total Income excluding Rates		Net Expenditure met from Rates and Balances	Net Debt at March 31, 1952
		Government grants	Receipts in aid		
	£	£	£	£	£
London County Council	59,369,584	17,651,284	13,421,757	28,296,543	157,380,446
Metropolitan Borough Councils	32,114,983	4,899,286	10,835,815	16,379,882	71,682,791
City Corporation ..	3,369,659	331,576	1,359,241	1,678,842	7,389,438
Metropolitan Police (rateable proportion)	9,639,146	4,289,504	727,679	4,621,963	865,068
Metropolitan Water Board (rateable proportion)	6,015,965	—	5,901,977	113,988	35,238,143
	110,509,337	27,171,650	32,246,469	51,091,218	272,555,886
Less expenditure included twice	825,634	—	825,634	—	—
	109,683,703	27,171,650	31,420,835	51,091,218	272,555,886
Percentages	100%	24·8%	28·6%	46·6%	—
Per head of population..	£32 13 3	£8 1 10	£9 7 2	£15 4 3	£81 3 4
Per £ of rateable value	£1 18 8	9 7	11 1	18 0	£4 16 1

We can see from this that metropolitan borough councils spend in the aggregate a little over half as much as the London County Council. Government grants provided about a quarter of the total gross expenditure, and local taxation yielded nearly half of the money spent. The outstanding debt was about two and a half times the annual expenditure out of revenue.

We may now examine in somewhat greater detail the finances of the London County Council. The total expenditure on current account of

TABLE II

Expenditure of London County Council, 1952–53

Service	£
Children's services (including approved schools and remand home) ..	2,469,867
Civic restaurants	371,543
Civil Defence measures	204,815
Education	29,843,403
Fire service	2,263,965
Housing	13,713,926
Judicial expenses	261,195
Local health services, etc.	4,911,400
Main drainage (including Thames flood prevention and contribution to Lee Conservancy Catchment Board)	1,733,825
Means of communication—bridges, ferry, tunnels, etc.	895,963
Parks and open spaces	1,539,685
Public control services	169,292
Royal Festival Hall	192,463
Town planning	857,644
War Damage Act, Part 1.—cost of repairs, etc.	503,771
Welfare services	2,604,349
Sundry services	487,473
Other items not included in services—	
Administration	1,568,603
Finance Committee—	
Surplus on Consolidated Loans Fund (Income Account)	*249,209*
Acceleration of redemption of debt	327,264
Capital expenditure met directly from rate	172,736
Contribution in lieu of rates on nationalized undertakings and government property	—
Miscellaneous	*176,898*
Local taxation licence duties	16,767
	£64,683,842

Figures printed in Table II in italics represent a surplus.

the London County Council (for the year ending March 31, 1953) was £65 millions. Table II shows the distribution of this among the principal services.

The figures given above include debt charges on outstanding capital

TABLE III

Sources of Income of London County Council, 1952–53

INCOME

Service (1)	Miscellaneous (2) £	Exchequer grants in aid of services (3) £	Total (4) £	Net Expenditure falling on rate (5) £
Children's services (including approved schools and remand home)	254,261	1,063,639	1,317,900	1,151,967
Civic restaurants	371,543	—	371,543	—
Civil Defence measures	9,978	227,731	237,709	*32,894*
Education	2,399,648	12,267,539	14,667,187	15,176,216
Fire service	236,028	501,159	737,187	1,526,778
Housing	9,004,642	2,079,728	11,084,370	2,629,556
Judicial expenses	11,888	162	12,050	249,145
Local health services, etc.	402,248	2,194,376	2,596,624	2,314,776
Main drainage (including Thames flood prevention and contribution to Lee Conservancy Catchment Board) ..	306,866	8,892	315,758	1,418,067
Means of communication—bridges, ferry, tunnels, etc.	104,799	—	104,799	791,164
Parks and open spaces	201,429	—	201,429	1,338,256
Public control services	45,065	—	45,065	124,227
Royal Festival Hall	158,353	—	158,353	34,110
Town planning	175,307	62	175,369	682,275
War Damage Act, Part I—cost of repairs, etc.	503,771	—	503,771	—
Welfare services	1,047,324	29,837	1,077,161	1,527,188
Sundry services	395,239	31,202	426,441	61,032
Other items not included in services—Administration	41,070	—	41,070	1,527,533
Finance Committee—Surplus on Consolidated Loans Fund (Income Account)	—	—	—	*249,209*
Acceleration of redemption of debt ..	—	—	—	327,264
Capital expenditure met directly from rate	—	—	—	172,736
Contribution in lieu of rates on nationalized undertakings and government property	1,690,267	—	1,690,267	*1,690,267*
Miscellaneous	341,083	—	341,083	*517,981*
Local taxation licence duties	—	64,081	64,081	*47,314*
	17,700,809	18,468,408	36,169,217	28,514,625
Less Reduction in balances during year				2,452,090
Amount raised by County Contributions				26,062,535

Figures printed in Table III in italics represent a surplus.

borrowings, but they do not include expenditure on capital account, except in so far as this is financed out of revenue.

Table III shows the sources from which the London County Council received revenue required to meet the expenditure in Table II.

'Miscellaneous' receipts shown in Column 2 of Table III consists of money received from charges, fees, repayments and contributions of many different kinds for services rendered or facilities provided by the council. The parks department, for example, receives money by way of rent for allotments[1]; charges for games, licences for chairs, boating, refreshments, etc.

The total expenditure to be met by the London County Council after deducting these miscellaneous receipts was £47 millions. Of this, nearly £18½ millions was received by way of Exchequer grants from the central government, leaving about £28½ millions to be provided from local taxes. (The sum actually raised was just over £26 millions.)

The finance of the London County Council's very extensive housing operations can be better understood from the following figures:

TABLE IV

London County Council Housing Finance—1951-52 and 1952-53

	1951-52 £	1952-53 £
Expenditure	11,994,901	13,713,926
Income other than subsidies	7,943,158	9,004,642
Exchequer subsidies	1,680,196	2,079,728
Net deficiency	2,371,547	2,629,556
Made up by:—		
(a) Statutory rate contribution	832,750	971,686
(b) Additional rate contribution	1,538,797	1,657,870

This shows that the proportion of the council's housing expenditure falling on local rates is 15–20 per cent of the total cost. It is greater than the proportion received from grants-in-aid from the Exchequer. The 'statutory rate contribution' is a compulsory subsidy from rates which must be made in order to qualify for the Exchequer subsidy. The 'additional rate contribution' is a supplementary subsidy from local taxation to meet any deficiency remaining after rent and the other two subsidies have been deducted from expenditure. The item shown as 'income other than subsidies' represents in the main rents and rates, electricity, gas and water charges recovered from tenants.

---

[1] Allotments are small plots of land let out to amateur gardeners for growing vegetables.

London

County boundary
Greater London Plan
Area of London
Electricity Board
Metropolitan
Water Board
Metropolitan
Police District
Built up areas:-
I. County of London
II. Adjoining Counties

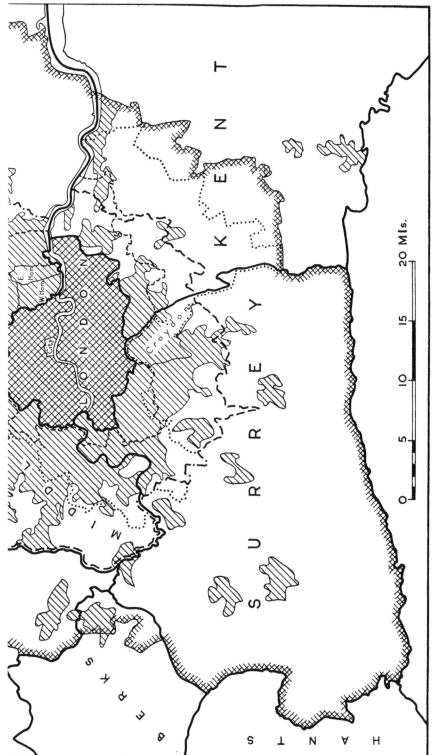

THE GREATER LONDON REGION. The hatched area shows the spread of urban population beyond the area of the County of London into the adjoining local government areas.

In London, as elsewhere in Britain, local authorities have at their disposal only one major form of local taxation known as rates. These are taxes levied on the rateable value of real property, such as land and buildings of all kinds. The rate is levied at so much in the £ of annual value. It falls on the occupier; and unoccupied land or buildings pay no rates. Industrial premises are liable to only one-quarter of their rateable value; and agricultural land pays no rates at all.

The London County Council is not a rating authority. In 1948 valuation for rating was transferred to the Inland Revenue, a central government authority, and the rates are levied and collected in the county of London by the metropolitan borough councils and the City Corporation. The London County Council demands from each metropolitan borough council and the City Corporation the rate required to cover expenditure on general and special county purposes. The latter bodies then add this county rate to their own rate demands and levy a consolidated rate on the local taxpayer.[1] In 1952–53 the London County Council raised a rate of 9s. 2d. in the £.

There are great inequalities of wealth among the London boroughs. The rateable value per head of the population varies from £113 in Westminster—and an enormously high figure in the City—to only £8 a head in Poplar, Deptford and Bethnal Green.

The Local Government Act, 1948, introduced a new grant known as the Equalization Grant. This provides that a sum is paid by the central government to each county and county borough council whose area has a rateable value per head below the national average. The London County Council receives no equalization grant.

The 1948 Act introduced, however, within the administrative county of London an equalization scheme designed on similar lines whereby the wealthier boroughs contribute towards the *local* expenses (as distinct from expenditure on county and police purposes) of the poorer boroughs. This scheme has considerably alleviated the disparities of taxable wealth between the metropolitan boroughs. In 1947–48 the rates levied in London boroughs ranged between 11s. and 23s. in the £ (a difference of 12s.); in 1952–53 they ranged between 15s. 0d. and 22s. 8d. (a difference of only 7s. 8d.).[2] The *average* rate levied in 1952–53 in the county of London

---

[1] The rates for special county and metropolitan police purposes are not chargeable in the City. These amounted to nearly 2s. in the £. The reason is that the services for which the rate is raised are not provided in the City of London.

[2] For details see *Rates Made* 1952–53 (published by the London County Council) pp. 2–5. Cf. H. Morrison: *op cit.*, pp. 144–5.

(excluding the City) was 17s. 3·38d. in the £.[1] Nearly two-thirds of the rates on the average were required for the expenditure of the London County Council and the metropolitan police.

The rateable value of the administrative county of London was (in April 1952) £57½ millions, which works out at £17·1 a head of the population.[2] The rateable value for outer London was £48½ millions, making £106 millions for Greater London as a whole. Since 1946 the average rate in the £ has been higher in outer London than in the administrative county. In 1952–53 the figure for outer London was 18s. 11·76d. in the £, which is about 1s. 10d. in the £ more than the average rate for the county. Prior to 1948 there was much inconsistency —some of it intentional—between the valuation of property for rating in different areas. With the transfer of this function to the Inland Revenue a much higher degree of consistency will be achieved.

The capital expenditure of the London County Council is often large. In the year ended March 1953 it was about £29 millions and it seems probable that the council's capital expenditure will continue at a high rate for some years to come. The aggregate net outstanding debt is about £178 millions.[3]

POLITICS IN LONDON GOVERNMENT

Party politics have for long played a prominent role in London government. This applies to the London County Council, the metropolitan borough councils and the out-county authorities. The City Corporation claims that party politics exert no influence on its proceedings, but the Guildhall has long been the stronghold of Conservatism in every sense of the word.

The two principal parties in London are the Labour Party and the Conservative Party. At one time the Liberal Party was important but it has now declined to a negligible power. The Progressive Party, an alliance of Liberals and Socialists, was in control of the London County Council from 1889 to 1907. The Labour Party has had a majority on the London County Council continuously since 1934.[4]

The Labour Party has also been in power on a majority of the metropolitan borough councils for some years. After the election of November

---

[1] *Rates Made 1952–53*, p. 12. [2] *Ibid.*, p. 10. [3] *Ibid.*, p. xxviii.
[4] See Brian Barker: *Labour in London. A Study in Municipal Achievement, passim,* for an account of the rise of the London Labour Party to power and what it has done in local government.

1945, 23 borough councils had a Labour majority, and 5 a Conservative majority. The election of May 1953 reduced the number of Labour councils to 19 and increased the Conservative councils to 9.

Both the main parties are well organized in London. There is a high degree of party organization at County Hall and at most of the town halls in the metropolitan boroughs.[1] We have already seen that the committees of the London County Council are appointed so as to reflect the relative strengths of the parties on the council, and the chairmen and vice-chairmen of committees are drawn from the majority party. The institution of a leader of the majority party and a leader of the opposing party has done much to increase the concentration of power, leadership, and responsibility exercised by the dominant party at County Hall.

The political parties are closely organized at County Hall. Each party holds regular private meetings of its council members and the party's attitude on all important questions is decided at these meetings. The members of each party on a committee usually meet regularly in private to determine the party attitude on matters coming before the committee.

It must not be thought that the differences between the political parties are irreconcilable, or that they extend to all matters with which the council is concerned. There are many matters on which there is no cleavage between the parties and where the members of the rival parties can co-operate.

The main drive of the Labour Party in London has been for more and better social services, particularly housing, education, health, and welfare. The Conservative Party has based its appeal to the electorate chiefly on criticism of Labour's shortcomings, the promise of a more effective housing programme, and wiser spending rather than the hope of reduced taxation. The Labour Party on the London County Council, like all parties which have been in power for a long period, has exhausted some of the enthusiasm and reforming spirit which it showed when it gained supremacy under the leadership of Mr. Herbert Morrison.

In general the existence of vigorous and well-organized political parties has done much to clarify the issues of London government, to stimulate public interest in the activities of the council, and to foster coherent policies and efficient administration. Neither party has attempted to introduce the 'spoils system,' or to profit by corruption or nepotism.[2]

From time to time the public interest has been disregarded in irrelevant bickering between the parties. An example of this was when a badly

---

[1] Morrison: *op cit.*, pp. 61–8.
[2] W. A. Robson: *The Government and Misgovernment of London* (2nd edition), pp. 344–51.

needed replacement of Waterloo Bridge, after a century of use, was delayed for many years by being made a burning political issue in Parliament as well as in the council, though there was no good reason why the parties should have so regarded it.[1]

A recent tendency is the increasing influence exerted by national politics on municipal elections in the metropolis. It is too early to say whether this tendency, which was manifest in the London County Council election of 1949, is likely to develop. It is undoubtedly detrimental to the best interests of local government. Local government elections should be fought on local government issues, not on national or international questions.

THE PLANNING OF LONDON

London has grown and developed for more than a century in a sporadic and uncontrolled manner. The splendid squares of Bloomsbury, Mayfair, Westminster and Belgravia, the noble parks which are the finest feature of central London, the secluded beauty and convenience of the Inns of Court, were early examples of civic design which might well have evolved into comprehensive town planning. Unfortunately the nineteenth century belief in *laissez faire*, combined with the degradation in taste of the Victorian era, destroyed any hope of fulfilment of this early promise. The destructive hand of the nineteenth and twentieth centuries fell like a blight on the metropolis. Nothing comparable to Nash's great scheme, which produced a unified and beautiful design for the whole sweep of Carlton House Terrace, Waterloo Place, Regent Street, Portland Place, Park Crescent and the terraces in Regent's Park, was ever again even conceived; and when Regent Street was rebuilt in the 1920's and 1930's, its unity and harmony disappeared and were replaced by a multitude of separate, undistinguished buildings.

As London spread, all the usual consequences of unplanned growth appeared. Urban sprawl and ribbon development became a common sight. The lovely countryside was submerged beneath a sea of suburban villas, unsightly bungalows, cafés, filling stations, arterial roads, cinemas, road houses, housing estates, shops and factories, There was no attempt to control industrial location or sporadic building. Improved transport and traffic facilities induced people to live farther and farther away from their work. In consequence, large numbers of citizens reside in one local government area and work in another. Their interest is divided and their civic allegiance is diminished.

[1] For details see D. N. Chester: *Central and Local Government*, pp. 150–1.

These deplorable results were due partly to the multitude of small and ineffective planning authorities which littered the metropolis—there were 143 in the Greater London region—and the absence of any regional planning organ except a completely futile advisory committee. They were also due to the exorbitant cost of compensation which local authorities were obliged to pay landowners whose rights were in any way restricted by planning schemes. Other contributory factors were the lack of adequate powers to control the location of industry, and the heavy migration of industry and population to London during the inter-war period.

In recent years the situation has entirely changed. The Barlow Commission[1] reported in 1940 that the continued drift of the industrial population to London and the Home Counties constituted a social, economic and strategic problem requiring immediate attention. They advocated control over the location of industry as the best method of restraint. Since 1945, increasingly drastic powers of regulating the location of factories and other industrial establishments have been acquired and exercised by the central government and by the local planning authorities.[2]

The planning outlook in Britain was transformed during the war of 1939–45. Under the stress of violent and massive destruction by enemy air raids, the planning of London came to be regarded as desirable, practicable, and even inevitable. Important Acts of Parliament were passed establishing new central and local planning authorities; very full powers of planning, control and development were given to them; and steps were taken to overcome the financial obstacle to planning arising from compensation obligations.

Three great plans have been prepared for the metropolitan region. There is the Abercrombie-Forshaw plan commissioned by the London County Council for the County of London; the Holford-Holden plan for the ancient City; and the Abercrombie plan for Greater London. The last-named plan covers an area of approximately 2,600 square miles, extending to a distance of about 30 miles in every direction from Charing Cross. A population of more than 10 millions is covered by these plans.

The plans envisage vast improvements in the capital city. The fundamental aim is to stop the haphazard growth of London, to effect a substantial measure of decentralization within the region, and to introduce

[1] Report of the Royal Commission on the Distribution of the Industrial Population, Cmd. 613/1940, par. 426. W. A. Robson: *The Government and Misgovernment of London* (2nd edition), pp. 439–55.

[2] The Distribution of Industry Acts, 1945 and 1950; Town and Country Planning Act, 1947. See also White Paper on Employment Policy, Cmd. 6527/1944, Ch. III, par. 26.

controlled development of housing, industry and communications. The county and Greater London plans together call for the decentralization of a million persons. About three-quarters of them will be rehoused in or near the region by means of additions to existing small towns and quasi-satellites. The remaining quarter will be dispersed outside the region. A striking proposal, now being carried out, is the creation of eight new 'garden city' towns outside the green belt. They are designed to absorb ultimately an aggregate population of 312,500, drawn so far as possible from the more congested portions of inner London.

The London plans[1] provide for the control and decentralization of industry; for a vast improvement in the highway system; for the electrification of all railway transport; for the protection of the countryside; for amenities such as parkways and footpaths, bridle and bicycle tracks, rest gardens, children's playgrounds, recreation and sports centres, urban squares, parks, playing fields, open spaces and riverside walks. They contain extensive proposals for civil airfields, inland waterways and markets; for redeveloping congested areas; for shopping and amusement centres. The administrative, economic and social problems involved in carrying out these plans are of great magnitude.

THE UNSOLVED PROBLEM OF LONDON GOVERNMENT

Satisfactory machinery to carry out these great schemes does not exist. The local planning authorities in Greater London consist of the London County Council, the three county borough councils of Croydon, East Ham and West Ham, and eight county councils whose areas are covered wholly or partly by the Greater London plan.

There are thus a dozen planning authorities with diverse and often conflicting interests, instead of a single organ responsible for the planning of the metropolitan region as a whole. In order to secure greater unity and co-ordination a standing committee of officials representing government departments interested in the London plan has been in existence for some time; but a centralized approach of this kind has many drawbacks and is no substitute for a proper regional solution. .

The problem of achieving unity of conception and consistency of execution in the sphere of regional planning and development is only one of several similar problems which have arisen in connection with other major services. The tendency so far has always been to deal with each matter on a piecemeal basis rather than face the difficult problem

---

[1] For further details see 'The Greater London Plan' by W. A. Robson in *Planning and Reconstruction 1946*, edited by F. J. Osborn, pp. 152–7.

of reforming London government as a whole. Hence *ad hoc* authorities have been created to administer the police, water supply, transport, hospitals, electricity, gas supply and the port, while the structure of London government has continued unchanged for more than fifty years. This is bad for London, and opposed to the interests of democracy, since these one-purpose bodies are never as democratic as a directly elected council with general responsibilities over a broad field.

A thorough-going reform of local government in London is long overdue. London is unquestionably a region, and needs an elected regional council in place of the London County Council, whose area is entirely obsolete. The lower tier of authorities also needs drastic reform throughout the region. If this were done, special bodies like the Metropolitan Water Board and the Commissioner for Metropolitan Police could be abolished and their functions transferred to the regional council.

Local self-government in London is on the whole honest, efficient and democratic. Considering the obsolete framework within which it is required to operate at present, its virtues far exceed its defects. Londoners have much to be proud of in their schools and remarkable facilities for technical education; their water supply and main drainage; their public housing achievements; their fire brigade and police forces; their local transport system; their public health and welfare services; above all, in the bold and inspiring vision of the plans which have been formulated for Greater London. If Parliament and the central government can be induced to modernize and rationalize the machinery of London government, the people of London may be relied upon to take full advantage of the opportunities for further achievement which would arise from a reformed constitution for the capital.

Los Angeles

WINSTON W. CROUCH
DEAN E. McHENRY

K*

WINSTON W. CROUCH

Born 1907. A.B. (Pomona College), 1929; M.A. (Claremont Colleges), 1930; Ph.D. (University of California, Berkeley), 1933.

Professor of Political Science and Director, Bureau of Governmental Research, University of California, Los Angeles; Commissioner, Los Angeles County Civil Service Commission.

Author of: *State Aid to Local Government in California* (1938); *The Initiative and the Referendum in California* (1938) (with V. O. Key, Jr.); *California Government—Politics and Administration* (1945) (with D. E. McHenry); 'Local Government Under Labour' (*Journal of Politics*, May 1950): 'Metropolitan Decentralization: Britain's New Town Program' (*The Western Political Quarterly*, June 1950); 'Administrative Supervision of Local Government: The Canadian Experience' (*The American Political Science Review*, June 1949).

DEAN E. MCHENRY

Born 1910. B.A. (University of California, Los Angeles), 1932; M.A. (Stanford University), 1933; Ph.D. (University of California, Berkeley), 1936; study and research in Europe, 1935–36; Carnegie Fellow in New Zealand and Australia, 1946–47; Fulbright Lecturer, University of Western Australia, 1954.

Professor of Political Science, University of California, Los Angeles.

Author of: *The Labour Party in Transition, 1931–38* (1938); *His Majesty's Opposition: Structure and Problems of the British Labour Party, 1931–38* (1940); *A New Legislature for Modern California* (1940); *California Government: Politics and Administration* (1945) (with W. W. Crouch); *The American Federal Government* (1947) (with John H. Ferguson); *The American System of Government* (1947) (with John H. Ferguson); *The Third Force in Canada* (1950); *Elements of American Government* (1950) (with John H. Ferguson).

Los Angeles

GREATER Los Angeles, third largest metropolitan area of the United States, is 'an island in the land'—3,000 miles from New York and 2,000 miles from Chicago. Because its greatest growth came after the advent of the motor car, its problems may indicate the shape of cities to come.

This vast sprawling metropolis of California stretches out over a semi-arid coastal plain, between a man-made harbour at San Pedro and the rugged San Gabriel mountain range. The 1950 federal census shows 1,970,358 persons living in a total area of 453·42 square miles within the city of Los Angeles. The city of Los Angeles, without its metropolitan over-spill, is the fourth largest city of the United States in population. Another 1,309,403 persons live within 44 incorporated cities immediately adjacent to Los Angeles, within Los Angeles county; and 871,926 additional live outside the cities under county government. The metropolitan area is largely coterminous with the boundaries of the county, but recent urban development has spread across the county limits and links many of the 15 cities in Orange County to the Los Angeles metropolitan area. Recent census figures indicate that 4,367,911 people live within the metropolitan area.

Since 1910 the increase in population both for Los Angeles and for its suburban fringe, has been a phenomenally rapid one. The metropolitan area population increased 48·8 per cent between 1940 and 1950. This huge influx might have overwhelmed the local authorities except for the fact that basic local works, e.g. water supply, streets, drains and sewers, electricity supply, had been constructed during the 1920's on a scale that could serve a population of the 1950 proportions. During the 1920's, Los Angeles and neighbouring cities had spread over a large area; from 1946 onward the vacant lands were built upon and the gaps were filled.

The population influx to Los Angeles has come from other states and to a very slight extent from the agricultural areas of California. Until the last decade the larger percentage of new Los Angeles residents came from middle western and western states. In more recent times large numbers have come from other sections of the nation as well. Thus, Los Angeles has largely escaped the big problem that has confronted

many American cities on the Atlantic seaboard and some in other sections
—that of assimilating huge numbers of foreign born. While Los Angeles
has some language-group communities, these have not been as tangible
nor as important in municipal politics as those in New York, Boston,
or even San Francisco. Los Angeles today has a large negro population
which has been growing by in-migration at a very rapid rate. While
Los Angeles does not have the industrial attraction of Detroit, climate
and social conditions have drawn many negro migrants. Los Angeles
has a large Spanish-speaking population that presents problems of cultural
and economic assimilation; this group has only recently exhibited signs
of political consciousness. Moderately large groups of Chinese and
Japanese ancestry are to be found. Both groups show a tendency to
spread throughout the city and a desire to absorb most of the dominant
culture patterns of the community. Neither has played any prominent
part in group politics.

One of the very important natural resources upon which the economy
of this area is based is petroleum. Large oil fields within the city, in
neighbouring Long Beach, the harbour area, and in nearby Orange
County have provided an important resource, both for local consumption
and for export. For the most part, however, Los Angeles is not a heavy-
industry town, although there are several manufacturing and industrial
centres in the suburbs and in county territory. The two largest industries
of the area, motion picture and aircraft production, are not dependent
upon the type of resources that normally are required by heavy industry.
Historically, both industries concentrated upon Los Angeles and southern
California because of the mild climate and the many days of favourable
weather. Today both industries are great users of hydro-electric power,
a resource that has been intensively developed in this area. In recent
years, radio, television, and other entertainment businesses have established themselves here. The tourist trade remains, as it has for many
years, one of the chief enterprises of the entire region.

Since 1941 other industries have come into the area partly because of
power resources and petroleum, partly because of other natural resources
available in the desert back country, and partly because this has become
a great residential area with many customers to be served. Los Angeles
has become a national centre for the manufacture of many types of
clothing; likewise some of the large insurance companies have decentralized their offices and established major, regional offices in Los
Angeles. Gradually the economic base of the community has been
broadened and diversified.

THE MECHANISM OF GOVERNMENT

The Los Angeles metropolitan area has numerous and diverse units of government, although not so numerous or so diverse as those in metropolitan Chicago or metropolitan Boston. The principal local governments operating in the area are: the County of Los Angeles and its many special districts, the City of Los Angeles, the satellite cities, the Los Angeles City School District, and the smaller school districts of the area.

The county's chief source of revenue today is the county tax upon land and buildings. The state gives grants-in-aid for care of the aged, the blind, and needy children, and for certain health services. It also gives the county annually a share of the state taxes upon gasoline and motor vehicles.

In California, only San Francisco has a combined city and county government, roughly comparable to the British county borough. There have been many proposals to consolidate Los Angeles city and county governments, or to detach the city from the county, but neither plan has ever proceeded very far.

The county is governed by a group of officers specified in the state constitution and laws. The board of supervisors is the general governing body, combining the legislative and executive authority in a manner unfamiliar in American state and national government. The standard pattern is five supervisors, each elected from a single-member constituency on a non-partisan basis for a four-year term. The simplicity and directness obtained by eliminating the separation of powers are, in practice, rarely used to full advantage.

In Los Angeles county the vast population and the smallness of the governing board make it difficult for the governors to keep in touch with the governed. Each of the five supervisorial districts has a population of almost 1,000,000. Obviously a supervisor cannot give much personal service to so many constituents. Field representatives of the supervisors investigate complaints and attempt to maintain contact with the electorate, but this is a difficult task with such large constituencies.

The board itself is a businesslike body. Although the Los Angeles supervisors, unlike those of small and rural counties, serve full-time, the board meets regularly only once a week. The atmosphere at these meetings is more like that of a corporate board of directors than a public legislative body.

It is difficult to describe county politics except against a background of state and national politics. Because all local government elections in

California are conducted on a non-partisan basis, candidates for super-visor do not parade their party labels, and party organizations (which are weak in California) do not participate in nominating and electing county officers. Into the vacuum caused by the withdrawal of political parties from local government have crowded organized pressure groups and the potent force of the local newspapers. The successful candidate for supervisor usually wins by combining support of local chambers of commerce, improvement associations, and the newspapers that circulate in his district. Because county officials are elected at the same time as state and federal officials, the ordinary citizen is rarely able to judge from his own knowledge the merits of the candidates for supervisor. The common American habit of having too many elective offices in state and local affairs actually diminishes rather than augments responsibility to the public. The supervisor must build his political fences with care. He secures the endorsements of organized groups that are influential in his area, tries to build up a personal 'organization' in his district, and seeks newspaper support—especially that of the metropolitan dailies. The merit system is so widespread in county service that he has no considerable patronage at his disposal.

It costs a great deal of money to campaign among voters in a large metropolitan area like Los Angeles. Lacking party help, the candidate must forage for funds where he can find them. The most likely source is from persons who, or interests which, want something that the county can furnish. Business interests certainly contribute the largest amount of money in county politics; in recent years trade unions have played a modest part in campaign finance. The supervisor has opportunities to reciprocate when the board is considering zoning changes, reviewing tax assessments, awarding contracts, or extending governmental services to a community within the supervisor's constituency.

A few California counties have moved to concentrate administrative authority in the hands of a professional manager. Los Angeles and San Diego counties have created a 'chief administrative officer' and have assigned to him budgetary and other authority over the functions that fall under supervisorial jurisdiction. Sacramento County alone has developed a real county manager.

The county may also have other elective officers: sheriff, district attorney, tax assessor, auditor, treasurer, superintendent of schools, and county clerk. The sheriff and district attorney are most important; several of the other officers may be appointed by the board of supervisors in counties that have 'home rule' charters. The traditional functions of the

American sheriff remain: he enforces state law and county ordinances, serves as an officer of the courts, maintains the county jail. In Los Angeles County, the sheriff heads a department that is similar in all respects to a city police department. The district attorney is the officer who secures indictments from grand jury or court and who prosecutes alleged violators of the law. Most of the work he performs is concerned with state law, but direct election by the county electorate is still considered a means of securing an expression of local will regarding law enforcement.

The city of Los Angeles, as we have stated earlier, is large in area, but embraces less than one-half the population of the metropolitan area.

Some fifty smaller cities of Los Angeles and Orange counties have an important role in municipal government of the area. The term 'city' is applied to all California municipalities, large and small. A community with five hundred or more population may, by following the prescribed routine and securing the required consents, obtain incorporation papers and become a new city.

California cities may be organized under the general law or they may have charters of their own making. The general law cities have councils of five members, elected either at large or by wards. The council elects one of its members mayor, but his duties are not executive in nature; he presides over the council and has ceremonial and other non-administrative functions. The city council appoints a city attorney, chief of police, fire chief, street superintendent and other employees. The city clerk, an elected officer in most general law cities, is the main executive officer; the clerk keeps the council's records, conducts city elections, administers the city's tax laws. Alternately, a general law city may choose to have a city manager if the council passes an ordinance creating the post and a majority of the voters approve in a plebiscite. When the city manager plan is adopted, the manager is given general administrative authority over all officers except the local treasurer and judge or judges, who are elected popularly.

Since the adoption of the state constitution of 1879 cities of over 3,500 population have been permitted, if they choose, to draft charters to suit their own needs. The charter is formulated by an elected board of freeholders. If approved by a majority of voters participating in a special election, the charter is forwarded to the state legislature for final authorization. Home rule cities, as they are called, have much variety in forms of government. Some have the 'strong mayor' type, with real executive power concentrated in an elective head. Most of those around Los Angeles have preferred the city manager plan. The home rule cities usually have a larger number of officers than is provided for under the general law. About one-third of cities in the Los Angeles metropolitan area have freeholders charters, and they include most of the larger municipalities—Los Angeles, Long Beach, Pasadena, Glendale, Burbank, and Santa Monica. Home rule cities have full authority to legislate, without state interference, in all municipal affairs.

In view of its preponderant size, it is not surprising that interest in Los Angeles city politics overshadows that taken in even the aggregate of other municipalities of the area. The home rule charter under which the city operates is a massive and somewhat archaic document that was adopted in 1924 and has been amended more than one hundred times. Legislative power is vested in a council of fifteen members, elected by districts (wards) for two-year terms. The charter requires the council to meet five days per week; a salary is paid for full-time work by councillors. The council meets in its elaborate chamber for daily sessions. The few citizens who bother to look in on the council sessions often are disappointed with the atmosphere and the quality of the debate. It is not an orderly deliberative body. Procedure is highly informal; council men talk at length and range over a wide variety of subjects, frequently remote from the business of the day.

The constituencies represented in council average in population around 125,000. The district boundaries are altered frequently to take into account the rapid population shifts for which Los Angeles is noted; communities widely separated both in geography and interests often are thrown together into a single district. Required to attend council sessions daily and tied down with an elaborate committee system (fifteen committees; every man a chairman!) members find rather little time left for personal contact with their far-flung constituencies. Complaints from the most distant districts are most vociferous. After seeing such a body in action, a student of comparative local institutions would probably conclude that the council is too small, too highly paid, that it meets too often, meddles in administrative matters too much, and has little contact with the people of the city.

The mayor of Los Angeles also is elected by the people, but for a four-year term. On the surface it appears that he has considerable authority. Actually, however, the mayor has relatively little direct power over the city departments, which are headed by citizen boards and commissions. The mayor appoints, with council approval, the members of these bodies, but they have staggered terms of office and ordinarily a mayor does not appoint a majority during his single four-year term.

Each department also has an administrator, chosen by civil service procedure, who serves under the direction of the board or commission.

No aspect of Los Angeles city administration is more controversial than the usefulness of the boards and commissions. Energetic mayors and department chiefs have chafed under the situation. They seek charter reform to abolish the plural bodies or to reduce them to purely advisory committees. They would establish a direct chain of command from the mayor to a single department head. Supporters of the 1924 plan assert, on the other hand, that no better way has been devised to ensure administration close to the people. They declare that if only citizens able and willing to devote time are appointed to commissionerships, the results will be better than through any other system.

Los Angeles operates some proprietary enterprises of importance. Its department of water and power has a virtual monopoly in provision of water and electricity services within the city. The harbour, one of the nation's largest, is municipally owned and operated. So is the Los Angeles international airport.

The municipal water and power department is self-supporting from the sale of services. It also contributes sums deemed equivalent to taxes for the support of the general city government inasmuch as the water and power facilities are exempt from municipal taxation. Earnings of the harbour and airport departments are devoted to those departments' programmes, although their earnings are not sufficient to cover the entire cost of operations and to repay the bonded indebtedness incurred to finance construction of harbour and airport works. The balances required are voted by the city council from the city's general revenues.

The following table will indicate the proportionate share of revenue contributed by the city's sources of revenue and the proportionate distribution of the city's expenditure:[1]

Los Angeles city income (exclusive of earnings of proprietary enterprises which are devoted to the use of those enterprises):

> 39·6 per cent—general property tax
> 15·1 per cent—subventions and grants
> 13·4 per cent—licences and fees
> 9·7 per cent—municipal sales tax
> 9·0 per cent—proprietary earnings paid into city
> 5·2 per cent—fines
> 4·0 per cent—reserve fund
> 2·7 per cent—franchise fees from private utilities
> 1·3 per cent—miscellaneous

[1] Los Angeles City. *The First Hundred Years of Municipal Government* (Los Angeles: 1950).

Los Angeles city expenditure:

24·9 per cent—public works and sanitation
20·9 per cent—police
12·7 per cent—bond interest and redemption
11·6 per cent—general government
10·7 per cent—fire protection
6·0 per cent—employee pensions
5·5 per cent—recreation and parks
2·8 per cent—health
2·5 per cent—building regulation
2·4 per cent—library services

Municipal politics resemble county politics. The absence of parties produces a situation in which personal, pressure group, and newspaper influences predominate. The mayoralty contest in Los Angeles usually is a keen one. Business and labour groups want representation on the boards and commissions. Church and morality groups do battle with vice and gambling interests over appointments to the police commission and consequent police policy. Council politics are more difficult to fathom. Local chambers of commerce and improvement associations play an important part, but the most basic factor is the individual candidate and the organization he develops to secure his election. Some businesses, even cemeteries, being concerned with zoning and other matters that may come before the council, take an interest in council elections.

As American ballots go, that of the city of Los Angeles is not long. Every two years a municipal election is held. Councilmen are elected each time; the mayor, city attorney, and city controller are elected for four-year terms. The regular California non-partisan election rules prevail. A preliminary ('primary') election is held in April; all candidates who pay the filing fee and file nominating petitions with sufficient signatures have their names printed on the ballot. If any candidate receives an absolute majority of the votes cast, he is forthwith declared elected. If there is no majority, the two highest candidates engage in a 'run off' election which is held several weeks later.

CITY EXPANSION AND METROPOLITAN INTEGRATION

The story of the territorial growth of Los Angeles is unique among the great cities of the world: most of its territorial growth came before the heaviest influx of population to the region. Most of the major cities of the United States began with a thickly populated core section and added a populated fringe area to become great metropolitan cities. Los

Angeles began with 28 square miles in 1850 when the city was incorporated under American law. (Los Angeles' earliest municipal government was a *pueblo*, under Mexican law.) After 1906, and especially between 1913 and 1925, Los Angeles annexed large tracts that were only sparsely populated at the time, but urban development has filled in the open spaces since. Analysis of the factors that caused that peculiar expansion is important for an understanding of the problems that face the city government today.

Water is written large in the politics of the entire south-western United States. It is important to any city population, but it is a doubly significant resource for a metropolis developing in a semi-arid locality. The size and shape of local governmental areas in the Los Angeles metropolitan region can be explained largely in terms of the struggle for water of an urban population, present and future.

Shortly after 1900, Los Angeles began to reach outside its existing municipal limits for an assured water supply; in 1910 it reached more than 150 miles northward into the eastern Sierra Nevada to obtain that resource. Some of the water desired was pertinent to public domain still controlled by the federal government and federal officials were persuaded to assist the city. The President withdrew 100,000 acres of land in the area from homestead entry, and the Congress permitted Los Angeles to develop water resources from the lands. Other waters desired were parcelled with land in private ownership and used for agricultural purposes. Over a period of years the city purchased the land and water rights, becoming as a result the largest single landlord in Inyo and Mono counties. These municipal lands were never annexed to the city, although there was serious consideration given to that idea. It would have been necessary to annex a connecting strip many miles in length and extending through three counties. For local governmental purposes the city's lands remained under the jurisdiction of other governments. Water from this area has been brought to Los Angeles by a great aqueduct and stored in reservoirs.

Even before the aqueduct reached Los Angeles in 1913, large areas surrounding Los Angeles petitioned for annexation. The first of these was the San Fernando valley, a large partially-developed agricultural area just north of the city.[1] Shortly after San Fernando valley applied, other areas, mostly those lying between the city and western beaches, applied for annexation; clearly they lay within the path of urban development.

[1] Between 1940 and 1950 this valley was the most rapidly growing residential and light manufacturing area in the city.

While Los Angeles' officials had an aggressive policy to persuade fringe areas to annex, elements in each area petitioned the Los Angeles City Council for annexation and the actions were ratified by popular referendum in the annexed area. The attraction in each instance was water, and the city undertook to put in supply mains even before urban development took place. By 1925, however, the city was forced to call a halt or at least to slow this annexation trend because it was over-committing itself to capital expenditures for water supply facilities for areas that were still so thinly populated that it would be years before they could yield a return on the city's investment. Many areas have been annexed since 1925, but the greatest development was in the 1920's.

Los Angeles grew also by absorbing a number of smaller cities, some of which were attracted by the aqueduct water supply and some because they were unable to support a sound city government from their tax resources. Hollywood, identified throughout the world with the motion picture colony, was originally a separate city, but it joined Los Angeles in 1906.[1] A fabulous real estate development, patterned after Venice, Italy, thrived for a number of years under a separate city government, but later joined Los Angeles, and today it is the demoded beach-front fun zone of the metropolis. Eagle Rock, Barnes City, Watts, and Sawtelle joined during the great expansion period.

Another absorption programme developed almost simultaneously with the water-annexation period, and it too contributed to the conditions that figure in the Los Angeles city problems of today. The original city of Los Angeles was twenty and thirty miles from either of the two poten-tial harbours, Santa Monica to the west or San Pedro to the south. The dominant political and commercial groups in Los Angeles favoured the San Pedro area. In 1907, negotiations were completed for the two small cities of Wilmington and San Pedro to consolidate with Los Angeles and for the latter city to annex a narrow 'shoe-string strip' of land to connect the newly acquired harbour with the city. In a spectacular court fight in 1910, the state government ousted several large private land-owners from the waterfront of this harbour area by quashing title to lands granted them by the state legislature in previous years in violation of the state constitution. The state legislature thereupon granted the recovered lands to the city of Los Angeles in perpetual trust to develop port facilities under municipal ownership and operation. Operation and management of the port facilities was assigned to a harbour commission

[1] Most motion picture studios are now in neighbouring Culver City and Burbank, and in other sections of Los Angeles.

appointed by the mayor. This commission became an influential element in the economy of the harbour district but it had no authority over general government; the mayor and council are responsible for general municipal government.

Residents of the harbour area, located thirty miles from the city hall, have retained a feeling that they are apart from the rest of the city; the issue of separatism is always a popular one in that area. At the time San Pedro and Wilmington were consolidated with Los Angeles, leaders of the latter city are alleged to have made promises that a type of borough government would be created to give this new area some autonomy over community affairs. The city charter has twice been amended to provide for a borough form of government, but considerable doubt has been thrown upon the legality of these schemes and the central city has given no support to the plans.

In the meantime the San Fernando valley has developed to be a very populous section, and similar signs of discontent have come to view there also. Some groups talk of pressing for borough status for the area, others urge separation and incorporation as a separate city. There is a pronounced feeling in both areas that the central city officials are not sufficiently alert to the needs of the communities that are far from the centre of government. Perhaps Los Angeles city has ceased to be a city in the basic sense of the word and is now really trying to cope with problems that are similar to those of the state.[1]

In an attempt to carry city services out to the communities, branch city halls or administrative centres have been built in three sections of the city; in San Pedro, Van Nuys in the heart of San Fernando valley, and West Los Angeles. A similar administrative centre is being developed in Hollywood. This is convenient for citizens using public library branches, paying water and electricity bills and for obtaining routine building permits. Precinct police stations and fire stations are located nearby, also. However, when real decisions of policy are to be made, the citizen still has to go to the city hall, which may be fifteen or thirty miles from his home. For example, if a variance in the zoned land use is contemplated or if building plans involve more than routine checking and approval, the matter must go to the city hall for decision. Location of traffic signals, parking limits and the like are also determined at central headquarters. This is a real source of irritation to the outlying areas that are now quite populous. Demand has been made several times for further decentralization both of some policy making and of administration.

---

[1] In terms of population this city is larger than one-third of the states in the Union.

City expansion has left some other very curious problems in its wake also. The problem of government in the metropolitan area is greatly aggravated by the existence of some thirty or more unincorporated (non-city) islands of territory, almost wholly surrounded by incorporated cities. Some of these, such as West Hollywood, are thickly populated; others are small in area and are uninhabited or are owned by a single landowner. One that is surrounded entirely by Los Angeles city, has been the object of contention for years because a sports arena, baseball field, and television studios are operated in this area and provide a considerable traffic and noise nuisance to the surrounding residential section that is wholly in Los Angeles city. These county islands resist pressures for annexation to adjoining cities and there is no law that requires annexation. These islands depend upon the county government for services. In a large area like West Hollywood the county can decentralize offices to serve that area, but many county islands must be served by county offices miles away, and county sheriff's deputies or county fire brigades must travel a considerable distance through city territory to reach the county island. In some instances the only practical solution has been for the county to contract with a city for the latter to provide services in a county island adjacent to the city.

Another problem has arisen from certain cities annexing a narrow strip of land which either connects two sections of the city or which curves around and encloses a county island. The first and best known of these is the 'Shoe-String Strip' connecting Los Angeles with the harbour area, a narrow strip approximately a mile wide and twenty miles in length. Long Beach and Compton have similar but smaller strips. These were annexed to prevent other cities, chiefly Los Angeles, from annexing areas covered by this city; also it has been hoped that property owners in the surrounded area could be persuaded in later years to annex to the city to obtain municipal services.

COUNTER-INTEGRATION TRENDS

The politics of water, plus the ambitions of real estate promoters, and a 'booster' spirit fostered by some who were interested in commercial development, and by others who had an emotional interest in a 'bigger and better' city, served to bring about some integration of the metropolitan area by incorporating a large part of the area into the city of Los Angeles. The very success of this programme, however, brought reactions that served ultimately to halt this integration effort. Four identifiable trends ran counter to city expansion. Two were especially prominent in the 1920's and two others have developed a major role more recently.

Other cities of the area, such as Pasadena, Glendale, Long Beach, and Compton, had territorial ambitions also, but were not able to offer as large a prize for annexation as Los Angeles. However, their combined territorial expansions served to block Los Angeles and confine it to the areas it had already won. Some cities, such as Santa Monica, successfully resisted consolidation with Los Angeles but did not themselves expand.

During Los Angeles' greatest expansion period, many suburban communities, not wanting to be swallowed up by this colossus, rushed to incorporate and thereby either escaped or put themselves in a more favourable bargaining position. Beverly Hills, which lay in the westward-growth path of Los Angeles, incorporated in 1911, and it has since been completely surrounded. On the eastern boundary, Maywood, South Gate, Huntington Park, Lynwood, and Bell form an almost complete chain of suburban cities. All of the forty-four cities in Los Angeles county today are very sensitive to any move that would bring them within the Los Angeles orbit, although as the reader will see, there are numerous instances of co-operation short of merger.

Many other suburban fringe communities that have not wished to establish a city government have sought to obtain particular municipal services by means of special districts. These districts are created under permissive state law at the request of local property owners and residents. Districts may be created to supply library services, garbage disposal, lighting, fire protection, cemeteries, and numerous other services. Nearly all these districts have the Los Angeles county board of supervisors as their governing board and hence their administration is co-ordinated by the county government. An excellent example of government by special district is Altadena, a thickly populated area north of Pasadena. Library service, fire protection, street lighting and drainage is provided by districts. Police and health protection is supplied by branch offices of the general county government. Both incorporation as a city and annexation to the neighbouring city of Pasadena have been rejected by Altadena voters.

State law has also permitted the county government departments to develop municipal services; hence, through county departments and through special districts that are under county supervision, the county serves many fringe areas with government almost comparable to that of a first-class city. Costs of special districts are collected from the property taxpayers in the district, whereas county costs are paid by the property taxpayers throughout Los Angeles county. Considerable competition has developed between the cities and the county, and the League of California Cities has recently taken a stand opposing county provision of municipal type services to the fringe areas, because it feels that city taxpayers are taxed twice, once for city services within the city and again for county services. For example, Los Angeles city has a health department supported by city taxpayers; the county has a health department supported by taxpayers throughout the entire county, but the county does not attempt to duplicate the city's department within the city. Los Angeles has a police department; the sheriff provides similar protection to suburban areas. The League of Cities demands that the fringe areas either incorporate as cities or that the cost of services to these areas be collected by special district taxes, that the county government as such confine itself to county-wide services. The county, however, has developed an efficient organization with a large staff of civil servants, and with a chief administrative officer or general manager heading the county's administrative work. The county has exercised an important unifying influence in the metropolitan area.

SPECIAL AUTHORITIES IN THE METROPOLITAN AREA

Special districts and *ad hoc* authorities have been used in some instances where a problem was regional in extent or where a large mobilization of resources was required. The most outstanding of these is the Metropolitan Water District. Leadership for formation of this district came from the principal policy makers in the Los Angeles water department; the district was designed to draw together the major cities in southern California for a joint effort to secure water supplies for the urban population that was estimated would emerge in the calculable future. Development of this plan relieved one of the pressures for Los Angeles' expansion, but it was recognized also that the existing territory of Los Angeles would need more water to satisfy the demands of population than the city could reasonably expect. Thirteen cities originally formed the district, which was created by special state law. Cities and districts may join.[1] The district obtained a portion of the Colorado river water under the terms of a compact between the federal government and the eight western states

[1] The Metropolitan Water District was concerned originally with supplying urban needs. Some districts that have been formed to obtain water from the Metropolitan Water District are composed of several cities; such districts are of course interested in urban water supply. In recent years, however, some districts that have organized and have been admitted to the district are composed chiefly of agricultural lands.

that have an interest in the river. A viaduct brings the water across the desert and through a mountain range to the coastal area. Cost of this construction was financed by bond issues and the costs of the district are paid by taxes upon property within the cities or districts that are members, hence newly admitted members must pay a retroactive share of the bonded indebtedness. Voting rights within the district board are allocated in proportion to the assessed wealth of each of the member cities or districts. Los Angeles, with the greater share of assessed wealth, is the major force on the board, although it may not have more than half the votes. Member cities draw water from the district's aqueduct to augment their other water resources; only one city is fully dependent upon the district for its supply at present. Los Angeles takes an average of about 19 per cent of its water supply from this source.

A recently developed special authority covering the metropolitan area is the County Air Pollution Control District. Industrial developments in the area since 1941 have resulted in the discharge of gases and smoke into the atmosphere; and this, combined with the meteorological conditions common to an arid coastal plain surrounded by a rim of mountains, produces an irritating haze or 'smog.' Sources of these fumes could not be controlled by any one city, nor was the county government an acceptable authority to all interested groups. Therefore the state legislature created a special district and designated a board of supervisors as the governing body. The district has powers to establish standards under state law and to inspect and prosecute establishments emitting smoke or gases into the atmosphere. The research and enforcement work has not yet produced completely satisfactory results.

Water pollution prevention has also been assigned to a special authority recently, although the regional board having jurisdiction in this area has been assigned to a watershed that is more extensive than the Los Angeles metropolitan area. Regional board members are appointed by the state governor and must be selected from among local sanitation officials and industrial groups. Such boards are to supervise local sewage disposal systems and industrial waste disposal or other practices that are likely to imperil the water supplies of the region. This system has only begun to function and is still developing standards and procedures. There is every intention that it shall work with local government.

The metropolitan area has been served for thirty years by two sewage systems. Each of these is controlled by a governmental organization which is an important illustration of inter-governmental relationships. One system is dominated by the city of Los Angeles and involves a series of inter-city contracts. The other is a special authority associated with the county government.

When Los Angeles spread, it often found it advantageous to arrange to run sewer lines through another city and thus connect a portion of its own area with a trunk line at gravity flow. It followed, then, that it was mutually advantageous to permit the other city to connect with Los Angeles' sewer lines. The private housing boom that followed the First World War brought the sanitation problem of the metropolitan area into a critical condition. The city of Los Angeles built a treatment plant and ocean outfall at that time, and with great optimism contracted to permit some eight cities to use the Los Angeles trunk lines leading to this plant. In one respect this arrangement was highly advantageous; it eliminated the possibility of several sewer systems operating in the metropolitan area, each with a separate treatment facility. It was not foreseen, however, that continued residential and industrial development would soon tax the capacity of this system, or that obsolescence would require reconstruction. Both factors developed to a degree that bathing beaches in Santa Monica Bay became so seriously polluted that the state board of health quarantined the beaches for three summers. Los Angeles voters rejected bond issues proposed for reconstruction; and some of the contracting cities adopted the attitude that Los Angeles alone was responsible under the contracts for correcting the sewer deficiencies. Los Angeles owned the trunk lines and treatment plant, the other cities merely paid a share of annual operating costs. In 1946, the state brought suit to compel all the associated cities to rebuild the system, and the state won its suits. Two smaller cities at first refused to accept this decision, and their city councils were in peril of going to jail for contempt of court. The other cities have joined Los Angeles in financing a new plant, and the state contributed over $13,000,000. This project has recently been completed and all contracting cities have re-negotiated agreements.

At the same time that Los Angeles was inviting other cities to contract with it for sewage disposal service, the state legislature authorized the creation of county sanitation districts, a type of district adapted to this metropolitan area. A district may be composed of two or more cities, a city and a non-city territory, or entirely of non-city territory. Costs of operation are collected from taxes upon property. Governing boards of these districts are made up of mayors of member cities and members of the Los Angeles county board of supervisors. Thus the county board co-ordinates the policies of the twelve districts that have been created. The districts have pooled their engineering and management staff and have employed a joint chief engineer. A single comprehensive trunk-line sewer system serves all the districts and feeds to a central treatment plant and ocean outfall. Two of the districts, because of their location and elevation, contract with Los Angeles to use the other disposal system. One district, composed of three cities, originally contracted with Los Angeles, but has recently built a new line with pumping stations, to connect it with the county system. The two systems serve almost the entire metropolitan area, and since the recent reconstruction of the city system, the area's needs are being effectively met.

Flood control is another function which has been assigned to a special authority. While much of the flood control work is performed by the federal Corps of Engineers, local responsibilities are supported by the Los Angeles County Flood Control District. This district was created by special state act after disastrous floods in three successive winters had destroyed much property. As a result of effective work by district and federal engineers, flood waters are tamed near their sources and spread out on settling grounds to percolate into the natural underground basins, and storm waters are channelled out to the ocean. The district centres its energies on check dams and spreading operations. As a result of the joint programmes, an area that formerly was flooded each winter has been reclaimed and developed as the central industrial area, a part of which is in Los Angeles and a part of which is in Vernon and other industrial suburbs. The area of the district is that portion of Los Angeles county south of the mountains drained by this watershed; most of Los Angeles city is included in this area. When the legislature created this district, it provided for informal integration by designating the county board of supervisors the governing board of the district. The district has a separate staff and a separate tax rate upon property to support its work. A few years ago arrangements were completed between the county and the district for the county civil service department to perform personnel work for the district.

Educational services are provided by 130 separate school districts in Los Angeles county. The largest of these is the Los Angeles City School District, which has about 350 schools—elementary, high school, and junior college. At the other extreme are some rural districts that operate a single one-room elementary school. Following the usual American pattern, California separates the education function from general local government; school districts are largely independent of city and county governments and possess the power to levy taxes for the support of the school system. An important share of school finance is provided by the state and paid as a grant to the local school district on a per pupil basis. The rapid population increases in the area keep local districts busy with an extensive building programme just to keep pace with the new children who need education. Not long ago, the Los Angeles city school system reported that in-migration required that the equivalent of two new elementary schools, of 750 pupils each, be ready every Monday morning of the school year. School board members are elected at large from the school district; the smaller districts elect three and the largest, seven. Each of the major school districts administers its schools through a superintendent of schools, who is appointed by the board.

INTER-GOVERNMENTAL RELATIONS

A relatively unique system of inter-governmental relations has developed among the cities of the Los Angeles metropolitan area and between them and the county government, in spite of occasional difficulties. The contracts of the Los Angeles sewer system have been described. A somewhat similar contract arrangement was made by Los Angeles, Glendale, Pasadena, and Burbank, whereby Los Angeles conducts specified amounts of hydro-electric power over its lines from generators which the city operates at Boulder Dam under federal permission. Instead of four power lines, one city brings a supply to the four distributing systems.

A less formal relationship may be illustrated in the sphere of fire fighting. Each city in the metropolitan area has its own fire department, and in addition, the county operates a trained force to suppress brush and grass fires in the mountains and supervises a series of county fire protection districts which provide fire fighting services for buildings in the suburban fringe area.

Whenever a fire breaks out near the Los Angeles city boundary and threatens to spread, or when a fire in an industrial suburb threatens to overcome local resources, the fire chief in the city or the county district may request assistance from the Los Angeles chief, who determines the amount of assistance that can be given. The Santa Monica range of hills extends between two thickly settled portions of Los Angeles and runs on into county territory. Fires in this area are frequently fought by city and county forces co-operatively.

Relations between Los Angeles city and county are extensive; sometimes they are harmonious, occasionally they are bitter. Complete functional consolidation was achieved for one function some years ago when enforcement of weights and measures standards was transferred voluntarily from the city to the county. For thirty years the city has contracted with the county to assess property for city tax purposes and to collect city property taxes. The city pays a charge for this service equal to about two per cent of the amount collected—a sum much less than the city would spend if it did the work itself. This contract may be cancelled at any time if the city or the county is dissatisfied. For several years the county has leased space in the city hall for county courts and paid a figure approximating $13,000 per month. Plans are now about ready for a new court building, and when it is completed, the city may then occupy all its own building.

A four-government joint project relates to wage and salary fixing. Los Angeles city, the county, the city school district, and the city housing authority budget and personnel staffs co-operate each year in conducting a survey of prevailing rates of wage in representative private and public employment in the area. One of the four takes responsibility for directing the survey; and the work is aggregated. Some fifty or more carefully selected classes of positions are first studied for use as specimen positions. This has tended to reduce diversity of practice among the four agencies. After the sample positions are chosen data on wages paid in comparable positions is obtained from a large number of employees, and refined to produce tentative wage scales. This co-operative survey has brought about several valuable results, although it has not yet resulted in uniformity in job classification or wage schedules among these four important employers. It has saved costs and reduced the confusion that would arise if each government went its own way.

One other city-county relationship sometimes produces friction. The state collects a tax upon all gasoline sold for highway use, and divides the proceeds with the cities and the counties. Los Angeles city receives about $4,600,000 per year directly from the state for streets and for police protection. Los Angeles county devotes a portion of its share to construction and maintenance of county roads, but it also distributes a portion to the cities in the county on the bases of population and street mileage. This division is not required by law, but is done on the theory that much of the county is served by city streets. Los Angeles city normally receives about $1,500,000 from the county. On several occasions the county has threatened to withhold portions of this money from the city if the latter did not set city employees' salaries nearer the level of county salaries, or unless an equitable rental price for the courts in the city hall was determined. Another source of friction between the city and county has been over health services. The county has a well developed health programme that operates in forty-one of the cities as well as in non-city territory. Los Angeles city also maintains a large, modern health department, although city property owners are taxed for county health purposes as well as city. Numerous efforts have been made to consolidate the two departments but each has failed. For many years the city health department has ranged far beyond the municipal borders to inspect dairies producing milk for the metropolitan market. The county also has an inspection service. Competition and duplication became so disturbing that the state passed milk and dairy standard laws and assigned separate areas to the city and the county inspectors. State inspection was not established because both local governments were well established in the function and the state was a newcomer, even though legally superior.

PROBLEMS OF THE AREA

The role of water supply as the paramount problem in the area's development has already been indicated. Los Angeles has long been unable to sustain life on the meagre surface and underground water of the region. Now, recognizing that the available water from the Colorado river may not be adequate for the future, civic leaders are pressing for research on economic methods of converting sea water into potable water, and for a survey of the possibility of bringing Columbia river water 1,000 miles from the Pacific north-west into California.

A casual visitor to Los Angeles often goes away with memories of the greatest traffic congestion he had ever experienced. One agency reported that in 1890 it took 10 minutes, 21 seconds for a horse and buggy to traverse Broadway (a main downtown thoroughfare) from First to Tenth Streets. In 1938 an automobile required 14 minutes, 12 seconds to cover the same course. Today even more time would be necessary. Alone among the great metropolitan areas of the world, Los Angeles lacks even the beginning of a rapid, mass transportation system. Two private companies operate street and inter-urban railways, largely with ancient rolling stock. Both are seeking to abandon electric lines and to operate motor buses in the streets. The only real progress that is being made is in the construction of freeways—great super-highways that eventually will criss-cross the metropolitan area. Judging from the results of the first one completed, the Pasadena Freeway, traffic can be speeded up considerably, but more people are induced to use their private motor cars, thus adding to congestion at the end of the freeway in the downtown area. Freeways are constructed by the state and paid for from state motor fuel tax proceeds.

A rather unique proposal of a "monorail" rapid transit system has stimulated much interest recently. Cars would be suspended from an overhead single rail. The initial proposal calls for utilization of the bed (usually dry) of the Los Angeles river, which winds from San Fernando valley through the metropolitan area to the harbour area.

From time to time proposals have been launched to create a metropolitan transportation district. Population has spread so widely that there is grave doubt whether there is sufficient density of population to support a rapid transit system. Privately owned lines have provided poor service for years and the companies plead poverty. Climatic conditions and the lack of an integrated transit system has made Los Angeles an automobile-riding metropolis. In addition, shopping and industry have decentralized to such an extent that instead of a stream of workers flowing in and out of a metropolitan centre, many cross-currents flow from residential areas to several sub-centres. Many persons have thought that any public transportation scheme would necessarily require generous subsidies. Thus far, all attempts to organize a special district have been defeated by the cities surrounding Los Angeles, who fear such a proposal is a threat to their own economic development and another effort to bind them to the Los Angeles 'planetary' system.

The transport problem has been complicated by many factors: the lack of a single local government with jurisdiction over the problem, the conflict between proponents of private and public ownership, the vastness of the area to be covered and its sparse population, and the disagreement among engineers over the feasibility of underground lines in soft earth and in an earthquake prone area.

Planning and zoning are employed both by cities and county to control development and land use. As in most large American communities they came too late to affect materially the basic layout of the metropolitan area. Sub-division of land plays a prominent part in the growth of Los Angeles. Each year new tracts are opened up farther and farther from the central city. So rapid has been this process that lots (sections) of previous sub-divisions are left unsold or unused. Los Angeles has been called the city of vacant lots; areas sub-divided fifty years ago still stand idle, yet require expensive municipal services. If local taxes on property emphasized the unimproved land value rather than the value of improvements, owners might have been induced to put their holdings to use rather than retaining them for speculation. Even now sub-dividers of vacant properties are not required to include in their plans space for schools, parks, and playgrounds.

The pattern of housing in Los Angeles differs markedly from that of most large cities. About 70 per cent of the population is housed in one- or two-family dwellings. The typical house is a one-story bungalow or cottage, usually of wooden construction, on an individual lot or section. Most of these dwellings meet the minimum standards of running water, indoor flush toilet, bathing facilities, proper ventilation, and good state of repair. The 1940 census showed, however, that the county contained over 100,000 sub-standard family dwellings. To meet the challenge of the 'shack slum,' city and county housing authorities have utilized federal funds to clear slums and build public housing projects. During and after the Second World War the local housing authorities also operated for the federal government a number of emergency housing projects. In 1950, just as the post-war programme was getting under way, a state constitutional amendment was adopted which will require in future an affirmative vote of the people of the city or county concerned before a local authority can launch a particular housing project. This restriction will hamper greatly the expansion of public housing programmes in California.

Law enforcement in the metropolitan area is exceedingly fragmented. Some fifty local police forces have jurisdiction over various portions of the area. Police officers, in hot pursuit of a lawbreaker, may pass through several jurisdictions in a few minutes. Inevitably, different local administrations vary in their interpretations of the state law; local ordinances on a given subject often are diverse. Laws against gambling are enforced vigorously in some cities of the area, while other municipalities openly permit virtually games of chance. Prostitution may flourish in county territory and be driven underground in incorporated cities. Vice and gambling interests can nearly always find some part of the metropolitan area where local politics is such that they can operate unmolested. The need for a unified police force becomes more apparent year by year.

Nature has provided the Los Angeles area generously with nearby recreational facilities: high mountains, the long coast line of the Pacific, long hours of sunshine. Man has both added and detracted; he has made the mountains and the beach accessible by means of good highways, but has permitted so much of them to pass under private ownership that their use by the public is limited; the sun is often obscured by an air pollution that causes great discomfort to inhabitants of the area. In acreage of parks and playgrounds Los Angeles ranks reasonably well among American cities, but Griffith Park alone accounts for a large share of the green spots on the map. Parks are unevenly distributed through the metropolitan

29. LOS ANGELES

A view of the downtown business centre.

[*Photo: Spence Photos*

30. MANCHESTER

A view of Wythenshawe garden suburb developed by the Manchester City Council.

[*Photo: Airviews*

area, and often are located in the areas of lesser need. Playgrounds are generally insufficient; many school yards are locked throughout the summer months while children play in dangerous streets.

PROPOSALS FOR REFORM

It may be seen that Los Angeles is ill-organized to deal with the acute problems that confront the metropolitan area. There are too many units of local government to get a consensus on most questions of regional policy. The task is to devise a new local government arrangement that will provide centrally those functions that affect the safety and welfare of the entire metropolitan area, yet allow diversity and local self-determination in matters that ought to be close to the people.

Some of the methods of solving these problems have already been used about as much as they are likely to be under present conditions. Annexation of unincorporated areas to existing cities, especially to the city of Los Angeles, remains a possibility, but the process has slowed down and will not bring in much territory in the future. Consolidation of several small cities into a single large city seems unlikely, given the persistence of separatism and local pride.

Other methods are in use now, and probably will be utilized actively in the years immediately ahead. The cities and the county have entered into a host of voluntary agreements and contracts regarding specific functions. These arrangements appear advantageous to both levels of government and certainly ought to continue.

Likewise the tendency toward the creation of special service districts like the Metropolitan Water Board will undoubtedly be continued. It is difficult to see how the transportation problem can be solved, given the present pattern of local government, without the creation of a transport authority for the whole region.

Finally, we come to proposals for reform that have been made but not tried. They vary in detail, but have in common the idea of establishing a metropolitan government for the whole area, either by vesting additional power in the county, or by combining city and county governments, or by creating a new level of regional government. The first stumbling block that inevitably will be encountered by any of these schemes is the necessity of securing the consent of a majority of the people of an existing city or unincorporated territory before they may be absorbed by a local authority. In the end a state constitutional amendment would probably be necessary to overcome objections of local

L

groups. The people of the state would be reluctant to adopt an amendment that would forcibly alter the status of a city with a home rule
charter, of which there are thirteen in Los Angeles county.

If this hurdle is surmounted, the knotty problem of keeping local
services close to the people will remain. One possibility would be to
retain the existing smaller cities, but with reduced status and a smaller
range of functions. The largest of these is Long Beach, which has about
270,000 population; the smallest are Vernon and Palos Verdes Estates,
each with about 1,000. Although some consolidation of the smaller units
might be advisable, most of the present incorporated cities are large
enough to justify continuance with borough status under a general
government for the area. Los Angeles city would provide the most
perplexing problem of all. It is too large to fit into a federation of municipalities without playing a dominating role; in addition there are many
separatist grievances within the existing city. In the circumstances, if a
metropolitan government were established, Los Angeles city might well
be divided into thirty to fifty community boroughs for purposes of purely
local matters and representation in the metropolitan governing body.

Because the obstacles imposed by special interests, narrow localism,
tradition, and inertia are so formidable, Los Angeles is not likely soon to
adopt reforms so sweeping as those provided for London in 1888 and
New York in 1898. Several competent studies have been made of one
or another aspect of this vast problem, but none so far has ended with
recommendations so bold that they commanded attention and attracted
the support necessary to put them into force. Los Angeles badly needs
a reconsideration of the whole problem of metropolitan government by
a group of authorities who have shown a capacity both to dream and
to get things done. But just another survey is not enough. The results
must be 'sold' to the public in a great campaign of education through
every channel of communication. The issue may well be whether urban
man can solve the problem of metropolitan living through governmental
institutions of his own choosing.

Manchester

SHENA D. SIMON

SHENA D. SIMON
(Lady Simon of Wythenshawe)
M.A.(Cambridge), Economics.

Member of Manchester City Council, 1924–33; Member of Consultative Committee of the Board of Education, 1929–36; Member of Royal Commission on Licensing, 1931; Co-opted member of Manchester Education Committee since 1936; Member of Departmental Committee on Valuation of Dwelling Houses, 1939; Member of Council of Manchester University; former Governor of Newnham College, Cambridge; Vice-President, Workers Educational Association; Vice-President of the Association of Rating and Valuation Officers.

Publications: *A Century of City Government—Manchester 1838–1938*; various pamphlets and articles on Education, Local Government, Rating and Valuation, and Local Income Tax.

Manchester

SOCIAL, POLITICAL AND ECONOMIC SIGNIFICANCE

MANCHESTER, the third largest city in England (population 715,000) is the core and pivot of one of the most highly industrialized regions in the world,[1] that of south-east Lancashire. It includes a population of 10 million people within a radius of fifty miles, and the proportion of the total population employed in manufacturing industry is nearly double that so employed nationally. A market town in the fourteenth century, the cotton industry was established in Manchester in the sixteenth century, and with the development of the mechanical inventions of the nineteenth century, the proximity of coal, an abundance of soft water, and the damp air, it grew rapidly. Gradually the city itself developed as the commercial centre of the cotton industry—the big warehouses were built and the merchants set up their offices there. The actual manufacture of cotton goods was concentrated more in the surrounding towns.

The construction of the Manchester Ship Canal (1894) which connected Manchester with the sea near Liverpool was a joint undertaking by private enterprise and the municipality. It revived the trade of Manchester which was then on the decline and made possible the development of the Trafford Park estate—not to be confused with its neighbour, the Old Trafford cricket ground! By its proximity to the canal and the Manchester docks, this became the centre for big engineering works of which Metropolitan Vickers is the best known. Trafford Park forms the largest concentration of heavy industry in the area. The joint enterprise which resulted in the Ship Canal was an example of bold and imaginative action on the part of the city, not, unfortunately, to be matched again until it undertook the development of the Wythenshawe estate in 1930.

In the last twenty years engineering rose to be one of the two most important industries. In 1947 engineering actually employed the largest number of people in Manchester, whilst in the north-west region[2] it remained second to textiles. In Manchester itself the manufacture of clothing has succeeded that of textiles and is now second in number employed to the engineering industry.

[1] Regional Planning Report, p. 47.

[2] I.e. the standard north-west region as defined for administrative purposes comprising Lancashire, Cheshire and the High Peak district of Derbyshire.

The Manchester Chamber of Commerce is one of the most important in the country and the various war-time and post-war controls of the cotton industry were operated from the city.

The position of Manchester as a regional centre is perhaps shown by the fact that for the last twenty years there were actually more people employed in the distributive trades in the city than in any form of industry, whereas in the north-west region, textiles maintained its lead through the period of the slump, war, and post-war readjustment.

THE CITY'S PROBLEMS

Pressed up against its neighbours, Salford and Stretford on its western boundary and hemmed in on the east, Manchester has had to develop north and south, with the result that it resembles nothing so much as a banana, over twelve miles long and not quite five miles broad; but the closely packed immediate region of which it is the centre includes fourteen local authorities, a population of 1,296,000 and an area of 73,000 acres.

As the largest of the manufacturing towns not only in the region, but in the whole of Lancashire, Manchester in its physical aspect shows the worst horrors of the uncontrolled industrial revolution side by side with the development of Wythenshawe, which is perhaps the best attempt so far made by any municipality to make a serious beginning to remedy them.

'As in most northern industrial regions, the intermingling of factories, mills, business premises and dwelling houses, without any semblance of order, results in a lack of light, air and good working conditions.'[1] Out of a total of 200,000 houses, 48,000 are more than eighty years old, which means that they were built before there were any effective building by-laws. Speculative builders were free to cram as many houses on to an acre as was physically possible, with no restrictions as to light, air or room space. Baths were an unheard of luxury; even in houses built under the 1890 by-laws they are rare. A billeting survey during the Second World War showed that only 55 per cent of all houses have baths. Since this survey included the outer suburbs where the well-to-do have lived for generations and where the inter-war housing estates were developed, it is obvious that the percentage of baths in the centre areas is much less, in fact it is 15 per cent. The Manchester Council, however, deserve praise for the fact that, unlike Birmingham and Leeds, it had abolished the worst legacy of the industrial revolution, the back to back house, by 1914.

[1] Regional Planning Report, p. 47.

The central part of the city is short of parks and open spaces, for these were not considered necessary when Manchester was developed. All except one, generously given by a cotton manufacturer in 1843, are in the suburbs. Although the city council realized this need at the beginning of the twentieth century, and bought various open spaces, Manchester still has only three acres per thousand of the population, whereas the recognized minimum standard nowadays is seven acres. The centre of the city has only two public gardens where city workers can spend their lunch hours out of doors, and children who are still condemned to live in the inner ring have at the least a mile to walk before they can find a park to play in. There are some fine buildings in Manchester, but they are mostly hidden amongst offices as if the city were ashamed of them, and those that can be seen, the old Town Hall, the Rylands Library, and the University, were all built at the time of the Gothic revival. Although they throw an interesting light on the relation which the successful Manchester business man of the nineteenth century saw between money and an ecclesiastical setting, they can hardly be described as functional, and one can only be thankful that the fourth building of the same period, the Assize Courts, was destroyed in the blitz.

Although Manchester is the seat of a bishopric, the cathedral has never aspired to be other than the former parish church of an immense parish, and is only visible from a distance now because the blitz destroyed some of the intruding buildings. There are still terraces of late Georgian houses to be found, but they are now either used for small industries, offices, or, at the worst, for tenement dwellings. The exodus of Manchester merchants from the centre of the city to the suburbs of Rusholme, Fallowfield, Withington, Didsbury, took place unfortunately at a time when domestic architecture was at its lowest ebb.

The inhabitants of central Manchester live under a perpetual smoke pall which not only reduces the health-giving property of the sun's rays and lowers the general vitality and power to resist infection,[1] but deposits soot over the buildings although increasing use of gas and electricity has certainly improved the atmosphere in the last twenty-five years. Under local powers obtained in 1946 the city council has recently applied to the Ministry of Health to establish a smokeless zone in the centre of the city, where it will become a punishable offence for occupiers of premises to emit smoke.

The schools in Manchester have suffered the same fate as the houses built at the same time over the last one hundred and fifty years. Only 25

---

[1] Manchester Plan, p. 1.

per cent of the schools now in use were built between the two world wars. A number have been on the Ministry of Education black list, as being past reconstruction, for years, and if the standard set by the Education Act of 1944 is to be reached, most of the existing buildings will have to be reconstructed. Two of the schools which serve the region, the Manchester Grammar School, a foundation dating from the sixteenth century, and the Manchester High School for Girls, had moved out from the centre of the city to the suburb of Rusholme before the war, and the new housing developments have necessarily brought with them new schools.

This grim aspect of the centre of Manchester must be seen in contrast to the housing estates developed between the wars. Mancunians, like most Englishmen, do not like flats as family dwellings, although some have been built in the inner ring of the city.

Between 1920 and 1939, 30,685 houses were built by the corporation, and 22,000 by private enterprise at a density of twelve to the acre on the outskirts of the city, and families from the grimy, over-crowded centre of Manchester moved into them. The most interesting experiment was the development of an estate in the south of the city, Wythenshawe, as a satellite garden suburb, with factories as well as houses, schools, shops, cinemas and other public buildings. This was begun in 1930, held up during the war, and now provides almost the only available land for houses within the city boundary. It is estimated that in order to house the population on a satisfactory basis at least 150,000 of its population —over 20 per cent—will have to be housed outside.

A few years before 1939 a beginning had been made to clear some of the slum areas, but the Second World War put a stop to that, and when building began again the post-war need for houses was so desperate that no demolitions have taken place except of houses that are actually unsafe.

MANCHESTER AS A REGIONAL CENTRE

But as we have seen, Manchester is both the cultural and the commercial centre of a wide region, which includes the industrial towns of Bolton, Bury, Rochdale, Oldham, Stockport and others. These developed originally as centres of cotton spinning and weaving, and those industries still play a major part in their employment. Their women come to Manchester to shop, their men to buy cotton on the Royal Exchange— which, before part of it was damaged in the blitz, could accommodate 10,000 people in its main hall—and both men and women come to Belle

Vue for fireworks and large meetings, and to the Free Trade Hall for Hallé concerts. (This historic hall, also destroyed by the blitz, was rebuilt for the Festival of Britain in 1951.) Thousands of citizens from outside Manchester use the excellent facilities of the Reference Library which includes the unique Watson Music Library. Manchester is the seat of the legal Court of the County Palatine,[1] and the assizes meet there four times a year.

The Co-operative Wholesale Society and several banking and insurance houses have their head offices in the city. During the Second World War Manchester was the headquarters of the North-Western Civil Defence Region, and all government departments which now have regional offices make Manchester the centre of the north-west region.

It is the proud home of the *Manchester Guardian*, the leading provincial daily paper which circulates far beyond the region and has a worldwide reputation, and it is the great northern centre where all the leading London papers which have northern editions print them.

Manchester University, the oldest of the provincial universities, celebrated its centenary in 1951. It was founded by a Manchester man who had made money in cotton and it has benefited since from the generosity of succeeding generations of Manchester business men.[2] It serves an even wider region, one which extends north to Cumberland and Westmorland, as well as east to Derbyshire and south to parts of Cheshire. Although the majority of the students come from the region, many, especially research workers, have always been attracted from all parts of England by the reputation of its professors. It is no exaggeration to say that Manchester possesses one of the best schools of the physical sciences in the country.

In the College of Technology Manchester possesses the only technological college outside London which provides a full-time university degree course as well as day and evening courses in all branches of technology. This institute is run jointly by the University and the city council, a typical example of British methods of compromise.

Although Manchester may be the ugliest and dirtiest of all the big cities in England—and few Mancunians would deny this charge—they

[1] The three counties palatine (of Chester, Durham, and Lancaster) were originally separate kingdoms or principalities in themselves, created by the king as a means of defence against the neighbouring Welsh and Scots. The county palatine of Lancaster was granted by Edward III in 1377 to John of Gaunt. The County Palatine had its own court, which was not an inferior court, as it possessed the full rights of administering royal justice to the exclusion of the king's writ.

[2] Today income from endowments is only five per cent of the total income. Grants from the Government form the main item of revenue.

L*

have the satisfaction of knowing that if mention of the word 'Manchester' in other parts of England brings an inevitable reference to fog and excessive rain, in far away countries it stands for first-class cotton goods and engineering products, the *Manchester Guardian* and the Hallé orchestra.

The city has suffered like all provincial cities from the drift to London of finance, the professions, and the arts. 'At one time Manchester was a cultural centre in its own right, living up to a tradition of leadership in scientific and political thought and making a not inconsiderable contribution to the common fund in music and the arts. Latterly, however, the tendency towards centralization has weakened the power of every provincial city to stimulate and satisfy the creative imagination of its citizens.'[1] But, more than Liverpool and Birmingham, it has suffered by the exodus of its leading business men to homes outside the city boundaries.

'I do not live in Manchester now,' said a leading citizen in 1866 when the term referred to what is now the inner ring of the city. 'No one does, who is at all in a position to live out of it,' and this movement increased as transport improved, as fortunes were made and as the municipality, largely owing to its geographical limitations, but still more to false ideas of economy, made no attempt to attract good residential building. London, Birmingham, and Liverpool have kept leading citizens within their own boundaries. This has not only enriched the cultural side of civic life at a period when only the well-to-do could cultivate and enjoy the arts, but has kept on their municipal councils men of affairs both in industry and the professions who have brought valuable experience to bear on local problems.

Whereas merchants and manufacturers formed more than half of the Manchester Council in 1838, they only formed thirteen per cent a hundred years later, and only one of the present forty-two directors of the Chamber of Commerce is a member of the city council. Their place in Manchester was taken by small shopkeepers and traders so that the Conservative Party until recently has been the party which has advocated low rates rather than of wise expenditure as at Birmingham under Neville Chamberlain. The Labour Party, which has succeeded the Liberal Party, has always stood for expenditure on the social services.

Between the wars the exodus to Cheshire was not confined to leading business men, doctors and lawyers, but included teachers, clerks, and men occupying managerial positions in industry. The lure of the countryside, of purer air and lower rates has meant that Manchester finances have not benefited from the increase in the city's wealth. The people needing

[1] The Manchester Plan. p. 99.

the social services most are those left behind who are less able to bear the burden of paying for them. When the last count was made, it was found that about 118,000 people travelled into Manchester each day. To extend the city boundary—as was done in the early days to include the residential areas—is no longer possible since the extension would have to include such a large part of the county of Cheshire that there is no hope that Parliament would agree. Also such an extension would make Manchester still more unwieldy both as a geographical and as a municipal unit. This problem will be referred to later when the problem of regional government is considered.

In 1838, when the collection of townships that then composed the borough received its charter of incorporation, the population was 242,357. Today, with a population of 715,000, it is the third largest town in England. In 1838, the area of the municipality was 4,293 acres; now it is 27,255. The rateable value increased in the first hundred years from £669,954 to £6,398,000. In 1950 it was £6,223,164.

THE GOVERNMENT OF MANCHESTER

Manchester's form of government is similar to that of all other county boroughs, of which there are eighty-three in England and Wales. The city council can exercise the full powers that are allowed to a local authority by law. Most of these powers are found in general legislation, i.e. Acts relating to education, housing, police, health, and town and country planning. In England, unlike Germany and the U.S.A., if a local authority wants to exercise any other powers it has to get a private Act passed by Parliament. Manchester has had her share of these private Acts, one of which gave her the right to develop Wythenshawe as something more than an ordinary housing estate.

The city is divided into thirty-six wards, each of which has three councillors and one alderman allotted to it. Every year elections are held and one councillor is elected in each ward for three years after which he may stand for re-election. Thus a third of the council is renewed every year. Aldermen are elected by the city council from among the councillors.[1] The custom has now developed of electing the senior councillor

---

[1] The legal qualification for an alderman is that he must either be a councillor or a person eligible to be a councillor. The practice among towns in Britain varies regarding the choice of aldermen. Some municipalities, like the London County Council, elect people who have never been councillors; others elect only members of the dominant political party, so that when the party complexion of the council changes, those aldermen who do not belong to the party in the majority, may not be re-elected. Manchester avoids the charge that aldermen have not had to stand for popular election by choosing those who have stood and have been re-elected for a number of times.

to fill an aldermanic vacancy, which, with a very few exceptions of resignation, only occurs by death. Although aldermen have to be re-elected by the council every six years, this is a pure formality, so a councillor who is successful in keeping his seat for eighteen or twenty years without a break of more than one year is sure of a safe seat to the end of his life.

The electors, who are men and women over twenty-one, are in general the same as those who vote for Members of Parliament, and the elections take place in May. The first business of the new council is to elect the Lord Mayor. The person has been chosen previously by whichever party has the nomination for the year. As a result of an informal agreement between the political parties, the Liberal, Conservative and Labour parties have the nomination in turn, but the Liberals having now practically disappeared—they are all aldermen who have already 'passed the chair'[1] —Labour and Conservative take it in turn to nominate the Lord Mayor, irrespective of the proportion of the respective parties in the council at the time of the election. The Lord Mayor serves for a year and receives an allowance for his expenses. He is the chairman of the council, chief magistrate, and the official head of the city, and he spends most of his time on public relations. He entertains all distinguished visitors to the city, presides at meetings of voluntary organizations, and represents the city at conferences with other local authorities. During his year of office he eschews all party politics; he represents the whole city.[2]

POLITICAL PARTIES

At the present time (1950) the council consists of 77 Conservatives, 61 Labour and 6 Liberals, the Conservatives having a small majority. Out of a total membership of 144 there are 13 women. But although council elections are fought on party lines and the true 'independent' has now no chance of election, party politics are not so important in the Manchester City Council as in others. For instance, the dominant party does not claim necessarily the chairmanships of all the committees, nor a majority on each of the committees. It has become the practice to have a chairman and deputy chairman of opposing parties, and the main committees, those dealing respectively with finance, education, and public health, are roughly made up of the parties according to their

---

[1] This idiom is a way of saying they have served as Lord Mayor.—*Ed.*

[2] For a full and critical account of the Lord Mayor's functions, see E. D. Simon: *The City Council from Within*, pp. 151–67.

representation on the council. It is very rare that a member is turned off a committee against his wish, but each election brings vacancies.

This lack of strict party government in one of the largest municipalities has a long history. Mrs. Sidney Webb remarked on it adversely when she visited Manchester in 1899 before there was any separate Labour Party. She thought that a two-party system brought greater efficiency. The Liberal Party certainly existed longer as a third party in Manchester than in Birmingham, Liverpool, London, Sheffield, and most of the big towns. In Manchester, too, outstanding Labour men were elected chairmen of the public health and the parks committee long before the size of the party membership on the council justified it, and the city certainly benefited from their services. There are certain issues that divide the parties—the acquisition and development of the Wythenshawe estate twenty-five years ago was one—but the recent vote on the establishment of a new town in Cheshire found several Conservatives voting with the Labour Party. The amount of expenditure from the rates[1] is another, though Conservatives nowadays do not fight elections (as they did before the Second World War) on the cry of 'reducing the rates' but on the appeal for 'wise spending,' and there has been recently an alarming tendency on the part of Labour members to avert a rise in the rates by co-operating with the Conservatives to reduce expenditure, even on the social services.

COMMITTEES OF THE COUNCIL

There are 23 standing committees which are appointed every year and are permanent. In addition, several other committees are appointed for special and temporary purposes. Apart from the general and parliamentary committee (which has 45 members) and the education committee (which consists of 30 members), the usual number of members is 15. The council has four main committees which exercise co-ordinating functions over the other committees. These are finance, establishment, central purchasing, and the general and parliamentary committee. This latter committee gives guidance to other committees on questions of major policy and helps to co-ordinate policy. Issues between committees can be discussed more freely and in more detail than in open council. Although the action of these four co-ordinating committees does limit the independence of the other committees, it provides to some extent the leadership that in other councils is provided by a more rigid party system.

The minutes of every committee have to be passed by the monthly

---

[1] Rates are a municipal tax on certain forms of property. See *post*, p. 341.

meeting of the council. This makes for some delay in matters of major importance and gives an opportunity for any member to raise any point on the minutes of any of the sub-committees which met during the preceding month, but it also means that action can be challenged by any member of the council. If this power is not abused—and it seldom is—it provides a safeguard against arbitrary action on the part of any committee. Although there is provision under standing orders for questions to be put to chairmen of committees, it is only rarely used, whereas it is the general practice in the London County Council, which follows the pattern of questions to Ministers in Parliament.

The council also has representatives on various joint boards, such as the Port of Manchester committee, and eleven directorships on the Ship Canal Company. The council has members on the regional hospital board, but they are appointed by the Minister of Health.

The council and the main committees meet once a month, always in the day time because so many Mancunians live outside the city that evening meetings are difficult to arrange. This limits membership to men and women who are either retired professional people, housewives, business men who are self-employed or in a position to take time off for public work, or trade union officials.

The education committee includes ten co-opted members; that is, members from outside the city council, but elected by it to serve on this committee. It is the custom in Manchester to include in these the Dean of Manchester, the chief functionary of the Manchester Cathedral, and the Vice-Chancellor or another member of the Senate of the University. Apart from these dignitaries, the co-opted members are appointed on a party basis, but by law they must be people with experience in education and acquainted with educational conditions in the area. There are co-opted members also on the art galleries committee who are experts and whose party politics are not known, and on several other committees and sub-committees. Although it may offend against the doctrine of pure democracy that people who are not directly answerable to the electorate should be able to influence policy and expenditure, they are nevertheless able to make a useful contribution to city government. It must be borne in mind that the co-opted members are always in a substantial minority on a committee, and also that the proposals of each committee must be approved by the council. Co-opted members are appointed by the city council for a period of one year, and because they confine their interests to one committee they are able to give more time to its activities than the average councillor.

For these reasons Manchester, under different political majorities, has always welcomed co-opted members, although she has not used the power conferred by law to co-opt members on all committees (other than those dealing with finance and the police force).

The councillors give their services voluntarily, but although they receive no payment by way of salary, they may be reimbursed for loss of remunerative time up to a limit of 10s. for four hours' absence, or 20s. per day. They also receive travelling and subsistence allowances to recoup them for out-of-pocket expenses when travelling outside the city area on council business. Even those, and they are the large majority, who are unable to devote all their time to council work, are expected to serve on three committees. When one of these is the education committee it involves attending at least three sub-committees, and there are altogether thirty sub- and sub-sub-committees of the education committee which have to be manned. In addition, schools and other institutions have to be visited. It is only because of the co-opted members on the main committees and on various sub-committees, such as those responsible for administering the college of technology, school of art, high school of commerce, youth service and juvenile employment, that the work can be done. For all its committees the council can draw upon men and women of integrity who are anxious to serve their city to the best of their abilities.

Since 1927 the chairmen of committees are not permitted to serve in that capacity for more than three years and cannot, unless by special resolution of the council, be re-elected except after an interval of one year. This was introduced to prevent the same man occupying the chair for years, sometimes after he had become almost senile, and being re-elected for sentimental reasons. It is a good system, and as no one is elected chairman until he has served for one or two years as deputy chairman, he knows the work of the committee well before he occupies the post. It also has the great advantage of ensuring that the committee will probably include a number of ex-chairmen who know more about the work than an ordinary member. On the other hand, the shorter tenure of the chairmen of committees weakens control of officials, reduces the influence of those representing towns on national bodies, and causes a chairman to vacate his post when he is just beginning to be useful. Probably five years would be a better period for the office of chairman than three years.

LOCAL GOVERNMENT OFFICERS

The committees have the services of their paid chief officials who carry out policy and control the day-to-day work of the departments. The officials are appointed by the committee concerned. Certain officials, such as the medical officer of health, the chief constable, and the sanitary inspector, have to be approved by the respective ministries on appointment. The chief education officer has to be chosen from a short list approved by the Ministry of Education. The appointment and dismissal of the town clerk—the chief official—is entirely in the hands of the council.

Inevitably and rightly, much is left to the officials and the members do not concern themselves with as much detailed work as appears on the agenda of the education committee of the London County Council, for example. It is much more difficult to draw the line between policy and its execution in local than in national affairs, but it can be done if the councillors trust their officials as they do in Manchester. The councillors realize that they are not experts and should not attempt to control the ordinary running of the departments. The officials know, or quickly learn, what issues should be brought for decision to the committee. The fact that any real or alleged injustice at the hands of an official is so easily brought to the notice of the councillors of a ward is an effective safeguard against any abuse of power. With the increasing complexity of local government the initiation of policy is now mostly in the hands of the officials, who are experts, but the committees approve and take responsibility for it. There are exceptions to this statement when certain councillors showed imagination and initiative. The Wythenshawe project was one instance of this, and the Ship Canal undertaking was an earlier example.

The municipal officials consist of 29,000 people, of whom 18,500 are manual employees, 4,000 are teachers, 1,500 police and firemen, 800 nursing and medical personnel, 4,200 clerical, technical, administrative and professional officers.

The entry to the clerical and administrative side is by examination at the age of fifteen or sixteen. Except in the case of technical or professional men, i.e. doctors, lawyers, architects, etc., university graduates are recruited direct to the service only to a limited extent. In this respect, Manchester is behind some other authorities, notably the London County Council, and undoubtedly suffers from the restriction. It is partly mitigated by the existence at the Manchester University of an evening course in commerce and administration, and by the encouragement, both moral and financial, given by the city council to its employees to take these degrees, a practice which is extensively followed. But such a large munici- pality could easily recruit more university graduates each year and use them with advantage, particularly in the town clerk's and city treasurer's departments, in the libraries, and in the education service. As it is, the town clerk has to rely on committee clerks who left school at fifteen or sixteen and on assistant lawyers for administrative work. This idea derives from the practice common to nearly all municipalities, that the town or county clerk must be a lawyer by training. Many people think that since his post is primarily an administrative one, a legal training is not neces- sarily the best preparation for it, neither do these same critics hold that a medical training is the best preparation for the post of medical officer of health, which is again an administrative post.

One of the differences between the municipal and the national civil service in Britain is that in the former the chief official is a technician— the chief education officer has been trained as a teacher, the medical officer of health as a doctor, the town clerk as a lawyer, the parks super- intendent as a gardener, the chief constable as a policeman—whereas in the latter the chief official of any government department is an adminis- trator with technical experts under him. It must, of course, be borne in mind that, to a much greater extent than in the case of government departments, the functions of the local authorities' departments are executive. The analogy of the civil service cannot, therefore, be pressed too far.

The efficiency of the government of Manchester, as has been said, depends more upon its chief officials than upon its councillors except that the councillors appoint the officials and realize the importance of securing first-rate men. The town clerk is not in a position of authority over the other officials, but as a co-ordinator and general adviser he can exercise great influence over them. The city treasurer is now generally accepted as a general financial adviser and is the man responsible for seeing that the cost of each service is not excessive—in so far as costs can be checked against national figures—but each committee is very jealous of its independence, and those like the education committee, which receive grants from government departments, have their costs checked by the Ministry concerned. The finance committee not only has to approve each year the estimates of each committee, but during the year its approval has to be sought for every project involving capital expenditure, and any new proposal for revenue expenditure. These applications, after being approved by the finance committee, are sent by them to the council with a full report. That this considerable power in the hands of the finance committee is usually accepted by the council is due largely to the wisdom as well as to the exceptional ability of the present city treasurer, whose aim is not so much to keep down expenditure, as to ensure that there is no waste.

THE SERVICES PROVIDED BY THE CITY COUNCIL

Manchester City Council provides all kinds of schools and other educational institutes. It has a central art gallery and six branch galleries which are accommodated in old houses set in the parks and it operates the Rutherstone loan collection of pictures to schools and schools of art in Manchester and the region.

There is a large reference library in the centre of Manchester which is used by people from all over the region and it includes, besides a technical and a commercial library, the Henry Watson music library from which amateur societies can borrow scores for orchestral and choral works. In the basement of this building is a small theatre to which professional as well as amateur companies pay visits. There are thirty-three branch libraries and seven travelling libraries which serve the new housing estates on which permanent buildings have not yet been erected. The library committee also operates a library service for the hospitals in Manchester and for the prison. Baths and wash-houses, buses and trolley vehicles, houses, markets, parks, cemeteries, main drains, water and police are all controlled by the city council. It town-plans the area, maintains highways, lights and cleans them, looks after old people, and the blind, deaf, dumb and crippled persons, and through its children's committee it takes responsibility for children deprived of a normal home life. Its health committee runs maternity and child welfare clinics, and employs health visitors and home helps. The city council controls the Wythenshawe estate as ground landlord and estate developer, as well as providing the services as a local authority.

Before 1948–49 Manchester owned and operated gas and electricity undertakings, but when these became nationalized they were taken away from municipal control. Manchester had gone further than some other authorities in municipalizing gas, water, transport and electricity, and therefore has lost in proportion from the policy of nationalization of public utilities.

Since the government had not reformed local government or tackled the problem of local government areas to make them suitable for modern conditions, this loss was inevitable and the country as a whole will gain from unified control, but in common with other local authorities Manchester has suffered from the centralizing policy of the central government. Its municipal hospitals have been taken away and put under a regional committee responsible to the Minister of Health and not to the elected representatives of the city. The relief of the poor, which had been a local responsibility since 1601, had before the war of 1939–45 been largely taken over by the National Assistance Board—a government agency—and now the only remaining duty of the city council is to provide residential accommodation for old people, unable by age or infirmity to look after themselves. This is now known as the welfare service. Finally, the valuation of property for rates, which had been a local responsibility for over three hundred years, has now been taken over by the Inland Revenue—a government department. Police, highways and education have been left in an unmutilated state, but education, by the 1944 Act, has now been put under much more governmental control and direction than before, although its administration is still left in local hands.

It is only fair to say that the governments which have taken away so much have also given some new powers to local authorities. The wartime power to provide restaurants has been made permanent provided the undertaking does not run at a loss for more than three consecutive years. Local authorities can now, subject to a rate limitation, provide theatres, concerts, cinemas, and other forms of entertainment.

The Wythenshawe estate, an area of about 5,500 acres, across the River Mersey in Cheshire, was incorporated with the city in 1930. The corporation owns 4,600 acres of the estate, which it has developed as a satellite garden suburb with factories, shops, public houses, cinemas, parks, schools, and, of course, houses. Before 1930 there were two garden cities in England, Letchworth and Welwyn, both controlled by non-profit making bodies which had no local government powers. On the other hand, local authorities dealt with the post-1918 building developments by providing housing estates with schools and some shops in their suburbs. Manchester was the only municipality to develop something between an ordinary housing estate and an independent garden city, financed from the city rates. The war held up development which is now nearly completed. There is now a population of approximately 51,000. When development is finished, in about four years' time, it will be 90,000. The initiative for this enterprise came from one or two members of the city council and a stiff fight against reactionary forces had to be maintained for many years in order to safeguard the development, although all parties now take pride in a really remarkable municipal enterprise.

Current proposals for the development of a new town in Mobberley, Cheshire, for part of Manchester's surplus population and for a new form of regional government, to both of which reference will be made later, show that there is plenty of leadership still to be found in the city council, although these later proposals perhaps owe more to the initiative of officials—at least in their early stages—than to that of individual councillors.

The power to develop new towns was withheld from Manchester and other large cities by Parliament when it enacted the New Towns Act, 1946. The many new towns under construction in Britain are all being developed by non-elected corporations appointed by the government, and responsible only to the Minister of Housing and Local Government. As the pioneer of Wythenshawe, Manchester had hoped to be able to deal with a part of her housing problems by developing a new town in Cheshire which, unlike Wythenshawe, would not be absorbed within the city boundaries, but eventually become an independent municipal borough within the county.

THE CITY'S FINANCES

The gross expenditure of the city was £13·6 millions in 1949–50. This does not include capital expenditure, or the expenditure of the trading departments, running public utility and similar services.

The net expenditure of the various committees of the council on the General Rate Fund during 1949–50 is shown on the opposite page.

The income of £3,080,871 received in respect of education consisted of £2,508,862 from government grants, and 'other income' of £572,009 received from fees and charges of various kinds. The income of £1,265,718 received in respect of assisted schemes of housing included £804,985 from rents, a government grant of £359,724, and the remainder from other sources. Government grants for specific services are thus shown as 'income' received in respect of that service.

The rate in the £ for the year under review was 20s. on the rateable value of the property subject to local taxes. That is, 100 per cent.

Houses, shops, offices, warehouses, factories and mills are assessed roughly on their rent, but industrial undertakings pay on only 25 per cent of their assessment. Agricultural land (of which there is little in Manchester) pays no rates at all.

	Expenditure £	Income £	Net £	Equivalent Rate in Pound d.
Art galleries	57,303	18,589	38,714	1·52
Baths and wash-houses ..	229,467	96,202	133,265	5·24
Children's	308,206	165,987	142,219	5·59
Cleansing	520,614	95,850	424,764	16·69
Establishment	1,649	—	1,649	0·06
Finance	384,301	213,923	170,378	6·70
Health	917,392	524,212	393,180	15·45
Highways	1,439,059	286,679	1,152,380	45·29
Housing (non-assisted schemes)	95,176	95,469	Cr. 293	Cr. 0·01
Libraries	213,970	24,102	189,868	7·46
Markets—diseases of animals ..	891	411	480	0·02
Parks and cemeteries	414,496	134,307	280,189	11·01
Rivers	305,767	56,851	248,916	9·78
Town hall	585,931	323,453	262,478	10·32
Town planning and buildings	207,228	47,471	159,757	6·28
Watch	1,192,646	571,250	621,396	24·42
Welfare services	233,032	126,457	106,575	4·19
Wythenshawe estate	150,068	111,483	38,585	1·51
Airport	119,839	87,367	32,472	1·28
Civil defence	50,230	48,799	1,431	0·06
General and parliamentary ..	12,831	11,144	1,687	0·07
Public safety	667	329	338	0·01
	7,440,763	3,049,335	4,400,428	172·94
Education	4,784,454	3,080,871	1,703,583	66·95
Housing (assisted schemes) ..	1,425,257	1,265,718	159,539	6·27
	13,650,474	7,386,924	6,263,550	246·16
Precepts from other authorities	8,239	—	8,239	0·32
	£13,658,713	7,386,924	6,271,789	246·48

The total expenditure shown above was met by revenue from the following sources:

	£
General rate	6,030,290
Government grants	3,842,649
Other income and balances ..	3,785,774
	£13,658,713

Some government grants are given for specific services, others for general purposes. Thus the Home Office pays a 50 per cent grant-in-aid of local expenditure on the cost of the police, 25 per cent on the cost of the fire service, and 50 per cent of the cost on the service for the care of children deprived of a normal home life.

The Ministry of Education assists the expenditure of the city council on education based on a formula which, in Manchester, provides 55 per cent of its total expenditure on this service. The Ministry of Health pays 50 per cent of the cost of the health services, and the Ministry of Housing and Local Government makes a grant of £16 10s. per annum for sixty years for every house built by the housing committee. This last-mentioned grant, together with a subsidy from the rates, enables the houses to be let below their economic rent. There are other special grants such as those in connection with road improvements and the civil defence services. Unlike some foreign municipal bodies, English municipalities are not allowed by the government to raise any local taxes—apart from rates—neither do they get the proceeds in their locality of the duties on alcoholic drinks, and the tax on entertainments. There is very little elasticity about their income, although the specific government grants are based on expenditure. A rate of more than 20s. in the pound is still considered—in Manchester—a financial enormity, although similar rates and even higher ones, are found in other parts of the country. The new revaluation of property which will come into force in the near future may reduce the actual poundage of the rate but only by increasing the valuation of property. Manchester, in common with all municipalities, suffers from the ban of the national Treasury on any competitors in the field reserved for national taxes.

RELATIONS WITH THE CENTRAL GOVERNMENT

There are close relations between the departments of the council and the central government departments. Those committees, such as those dealing with education, the police, and health services, which receive a percentage grant based on their expenditure, have to submit preliminary estimates to the government department concerned. The latter can question the amount of any item and if they disapprove of any proposed expenditure, can refuse to pay grant on it. This does not mean that the council cannot go on with the expenditure—if it is within its legal power to do so—but that the whole amount will have to come out of the rates. This form of government control (which is under review at the present time as the result of an enquiry into manpower in local government) is usually exercised with discretion, especially in the case of a large authority like Manchester, which has Members of Parliament who are ready to take up any corporation grievance with a government department. The percentage grants are given subject to the service being regarded as 'efficient'; and in addition to the check on expenditure there are inspectors appointed by the government who regularly inspect schools, the police force, and the public health services. There are separate inspectorates for each of these services.

There has been an increasing tendency in latter years of imposing expenditure on local authorities over which they have little control. In the first place, higher standards, such as those for schools required by the Education Act, 1944, are enforced by the Ministry as a condition of grant. In the second place, salaries of many employees are now settled either by national joint negotiating bodies, such as those which determine the scales of remuneration of teachers, policemen and local government officials, or by regional Whitley Councils. Local authorities are jointly represented on these bodies, and are obliged to pay the scales agreed upon.

PLANNING AND REDEVELOPMENT OF MANCHESTER

Manchester, like other towns, has been seriously tackling its housing problem since 1919, when municipal building began on a large scale. Manchester was also the chief of fourteen local authorities which formed the area of the Manchester and District Regional Planning Committee, and by 1939 had prepared a draft scheme for the area. But the powers of such committees were very limited under pre-war legislation, and it was not until the Town and Country Planning Act, 1947, that more adequate powers were given and the present planning was instituted. Manchester was, therefore, the centre of two new development plans; one for the city alone, and the other for the immediate district. The fact that the city surveyor, treasurer and town clerk acted in an honorary capacity for the Regional Town Planning Committee prevented any conflict between the plans, and two separate reports were published in 1945.

Main roads—especially the outer of three ringroads—extend beyond the city's boundary and so, to be effective, does zoning of industry and open spaces. But the greatest interest to Manchester residents is the plan for the redevelopment of the city itself, which is based on a most comprehensive survey, physical, economic, and sociological. Although Manchester did not suffer so seriously from bombing as did some other cities, such as Liverpool, Bristol, Hull, Southampton and Coventry, an area in the commercial centre of the city was destroyed, and she had every reason to share the impetus to replanning given by the war.

The main object of the plan is to enable every inhabitant of the city to enjoy real health of body and health of mind. That to attain such an object involves drastic changes is obvious from the descriptions given of the inner ring of the city, re-inforced by the following account. 'Industry and housing are jumbled together in many districts on the fringe of the central area of the city and among the inner wards. Narrow streets lined with terraced houses lead up to the very gates of old, unsightly, cramped and ill-planned factories. Loading facilities are often crude and confined; accommodation for newer processes is restricted. In some areas industries still carry on in rows of houses hurriedly converted into workshops a century ago. Amid these disordered industrial slums are more recent factories, rising above their outworn neighbours, but often occupying every available square inch of ground, and thus adding to the general congestion. Sometimes, too, noxious industries are found near those engaged in food production.

'Such is the aspect of industry in Ardwick, Bradford, and Miles Platting, where succeeding generations of families have worked together all their lives. Conditions in the clothing section are equally bad; sewing machines, guillotines and presses are packed closely together in the upper floors of retail shops and other unsuitable premises. Book-binding and printing in the city centre are often similarly housed.'[1]

The plan proposes a radical redevelopment of the city to be carried out in stages. The centre is allocated to commerce and administration, with a large cultural area which includes the University and its hostels, the principal hospitals, a new college of art, and other institutions of higher education. There are to be more open spaces and roads, planted with trees brought right into the heart of the city. Industry is to be restricted to special zones. The residential areas are planned in neighbourhood units which will contain between 5,000 and 10,000 people, combined into district units of about 50,000. These will have the schools, health centres, libraries, shops, churches, public houses, cinemas, and community centres usually associated with neighbourhood planning. Open spaces will be provided in what is now the heart of the crowded city and for the most part houses and not flats will be built. This increases the size of the overspill, which it is estimated will be at least 150,000 people, or nearly 20 per cent of the present population, but is in conformity with the wishes of the citizens. Here there is a fundamental

[1] The Manchester Plan, pp. 87–8.

difference between England, on the one hand and the continent and the U.S.A. on the other. The Englishman loves his garden and only a minority would willingly forgo it even to live nearer to his work. 'In the countries where workers' flats have been most generally popular, housing standards are markedly lower than in Britain, the climate encourages a gregarious outdoor life, and public transport is not sufficiently developed to permit the employees of concentrated industries to live in open surroundings. It would be a profound sociological mistake to force upon the British public, in defiance of its own widely expressed preference for separate houses with private gardens, a way of life that is fundamentally out of keeping with its traditions, instincts and opportunities.'[1]

Under the new plan, industrial zones will be within the reach of all houses, even if they are separate cottages as proposed. How to house the overspill has not yet been solved. A proposal of the city council to establish a new town in Cheshire, twelve miles from Manchester—not to be incorporated into the city like Wythenshawe, but to develop into a local government unit which would eventually become independent—was turned down recently by the electors. One of the Government's new towns may be built there instead, but in that event preference will not be given to the migration there of Manchester citizens as against those from other overcrowded towns in the area.

Another proposal for the overspill from Manchester, Salford, Stockport and Stretford, is to fill in the gaps and extend various country towns in north Cheshire and to a small degree in south Lancashire. Both plans would be necessary to solve the problem of the overspill from these areas, but both are at present in abeyance.

As soon as the rate of new building has caught up with the wartime shortage and the present overcrowding, the demolition of old houses can begin. This is planned in five yearly periods, and when the plan was published, it was hoped that building would be at a much more rapid rate than has proved possible, owing to the severe post-war shortages of building labour and materials, and the restriction of capital development. It was hoped that over 5,000 houses a year would be built, but, in fact, the yearly average has been only 1,800, so that the date for demolition of the first 60,000 houses now over seventy-five years old, which was fixed for 1961, will be considerably delayed. But the remaining land in the city zoned for residential purposes will be built up in about another six years' time and if other land for expansion is not found in the next year or so, building will have to stop.

[1] The Manchester Plan, p. 7.

A plan of this magnitude could not be carried out if the pre-war powers of local authorities to acquire land had remained unaltered, but the 1947 Act was 'both a planning enactment and a land measure of great importance. The compensation problems which had handicapped earlier planning acts were partly overcome by vesting in the state all future development rights in land and buildings; and by making a development charge when their use was changed to the advantage of the developer. At the same time, the sum of £300 million was set aside to compensate property owners whose interests suffered unduly as a result of these provisions. In spite of teething troubles and anomalies, planning control—particularly of virgin land—has been largely freed from the burden of compensation, while the acquisition of land to carry out plans has been made easier. Moreover, a generous scale of grants for redevelopment has been laid down.'[1]

A development plan for Manchester, in common with those of all other authorities, had to be sent to the Ministry of Housing and Local Government by setting out the proposals the city council considers can be carried out within the next five years, with proposals for the future use of this land during a further fifteen years. This development plan, which is to be reviewed every five years is, in fact, a continuous process. The plan is used by the town planning committee as a basis for making decisions on current applications for planning approval to proposals submitted by individual citizens or firms, by other committees of the council, and by bodies like the University. Limited approval is given where redevelopment is likely to take place within twenty or thirty years. The congested parts of the city, will, so far as industrial and commercial buildings are concerned, be developed first, and permission to use any existing building there for which planning approval is required may only be given for five years, so that there will be no difficulty in clearing it away or altering its use in accordance with the progress of the redevelopment in the city.

Development is continuous and it is only by keeping control of it in the town planning committee that redevelopment can ever take place in accordance with the plan. It was estimated that, given normal conditions, and a determination on the part of Manchester citizens to do it, the whole plan could be carried out in fifty years, but in the changed economic situation it is not possible to give a date. Meanwhile the acquisition of blitzed and adjacent sites is occupying the attention of the council.

It is not possible to calculate the cost in pounds, shillings and pence

[1] *Town Planning and the Public.* P.E.P., p. 3.

of redeveloping the city on the lines proposed, but the assumption underlying all town planning and especially the redevelopment of congested areas is that it pays good dividends, if not in actual cash, in improved health and happiness, time saved in travelling, shopping and getting to places of recreation, and in the general well-being of the community. Industrial cities in England—and Manchester as the oldest is one of the most outstanding examples—are ugly, dirty and ill-planned. They now have a chance to transform themselves into beautiful, clean and attractive places for living. By general consent Manchester has prepared the most thorough and comprehensive plan for redevelopment. Its citizens, through the town planning committee of the council, can be trusted to carry it through.

THE FUTURE OF LOCAL GOVERNMENT IN MANCHESTER

One of the main problems of local government today is that of areas. It is a commonplace that the development of modern forms of communication which have reduced—though by no means eliminated—distances, and the need for large areas for the modern development of transport, water, gas, electricity, and hospital services, have made the old administrative areas obsolete. As we explained in the first part of this essay, Manchester is the economic centre of a large industrial region of 10 million people. But 72 local authorities with a total population of $2\frac{1}{2}$ millions exist within a radius of about fifteen miles of the city. Manchester's local government services spill over its boundaries. It receives and disposes of sewage from twelve adjoining authorities and its higher education institutions are used by residents from outside. Its municipal reference library serves a wide area. It is the second largest water undertaking in the country, bringing water from the Lake District one hundred miles away, and supplying in bulk 26 other districts in addition to itself and Salford. Many of the services it provides, such as roads, cleaning, lighting, etc., are used by the large number of men and women who work in Manchester but live and pay rates outside.

The redevelopment of the city and the rehousing of its congested population involves, as we have seen, an overspill of 150,000 people since there is no more land for housing within the city's boundaries. In the past, the solution which each town applied to a similar situation was to extend its boundaries. The last extension of Manchester's boundaries took place in 1930, when Wythenshawe was taken from Cheshire and incorporated with the city. But the limits of such action have been reached. To extend Manchester still further south, even if the government's policy of building new independent towns to solve the housing problem does not stand in the way, would make it too unwieldy a unit for efficient government. There is, moreover, no agreement as to the areas that should supersede existing ones, and the government, in default of such agreement, has nationalized some municipal services, and regionalized their administration. Even so, the close proximity of so many local authorities in this area leads to overlapping, and the rigidity of local government boundaries confines the financial powers of each authority to its own area. Thus, although Manchester and Salford touch along a main street, the former is the richer authority and both are poorer than the borough of Stretford, whose rateable value benefits from the industries of both Manchester and Salford, but whose population, being of a higher income group, does not make such heavy demands upon the municipal services.

Similar irrational and inefficient situations occur elsewhere in the region, and the problem of redevelopment of the city and of the overspill of so many of its inhabitants makes the problem of areas even more difficult. As we have seen, the only land available for rehousing the congested populations of Manchester and Salford is outside the boundaries of both. Another consideration is the growth of community of interest and of economic interdependence between the parts of the greater Manchester region which is not reflected in the local government arrangements.

A REGIONAL SOLUTION

In 1945 the Government appointed a Local Government Boundary Commission[1] to enquire into the problem of local government areas of England and Wales, and in 1946 Manchester called a conference of representatives of the 72 local authorities of all kinds which comprise the region mentioned above. Of those which accepted the invitation, 33 agreed to pursue the matter further and a sub-committee was set up to prepare a scheme. This scheme which envisaged the establishment of a Greater Manchester County Council, has caused much discussion in local government circles because although no two conglomerations of populations and local authorities in England are alike, some features of the Manchester scheme might be applied to other parts of the country. It proposes to set up a Greater Manchester County Council to include the 72 existing local authorities in Lancashire, Cheshire and Derbyshire,

[1] The Local Government Boundary Commission was abolished in 1951, after presenting some valuable reports on which no action has been taken.

with a population of $2\frac{1}{2}$ millions, an acreage of 450,000 and a rateable value of £219 millions. Existing local authorities, whether county boroughs, boroughs, or urban district councils would be known as district councils. The municipal services were to be divided into the following three classes:

(1) Those that should be provided, financed and administered solely by the county council, namely, water supply, police, fire service, sewage disposal, and welfare of the blind.

(2) Those that should be financed, planned or organized by the county council but administered by the district councils, such as education, health, maternity and child welfare, town and country planning, public libraries, art galleries, and some parts of housing and slum clearance.

(3) Services of an essentially local character which the district councils should both provide and administer. Examples of these are sanitary services, maintenance of local and district roads, street cleansing and lighting, removal and disposal of house refuse, local parks, community centres, markets, and the collection of rates. Although to start with the district councils would be the existing municipal authorities, it was assumed that the Local Government Boundary Commission would review all boundaries, and readjust them so as to ensure that all district councils would be able to carry out the functions allotted to them, which would not be the case at present.

The plan proposes that the Greater Manchester County Council should consist of representatives of the district councils, that is to say, the members would be indirectly elected. This is a very debatable question because indirect election has been found in the past not to be consistent with full responsibility. On the other hand, the system by which the same electors elect a district council to carry out certain functions, and a county council to carry out others, is not satisfactory. If, as the authors of the scheme hope, the main issue at district council elections would be the suitability of people to represent it on the Greater Manchester County Council, the larger questions of the whole region would attract the attention of the electors. To avoid any danger of the county council being undemocratic no co-opted members or aldermen of district councils could be members and all members would have to submit themselves for periodic re-election as members of their district council.

With regard to finance, it is proposed that the Greater Manchester County Council should decide the amount it needs to carry out the duties put upon it and that it should then make a levy upon the district councils.

They would also raise the money they need to carry out their specific duties.

The financial aspect, in some ways the most important of all, has not been worked out in any detail, nor has the delicate question of the delegation of powers to the district councils to enable them to administer the class (2) services, been considered. Experience since 1945 of the delegation of powers by existing county councils to divisional executives for the administration of education in counties, shows how difficult it has proved to be to get a major authority to delegate fully and wisely to minor authorities. But if autonomous county boroughs like the present Manchester, Salford, Bolton, Rochdale, Oldham and Bury are to sacrifice much of their sovereignty to a Greater Manchester County Council, they must be given considerable freedom in return and adequate financial resources to carry out the functions left to them. It is likely, however, that just as some of the existing districts should be enlarged, so the big towns might be cut up into a number of separate district councils to make for a greater uniformity of size in the lower tier. At present the scheme is no more than a scheme. It cannot be implemented without an Act of Parliament and this would certainly be opposed by those authorities, the majority in the area, that have not given their agreement. But if some such scheme could be carried out in this region and similar ones to fit other parts of England and Wales where similar problems have arisen, it is perhaps not too much to hope that these new and much larger local government units could take back from the central government the provision of some of those services that have been taken away on the grounds that local government units were not suited to modern conditions. Those who believe that extreme centralization by the national government is incompatible with a healthy democracy and that local self-government is essential to our British form of government are looking at the Manchester scheme with interest. The reform of local government is urgent, and if no one plan will fit the large urban concentrations and the sparse rural areas, there are fundamental principles which are applicable to both.

Montreal and Toronto

K. B. CALLARD

KEITH BRENDON CALLARD

Born 1924. Educated at Whitgift Middle School, Croydon, and the London School of Economics and Political Science. Served in Second World War with the Queen's Regiment and the Madras Regiment, Indian Army. Returned to the L.S.E. and graduated in 1948 with 1st class honours in Government, B.Sc.(Econ.).

Lecturer in Political Science at McGill University, Montreal, 1948–50. Junior Fellow, Harvard Society of Fellows, 1950–52. Returned to McGill University in 1952 and is now Associate Professor of Political Science.

31. MANCHESTER

The Civic Centre. The circular building is the Public Library. On the right are the municipal offices, and immediately behind them is the clock tower of the Town Hall.

[Photo: N. S. Roberts, Manchester

32. MONTREAL

A view of the business section of the city. The St. Lawrence river is in the background.

[Photo: Canadian National Film Board

Montreal and Toronto

LOCATION, HISTORY, IMPORTANCE

MONTREAL and Toronto are, by considerable margins, the two largest cities of Canada. Each metropolitan area contains more than one million people, and jointly they comprise 18 per cent of the Canadian population. Both are cities of great industrial and commercial importance, situated along the St. Lawrence–Great Lakes water route that has been the mainstay of the continued existence of the Canadian state. Consequently each has a hinterland that stretches eastward to the Atlantic and westward to the Rocky Mountains. To the people of the Maritimes and of the Prairie Provinces they represent the twin colossi of financial might.

It is not surprising that the two cities should be rivals for prestige, influence and commerce; but the visitor who is not aware of their history finds it remarkable that they form almost as great a contrast as Paris and London. The differences are apparent as well as real. The new arrival in Montreal can see from a distance of twenty miles the illuminated cross on top of Mount Royal ('royal' in honour of a Bourbon king), and will soon notice the blue and white banner with the *fleurs-de-lis*, which is the flag of the Province of Quebec. The careful alternation of street names (Papineau, Amherst, St. Hubert), shows the balance of the two cultures. Only one culture is apparent to the visitor to Toronto. Each road which leads to the city is labelled 'the King's Highway,' and the seat of the provincial government lies at Queen's Park.

The linguistic aspect of this contrast between the two cities is demonstrated by the census figures [1] of languages spoken.[2]

Greater Montreal

Persons speaking English only	324,951 (23 per cent)
Persons speaking French only	501,649 (36 per cent)
Persons speaking English and French	557,064 (40 per cent)

Greater Toronto

Persons speaking English only	1,065,530 (95 per cent)
Persons speaking French only	1,330 (0 per cent)
Persons speaking English and French	38,821 (3 per cent)

[1] Unless otherwise stated all statistics of population are based upon Ninth Census of Canada (1951) *Report*, Vols. 1 and 2.   [2] *Census*, Vol. 1, Table 58.

M

As might be expected from the linguistic division of the inhabitants, the religious balance is also a point of contrast between the cities.[1]

Greater Montreal
> Total, 1,395,400, Roman Catholic, 1,050,712 (75 per cent)

Greater Toronto
> Total, 1,117,470, Roman Catholic, 187,262 (17 per cent)

Montreal is an old city by the standards of North America. It was founded by Maisonneuve in 1642 as a religious and administrative settlement on the upper St. Lawrence. The influence of the French régime is still much in evidence in the contemporary city. The whole of the land was granted to the Sulpician Order which still possesses considerable holdings. The early form of land sub-division, with the typical 'long-lot farm' fronting upon the river and extending deeply inland, continues to complicate the task of the present-day planner.

During the early nineteenth century Montreal surpassed its rival, Quebec City, in population, trade and influence. Its growth had continued rapidly after 1760 when Canada passed from French to British rule, and its character as a centre of dual cultures began to form. It was incorporated as a city in 1832; the charter lapsed in 1836, but a new charter was approved four years later. The form of municipal administration has never been stable, and minor changes in the council, the office of mayor and the method of election have been made at frequent intervals. A new charter was introduced in 1899, and was completely revised following a referendum in 1909. In 1918 the provincial legislature placed the administration of the city government in the hands of an appointed commission. Another plebiscite soon ended this experiment and a more orthodox council-and-executive-committee system took its place.[2] In turn this has been modified by substantial amendments in 1940 and 1949.

These oscillations in administrative history are explicable in part by the uncertainties of its internal politics, which have been great,[3] and in part by the frequent hostility of the provincial government.

Toronto is a city of the nineteenth and twentieth centuries. The site was selected by Lord Dorchester and the outline of the town was laid

---

[1] *Census*, Vol. 1, Table 43.

[2] *Charter of the city of Montreal*, 1942 ed. Historical notes, p. iv.

[3] E.g. Royal (Cannon) Commission on Montreal, *Report* (1909). The report showed that '25 per cent of the annual revenue of $5 million had been spent in bribes and malversations of all kinds.' The citation is given in S. Leacock: *Montreal* (1945), p. 224.

out by the Lieutenant-Governor of the province of Upper Canada in 1792. From the very beginning it acquired a military and official air.[1] Concession lines were drawn in chessboard fashion $1\frac{1}{4}$ miles apart, disregarding entirely the natural topography, especially the contours of the lakefront, river valleys and ravines.[2] The town was granted a form of self-government in 1817, and having grown to a size of 9,000 inhabitants by 1834, it was incorporated as a city by Act of the province of Upper Canada.[3] Eight years later Charles Dickens had little difficulty in assessing the salient points of its character. He remarked the 'motion, bustle, business and improvement,'[4] of its commercial life but on political matters he could only conclude that 'the wild and rabid Toryism of Toronto was appalling.'[5]

The administrative experience of Toronto has been less erratic than that of Montreal. Changes have taken place, but their most frequent subjects have been the number and boundaries of the wards, and the size of the council. The police force was removed from the direct control of the council in 1854 and placed under a board; and a separate Board of Waterworks Commission was established in 1872. These were changes of 'prudent housekeeping' and serve to reveal the city's character. A trace of Scottish ancestry may perhaps be discerned in its own statement: 'The City of Toronto is proud of its reputation of never having defaulted nor asked for a renewal of a debt.'[6] And the official motto of the city is, 'Industry—Intelligence—Integri.y.'

The site of Montreal was determined by the twin necessities of communication and defence. Montreal Island lies immediately below the rapids that have traditionally marked the limit of main navigation, and extends to the junction of the St. Lawrence and the Ottawa rivers. Mount Royal rises sharply from the middle of the south shore[7] of the island, and apart from the St. Lawrence, forms the dominating topographical feature of the area. The site of the original settlement was, not unnaturally, between the mountain and the river, and the centre of the present city is crowded into successive 'strata' on the lower slopes of Mount Royal. Growth could take place only by expansion east and west along the river

---

[1] D. C. Masters: *The Rise of Toronto* (1850–90), (1947), Chap. 1, *passim*.

[2] Griffith Taylor: 'Topographical Control in the Toronto Region,' in *Canadian Journal of Economics and Political Science*, Vol. 11 (1936), No. 4, p. 493.

[3] City of Toronto, *Municipal Handbook* 1949, p. 19. [4] Cited *ibid.*, p. 22.

[5] Cited in Middleton: *Municipality of Toronto*, Vol. 11, p. 679.

[6] City of Toronto, *Municipal Handbook* 1949, p. 11.

[7] Conventional directions in Montreal are derived on the assumption that the river flows from west to east. Thus 'conventional' north is in fact somewhat west of north-west.

banks and over and around the mountain. The result is a magnificent natural park in the centre of the city, and an almost insoluble traffic problem.

Toronto lies on the north shore of Lake Ontario, almost opposite the mouth of the Niagara river. It is situated at the base of the peninsula of southern Ontario, which contains some of the most fertile land and the highest concentration of population in Canada. It has a fine natural harbour, protected by an island, and is the focal-point of a network of railway lines radiating to New York, Chicago, Montreal and western Canada. The city is concentrated in a semi-circle from the harbour, on the gently rising shore of the lake. There is no great topographical obstacle to expansion in any direction.

POPULATION AND LOCAL AUTHORITIES

Each province in Canada has its own system of local government, but both Quebec and Ontario have classified urban units as: cities, towns and villages. In Ontario, the city is not in any way subject to the authority of the county, while the town and village municipalities are represented on the councils of their appropriate county, which has responsibility for some local government functions. In Quebec there exists a similar situation except that the town is classed with the city, and not subordinated to the county.

The census designation 'urban'[1] does not necessarily imply the presence of a large population; the 'town' of Ile Dorval in Greater Montreal, contained in 1951, 17 persons. Nor may it be assumed that the hierarchy of status is in any exact fashion correlated with size;[2] in Greater Toronto the 'village' of Forest Hill has more inhabitants than each of four neighbouring 'towns.'

'Rural' areas are divided into townships (Ontario) or parishes and townships (Quebec), and these are grouped into counties. In the vicinity of large cities the word 'rural' has become completely ambiguous. For example, the township of York, adjacent to the city of Toronto, contains 101,582 persons in an area of 5,050 acres,[3] giving an average density of 20 persons per acre.

[1] For a discussion of the definition of the term 'urban' see: Seventh Census of Canada, 1931, Monograph No. 6, *Rural and Urban Composition of the Canadian Population.*

[2] K. B. Callard: 'The Present System of local government in Canada: some problems of status, area, population and resources,' *Canadian Journal of Economics and Political Science,* Vol. XVII, No. 1.

[3] Ontario Municipal Board, *Decisions and Recommendations Dated January 20, 1953,* pp. 11, 15.

No Canadian province has established any scheme for the regular review of local government boundaries. In central Canada the original boundaries, for the most part, had been drawn by the early years of the nineteenth century. Since that time alteration has been intermittent and piecemeal. New incorporation or promotion in the hierarchy has normally been granted in response to a petition from the inhabitants, provided that certain statutory minimum standards have been attained.[1] There has been little indication that the provinces have regarded themselves as responsible for the establishment and maintenance of an effective pattern of local authorities. In such matters the role of the province has been that of an arbiter of disputes rather than that of an interested party.

A city may enlarge its own boundaries by process of 'annexation' of adjacent areas. Section V, para. 8 of Montreal's Charter[2] gives the council the power, by by-law, to annex any city, town, village or municipality on the Island of Montreal. Such a by-law must in normal circumstances be approved by the council and the property-owners of the area to be annexed. Final effect is given to such a measure by the approval of the provincial government. In Ontario it has been more common for small tracts of land to be separated from a neighbouring municipality and annexed to a city. This may be effected by the Municipal Board (the organ of provincial control), with the consent of the electors of the area concerned.[3] An entire town or village may be annexed with the consent of its council and electors.[4]

During the early years of their incorporation both cities enjoyed some room to expand within their original boundaries. But between 1883–1914 Toronto effected over 30 acquisitions and grew to an area of 40·57 square miles. Montreal's period of extension also occupied the years 1883–1914 and 23 municipalities and some lesser areas were added to the city. The present area of Montreal is 50·39 square miles. After 1918 the idea of indefinite expansion of the city limits was abandoned together with the wilder assumptions regarding the probable increase of the population. The larger adjacent municipalities had grown strong enough to desire to maintain their independence and the two big cities had no wish to assume the debt burdens of their weaker neighbours.

Annexation is essentially a haphazard process. It results in the addition of arbitrarily designated pieces of territory at one point of a city's peri-

---

[1] Seventh Census of Canada, 1931, *Monograph No. 6*, pp. 95–9.
[2] Province of Quebec, 62 Vict. C. 58 (1899) as amended.
[3] Province of Ontario, *Revised Statutes* 1937 C. 266 (*The Municipal Act*) s. 20.
[4] *Ontario Municipal Act*, s. 23.

meter, and may even result in the creation of an enclave of independent
jurisdiction entirely surrounded by the major authority. Since the
consent of the minor municipality is required, the negotiations which
precede annexation often assume the character of a hard-fought commercial bargain rather than an agreement for mutual benefit.

The year 1914 may have marked the end of one phase of the extension
of city boundaries, but it did not indicate any decline in the rate of
accumulation of population or of territorial extension of the metropolitan
areas. With the coming of the motor car and the improved street-railway
it became possible for large numbers of city workers to live in dormitory
areas many miles from their place of work. Statistics of the growth of
metropolitan areas require to be used with some care, but those given
in Table 1 reveal clearly the main trend.[1]

TABLE I
The Growth of Montreal and Toronto

Date	Montreal city population	Percentage increase by decades	Greater Montreal population	Percentage total in suburbs
1871	129,822	—	—	—
1881	176,263	36	—	—
1891	254,278	44	—	—
1901	325,653	28	—	—
1911	490,504	51	—	—
1921	618,506	26	—	—
1931	818,577	32	1,023,158	20
1941	903,007	10	1,145,282	21
1951	1,021,520	13	1,395,400	27

Date	Toronto city population	Percentage increase by decades	Greater Toronto population	Percentage total in suburbs
1871	59,000	—	—	—
1881	96,196	63	—	—
1891	181,215	88	—	—
1901	218,504	20	232,346	6
1911	381,833	75	405,061	6
1921	521,843	37	603,858	14
1931	631,207	21	810,467	22
1941	667,457	6	909,928	26
1951	675,754	1	1,117,470	40

The city of Montreal is still growing rapidly, but more than one-
quarter of the inhabitants of the metropolitan area live in the suburbs,

[1] Statistics of population shown are based on Eighth Census of Canada (1941) Vol. ii,
Tables 10, 17 and Ninth Census of Canada (1951), Vol. i, Table 12. Figures for the population
of Greater Toronto 1901–21 are taken from Toronto City Planning Board, *Second Annual
Report* (1943), p. 9.

and this is a proportion that is expanding. The city of Toronto has reached its maximum size, and may even lose some of its citizens to neighbouring areas, where the growth of Greater Toronto is now concentrated.

Both areas have to contend with the problem of local administration by one major unit, and a large number of autonomous smaller units, many of them ill-provided with powers and resources. The census of 1951 listed 50 municipalities (exclusive of the county authorities), as part of Greater Montreal and it may be maintained that additional areas should have been included.[1] No fewer than 35 of these units are smaller, in

TABLE II
Municipal Units, 1951

	Number	CITIES Population Range Large	Small	Number	TOWNS Population Range Large	Small
Montreal	7	1,021,520	8,615	27	22,450	17
Toronto	1	675,754	—	4	16,233	8,677

	Number	VILLAGES Population Range Large	Small	Number	PARISHES OR TOWNSHIPS [2] Population Range Large	Small
Montreal	4	1,322	411	12	6,899	57
Toronto	3	15,305	8,072	5	101,582	53,779

point of population, than the least of the municipalities of Greater Toronto. Toronto is not faced to quite the same degree with the problem of a very large number of small authorities. There are 13 municipalities included in the census definition and these form the new metropolitan authority. However, the Committee on Metropolitan Problems[3] suggested that a more adequate definition of the metropolitan area should include 10 additional municipalities with a combined population of 46,691.

CONSTITUTION AND ADMINISTRATION

Under Section 92 (8) of the British North America Act[4] the legislature of each province is given exclusive powers to make laws governing

---

[1] A count of all municipalities in which the physical impact of metropolitan extension was apparent would yield a total of over 100.

[2] For Greater Montreal this category includes also municipalities without other designation.

[3] Civic Advisory Council of Toronto, Committee on Metropolitan Problems, *op. cit.*, pp. 1, 29

[4] 30–1 Vict. c. 3 (1867).

municipal institutions. In the province of Quebec the two basic local government measures are the Cities and Towns Act[1] and the municipal code[2] which governs the affairs of the lesser authorities. The city of Montreal has received individual legislative attention and its charter takes the form of a special Act.[3]

The authority of the corporation is exercised by a council composed of a mayor and 99 councillors. Councillors are grouped into three categories 'A,' 'B' and 'C.' Each of 33 Class 'A' members is elected by the elector-proprietors of 11 electoral districts. Class 'B' is chosen for the same districts but by all local government electors.[4] The remaining 33 councillors are appointed by 13 designated organizations connected with the life of the city, including the two universities, the Canadian Manufacturers' Association, and the trade unions. As one result of this unique method of selection, Montreal has a council that is several times as large as any other in Canada.

A councillor in class 'A' must hold, in his own name or that of his wife, immovable property of the value of at least $1,000. The mayor, who is popularly elected, and all councillors are chosen every three years. Each councillor receives $600 per year, minus the sum of $20 for every absence from a council meeting, or for any meeting in which he did not vote on every matter put to the vote. The council is required to meet at least four times a year, but in practice it meets at least monthly.

Within thirty days following a municipal general election the council is required to appoint the executive committee. Each category of councillor appoints two of its members to the executive, and the entire council chooses one of these men as chairman of the committee. The chairman has a casting vote, in addition to a regular vote, and he receives an annual indemnity of $10,000. The other members receive $7,000.

The executive committee is to a great extent the effective governing body of the city. Its functions are listed in the charter. It must prepare and submit to the council the annual budget, all by-laws, and demands for the appropriation of loans or the raising of credit. Every matter within the jurisdiction of the council, unless expressly provided to the

---

[1] Province of Quebec, *Revised Statutes* (1941), c. 233.

[2] R. Tellier (ed.): *Code Municipal de la Province de Québec* (1947).

[3] Province of Quebec, *Statutes*, 62 Vict. (1899) c. 58 as amended, referred to hereafter as 'charter.'

[4] A local government elector is, broadly speaking, one who will be affected as owner or tenant by property taxes, or the wife of such person. City electoral rolls for the election of December 1950 list 284,175 electors and 77,186 elector-proprietors, out of a total city population of 1,181,955. Montreal *Gazette*, September 21, 1950.

contrary, must be submitted to the executive committee for report. The committee is responsible for the enforcement of the law, the carrying out of the city's contracts, and a large number of administrative duties. The committee appoints all heads of departments and tax assessors; the list of names is laid before the council which may only accept or reject it as a whole. The executive determines the salaries of all officials, except those whose appointment is vested in the whole council.

The director of departments is, so to speak, the general manager of the city administration. He is appointed by the council without report from the executive committee, of which he becomes a non-voting member. The executive is not functionally organized and the director of departments serves as the principal channel of communication between it and the departments and agencies of the city administration. There are eleven main departments, law, health, assessment, police, purchasing and stores, public works, social welfare, fire, planning, finance, and city clerk's department.[1] Positions in these departments are normally filled through the Montreal Civil Service Commission, composed of three members appointed by the council on the recommendation of the executive. The jurisdiction of the commission covers all continuing employees except the directors of departments and their assistants.[2] Municipal employees are not permitted under provincial law to strike, and certain categories, such as the police, may form only local employee associations.

The mayor is the first magistrate of the city and he represents it upon ceremonial occasions. He presides over the council and may submit recommendations to the executive and to the council. He is paid $10,000 per annum. Prior to 1949 the mayor was not entitled to attend meetings of the executive committee and if he failed to sign within forty-eight hours any document which by law required his signature, the chairman of the executive might act in his stead. The most recent amendment of the charter,[3] which was not formally requested by the city, gives the mayor powers of superintendence, investigation and control over all the departments and offices of the city, as well as the power to ensure 'that the provisions of the laws, by-laws, and ordinances of the council are faithfully and impartially enforced.' More specifically he may, at any time, suspend any employee of the city, though he is required to report the occurrence immediately and in writing to the council, which is entitled to rescind his decision. If the mayor refuses to sign any by-law,

---

[1] K. G. Crawford: *Local Government in Canada* (mimeographed) (1949), p. 138.
[2] *Ibid.*, p. 152.   [3] Province of Quebec, *Statutes*, 13 George VI, c. 73.

M*

resolution or contract it is to be presented to the council at the next meeting as a matter of urgency and privilege, and, if passed, becomes law with or without his signature.

Such, in outline, is the complex structure of Montreal's government. Within it may be found elements of a 'mayor and council' system, a 'weak' mayor, a 'strong' mayor, government by commission and of a city manager system. Simplification is not introduced in choosing the council by a mixture of appointment, a narrow property franchise and a wider property franchise. This remarkable amalgam is intelligible, if at all, only against the background of municipal and provincial politics, of which more will be said later.

It would be impossible, within the confines of a single chapter, to describe the forms of government of the remaining forty-nine units that make up Greater Montreal. Table II serves to indicate the variety of status and of size that is to be found. In organization and government none of them is as unusual, by Canadian standards, as the city of Montreal. Municipal councils are small, and six cities and three towns have appointed city managers who are responsible, to a greater or less degree, for day-to-day administration. Under the Cities and Towns Act the appointment of a clerk (*greffier*) and a treasurer[1] is mandatory, but in the smaller municipalities these offices are often held by one person.

The first major distinction to be drawn between the governments of Montreal and Toronto is that the latter city finds its constitution in the provincial Municipal Act.[2] The government is vested in a council of twenty-three members, including a mayor, four other members of the board of control, and eighteen aldermen. Elections are held annually, the mayor and controllers being selected by the voters at large, while the aldermen are chosen two from each of nine wards.[3] The mayor receives $6,500 plus $5,000 in his capacity as controller, the other controllers receive $5,000, aldermen $1,200, with an additional $100 to chairmen of committees.

The right to vote in municipal elections is, as in Montreal, limited by a property qualification. The requirement is that the elector shall be rated for at least $400 on the last assessment roll, for land held as owner or tenant, or shall be the husband or wife of a person so rated. Again, as in Montreal, there is a special category of elector-proprietors which alone

---

[1] Province of Quebec, *Revised Statutes*, 1941, c. 233, ss. 82, 93.

[2] Province of Ontario, *Revised Statutes*, 1937, c. 266.

[3] The size of the wards varies from 46,908 to 118,280 inhabitants. *Vide* K. G. Crawford, *op. cit.*, p. 68.

is entitled to vote on a 'money by-law';[1] this class of elector includes the nominees of tax-paying corporations.

The council meets every second week, except during the summer recess. It is organized into the board of control and five standing committees.[2] The board of control corresponds to the executive committee in Montreal and is responsible for the general administration of the affairs of the city. In particular it prepares the annual estimates; supervises revenue, expenditure and investment; awards municipal contracts; considers all by-laws; and nominates to the council heads of departments and sub-departments. All reports of committees are submitted to the board which transmits them to the council 'with such amendments as the board may deem advisable.'[3]

The council's power over financial matters is limited specifically by the Municipal Act. A two-thirds majority is required before any appropriation or expenditure is made which has not been recommended by the board of control. A similar restriction applies to the awarding of municipal contracts, and to the reinstatement of any head of department who has been dismissed by the board.

The city administration is divided into fourteen departments: mayor, city clerk, assessment, legal, treasury, audit, works, city planning and surveying, property, street cleaning, buildings, parks, health, public welfare, and the fire department. The city gives full-time employment to some 7,500 persons.[4] Each department selects its own employees under rules established by the director of personnel. Examinations are held to determine eligibility for entrance to the clerical field and for promotions to fill vacancies.

The director of personnel confers with employees regarding grievances or with union representatives on more general matters. He considers and submits any recommendations to the advisory personnel committee, which is composed of three members nominated by the heads of departments, one member of the board of control and three representatives nominated by the employees. The committee has no executive responsibility, but submits its decisions, through the director, to the board of control.

[1] A money by-law is one which authorizes an increase in the long-term indebtedness of a municipality. Such a by-law normally requires the assent of the provincial authorities and of a majority of the owners of property by number and according to the value of property.

[2] The Committees are: Works, Property, Parks and Exhibitions, Legislation and Public Welfare. City of Toronto, *Municipal Handbook* 1949, p. 29.

[3] *Ibid.*, p. 31.   [4] *Financial Post*, April 22, 1950.

The twelve remaining municipalities of the Toronto area have a more uniform system of government than their Montreal counterparts. They are all represented on the Metropolitan Council, which will be discussed later. Their own councils range from four to nine members, and with one exception, are elected annually. In the towns the head of the council is called a mayor, in the villages and townships, a reeve.

THE SCOPE OF MUNICIPAL SERVICES

Lists of municipal operating departments have already been given for both Montreal and Toronto, and these serve to indicate the main operating divisions of the city authorities. They fail, however, to give a sufficient picture of scope and importance of the services rendered by the municipality. Both cities maintain the streets in repair, see to their lighting and their cleanliness; as much as 120 inches of snow may have to be removed from the streets of Montreal in a single winter. Water is obtained and distributed for home and industrial purposes, and public baths are provided. The cities have offices concerned with attracting both tourists and new industries to the area. Civic patronage is provided for cultural activities; concerts and exhibitions are organized or aided, and large free public libraries are maintained. The appearance of the city is improved by planning, construction of civic buildings and the provision of parks and playgrounds. Sports facilities—ice rinks, golf courses, baseball lots, a gymnasium, ski-tows and toboggan runs—are available for children and adults. Montreal owns markets and extensive botanical gardens; Toronto operates a municipal abattoir. Health services on an extensive scale exist for the prevention as well as the cure of disease; special clinics deal with tuberculosis, with venereal diseases, and with medical and dental hygiene for children. Food offered for sale is inspected for quality and purity, and municipal licences control the practice of certain occupations. Sewers and garbage disposal services remove waste matter and some efforts, though without great effect, are made to limit smoke nuisance. Municipal orphanages care for the helpless young, while old people's homes do like service for the old. Reform schools are maintained for juvenile delinquents, and Montreal's police force even contains an anti-subversive squad to protect the citizens against Jehovah's Witnesses and communists. With all these, and many other services, is the Montrealer and the Torontonian provided by his city.

In addition, both cities have a wide range of private health, welfare and recreational agencies federated into and largely supported by community chests (fund raising and distributing bodies). Furthermore the 'service clubs,' Kiwanis, Rotary, Lions, etc., undertake philanthropic functions. In Montreal, as in Quebec generally, much of the social service work is done by the Catholic Church, or by church agencies of one kind or another;[1] and there is less municipal assumption of responsibility for such services than in Protestant Toronto. Strict comparison of municipal social service expenditure is therefore misleading.

The existence of three main tiers of government (federal, provincial, local) has led to a different distribution of functions than would be likely to prevail in a unitary state. In the period prior to the Second World War the municipalities carried a major part of the burden of relief and welfare responsibility, though considerable financial aid from superior governments was found to be unavoidable. During the worst years of the depression, when tax revenues from real property had sharply declined, this burden proved too much for many municipalities which defaulted upon their obligations or failed to provide even minimum standards of assistance to the necessitous. In 1940 when the immediate need was already passing, the federal government introduced a national scheme of unemployment insurance, with, however, only limited periods of benefit. Simultaneously the impact of the war raised revenues and decreased the opportunity of spending. For many years and especially following the war, the evident incapacity of the weaker municipalities to achieve minimum standards of adequacy has led to increased provincial performance and supervision of social services. Thus the Toronto Committee on Metropolitan Problems could report that the problem of aid to the indigent has now shrunk to almost insignificant proportions.[2] The committee went on to point out that if severe unemployment should recur the cost of relieving unemployed 'employables' who had exhausted their rights to insurance would again be prohibitive. Relief costs for unemployables are already shared in equal proportions by the provincial government and the municipality.

Provincial influence has been widened also in the field of public health. The Toronto Board of Health is a statutory body consisting of the mayor, the medical officer of health and five others appointed by the council, of whom one is a nominee of the Academy of Medicine.[3] Two of the

---

[1] E. Minville: *Labour Legislation and Social Services in the Province of Quebec* (mimeographed). Published as an appendix to the Report of the Royal Commission on Dominion–Provincial Relations (1939).   [2] Civic Advisory Council of Toronto, *op. cit.*, p. 151.
[3] City of Toronto, *Municipal Handbook*, pp. 32, 33.

other municipalities in the Toronto area have combined to form a single health unit whose board contains a provincial representative.

In the past the provision of housing has not been accepted as a municipal duty in Canada and the National Housing Act (as revised in 1949) still provides for federal and provincial responsibility for the major cost of each public housing project. Earlier provision had been made for local authorities to co-operate with the federal government in housing schemes for veterans, but neither Montreal nor Toronto was induced to participate on a large scale, though the Toronto Housing Authority has undertaken one major redevelopment project. Housing assistance in North America generally, has more frequently taken the form of cheap loans or of guarantees to prospective home owners, rather than direct municipal or state construction of low-cost housing for rent.

All the larger authorities in both areas maintain a separate police force. The police department and the Recorder's court are administered in Montreal in the same manner as other municipal departments. In Ontario a new Police Act, which did not alter the structure of the Toronto organization, was put into force in 1949. Under its provisions the police force of the city of Toronto is controlled by a board of commissioners consisting of the mayor, a judge of the county court and a magistrate, designated by the provincial government. A grant of 10 to 25 per cent of expenditures, decreasing with the size of the municipality, is made by the province, conditional upon the maintenance of satisfactory standards. The financial estimates of the board require the approval of the council.[1] Similar provisions govern the remaining police forces of the area. However, as a sign of lack of co-ordination it is said[2] that, of the twelve suburban police radio transmitters, only one operates on the city's wavelength.

Transport facilities and traffic congestion are two problems that beset all large cities. A network of major highways is maintained by the provincial authorities. The province of Quebec has established six road districts comprising thirty-five divisions, which, however, exclude the larger cities. On the Island of Montreal, certain powers regarding roads have been given to the Metropolitan Commission (of which more will be said later). It has completed much preliminary work on the Metropolitan Boulevard, an east-west artery that would by-pass the city. A later amendment to the powers of the commission appeared to empower it to plan suburban highways to link effectively with the main roads leaving the

---

[1] Civic Advisory Council of Toronto, *op. cit.*, p. 122 *et seq.*
[2] *Financial Post*, April 22, 1950.

city.[1] This function appears now to have devolved upon the city of Montreal. By the normal Quebec method of a new special act,[2] the city was empowered to set up a Board of Research on Traffic and Transportation Problems which might take into consideration the whole metropolitan area.

The board's report[3] included three main recommendations. First, it was proposed that a single transportation authority should be established to construct and operate a comprehensive transportation system for the metropolis. Such an authority would be given complete administrative independence, but would owe a general responsibility to the municipalities of the area. It would take over the surface transport facilities then operated by the privately owned Montreal Tramways Company. Second, the board recommended the construction of a subway (underground railway) for the centre of the city. Third, the building of some new major roads and the improvement of others were advocated. The council approved these recommendations and submitted them in the form of a private bill to the provincial legislature. The Premier objected to some of the clauses, but after a deadlock between the two chambers, the measure was finally carried establishing the transportation authority but omitting any provision for the building of the subway.[4] A new report on the transport problems of the metropolis was presented in June 1953.

Until December 1953 many of the roads in the metropolitan area fell within the jurisdiction of the Toronto and York Roads Commission. Three types of local road can be distinguished: municipal roads, maintained entirely by the municipality; county roads, built and maintained by the commission with costs shared equally between the province and the county; and suburban roads which are specially designated county roads, the costs of which are shared in the ratio of 2: 1: 1 by the province, the county and the city.[5] In 1954 the metropolitan council has assumed responsibility for county roads within its area and now treats such roads outside its boundaries in the same manner as Toronto used to contribute to suburban roads.

Toronto has had a municipal Transportation Commission since 1920. There are three members appointed by the council for terms of three years. The commission employs approximately six thousand persons and

[1] Taggart Smyth, 'Metropolitan Reorganization in Montreal and District,' in F. Wright: (ed.): *The Borough System of Government for Greater Montreal* (1947), p. 29.

[2] Province of Quebec, *Statutes*, 13 George VI, c. 73.

[3] Montreal Board of Research on Traffic and Transportation Problems, *Report* (November 1949).   [4] Montreal *Gazette*, March 17, 1950 and April 5, 1950.

[5] Civic Advisory Council of Toronto, *op. cit.*, p. 246.

operates ferry boats as well as buses, street cars and long-distance motor coaches. The construction of a subway, the first in Canada, was begun during the summer of 1949, and is expected to be in operation in 1954. Several of the adjacent municipalities have contracted with the Toronto Transportation Commission for transport facilities, although the commission has no statutory jurisdiction beyond the city.

The generation of hydro-electric power is performed by two provincial commissions which may sell it to municipal organizations for distribution to the consumer. There are eleven such municipal commissions in Greater Toronto. In Montreal both generation and distribution used to be performed by the privately owned Montreal Light, Heat and Power Company, which supplied the greater part of the area. This organization has been taken over by Hydro Quebec. The city of Montreal supplies water to a total of fourteen municipalities, while Toronto has made similar contracts with six of the adjoining authorities.

THE CONTROL OF EDUCATION

One of the more complex aspects of local administration is the control of education. Canada has to contend with the North American requirement that control of education should be democratic but 'non-political,' and should therefore be vested not in municipal councils but in separately elected school boards. In addition there is a constitutional guarantee for Ontario and Quebec that the main religious groups shall administer their own schools.[1] In the province of Quebec, all children are entitled to attend 'common' schools, but other denominations may establish 'dissentient' schools. In the city proper both Catholic and Protestant schools are 'common,' and contracts have been made between the Montreal Jewish School Commission and the Protestant Boards of Montreal and Outremont (an adjacent municipality).[2] There is one Catholic School Commission for the city and a considerable number of authorities for the surrounding area.

Owing to the disastrous financial position of some of the Protestant school boards in 1925 the Montreal Protestant Central School Board was established covering seventeen municipalities which had been grouped into eleven school boards.[3] The local boards continue to exist and they appoint the twelve members of the central board. The central board receives all tax moneys and is required to approve the budget of each

[1] British North America Act (1867), s. 93.

[2] The account of the educational system of Montreal is based upon L. P. Patterson: *Protestant Education in Montreal* (1947), Columbia University Teachers College thesis.

[3] Province of Quebec, *Statutes*, 15 George V, c. 45.

33. MONTREAL

As seen from Mount Royal.

34. TORONTO

A downtown view of the commercial and financial section showing the dominating influence of the New York skyscraper.

[Photo: *Canadian National Film Board*]

local board in order to balance the total budget. It sets limits to the number of teachers that may be employed and to the salaries that may be paid. It may acquire and hold property as well as control the building and reconstruction of schools. By a new provision of 1944 any local board may hand over all or some of its powers to the central agency.

The local boards, apart from the city of Montreal, are elected by the voters for three-year 'staggered' terms, although the elections are frequently uncontested. Three members of the city board are appointed by the province and three are selected by the council.

The most remarkable and complex features of the administration of education are those that relate to finance. A small income is derived from fees charged to some pupils. In 1946 the province assumed responsibility for the debt service charges of the school boards, and a special fund with a revenue from designated taxes was established for this purpose. The approval of the Quebec Municipal Commission is now required for new borrowing, increased taxation or increases of salary scales. Furthermore, in 1949 an additional 1 per cent was added to the sales tax in the Montreal area to be used for school purposes.

The main source of revenue for education is the real property tax. Each Catholic school board may 'precept' upon the appropriate municipality for its needs, subject to the 1946 limitation on tax increases. But the central Protestant board has no power to vary the tax rate which is fixed by provincial legislation at 15 mills[1] for Protestant and Jewish property and 16·5 mills for property on the neutral panel. Thus the only means of varying the revenue is to vary the assessment. Assessments are performed by each municipal authority, and although the Metropolitan Commission may exercise some supervision, opportunity for local discrepancies seems to exist. The money is collected by the municipal authorities and divided into three panels, Catholic, Protestant and neutral, according to the property from which it has been levied. The neutral panel is then in turn divided between the Catholic and Protestant panels in accordance with an elaborate formula which is not uniform throughout the metropolitan area. Three methods are employed to allocate the cost of Jewish education.

In the Toronto area ten municipalities, including the city, have established boards of education and three municipalities have united to form a combined board. The city board consists of eighteen elected members, two from each ward, and two members appointed by the Toronto and

---

[1] Local taxation in Canada is expressed in terms of 'mills' (one-thousandth part) per dollar of assessed total value.

Suburban Separate School Board. As in Montreal, the main basis of school finance is found in real property taxation, which supports both the public (Protestant) and the separate (Catholic) schools. Financial inequality between the various boards has been the root of the wide discrepancy between the amount spent per pupil. Average net expenditure per pupil in the public schools ranged from $46 (Scarborough township) to $140 (village of Forest Hill) in 1946.[1]

As a parallel development to Montreal, it is in the case of the schools of the minority group that central administration has been set up for the metropolitan region. The Toronto and Suburban Separate School Board is composed of twelve trustees, elected by wards for a term of two years— nine trustees representing the city and three the suburban area. This board operates forty-three schools with a 1952 registration of 15,044 pupils.

MUNICIPAL FINANCE

During the depression years the entire system of local government was in danger of collapse because, at a time when its responsibilities had expanded enormously, the possibilities of raising revenue had sharply diminished. The remedy, in so far as one has been found, has been to remove certain duties from municipal to provincial and federal shoulders, rather than to ensure adequate and flexible revenues to the lower tier of authorities.

The main sources of revenue for the city of Montreal are given below:

TABLE III

Revenue: City of Montreal
Year ended April 30, 1952[2]

Source of revenue	Amount $000	Percentage of revenue
General property tax (including school taxes)	43,848	45·3
Sales tax	12,253	12·7
Business tax	7,272	7·5
Water service	16,576	17·1
Share of provincial amusement tax	1,005	1·0
Other sources	10,051	9·3
Total ordinary revenue	83,706	92·9
Surplus from previous years	6,880	7·1
	96,885	100·0

[1] Civic Advisory Council of Toronto, *op. cit.*, p. 105.
[2] City of Montreal, *Report of the Director of Finance for the year ended April 30, 1952.* Statement No. 1, pp. 12, 13.

The sales tax to the amount of 2 per cent on most retail sales is received by the provincial authorities together with the provincial sales tax. Two per cent is retained by both the retailer and the province to cover the cost of collection.

The business tax is imposed at the rate of 10 per cent of the assessed rental value of any establishment operating for profit, but there are special rates imposed for hotels, restaurants, distilleries, breweries and banks. The tax is now subject to a surcharge of 8 per cent.

The provincial amusement tax amounts to 10 per cent of the price of admission to all places of amusement, plus a surtax of 25 per cent. The city undertakes the collection and retains one-half of the tax and 2 per cent of the surtax. There are no other direct grants from either the federal or provincial governments. The federal government makes grants only where special hardship is caused by the non-liability to taxation of its property, or where special services are rendered by the local authority. The province of Quebec has preferred, as with the school debt, to remove a service entirely from municipal control than to aid the city's revenue. The water tax is levied as a percentage rate upon rental value of the property; the present rate for residential property is 8 per cent with a surtax of 8 per cent in the territory of the city.[1]

The remaining sources of city revenue include fines, rentals, licences, sale of property and a tax on telephones. Nearly half of its income is drawn from the general property tax. All land and immovables are valued by the city assessors at their 'full' value, though the figures reached do not correspond to contemporary market prices. Before the tax rate is calculated the total is reduced by the amount of exempt property. Exemption is granted to property owned by the federal, provincial and city governments, churches and accredited educational and welfare institutions, as well as farm-land. Exempted property amounted in 1951–52 to 22 per cent of the total,[2] though some categories were liable for school and some other taxes. On property bearing full liability the tax rates were, for Catholic property 29·652 mills, Protestant and Jewish 32·652, and for the neutral panel 34·652 mills.

For the same period the principal items of city expenditure are given in Table IV. The functions of the city having already been

---

[1] City of Montreal, *Report of the Director of Finance for the year ended April 30, 1952*, Statement No. 1, p. 3. [2] *Ibid.*, p. 55.

discussed it would not seem necessary to add any further explanation to the table.

TABLE IV

Expenditure: City of Montreal
Year ended April 30, 1952

Expenditure	Amount $000	Percentage of current expenditure
Interest	5,293	5·9
Repayment of loans	12,362	13·7
Schools	17,015	18·9
Public works department	17,485	19·4
Health and social welfare departments	7,970	8·2
Police department	8,112	9·0
Fire department	5,309	5·9
Other current expenditure	10,680	11·37
Total current expenditure	83,326	92·37
Surplus ordinary revenue over current expenditure	6,680	
Expenditure from surplus—additional reimbursement of loans and capital expenditures	6,880	7·63
	96,885	100·00

The sources of revenue of the city of Toronto are broadly speaking similar to those of Montreal, except there is no sales tax. For 1951 the tax levy was 39·30 mills which yielded $83 *per capita.*

TABLE V

Revenue: City of Toronto, 1951[1]

Source of revenue	Amount $000	Percentage of ordinary revenue
Property and business tax	54,566	77·9
Grants and subsidies	4,522	6·5
Debenture debt charges recoverable	5,659	8·1
Other ordinary revenue	5,260	7·5
Total ordinary revenue	70,016	100·00

Apart from a share in the revenue of the Liquor Control Board, which conducts the retail sale of wines and spirits, the grants from the province

[1] Province of Ontario, Department of Municipal Affairs, *Annual Report of Municipal Statistics for* 1951, pp. 4–5.

are related to expenditures on roads, relief and health, as well as a small general municipal subsidy, amounting to the product of a tax of one mill.

A new scheme of unconditional grants to municipalities has been established[1] and is to start in 1954. The scale of grants varies according

TABLE VI

Expenditure: City of Toronto, 1951

Expenditure	Amount $000	Percentage of current expenditure
Debt charges	11,954	17·1
Schools	18,881	27·0
Public works	1,543	2·2
Sanitation and waste	5,224	7·5
Social services	10,685	15·3
Protective services	11,358	16·2
Recreation and community services	3,287	4·7
Capital expenditures out of revenue	1,280	1·8
Other expenditure	5,794	8·3
Total expenditure..	70,016	100·1

to the size of the municipality and its rate of growth. The Toronto metropolitan area will be treated as one unit and will receive a grant equal to $4 per inhabitant. Certain fire and police grants will terminate, but other provincial aid to municipalities will continue. The net increased contribution to all the municipalities of the province is expected to total about $8 million.

PROVINCIAL-LOCAL RELATIONS

The reader who is accustomed to the English pattern of local government may well be surprised at the small amounts contributed by the superior levels of government in Canada. In reality the province is playing a greater part in the provision of local services than appears from the study of municipal finances. In some cases, for example vocational schools in Montreal, the province has assumed complete control. More frequently when an agreement has been reached to share the cost of a service no formal grant is made though a portion of the expenditure is paid from provincial funds.[2]

[1] Province of Ontario, *Statutes*, 2 Elizabeth II, c. 72.
[2] K. G. Crawford: 'The Independence of Municipal Councils in Ontario,' *Canadian Journal of Economics and Political Science*, Vol. VI (1940).

Montreal is not the seat of the provincial government, but the metro-politan area contains almost 40 per cent of the population of the province. It presents a unique problem to the senior government and has received individual treatment. The city's charter is a private Act and provincial control has been exerted by amending the charter or by enacting further private Acts. The charter itself gives the province the power to disallow any by-law within two months of the receipt of an official copy. For municipalities outside the Montreal region the Quebec Municipal Com-mission, an 'independent' authority associated with the provincial Depart-ment of Municipal Affairs, exercises strict financial control, especially over the raising of new long-term loans. The Montreal Metropolitan Commission performs a similar function for the city and fourteen other municipalities on the Island.

The Metropolitan Commission was established in 1921, and its purpose was stated in the preamble to the Act:

Whereas, while still preserving the autonomy of the municipalities on the Island of Montreal, it is expedient that a system of financial control by a central authority repre-sentative of such municipalities, including the City of Montreal, be established for the future.[1]

The commission consists of fifteen members, eight chosen by the city, one by the province and six representing the remaining local authorities. It is empowered to assume control of defaulting municipalities and meet their debt obligations. In 1948 there were three 'aided' municipalities and the commission levied an assessment of $118,000, the lowest in its history;[2] $55,000 of this sum was for administrative purposes. The assessment, which is under the control of the commission, is levied on each member in proportion to its real-estate values.[3] The commission also borrows on behalf of solvent members, except the city of Montreal, in order that they may obtain more favourable credit terms.

While later amendments to the Act which created the commission appear to have envisaged much wider functions,[4] very little beyond preliminary consultation seems to have resulted. And the financial aspect of its duties, apart from the value of uniformity in metropolitan book-keeping, is decreasing in importance.

It would be unsatisfactory to leave the issue of the relations between Montreal and the province of Quebec without some mention of the

---

[1] Cited in F. Wright (ed.): *The Borough System of Government for Greater Montreal* (1947), p. 9.   [2] Montreal Metropolitan Commission, *op. cit.*, p. 9.
[3] F. Wright, *op. cit.*, p. 9.   [4] Taggart Smyth, *op. cit.*, p. 29.

political situation which has frequently dominated that relationship.
Greater Montreal is urban, contains a large number of persons who speak
English either as a first or second language, and has great interests with
the federal government in Ottawa and with business affairs outside the
province. Representation in the provincial legislature is weighted in
favour of the rural areas, and whichever party is in power concessions
have to be made to a sentiment of French-Canadian nationalism. The
city has tended to follow the changes in the political balance at Ottawa,
while the provincial politician finds his strength in opposition to the
federal government. While city elections are not conducted formally by
political parties, there is a general assumption that in recent years the
city administration has been controlled by those who favour the Liberal
Party. The amendment to the charter in 1940 was made by a provincial
Liberal government, and it has been seen as no mere coincidence that the
powers of the mayor should have been markedly diminished. The mayor,
who was certainly no Liberal, was at the time suffering internment under
certain wartime regulations; this disadvantage did not prevent his later
re-election.

In 1949, under a government of l'Union Nationale, some of the mayor's
powers were restored, though not at the request of the city council.
The following year the provincial premier attempted to insert into the
Montreal Transport Bill, a clause requiring a referendum to be held
asking the municipal electorate whether or no the present system of city
government had its approval. If the vote were to be in the negative, the
provincial legislature undertook to frame a new charter. This clause was
rejected by the provincial second chamber (Liberal controlled), and was
not incorporated in the Act.[1]

The city of Toronto, which is the capital of the province of Ontario,
is nevertheless governed under general municipal law. The provincial
organs concerned with supervision are the Department of Municipal
Affairs and the Ontario Municipal Board. The department is concerned
with the general working of the Municipal Act, and has power to regulate accounting procedures, estimates and auditing. The Municipal Board
is an administrative agency appointed by the province to supervise the
individual municipalities. Its approval is required for debenture issues
and zoning by-laws; it conducts assessment appeals and arbitrations
regarding expropriation of land; and it supervises the operation of
municipal public utility undertakings.[2]

[1] An account of the proceedings in the Quebec legislature is given in Montreal *Gazette*,
March 17 and April 5, 1950.   [2] City of Toronto, *Municipal Handbook 1949*, p. 194.

Beside these general controls a number of more specific checks is included in legislation regarding particular municipal services. Many of these have laid down conditions for the appointment and dismissal of categories of municipal employees. Since 1938 the appointment of a medical officer of health has required the approval of the provincial department of health, and a similar regulation has been applied to sanitary inspectors. The province has prohibited the removal of a municipal auditor without the assent of two-thirds of the council.

THE METROPOLITAN GOVERNMENT OF TORONTO

Since 1945 the attention of the municipalities and especially of the Province of Ontario has been focused upon problems of metropolitan organization. In 1946 the legislature added an extensive section to the Municipal Act conferring upon the Municipal Board power to create Inter-Urban Administrative Areas. The city of Toronto began to press for amalgamation of the whole area into one administrative unit, but was strongly opposed by all the remaining authorities.[1]

A committee of the (provincial) cabinet, with the premier in the chair, was constituted to study the issues. Its conclusions were finally written into law in April 1953[2] and the new scheme came fully into operation in January 1954.

Under the Act the thirteen existing local governments, or area munici-palities' retain residual powers and responsibility for services of a local nature such as the distribution of water, district sewers, streets. They exercise all licensing and regulatory powers and they will continue to control fire and police services, which are not being transferred for the present, owing mainly to the increased cost which would attend the establishment of a common standard of services and salaries throughout the area.

The metropolitan government has assumed the management of filtration plants, trunk water mains, pumping stations, sewage treatment plants, trunk sewers, arterial roads, the public transportation system (to be operated by the Toronto Transit Commission), homes for the aged, court houses and jails. The metropolitan government shares concurrently with the local governments broad powers with respect to housing, redevelopment, parks and recreational areas. It seems probable that these facilities will devolve increasingly upon the metropolitan authority.

[1] Ontario Municipal Board, *op. cit.*, pp. 26–32.
[2] Province of Ontario, *Statutes*, 2 Elizabeth II, c. 73.

The metropolitan government is made responsible for basic planning throughout the whole 245 square miles of the area and, at the discretion of the Minister of Planning and Development, in the adjacent territories. It has assumed the statutory liability of the local governments for payments towards services such as children's aid, hospitalization of indigents, and the administration of justice.

The borrowing power of the whole area is to be centralized, and the metropolitan authority may raise capital for the area municipalities, the school board, and the Transit Commission, as well as on its own behalf. The central authority will carry out a uniform assessment throughout the area. The area municipalities will continue to collect revenue from ratepayers, but will contribute sums fixed by the metropolitan authority in proportion to assessed rateable value.

The council of the metropolitan authority consists of 24 members. These are the mayor of Toronto, the two city controllers who receive the largest number of votes at each election, the alderman polling the largest number of votes in each of Toronto's nine wards and the mayor or reeve of each of the 12 remaining municipalities. The council may then select a chairman either from among, or outside, its own members. The first chairman, who holds office until December 31, 1954, was appointed by the Government of Ontario. The chairman receives a salary of $15,000, and the members receive $1,800 in addition to their remuneration as members of the local councils.

In conjunction with the Metropolitan Council was established the Metropolitan School Board, consisting of 10 members from the city school board and ten from suburban boards. In addition to these there are two members appointed by the metropolitan Separate School Board, but they will not participate in proceedings exclusively affecting public schools. Local school boards continue to exist and will receive from the central board uniform maintenance assistance grants ranging from $150 to $300 per pupil per annum. (After 1956 the Metropolitan School Board will determine the amounts to be paid.) The Metropolitan Board is charged with the preparation of plans for the expansion of school facilities and for the co-ordination of those at present existing.

TOWN PLANNING

The province of Ontario enacted in 1946 a general Planning Act which gives municipalities power to designate land use, regulate sub-division, acquire land for housing projects and other 'official plan' purposes, and to plan highways and other public works. Following the passage of the

Planning Act the Toronto and York Planning Board was established with five members from the city and four from the county.[1] An official plan was approved by the city council in September 1949 and confirmed by the provincial Minister of Planning and Development in August 1950.

The province of Quebec has not enacted a planning law. Planning powers therefore have to be found within the charter, in the general municipal capacity to expropriate land for public purposes and to regulate streets, buildings and public spaces. The city is authorized, 'To draw plans of streets extending from the city limits to . . . any place on the island of Montreal for the purpose of having a general plan of the streets throughout the island.'[2] Within its own limits the city may regulate the type of building to be erected on its streets and the uses to which such buildings are put. Under article 421 of the charter the city may acquire land (with certain exceptions), within its territory or outside it, by agreement or by expropriation.

To promote the development of planning administration a provincial statute was passed in 1941 to permit the city to create a department of city planning accompanied by an advisory commission of seven to fifteen members.[3] The department, which includes a building inspection division, has expanded its functions considerably since its establishment.[4] The department was closely concerned with the production of the traffic and transport survey[5] mentioned above. It is also authorized to prepare regional plans for the island of Montreal, and to submit them for the approval of the Metropolitan Commission, or to the municipalities concerned after they have been approved by the executive committee.

Until very recent years the issues involved in the physical planning of large urban areas had received little or no consideration in Canada. The isolated problems of traffic congestion, provision of open spaces and amenities, and the maintenance of real-estate values in 'blighted' areas had arisen but piecemeal solutions were proposed. The zoning by-law was the instrument of planning most frequently employed, and its use was essentially negative and regulatory. And in a metropolitan region, such as Montreal or Toronto, even this degree of control could not be exercised over the areas of new growth outside the city where small municipalities were competing for industry and enhanced land values.

[1] Toronto City Planning Board, *Third Report and Official Plan*, June 21, 1949.
[2] Montreal, *Charter*, s. 300 (119).
[3] H. Spence-Sales: *Planning Legislation in Canada* (1949) (mimeographed).
[4] City of Montreal, *Budget Estimates* 1950–51, p. 57.
[5] Montreal Board of Research on Traffic and Transportation Problems, *Report* (November 1949).

It is difficult for a municipality to undertake a programme involving a financial burden that would place it at a disadvantage *vis-à-vis* its rivals. In a federal system this is equally true of the provinces. Consequently the expense of a major extension of community planning can hardly be contemplated until there is some general agreement on the problems of compensation and betterment that are bound to arise.

CONCLUSION

Both Montreal and Toronto have, at the turning point of the century, outgrown their political and administrative bounds. Both cities are aware of this fact and also that the main problems it poses are similar. There are too many municipalities administering local services, and in consequence some are too small to provide even the minimum requirements of the inhabitants of a large urban area. In each instance tension arises from the relationship between the desire of the one predominant unit to expand and the struggle of the lesser authorities to justify and preserve their independence. The suspicion aroused by this tension serves to prevent the co-operation that might be more easily achieved between real equals. Some of the smaller units have a substantial pecuniary interest in upholding their present status. For the most part they are suburban dormitory areas, with a high property valuation and a high standard of provision of services, and they are reluctant to be taxed for the benefit of poorer localities. Each unit tries to plan and zone its own area so as to raise property valuations, to attract new industry but to provide the minimum of low-cost housing, the tenants of which pay the least in taxes and demand the most in services. These reasons for antagonism are cumulative and lead toward complete deadlock on many vital issues.

The solution, if one is to be found, will require the initiative of the province. Any rational settlement will be to the advantage of the metropolitan area as a whole, but any settlement will be contrary to some vested interest. It is the superior level of government that should take the responsibility for advancing the wider viewpoint. To do so by direct fiat might have unfortunate political repercussions, but this is a risk that the Ontario government seems to have undertaken, and the operation of the new Toronto scheme will certainly be watched with great interest by her sister metropolis. The Quebec government has established (1953) a Royal Commission to enquire into inter-governmental relations and it is certain that metropolitan problems will be discussed at length. To achieve lasting success, a plan for metropolitan reorganization would have to make a radical departure from the existing pattern. A basic choice which has already been outlined in the case of Toronto, has to be made between the federal and the unitary type of organization; whether there shall be one big city or a series of urban units with arrangement for the unified administration of certain functions. The metropolitan area of Toronto is sufficiently homogeneous in point of population, politics and administration, to make the unitary solution practical. This, however, is not to affirm that it is politically or administratively desirable. In Montreal such homogeneity does not exist, and the only solution which seems possible is some form of federal structure.

To build a federal scheme of reorganization upon the pattern of the present local authorities would be most unwise. The new subordinate units would need to be comparable in size, in resources and in function. The elimination of many of the smaller municipalities and the dissection of the leviathan, would be essential.

The twin Canadian metropolises are cities of great and real contrasts, yet their problems are in essence the same. At the present stage of their evolution it is impossible to say that they will be solved, but it is significant that they have, at length, attracted the keen interest of a large number of their citizens.

Moscow

ROGER SIMON and MAURICE HOOKHAM

MAURICE HOOKHAM

B.Sc.(Econ.), London School of Economics and Political Science, 1935. Lecturer in Government, University College, Leicester.

Has had five years' experience in the employment of local government authorities. Visited Soviet Union in 1935 and 1941. Is engaged in research into the 'Methods by which the Local Soviets in the U.S.S.R. plan the municipal economy.'

Author of: 'A Plea for Local Government,' *The Political Quarterly*, Vol. 19, 1948.

ROGER SIMON

B.A. Economics, Cambridge, 1935. Qualified as solicitor, 1939.

Assistant Solicitor, Kensington Borough Council, 1939–42; Assistant Solicitor, Hull City Council, 1947–48; Assistant Solicitor, Ealing Borough Council, 1948 to date. Visited Soviet Union in 1935 and 1936.

Author of: *Local Councils and the Citizen* (1948).

Moscow

MOSCOW is the foremost industrial, political and cultural centre of the Soviet Union. It is also the historic capital of Russia. Ever since 1147, when the name of Moscow first appears in the Russian chronicles, the natural advantages derived by the city from its central situation enabled it to play a leading part in the building up of a centralized state in Russia, and on three famous occasions to act as the centre of national resistance to the foreign invader: in the fourteenth century when the Moscow prince Dmitri Donskoi defeated the Tartar hordes at Kulikovo; in 1612 when the Polish-Lithuanian army reduced the greater part of the city to ashes before being driven out; and in 1812 when the city was again largely destroyed in resisting the Napoleonic invasion. Even after Peter the Great had made St. Petersburg his capital in 1713, Moscow remained the principal commercial and cultural centre of Russia, and an important textile industry grew up from an early date. The historical traditions of the city are preserved in the beautiful churches, palaces and fifteenth-century walls of the Kremlin; in the severely classical buildings of the eighteenth century; and in the 'Empire' style mansions erected after the Napoleonic wars.[1] Thus the transfer of the capital back to Moscow immediately after the 1917 revolution restored the city to the position it had always occupied in the eyes of the Russian people.

Moscow is today the capital, not only of the U.S.S.R., but also of the largest of its sixteen constituent republics, the Russian Socialist Federative Soviet Republic, which contains about half the total population of some 200 millions living in the Soviet Union. The city plays a leading part in the scientific and cultural life of the country, and there are some 30,000 students in the university alone, quite apart from the many other higher educational institutions in the city. It is the headquarters of the Soviet Academy of Sciences, the Union of Soviet Writers, and of numerous other scientific and cultural organizations which regularly hold their congresses there. Above all it produces a greater volume of industrial output than any other city in the U.S.S.R., and together with the

[1] A good account of the historical buildings in Moscow is given in *Moscow Correspondent* by Ralph Parker (1949), Chapter 6.

surrounding towns which make up the Moscow Region, constitutes one of the major industrial centres in the Soviet Union.[1]

Before the 1917 revolution Moscow was mainly a centre for light industry, and the textile, leather and food industries accounted for three-quarters of its total output. But the industrial character of the city was fundamentally transformed in the course of the first five-year plan (1928–1932); it became a great engineering centre, producing machine tools, a wide variety of instruments, motor-cars and lorries, electrical equipment and equipment for the mining and oil industries, while important metallurgical and chemical plants were erected as well. A great range of consumer goods are produced, clothing and footwear, motor cycles, radio sets, watches and household utensils, and Moscow is the biggest source of supply of fabrics, footwear and clothing in the Soviet Union. Small scale industry, organized in industrial co-operatives, also produce considerable quantities of clothing, furniture, china and other articles. Moscow has been the scene of an unparalleled increase in production during the past twenty years. Already by 1940 the city was producing twenty-one times more than it produced in 1913, and twice as much as the entire industrial output of Russia in 1913.[2] It is not surprising to find, therefore, that the population of Moscow has grown with extraordinary rapidity, and almost continuously since the end of the Civil War, with the exception of a brief period of stability from 1933 to 1936:

Population in Millions

1918	1·7
1920	1·0
1928	2·2
1931	2·8
1933	3·6
1936	3·6
1941	4·1
1950	5·0 (Approx.)

The striving for an ever higher industrial output is one of the dominating features of soviet life. Many of the most outstanding Stakanovites and workers who have started new movements for improving technique and reducing costs of production, which later spread over the rest of the country, come from Moscow factories; and the same workers play a

[1] In 1940 Moscow alone accounted for 14·7 per cent of the total industrial output of the U.S.S.R. See Ranevsky: *Moscow Industry* (1947), p. 32. It appears that since 1940 the total industrial output of the U.S.S.R. has risen more than that of Moscow.

[2] Ranevsky, *op. cit.*, p. 32.

35. TORONTO

A general view of the city showing the Ontario Parliament building in the right-hand foreground. The University campus is in the centre.

[Photo: *Canadian Pacific Railway*]

36. MOSCOW
The Moskvoretsky Bridge.

[Photo: Society for Cultural Relations with the U.S.S.R.]

leading part in the municipal government of the city and form the majority of deputies to the city soviet. Thus the city occupies a special position in the life of the country: its citizens are expected to be the advance guard of the whole of the Soviet Union in the application of socialist principles, not only in the field of industry, but also in the sphere of housing, architecture, municipal services and in the planned reconstruction of a big city.

Moscow is now in the midst of carrying out the 'General Plan for the Reconstruction of the City,' adopted in 1935, and intended, in the words of the preamble to the plan, to create a city which shall fully reflect 'the grandeur and beauty of the socialist epoch.' A substantial part of this comprehensive plan has already been carried out and the immensely broad streets and impressive new buildings constructed since 1935 have already transformed the appearance of many quarters of the city. In order to achieve the entire reconstruction of a great city as rapidly as possible, a form of municipal government is needed which can organize and co-ordinate the work efficiently, arouse the enthusiastic participation of its citizens, and provide adequate opportunities for public criticism without delaying the work. Let us now see to what extent the Moscow City Soviet, which is responsible for carrying out all aspects of the general plan, fulfils these basic requirements.

THE SYSTEM OF GOVERNMENT IN MOSCOW[1]

A local soviet in the Soviet Union occupies an entirely different position, and performs a far wider and more important range of functions than a local authority in a capitalist country. The Moscow City Soviet, like any other local soviet, is a local organ of the state power; it is considered to be the representative of the national government within Moscow, and as such it has a general responsibility to give effect to *all* the functions of the Soviet state. There are therefore no specific limitations on the powers of the Moscow City Soviet, except that it is subordinate to the higher authorities and must conform with their directives. According to Article 97 of the Union Constitution 'The soviets of working people's deputies direct the work of the organs of administration subordinate to them, ensure the maintenance of public order, the observance

---

[1] The only comprehensive account of the municipal government of Moscow that exists is *Moscow in the Making* by Sir Ernest (now Lord) Simon and others (1937). Although there have been many substantial changes since this book was written in 1936, it contains a great deal of valuable information, especially Chapter 1 on city government by Professor W. A. Robson, and Chapters 6–8 by Lord Simon on the ten-year plan.

N

of the laws and the protection of the rights of citizens, direct local economic and cultural affairs and draw up the local budget.' This means in practice that the Moscow City Soviet is responsible, not only for the administration of the social services,[1] passenger transport, and the distribution of electricity and gas, but also for a wide range of local industries (i.e. industries producing goods for the local market), wholesale and retail trade, public restaurants, the administration of the law courts, police and fire brigade, and for cultural institutions such as theatres, cinemas, libraries, museums and art galleries.

The government of Moscow is organized on a two-tier system: the Moscow City Soviet exercises authority throughout the city, and there are twenty-five district soviets which are subordinate to it. There does not appear to be any constitutional division of functions between the city and the districts. Each service is planned and administered by the city soviet on a city scale, and by each district soviet on a district scale. Within particular services from time to time divisions may be made by agreement. Thus the city soviet plans, finances and erects certain houses, and administers them, while these houses will be situated in the districts where district soviets may also plan, finance, erect and administer other houses. Detailed divisions of immediate administrative responsibility are decided in each branch of a service, but as a general rule both the city and district administrations are involved. The city soviet as a superior administrative authority has power to direct the district soviet on any branch of administration, with the right of appeal by the district to the region, republic or union soviet.[2] The same principles apply, for example, where a union, republic or region service or institution exists within the area of the city; both the city and district soviet concerned will be involved in some aspect of its administration. This matter is referred to again below in connection with the problem of the settlement of conflicts between the various organs of administration.

Each soviet is directly elected by the inhabitants, and appoints a smaller executive committee to act as its executive and administrative organ. Thus the Moscow City Soviet, consisting of some 1,400 deputies, elects an executive committee of 60 deputies; the district soviet, with about 200 deputies, has an executive committee of about a dozen members.

---

[1] Social services include: public health, education, housing, roads, bridges, drainage, sewerage, street lighting, town planning, parks and open spaces, physical culture and sport, and the welfare of children and old people.

[2] The administrative hierarchy is as follows: Supreme Soviet of the U.S.S.R. (Union Soviet); Supreme Soviet of the Russian Republic (Republic Soviet); Moscow City Soviet; District Soviets.

The relationship between higher and lower soviets is governed by the principle of 'democratic centralism,' which is expressed in Article 101 of the Union Constitution as follows: 'The executive organ of the soviet of working people's deputies are directly accountable both to the soviets of working people's deputies which elected them and to the executive organ of the superior soviet of working people's deputies.' How is it possible, it may be asked, for the executive committee of the Moscow City Soviet to be responsible both to the full meeting of the city soviet and to the council of ministers of the Russian Republic[1] at one and the same time? The soviet answer is that every soviet, from the lowest to the highest, has been directly elected by the people and is animated with the common aim of building a communist society. The basis for fundamental conflicts of interest, it is claimed, does not exist, and the practical problems that arise in co-ordinating the activities of the different soviets can be solved in the light of public criticism and discussion.

In the soviet press there are many examples of criticism levelled by representatives of one organ of government against others, higher, lower or at the same level in the soviet hierarchy. It is noticeable, however, that the relations between the Moscow City Soviet, the Moscow district soviets, with the republic, regional and union soviets, appear to give rise to much less of this sort of criticism than in other parts of the union. The criticism is raised in the press in terms of a lack of co-ordination (i.e. a conflict arising through bad administration or bureaucracy) rather than from a conflict of interest. Instances are quoted for other areas in the union in, for example, reports in *Izvestya*, of higher organs of government issuing instructions to lower organs without consulting them beforehand, or of the arbitrary siting of buildings for organs under the control of higher soviets in such a way as to conflict with the detailed plans of the local soviet.

The criticism is invariably directed at removing the basis for this bad administration and the main correctives insistently referred to are:

(a) the need for all administrators, deputies, and organs of government to draw in to discussion and at all stages of actual execution, large numbers of the ordinary population.

(b) the constant need to employ detailed checks by higher level soviets of the work of subordinate soviets and by responsible officials of each administrative unit of the execution of decisions by that

[1] The council of ministers of the Russian Socialist Federative Soviet Republic is the executive organ of the Supreme Soviet of the Russian Republic; its relation to the Supreme Soviet is comparable to the relation of the British cabinet to Parliament.

unit. Great emphasis is placed on budget and accountancy checks, and the development of a sense of responsibility throughout the administrative apparatus.

These might be called the cardinal principles of soviet administration, but their successful operation depends, in the opinion of soviet press commentators, on the consciousness of the population of their common purpose.

A frequent source of disagreement between the various organs of government is the allocation of housing to workers employed by organizations controlled by higher level authorities. Thus the Dzerzhinsky district in Moscow has ten factories fully controlled by the district soviet. Also located in the district are undertakings controlled by the city, republic and union soviet administrations. Each factory directorate will naturally be concerned that its workers are adequately and conveniently housed. It may be tempted to apply pressure through the apparatus of its controlling soviet on the local soviet to this end. Among the employees in these various undertakings are, in almost every case, deputies who have been elected by local constituencies to the various soviets. They may be tempted to press the claims of their fellow employees at a higher level. In every case, however, the allocation of housing is controlled by a local soviet committee made up of local deputies and a large number of voluntary assistants (in one district in Moscow the standing committee concerned consisted of 20 deputies and 70 voluntary assistants). Any attempt to direct bureaucratically from above will be resisted by this popularly based committee. Instances have been quoted where it has been necessary to carry the battle quite high up in the hierarchy, or into the press, before the bureaucracy is exposed and agreement reached.

In each local soviet area there will be factories controlled by, say, the republic soviet, and the technical direction of that factory will be the concern of the republic ministry. There will also be factories controlled by the all-union ministries which are responsible solely to the federal government. The local soviet will, nevertheless, be responsible for the provision of a large number of amenities for the workers employed in the factory and for the generation of enthusiasm amongst them for the fulfilment of the factory plan. The local soviet is therefore in a strong position to influence the work of the factory, and its success will depend on the degree of harmony which exists between the higher and lower authorities. One of the main functions of the Communist Party is to ensure that the higher and lower soviets work together in a spirit of mutual co-operation. The work of the Moscow organization of the Communist Party is considered in the soviet press to be generally of a high level and therefore few disputes between the soviets have arisen. In other parts of the country where disputes between soviets have arisen these are often mentioned in the press as self-criticism by party members of the unsatisfactory political leadership of the party in those areas.[1]

It would be wrong to imagine that this absence of conflict is secured by deciding everything at the top. On the contrary, the lower soviets are encouraged to exercise their initiative to the maximum extent so long as they comply with the directives and standards laid down by the higher soviets. For example, the five-year plan which regulates the work of the Moscow City Soviet is prepared in detail by the city soviet itself. In preparing the plan the city soviet must comply with certain preliminary directives issued by the republic soviet and then, after thorough discussion, the plan receives the approval of the republic and union soviets, is incorporated into the financial and economic plan for the whole country and thereafter must be strictly observed. Owing to its outstanding importance as the capital city of the Soviet Union as well as the Russian Republic, Moscow occupies an exceptional position in that, although it is subject to the direction of the republic soviet in most matters, it has direct relations with the union soviet as well in respect of certain matters. For example, the union soviet finances the reconstruction of some of the principal streets in the city.

There are six republic ministries which are directly concerned with the city administration: Public Health, Education, Local Industry, Internal Trade, Finance, and Communal Economy. The Ministry of Communal Economy deals with a wide range of functions including housing, town planning, public utilities, drainage, parks and open spaces; it may be regarded as a residuary legatee of all services other than those supervised by other ministries. The degree of central control exercised by each ministry varies, as in England, with the nature of each service. Thus the control exercised in the sphere of education appears to be greater than in public health.[2] The Ministry of Education settles the cost per child in the various types of school, lays down the specifications for school buildings (leaving the local soviet free to make its own designs so long as they comply with the specifications), and determines the teaching

---

[1] This is not the place to consider the role of the Communist Party in the Soviet Union. The reader is referred to the full treatment of this subject given in such standard works as *Soviet Communism* by S. and B. Webb, or *Political Power in the U.S.S.R. 1917–47* by Towster. The membership of the Communist Party for the City of Moscow was given in February 1949 as 407,013.   [2] *Moscow in the Making*, pp. 39–41.

methods to be adopted. The Ministry of Public Health inspects the work of the Moscow public health department, issues directives on matters of policy, and formulates standards such as the normal cost per hospital bed.

In the sphere of local industry, where a large measure of decentralization was effected in 1941, the Ministry of Local Industry leaves the initiative in determining the amount and character of production mainly to the local soviets.[1] Each department of the city soviet works on the basis of a five-year plan, and the ministry concerned assists in the formulation of the plan and approves the annual instalment to be achieved each year. In principle there is no sphere of the activities of the Moscow City Soviet which is immune from interference by a higher soviet, but in practice the control of the republic soviet is confined to the methods described.

BOUNDARIES

Co-operation between the different soviets has permitted a smooth adjustment of boundaries when circumstances demanded. When the two-tier city government was set up there were only ten districts in Moscow. In the early nineteen-thirties the boundaries of the districts were re-drawn and twenty-three districts created, as experience had shown that the previous size of the ten districts was too large for good administration and that the district soviets were too remote from the inhabitants. After the Second World War, the population and area of the city having grown further, it was found necessary to split one large district into half, and to amalgamate parts of other districts into a new district, thus creating two additional districts.

ELECTION OF DEPUTIES AND THEIR WORK

The Union Constitution, promulgated in 1936, provides in Article 94 'the organs of state authority . . . in cities . . . are the soviets of working people's deputies.' Article 95 provides that these soviets shall be elected for a term of two years. In common with many other countries involved in the last war no elections were held during the period of the war. The basis of representation, as laid down in the Republic decree of October 1947, is one member for each 3,000 electors in territorial constituencies. Chapter XI of the Union Constitution provides for universal, direct and equal suffrage by secret ballot, with a minimum age of eighteen years for voting and candidacy. The constituencies are determined by Moscow

---

[1] Dobb: *Soviet Economic Development since 1917* (1948), p. 344.

City Soviet for the city and it appoints an electoral commission to supervise the elections.

Electoral commissions for each constituency, consisting of representatives of the local branches of the Communist Party, trade union branches in factories and offices, co-operatives, youth organizations, and cultural societies, are also appointed. Each constituency returns one candidate and any of these organizations may nominate a candidate; the Communist Party itself does not, however, nominate candidates. No candidate may be nominated for more than one constituency for any one Soviet. For some two months prior to the election agitation points are established throughout the constituency by the various organizations and the electorate generally are informed about the details of the election.

Long before the election campaign is formally opened the Communist Party, in the exercise of its function of leadership, initiates discussions of a far-reaching character about desirable candidates throughout the constituency and especially in the factories. It may sometimes happen that, where it has become widely known that a certain candidate with an outstanding record is to be nominated by some factory, other organizations who had been provisionally favouring their own nominees will withdraw them before nomination day. Usually, however, several candidates are nominated. When this occurs a selection conference (or 'consultation' in Soviet terminology) is summoned by the electoral commission for the constituency and delegates in proportion to the membership are summoned from organizations entitled to nominate candidates for the election. Each candidate nominated is thoroughly discussed by the delegates at the conference and ultimately, in practice, one candidate is selected—if necessary by an exhaustive vote. Subsequently a public adoption meeting may be held at which the candidate is formally adopted. Assuming the local branch of the Communist Party has done its job properly, therefore, the merits of the candidate and his ability to carry out the instructions of the electors will have been discussed in all the factories and other organizations in the constituency.

On October 1, 1950, it was announced in the press that elections for the city and district soviets would be held on December 17, 1950. It was reported in *Izvestya* that by October 13th tens of thousands of agitators had started already to conduct meetings in Moscow in all the districts. On October 3, 1950, *Izvestya* devoted nearly three out of its four pages to a full text of the republic electoral regulations setting out the procedure for the elections.

From the many discussions held in connection with the elections are prepared the long and detailed lists of instructions to the elected deputies from the electorate. For example, in the 1947 elections, the electors in the Dzerzhinsky district of Moscow directed their deputies, among thousands of other things, to reorganize the administration of the local housing department. The elections are held on the same day both for the district and the city soviets. Voting is from 6 a.m. to midnight and the day is declared a public holiday.

The city soviet elected in December 1947 (the second to be elected under the constitution of 1936—there were no elections during the war) consisted of 1,392 deputies.[1] Article 142 of the Union Constitution defines the duty of the elected deputy. 'It is the duty of every deputy to report to his electors on his work and on the work of the soviet of working people's deputies, and he is liable to be recalled at any time in the manner established by law upon decision of a majority of the electors.' In practice the deputies report back regularly. They have fixed visiting hours for their constituents every day, and their constituents come to them with suggestions, demands and criticisms, both then and at other times in the constituency and wherever they may happen to work. This applies to the individual deputy and to the deputies appointed to represent the various committees of the soviet. Each deputy has throughout his constituency groups of inhabitants organized to help him in his work. Every block of flats has a house committee and members of this committee are appointed by it to keep in touch with the deputy and help him on behalf of the residents. Groups of volunteers also act in a similar way for hospitals, schools, etc., and recently there has been a great campaign for beautifying the streets, gardens, and open spaces by such volunteer groups. Many of the old untidy spots left over from the war period have now been filled with a mass of greenery and flowers. These groups also help

---

[1] The report of the mandate commission elected at the first meeting of the city soviet after the elections gives the following details about the deputies.

727 were workers, of whom:

  400 were directly engaged in production (manual workers).

  178 held leading positions in U.S.S.R. and R.S.F.S.R. undertakings.

  167 were employees of the city soviet.

  176 were officials of the Communist Party and other organizations.

812 were members of the Communist Party.

580 were non-party members.

503 were women.

75 per cent had either middle school or higher education.

20 were Heroes of the Soviet Union, 16 were Heroes of Socialist Labour, 52 were Stalin Prize winners.

1,315 held some medal or award for distinguished work in some field.

with advice and work in the reconstruction and development of schools, hospitals, and other institutions.

A FULL MEETING OF THE CITY SOVIET

Some idea of the business transacted at a meeting of the city soviet may be obtained from the full reports given in the national press. In 1948 there were four meetings of the full soviet.[1] The first meeting of the newly elected soviet began on January 23, 1948, and lasted three days. It was attended by prominent figures of Soviet society, and was opened by one of the oldest deputies of the city soviet, Academician Shchusyev, who 'reminded the deputies of the wise remarks of Stalin' that 'a deputy is a servant of the people and that he has (and can have) no greater concern than for the welfare of the workers. In all our work we must rely on the broad masses and continually strengthen the widest links with the electors, quickly responding to their needs and requests.'

The deputy chairman of Moscow Soviet then made a report on housing construction in Moscow, stressing the need to develop new ways of mass construction based on industrial methods with the greatest possible use of prefabrication in factories. This report was discussed by the members the following day and it was emphasized that the housing question was the most important task for the city soviet. Sharp criticism was made of various administrative organs, such as the Ministry of Aircraft Production of the U.S.S.R. which had failed to take measures for the fulfilment of the plan for the manufacture of housing components, and of the Architectural Academy of the U.S.S.R. for lack of interest in designing plans for speeding up building construction. The amplitude of the proposals discussed may be gauged by the demand for 150,000 workers required for this task and for a capital sum amounting to one-third of the city annual budget.

The second session of the soviet held on April 14–15, 1948, had two items on its agenda, the rehabilitation and development of Moscow city economy in 1948, and the city budget for 1948. The third session held on August 11, 1948, was devoted almost entirely to a discussion of the report of the director of the health department on the work of the department. In the 198 hospitals and 424 polyclinics maintained by the department more than 100,000 patients daily received advice and treatment and more than 10,000 were daily visited in their own homes. One third of the Moscow city budget is spent on health services. In her con-

[1] Reported in *Izvestya* for January 24 and 25, March 17, August 13 and December 10, 1948.

N*

tribution to the discussion the chairman of the standing health committee
reported that, in preparation for this session of the soviet, 23 brigades
of deputies and voluntary assistants had, under the leadership of health
specialists, made a thorough inspection of the work of the city and district
health services. The deputy chairman of the executive committee of the
soviet, in winding up the debate, strongly criticized, among others, the
work of three district health departments which had failed to check up
sufficiently on bad work in health centres in factories in their districts.

The fourth session of the soviet was held on December 6, 1948, and
had three items on its agenda: the improvement in quality and variety
of goods produced by city industry; the election of city judges, and the
election of people's assessors for railway and river transport courts. There
was much criticism of the failure of those responsible for the production
of goods to pay attention to the comments of the customers. Particular
emphasis was placed on the need to build up a reputation for the trade
mark of each factory, to develop competition between the factories and
to introduce greater mechanization in production. Among those who
took part in the debate were the Director of the Co-operative Industry
of the Council of Ministers of the Republic, the Minister of Light Industry
of the U.S.S.R., and the Minister of Local Industry of the Republic, all
of whom are elected members of the city soviet.

A few observations may be made on the material contained in these
reports of meetings of the full soviet:

(1) There are some members of the highest organs (such as the Council
of Ministers of the U.S.S.R.) who are also deputies of the Moscow City
Soviet, and this may be of some value in establishing good relations
between the central government and the city soviet.

(2) The purpose which the full meeting of the soviet is intended to
serve is to review the achievements and failings of past activities and to
discuss future plans. With only one or two key items on the agenda for
each meeting this can be done successfully by such a large gathering.
The opportunity which this provides for a thorough criticism in public
of the work of the executive committee is considered very important.
A leading article in *Izvestya* (March 11, 1947) stressed the need for the
executive committee to submit itself to the searching criticism of the full
meeting of a soviet. The Council of Ministers of the Republic had ordered
that this must be done systematically throughout the republic. The article
points out that 'the strength of the soviets lies in their indissoluble connections with the people. The soviet deputy, elected by the people,
knowing well the needs and desires of his electors, should systematically

influence the work of the executive committee and other executive organs of the soviet and its administration and point out the correct policy to be adopted in economic and cultural work. This can be done both in meetings of the soviet and through the standing committees.' The article criticizes those executive committees which gave 'a merely formal account of what was being done by the executive committee, shewing how active it was in various fields, but without any systematic deep analysis of the failings of the committee and its various executive organs, or of the leading members of them.' The condensed reports in *Izvestya* of the meeting of the Moscow City Soviet supply evidence that the city soviet is fulfilling its main purpose in these respects.

(3) Many of the leading full-time officials are also deputies of the city soviet. There is nothing corresponding to the rule, which is so marked a feature of British local government, which prohibits the paid employee of a local authority from becoming an elected councillor for that authority. The soviet system enables the head of a department who is also a deputy to take part in a meeting of the city soviet—as, for example, did the director of the health department in the meeting mentioned above.

THE EXECUTIVE COMMITTEE

Article 99 of the Union Constitution provides that the executive and administrative organs of the soviets in cities are the executive committees elected by them, consisting of a chairman, vice-chairman, a secretary and members. The present executive committee of Moscow City Soviet consists of sixty members.[1] It is primarily responsible for the preparation of annual, quarterly and monthly plans for the whole work of the soviet and for the detailed scrutiny of the day-to-day fulfilment of the plans.

Reports of its meetings are published in the local and national press. The agenda for the meetings appears to vary from one item (on February 17, 1948, *Izvestya* reported a meeting entirely devoted to a review of the level of education in Moscow) to a larger number of items where particular difficulties have been confronted. Thus on March 6, 1948, *Izvestya* reported an agenda of five items as follows:

(i) quality and variety of clothing produced in city factories,

---

[1] Before the adoption of the 1936 constitution the executive committee elected a smaller body known as the presidium which consisted of only fifteen members. This presidium was the principal executive and administrative organ of the city soviet. The executive committee played a much less important role. The presidium has now been abolished. One possible explanation of this is that the concentration of all executive and administrative power in such a small body has become unnecessary, and consequently undesirable, as the number of persons with the requisite administrative capacity has grown larger.

(ii) collection of the turnover tax from undertakings which were in
    arrears,

(iii) plans for the production of local building materials,

(iv) plans for the more thorough utilization of resources in production,

(v) measures to be taken by city and district departments for dealing
    with the spring thaw.

STANDING COMMITTEES OF MOSCOW CITY SOVIET

The major part of the detailed direction of day-to-day administration
is done by the standing committees of the city soviet. Their particular
functions, broadly defined by their titles, are:

1. Budget.
2. Industry (co-operative and local industry).
3. Housing (new construction).
4. Housing (repair and administration).
5. Heating supply (central heating, electricity and wood).
6. Gas (a committee recently established to exploit the supply of
   natural gas brought by pipe line from Saratov).
7. Traffic and communications, including transport and telephones.
8. Trade and food supply.
9. Health protection.
10. Education.
11. Planning and reconstruction of the city, including open spaces.
12. Cultural.

It is in these fields that the deputies and the electorate are most interested
and where it is considered necessary they should be most actively drawn
into the work of administration. The other spheres of the soviet's work
have special departments, but it is not considered necessary to have special
standing committees to supervise and direct that work.

One of the most interesting features of local government in the Soviet
Union is the large number of volunteer assistants who are co-opted to
help the deputies in their work on the standing committees. These assistants are called 'activists' and together with the deputies they bring the
total membership of all the committees of the city soviet up to about
4,000 persons. Activists do not merely attend meetings of the committee
to which they belong, but are also expected to take part in the work of
administration in various ways, such as by inspecting the institutions for
which the committee to which they belong is responsible, or in dealing
with correspondence and suggestions from the constituents.

THE DISTRICT SOVIETS

The constitution and methods of working of the 25 district soviets are on the same lines as the city soviet. Each district soviet elects an executive committee and a number of standing committees, the size of the executive committee and the number of standing committees being left to the district soviet to determine in accordance with local conditions. There are over 3,000 deputies elected to the 25 district soviets. The total number of standing committees appointed by the district soviets adds up to 236 (this averages just over nine committees for each district), and the deputies who are members of these committees are reinforced with some 10,000 activists in all, of whom 6,000 are women. The standing committees usually meet fortnightly.

The executive committee of the Dzerzhinsky District Soviet had in November 1950 13 members, 6 being employed full-time (chairman, 4 vice-chairmen, and a secretary). The total administrative staff consisted of 517 people. The total administrative costs amounted to 1·9 per cent of the district soviet budget. The administrative staff were grouped in 14 departments and two trusts. The titles of the departments were education, health, municipal enterprise, housing administration, road trust, industrial trust (administering ten factories), trade, state bank, restaurant trust, physical culture and sport, cultural institutions, district planning, finance, allotments and pensions, and there were two trusts for housing construction. The nine standing committees are industrial, trade, public feeding, municipal enterprise, culture, health, schools, budget, and housing.

The organization of the district soviet is clearly described in the following verbatim accounts. In *Moscow News*, January 17, 1948, this report on the Bauman District Soviet was given. 'The Soviet consists of 202 deputies, of whom 71 are workers, 68 engineers and employees of various district institutions, and 45 professional and intellectual workers; 117 are members of the Communist Party and 40 per cent are women.'[1] At its first session the district soviet elected an executive committee of 13 deputies, three of whom, the chairman, deputy chairman and secretary are full-time officers. The rest combine their duties with their basic occupation. The executive committee supervises the entire economic and cultural life of the district with the aid of ten standing committees. The deputies are free to select the committee on which they wish to serve

---

[1] This figure of 40 per cent appears to be the average for the district soviets and is somewhat higher than the proportion of women deputies in the city soviet.

and the report points out this may have no particular reference to their own occupation. An example is quoted of a school principal who asked to be appointed to the health committee and was ultimately made chairman of it. The health committee consisted of 12 deputies and they planned to begin their activities by visiting district health institutions and checking up on their work. The chairman reported that he had promised his electors to look into the need for the expansion of one of the polyclinics. If the committee decided to recommend its expansion then their recommendation would go to the executive committee. In case of disagreement appeal would lie to the next meeting of the full soviet.

The chairman of the Kirov District Soviet was reported in *Soviet News* (May 16, 1947) as follows:

'The whole council meets once in two months. In the intervals between these meetings the current work is conducted by the executive committee, with the assistance of a number of standing committees. In our soviet we have eight such committees. As a rule, these committees meet monthly. Each is headed by an experienced chairman. The chairman of our industrial committee, for example, is the power station engineer. A school teacher presides over our education committee. These committees play an important part, not only in conducting regular business, but in initiating new measures. Our industrial committee, for instance, recently launched a scheme for using the by-products of the textile mills and clothing factories in the district. All the materials for the full meeting of the soviet are prepared in advance by the committees, which conduct the necessary investigations, formulate proposals, and so on. The full meeting of the soviet decides upon the major questions affecting the district, such as the plan of work for the year, the district budget, and so on. The fulfilment of the district plan as a whole is the main concern of the executive committee. This executive, which is elected by the full meeting of the soviet, consists at present of eleven members. It meets weekly and the questions which come up at its sittings are as varied as life itself. As an example I will quote the agenda of our last meeting. We dealt with: the question of guardianship over orphan children, allowances for mothers of large families and unmarried mothers, applications for flats, arrangements for sale of non-alcoholic drinks during the summer, and the fulfilment of the plan for installing gas in apartments which had previously not got it. The executive committee direct the work of municipal industrial and trading enterprises. Under its jurisdiction are the district industrial trust, the municipal enterprise trust, the housing administration, and so on. The doors of the executive committee's office are wide open. Everybody may apply to the executive committee on personal matters and the officials must always be accessible to the public.'

FINANCE

The system of raising revenue for the soviets differs from anything known elsewhere and thus makes a comparison of the taxation system with others very difficult. Contemporary estimates of the value of the rouble most commonly equate it with sixpence. The city budget for 1949, and again for 1950, provided for an income and expenditure of about 3,500 million roubles (or about £100 million). The budgets for the 25 districts together amount to about half of this. The revenue appears to be derived as to 60 per cent from profits on industry, the remainder coming from income tax (which is entirely at the disposal of the local soviet) and profits on services. The collection of all revenue coming under the republic and city budgets is the responsibility of the local soviet closest to the citizen and most likely, therefore, to exercise the greatest initiative in its collection. The two main items of expenditure in 1949 were health and education which took 49 per cent, and housing which took 41 per cent of the budget.

MOSCOW'S MUNICIPAL GOVERNMENT

Leaving aside the fundamental role of the Communist Party in the Soviet Union, it may be useful to summarize briefly the principal features governing the relationship between municipal authorities in Moscow and its inhabitants:

(1) The full meeting of the Moscow City Soviet has a very different function from that of a council meeting in English local government, where the committees must await the approval of the council before action can be taken on many matters. In Moscow, the full delegation of executive power to the executive committee facilitates rapidity of action, and the full meetings of the city soviet provide opportunities for thorough review and public criticism of the most important aspects of the work of the executive committee.

(2) The active interest of the citizens of Moscow is aroused, not only by an election every two years, but also by the system of activists and the voluntary committees in each constituency, which work with the deputies. It is estimated that altogether 100,000 citizens of Moscow

have thus been to some extent enlisted into the work of administration of the city and district soviets, whether as deputies, activists or members of local committees.

(3) There is no sharp dividing line between officials and elected councillors, so that heads of departments, teachers and doctors, can and do get elected as deputies of the soviet which employs them.

(4) Women play an exceptionally prominent part in the municipal government of Moscow.

Does this large and elaborate municipal organization in Moscow work efficiently? It is impossible to answer this question completely without a thorough examination of the various social services which cannot be undertaken here; but a good measure of the efficiency of the city soviet and of its capacity for getting things done can perhaps be obtained from an examination of the extent to which the plan for the reconstruction of the city has been carried out.

PREPARATION OF THE RECONSTRUCTION PLAN

By 1931, the population of Moscow had increased by over a million since 1917 and was still rapidly growing as the large factories constructed during the first five-year plan were completed and came into production. In consequence of the concentration of resources on the construction of new factories and machines the development of the municipal services and utilities was failing to keep pace with the growth in population, although substantial improvement had been made on the primitive conditions inherited from the Tsarist régime. In June 1931, the central committee of the Communist Party of the Soviet Union frankly declared that this backwardness had become especially acute 'as a result of the great deficiencies in the working of the Moscow municipal services.' There had been 'blunders in the tramway system, shortcomings in housing, poor execution of street and underground work' and the sanitary condition of the city was 'extremely unsatisfactory.' Some figures quoted at the meeting graphically illustrated the magnitude of the tasks facing the city soviet at that time. Moscow required 4,000,000 cubic metres of wood fuel per annum and there was liable to be an acute shortage of fuel in winter. The number of houses supplied with gas was very small indeed. Only 42 per cent of the houses had a water supply laid on, while in one of the new districts only 12 per cent of the houses were connected to the water mains, the remainder taking their supply from street pumps. Only 627 kilometres of streets, out of a total length

37. MOSCOW

Gorki Street. Note the width of the street.

38. MOSCOW

Modern apartment houses on Bolshaya Kaluzhskaya Street.

[Photo: Soviet Weekly

39. MOSCOW
A tall building on Smolenskaya Square.

[Photo: Soviet Weekly

40. NEW YORK

Manhattan Island, the home of the skyscraper. Many well-known buildings, such as Chrysler and the Empire State, can be identified.

[*Photo: Fairchild Aerial Surveys, Inc.*]

of 1,000 kilometres, had a drainage system; and 91 per cent of the streets were paved only with cobble stones.

The city had grown through the centuries without any kind of plan, reflecting 'even in the best years of its development, the barbaric character of Russian capitalism.' The narrow and crooked streets were hopelessly inadequate for the needs of modern traffic, and the centre of the city was encumbered with warehouses and small workshops. The large numbers of small, single-storied wooden houses had earned Moscow the title of 'the big village.' Building was proceeding in an uncontrolled fashion and without the guidance of any general plan. Radical reconstruction was urgently required. Accordingly the central committee of the Communist Party, after dealing with the steps needed to secure better housing, transport and other services, gave instructions for the 'elaboration of a serious, scientifically developed plan for the building of Moscow.' At the same time the erection of any further factories in Moscow was prohibited.

Two major projects required for the reconstruction of the city were begun in the same year—the underground railway (known as the Metro) and the Moscow–Volga canal. But the preparation of a detailed plan took four years. A vigorous controversy took place at first on the basic principles of the plan. It was proposed that most of the city should be pulled down and replaced by a city of skyscrapers; that the old city should be preserved as a museum piece and an entirely new city built on a new site; and that Moscow should be made into a garden city, with a series of satellite towns composed largely of cottages. All these proposals were ultimately rejected in favour of a plan designed to retain the historical outlines of the city, but radically to reconstruct it on the basis of a proper disposition of industries, railways and houses and the widening and co-ordinating of the existing network of streets and squares.

July 10, 1935, has been described as 'the most decisive date in the shaping of Moscow's modern history,' for that was the day on which the plan, now known as the Stabi 'General Plan for the reconstruction of Moscow'[1] was adopted by the government.[2]

THE PRINCIPAL FEATURES OF THE PLAN

The plan covers the development of the entire 'municipal economy' of Moscow; the first section deals with population, area, density, and

[1] The plan is sometimes called the 'Ten Year Plan for the Reconstruction of Moscow' as it was originally intended to be carried out in ten years.

[2] It was approved jointly by the central committee of the Communist Party of the Soviet Union and by the council of people's commissars for the Soviet Union (now the council of ministers).

other matters relating to the town planning of the city; while the second section contains the plan for the development of housing, transport, education and all the other municipal services. In all essentials the plan remains in full force, although certain modifications have been made as experience has shown them to be necessary. It is only possible here to summarize briefly the most interesting features of the plan.[1]

(1) The population of the city shall be limited to 'approximately five millions.' The area of Moscow, which had reached 70,000 acres in 1935, is to be extended gradually to cover 150,000 acres. The area reserved for expansion will be included within the administrative boundaries of the city as building proceeds. In order to safeguard the future carrying out of the plan, all construction work both within the existing city limits and the area reserved for expansion, and in the green belt, is placed under the full control of the city soviet.

(2) A protective green belt of forests and parks up to ten kilometres in width shall be created outside the area reserved for future expansion to serve, in the words of the plan, 'as a reservoir of fresh air for the city and a place of recreation for its inhabitants.'

(3) The centre of Moscow had grown up as a rough approximation to a spider's web, with two inner ring roads on the sites of old fortifications. Taking this historical circular and radial system as a basis, the plan provides for the straightening and widening of a big proportion of the existing main thoroughfares to not less than 30–40 metres, the construction of two new ring roads further from the centre and of a number of new streets including three great diagonal boulevards, and the reconstruction of the twenty-six principal squares, including the Red Square which is to be doubled in size. In addition, the facing of both banks of the river Moscow with granite (already begun before 1935) was to be completed.

(4) The residential areas are to be laid out in large blocks of from 22 to 38 acres consisting of apartment houses of not less than six stories, while higher blocks of flats will be erected on the embankments, squares and main streets. In 1935, the average density of the population was 140 persons per acre of residential block, but it rose to 400 in the centre of the city. The plan provides that the population should be evenly distributed throughout the city at 160 persons per acre, except that it might rise to 200 in districts particularly suitable for residential development. In this connection it should be remembered that Moscow applies a non-differential rent system based on area of living space irrespective

[1] The full text of the General Plan is printed in *Moscow in the Making*, pp. 184 *et seq.*

of location. The basic conception of the plan is that the city's amenities should be equally accessible to all citizens wherever they live. There should be no exclusive residential areas with superior amenities.

(5) No special industrial areas are reserved in the plan, as almost all the factories already exist. They are mainly concentrated in the east and south-east, although they are distributed irregularly throughout the town. This is not considered a disadvantage, as it enables the people to live near their work, and the best of the big Moscow factories are spaciously laid out and surrounded by gardens and apartment houses for their workers, each factory having its own club, children's playground, nursery, health centre, night sanitorium, and so forth. Moreover, as the principal sources of power are gas and electricity, Moscow is a remarkably clean city, free from smoke and grime.

The second section of the general plan dealt with the development of municipal services during the next ten years and perhaps the most interesting feature of this section, apart from the sheer size of the various targets, is the striking illustration it gives of the wide range of services for which the city soviet is responsible: thus it covers not only water, sewage disposal, central heating, new schools, houses and hospitals, but also new department stores, warehouses and cold-storage plants, grain elevators and bread factories, as well as cinemas and palaces of culture.

THE REALIZATION OF THE PLAN, PRE-WAR PERIOD, 1935–1941

The plan seems to have all the necessary ingredients, population limit, a defined area, green belt, modern street plan and so forth, which are required to lay a sound basis for the reconstruction of a big city. It is not hard, however, to work out a bold and comprehensive plan of this kind. What gives the Moscow plan its exceptional interest is that a substantial part of the plan has already been carried out, even though the conditions which prevailed during the greater part of the period since the inception of the plan have been distinctly unfavourable.[1]

One of the outstanding characteristics of the Soviet method of planning is the ability to maintain an inflexible system of priorities, concentrating resources in the first place on the basic items of capital construction, and

---

[1] The threat of war was compelling a substantial diversion of resources to defence preparations even before the plan was adopted. The percentage of the national budget devoted to defence rose from 3·5 per cent in 1933 to 12 per cent in 1935 and 17·2 per cent in 1936. See p. 32 *Man and Plan in Soviet Economy* by Andrew Rothstein (1948).

this is clearly illustrated by the history of the Moscow plan. Acute though the housing shortage was in 1935, this was not allowed to divert attention from the urgent need to develop the basic services such as water and gas supply, transport and the widening of the main streets in the central areas. Already by 1935, the first section of the Metro of 12 kilometres was opened to the public, and the second line of 14·9 kilometres was completed in 1938. The great Moscow-Volga canal was finished in 1937. In the space of two years (1936–38) eight new bridges were built over the river Moscow (roughly the same width as the Seine) and by 1941 seven new bridges had been constructed across the smaller river Yauza. By 1941 the appearance of the Moscow and Yauza rivers had been greatly improved by the construction of a total length of 30 miles of granite embankments along both banks of these rivers.

The building of new suburbs proceeded side by side with the reconstruction of the central part of the city, the appearance of which was considerably altered in the pre-war period. The greater part of the ancient commercial district of 'Cathay town,' containing a mass of dilapidated old buildings and small workshops, was demolished and two broad avenues driven through it. Two central squares and four major streets were largely reconstructed, widened and flanked with imposing new buildings.[1] In the four years from 1937, a two kilometre stretch of Gorky street, one of the principal radial highways leading out of the city, was widened from an average width of 56 feet to a width of 200 feet.[2] Nevertheless, this high rate of reconstruction could not be maintained as the approaching shadow of war forced the diversion of more and more resources to the strengthening of the country's defences. The ambitious housing programme, for example, could not be fulfilled.[3] On June 22, 1941, the whole programme of reconstruction was halted by he Nazi invasion.

[1] In the course of widening streets considerable use was made of the technique of moving back existing buildings on electrically operated rollers, and about fifty buildings were rolled back in this way, including the Moscow soviet building and a hundred-year-old eye hospital, which changed its address from one street to another during the night.

[2] This was done by demolishing the buildings on one side or rolling them back to the required distance and building a series of new and higher blocks of offices and flats on that side, while the height of the old buildings on the other side was raised by the addition of a couple of stories.

[3] This provided for 15,000,000 square metres of new housing space in ten years. Actually only 1,800,000 square metres were completed by 1941. But great attention was paid to other social services. In a special drive for new schools, for example, 152 new schools were built in 1936, each having accommodation for 800 pupils.

THE REALIZATION OF THE PLAN, POST-WAR PERIOD, 1945-1950

The fierce Moscow winter wreaks havoc with buildings unless constant attention is given to repairs. This was impossible during the war, and frost, snow and incendiary bombs did incalculable damage to the buildings, while most of the mechanical equipment of the building trade, and even spades and wheelbarrows, had been mobilized for the front. Nevertheless, as early as 1943, just as the red army began its great drive to the west after the victory of Stalingrad, work on the third section of the Metro was resumed, and in 1944 the construction of the pipe line bringing natural gas from Saratov to Moscow was begun. But the first two years after the war had to be mainly devoted to repairing all the damage to the city's buildings done during the previous four years of neglect. The appalling devastation in the areas of the U.S.S.R. occupied by the German armies must be borne in mind.[1] The reconstruction of Moscow, which had escaped lightly in comparison with towns in the occupied territories,[2] inevitably had to take second place to the enormous task of restoration in these areas. Consequently it was not until 1947 that wartime dilapidation had been made good and the work of reconstruction and housing could be resumed on a large scale.

In 1946, the Moscow City Soviet adopted the five-year plan for the rehabilitation and development of Moscow's municipal economy (1946-50), and in pursuance of this plan further big strides have been made towards carrying out the general plan. All the main features of the plan have been retained, and although the population now appears to have grown a little beyond the five million limit, the five-year plan repeats the prohibition on the building of new factories in Moscow and the city soviet has every intention of keeping the population down to the limit of 'approximately five million' fixed by the general plan. First priority has been given to new housing construction and over one and a half million square metres of new housing accommodation have been built. There has been large expenditure on public utilities; construction of a new waterworks; completion of the Saratov gas pipe line and the construction of a large central heating network (these two measures together having eliminated the need to bring firewood to Moscow); tens

---

[1] Twenty-five million people were rendered homeless, 70,000 villages and nearly 2,000 towns wholly or partly destroyed, 32,000 industrial establishments employing 4 million workers destroyed, 40,000 miles of railways wrecked, 500,000 collective farm buildings destroyed. For further details see Rothstein, *op. cit.*, pp. 40-52.

[2] Although 42 per cent of the Moscow region was occupied.

of miles of new tramway track laid, mainly in the outer suburbs; and the Metro extended to a total length of thirty miles, involving the construction of eight new stations. Bridges and roads have been constructed, widened or newly asphalted to the extent of three million square metres, and the pattern of the new city is already clearly emerging—very broad main streets (up to 500 feet wide) lined with buildings which are generally from seven to ten stories high, though rising to much greater heights on a few specially selected sites.

There is great emphasis on the extension of parks and green amenities, such as gardens down the middle of the broadest streets, and in the previous ten years over 600,000 trees and nearly 3 million shrubs have been planted in Moscow. The Gorky Park of Rest and Culture has been extended to an area of 750 acres, and a new botanical garden of 1,000 acres is planned. The 10 kilometre wide green belt surrounding the city is being preserved intact. It is in fact impossible to give an accurate account of the progress of the plan, since any statement becomes out of date almost as soon as it is made; but an indication of the extent to which the plan has been carried out was given in a statement by the council of ministers of the U.S.S.R. in 1949 that 'the basic tasks of the ten-year plan would be fulfilled within the next three or four years.'

THE MOSCOW–VOLGA CANAL

This was one of the biggest construction projects completed before the war in the Soviet Union. It is 84 miles long and joins the river Moscow to the Volga by way of a chain of lakes. The weight of soil excavated was 148 million tons, only slightly less than the Panama canal (160 million tons). It links Moscow by water with the Baltic, the White Sea and the Caspian, and at the same time has solved the problem of the capital's water supply. The canal celebrated its tenth birthday in 1947, when the volume of shipping using Moscow as a port had risen from 890,000 tons to 2,350,000 tons, while the daily consumption of water in Moscow had reached 66 gallons per head.[1]

THE SARATOV PIPE LINE

Begun in 1944 and completed in 1946, this pipe line brings natural gas to Moscow over a distance of 525 miles, and has made a big contribution to the task of supplying the city with cheap gas. Over 70 per cent of the population have now been supplied with gas for domestic

---

[1] This includes industrial as well as domestic consumption.

purposes; the city soviet installs the gas supply free of charge, and the average cost to the consumer is two roubles a month.[1]

THE METRO

Of all the major works involved in the reconstruction of the city which have been undertaken, the Metro is the most spectacular. Begun in 1931, in the face of exceptionally difficult soil conditions, it had reached a length of 30 miles on January 1, 1950, and when the circular line has been finished, the underground system will be complete. The project aroused the enthusiasm of Moscow citizens to such an extent that many thousands volunteered to help in their spare time. It is intended to be a work of art as well as a triumph of engineering, and each station is designed as a separate architectural unit and lavishly decorated with marble, sculpture and bas-reliefs.[2]

HOUSING

The shortage of housing is unquestionably the greatest problem that the Moscow planners have yet to solve, as the following table shows:

TABLE I

Year	Total housing space square metres ([3])	Housing space in square metres per head
1913	12,000,000	([4])
1936	16,400,000	4·5
1941	18,200,000	4·4
1950	19,700,000	4·0 approx.

In considering the housing position certain factors affecting Moscow need to be borne in mind. Immediately after the 1917 revolution about half a million people were transferred from the terribly insanitary cellars, doss-houses and barrack houses in which they had been living, into the mansions and flats previously occupied by the wealthy, so that the available housing space was more evenly distributed. Although many of the

[1] It is difficult to express accurately the value of the rouble in other currencies. The average wage earned in Moscow is currently estimated at 900 roubles per month.

[2] A Russian writer puts it thus: 'The Moscow Metro represents something far greater than the normal achievement of technical construction. Our Metro is a symbol of the construction of a new society.' Savitsky: *Moscow. An outline of its Architectural History* (1947).

[3] Soviet housing figures are given in square metres of living space, excluding kitchen, bathroom, lavatory, corridors, and staircase. Thus 48 square metres (480 square feet) is about the size of a typical pre-war British municipal house.

[4] Large for the better off, very small for the workers.

old wooden houses built in Tsarist days can still be seen on the outskirts
of the city, large numbers have already been demolished, and in so far as any
of the old cellars remain, the Moscow City Soviet does not permit anyone
to live there. Rent averages only 3 to 4 per cent of incomes and is limited
by law to a maximum of 8 per cent of the chief wage earner's income.
The citizens of Moscow are exceptionally well provided with amenities
such as libraries, clubs, special children's clubs and so forth, which go
some way to mitigate the shortage of housing space. Lastly, it should
be recalled that the formidable task of restoring the national economy to
the 1913 level, after the devastation caused by the civil war, was not
achieved until 1927. Although the work of housing construction then
began, it had to compete with a huge programme of industrial construction as well as a rapid increase in population. Just as the housing programme was getting into its stride the Nazi invasion began, and before
they left the Germans had succeeded in destroying about one-third of all
the housing space in the cities they had occupied, amounting to 70 million
square metres, the restoration of which has had priority over the resources
of the building industry since the war.

All the same, the acuteness of the housing shortage is in no way
minimized by the Moscow planners. Housing was declared to be the
first priority in the post-war five-year plan, and it is noteworthy that the
pace of construction is rising rapidly and has been over twice as great as
before the war. The technique of the building and building materials
industries has made rapid strides in the direction of developing the mass
production of standardized prefabricated parts and the mechanization of
construction work on the site. There can be no doubt that many of the
citizens of Moscow will have to continue for some years to share a flat
between several families; but the task of providing the basic municipal
services is substantially complete and the way is now clear for a much
bigger concentration of resources on housing.

AREA AND DENSITY

Although the general plan provided for an expansion of the area of
the city from 70,000 up to 150,000 acres, the built-up area has so far been
extended only to 82,000 acres, and it appears unlikely that the whole of
the area reserved for future expansion will be used, in consequence of
a decision by the city soviet to build higher than had been originally
intended. The ultimate density of the population is likely, therefore, to
be greater than the limits set by the general plan in 1935. It is claimed

that the development of more highly mechanized methods of construction have made higher and larger buildings more economical to erect than smaller ones, that the cost of transport is reduced by raising the density—and the very broad streets are capable of taking a heavy increase of traffic in the centre of the city—and lastly, that bigger buildings offer better opportunities for architectural beauty. The planned architectural design of the city is receiving particular attention. On carefully selected sites a number of tall buildings of skyscraper variety are being erected, varying in height from 16 to 32 stories. Their construction is not influenced by excessive ground values, and they are accordingly well spaced apart, each building being designed so as to spread out wide at the base and merge into the adjoining buildings. The most ambitious building is the new Moscow University on the outskirts of the city which is designed to provide accommodation for 6,000 of the students, in addition to lecture rooms and laboratories. Rising to a height of 26 stories, it will be surrounded by 275 acres laid out as botanical gardens, parks and playing fields. The only real skyscraper in Moscow seems likely to be the Palace of Soviets, planned to be the largest building in the world and to stand impressively at the end of a very broad avenue. In spite of all this new building the greatest care is taken to preserve the best of the old buildings and to reveal them to the best advantage. The chief architect of the city has written: 'In Moscow you meet, at every step, a most rare fusion of the antique Russian style with modern architectural forms.'[1]

CONCLUSION

Although the foregoing account is necessarily brief and incomplete it is undeniable that striking progress has already been made in the reconstruction of Moscow. There are three advantages in the sphere of town planning possessed by a socialist municipal economy.

First, the social ownership of all the land in Moscow, the absence of private landlords and of high land values, means that the cost of reconstruction is limited to the cost of demolishing old buildings and erecting new ones. There are no vested interests to hold up the plans. The citizens of Moscow are conscious that the whole city belongs to them, and this enables the city soviet to enlist the enthusiastic participation of a significant proportion of the population in the work of reconstruction, from the architects, technicians and building workers down to the thousands of activists associated with the city administration and the members of the

[1] Chernishev, quoted in *Soviet News*, June 24, 1946. Ralph Parker, *op. cit.*, p. 239.

numerous voluntary commissions for the blocks of flats, schools and other institutions in the city.

Second, the social ownership of industry and all other resources within the city has great advantages for town planning: it enables the city planners to maintain a firm system of priorities for basic construction tasks so as to lay a sound basis for future development; it makes possible the adherence to a fixed population limit through the power to prohibit the erection of new factories, and it enables the city soviet to plan the architectural design of streets and squares as a whole.

Third, the Moscow City Soviet has control of all building operations and municipal services within the city, its boundaries are extended as the city expands, it has authority to prevent building in the green belt, and the twenty-five district soviets work under its direction and in harmony with it.

Only the future can show whether the city planners will succeed in preserving the planned population limit of about five million and in overcoming the powerful magnetic effect of the capital city of a large and rapidly developing country. If this can be achieved, and if the present rate of building is not slowed down by the threat of war, it seems reasonable to conclude that within the next few years—although the provision of adequate housing space for all the inhabitants may take longer— Moscow will have become the best-planned big city in the world

New York

REXFORD GUY TUGWELL

B.S.(Economics), Wharton School of Finance and Commerce, University of Pennsylvania (1915); A.M. (1916); Ph.D. (1922); Litt.D., University of New Mexico (1933). Instructor in Economics, University of Pennsylvania, 1915–17; Assistant Professor of Economics, University of Washington, 1917–18; Manager, American University Union, Europe, 1918; Instructor in Economics, Columbia University, 1920–22; Assistant Professor, 1922–26; Associate Professor, 1926–31; Professor, 1931–37; Assistant Secretary, U.S. Dept. of Agriculture, 1933; Under-Secretary, 1934–37; Chairman of Planning Commission and head of Dept. of Planning, City of New York, 1938; Chancellor of the University of Puerto Rico, 1941; Governor of Puerto Rico, 1941; Professor of Political Science and Director of the Program of Planning, University of Chicago since 1946.

Author: *The Economic Basis of Public Interest* (1922); *The Trend of Economics* (with various economists, also editor) (1924); *American Economic Life* (with Thomas Munro and R. E. Stryker) (1925 and 1930); *Industry's Coming of Age* (1927); *Soviet Russia in the Second Decade* (with various economists, also editor) (1928); *The Industrial Discipline* (1933); *Battle for Democracy* (1935); *Redirecting Education* (with L. H. Keyserling) (1934); *Puerto Rican Public Papers* (1945); *The Stricken Land* (1946); *Seven Lectures on the Place of Planning in Society* (1953). Also various articles and essays in learned journals and other publications.

New York

NEW YORK CITY was the capital of the confederated American colonies at the time of the Revolution; and it was the first capital of the United States of America. But not since 1790 has it served as the political head-quarters of the nation. It has, however, from the very first been an increasingly important commercial centre, and has never been challenged as paramount in financial affairs. For this there was the very good reason that the harbour was superb, and was superbly located in a strategic position on the Atlantic coast. Access to the hinterland was easier from New York, also, than from any other port. The New York Central Railroad, which in the mid-nineteenth century thrust westward first to Buffalo at the eastern end of Lake Erie, and then on to Chicago, has always advertised itself as 'the water-level route'; and, indeed, it paralleled the Erie Canal[1] which, by using a few locks, had by this time firmly established the port of New York as the gateway to the interior.

The lead furnished to New York by the canal and the water-level railway was never overcome; and down to the air age, commerce from and to the West continued to funnel through the port. Even then, thanks to the foresight of Mayor LaGuardia, the city was ready with the first full-scale international airport; and at the present time (in 1954), airlines from the other side of the Atlantic have their terminals there.

Because of this commercial supremacy, the city also became the financial centre of the United States. This function, once begun, developed by momentum; and, although other centres naturally developed in time, none ever thrust aside New York. The city remains the place where most large transactions occur. It is the home of the great underwriting firms, the private banks which gather up and allocate the nation's financial resources; and its stock exchange is the market on which the prices of securities are fixed. The pulse of the nation's economy has still to be taken in New York.

---

[1] De Witt Clintons' campaign for the governorship of New York State had as its main issue the authorizing of the canal. He won; the legislature passed the enabling act in 1817, and on July 4th of that year the first work was begun near Rome, N.Y. The canal was officially opened in 1825. Freight rates from New York to Buffalo immediately dropped from about $100 a ton to some $6.

It may be something of a commentary on the political trading which took the seat of government away from New York, that the United Nations should have chosen to locate there its permanent seat. For New York can be thought of now as a world pivot, as it could be thought of as a national pivot in 1790. It is no longer a suitable location for a national capital, being on the rim of a country now stretching out to the Pacific, but its suitability as a world capital is generally acknowledged.

Not the least of New York's physical and social problems arise from its location and surroundings. The mouth of the Hudson River made a magnificent harbour, but its several islands and numerous confluent waterways furnished difficult problems of transport, communications, and administration. These have tormented planners and administrators throughout its history. Manhattan Island, on which the first crowded business and dwelling districts grew up, is thirteen and a half miles long and two and a quarter across at its widest. For most of its length it is much narrower. City planners of the early nineteenth century believed that the rapidly increasing population would require more provision for lateral than lengthwise streets. The conception was that the Hudson River on one side, and the East River on the other, would supply a good deal of the up and down communications. And lower Manhattan was laid out on a grid system with this basic principle. Even in the later days of horse-drawn vehicles, this scheme proved unsatisfactory; but it grew intolerably restrictive in the age of automotive transportation. And many costly devices have been resorted to for relief. Highways have been built, with enormous capital expenditure, both on the east and west margins of Manhattan. Besides this, access to the metropolitan centre by water across the Hudson from New Jersey and across the East River from Brooklyn and outlying Long Island became insufficient. Numerous bridges, beginning with the famous Brooklyn Bridge in 1883, have relieved this difficulty. But these too have had to be provided at a capital cost which continues to be a very heavy burden for the city's inhabitants. Many of these structures are today rather marvellous monuments to the engineering arts, but they are, in fact, artificial substitutes for the easy communications possessed by other comparably great cities.[1]

The physical separation of the old boroughs accounts, also, for the kind of growth the commercial and government structure has had, and

[1] The census metropolitan region had in 1950 more than 20,000 miles of traffic arteries, besides its local streets; it had more than 2,000 miles of railway; it had 74 airports of various sizes, including half-a-dozen major ones; and there were $3\frac{1}{2}$ million (approximate) motor vehicles used within the region. There were nearly $4\frac{1}{2}$ million housing units. There were 300 varied types of industry, employing more than six million workers.

for the heroic measures taken to overcome some of the difficulties. In 1898, the five boroughs were fused into Greater New York (Manhattan, Brooklyn, the Bronx, Richmond, and Queens) but this still did not bring into the united family the populous and busy area across the Hudson in the State of New Jersey; and, as time went on, it was apparent that the unification scheme had not reached nearly far enough out into the Long Island and Westchester County suburbs, to say nothing of other areas in upper New Jersey and Connecticut. The result is that now of the metropolitan area's fourteen millions of population, only some eight millions are actually within the city itself. The rest live in 550 surrounding cities, towns and villages.[1] Nevertheless, the whole area of 650 square miles has more problems in common than it has as separate municipalities, and these common problems have forced resort to joint arrangements of several sorts—such as the Port of New York Authority and the Triborough Bridge and Tunnel Authority. Such arrangements have served to mitigate some of the most serious problems of division. It is generally agreed, however, that more unification will somehow have to be achieved before the operations of the metropolitan area can be brought to the necessary standard of efficiency. The physical difficulties are sufficient without the added governmental cross-purposes inevitable in separation.

The obstacles to further progress in governmental consolidation are two. The metropolitan area is divided between two sovereign states—New York and New Jersey (with Connecticut on the outskirts); and those who have escaped to low tax areas on the periphery object strenuously to assuming part of the expense of maintaining the central services. These problems are not unknown elsewhere. They are, in fact, similar to those of many other large cities, but they are intensified by the federal structure of government in the United States. This not only makes union actually difficult to design, but furnishes a convenient bulwark for those whose interests run contrary to the general interest.

There is another problem, not unique in New York, but particularly bothersome there, which arises from the sheer size of the metropolis in comparison to New York State. Municipalities, no matter how large, are legal creatures of the states; that is to say, they owe their corporate existence to a charter issued by the state. This charter originates in a legislative act, and it may or may not provide the freedom necessary to the efficient functioning of the civic organism. Where, as in New York, the city grows faster than the state, there may be a serious lag in the apportionment of representation; and in that case, the city may find

---

[1] The census figure for New York City population in 1950 was 7,891,957.

itself at the mercy of a legislature whose majority consists of representatives elected from outside the city. Not only the charter is involved in this, but also the current legislative acts affecting taxation, the police powers, welfare, public works, municipal ownership and many other issues. In New York State, taxation and the distribution of funds have always favoured the rural areas because of their weight of representation. And as programmes of federal aid have developed, the tendency to administer them through state agencies has made more problems of a similar sort. For, in general, there are no direct relations between cities and the federal government.[1] This has led to a situation in which a less populous organization—the residue of the state—has exercised control over a larger one—the city—with all the consequences to be expected. There has, in fact, been for many years a conflict between most big American cities and the states, strengthening movements for home-rule. There have been compromises as time has gone on, but the situation remains anomalous, in spite of some gains for the cities; and nowhere is the issue more acute than in New York.

The physical and governmental difficulties have not prevented New York from growing; but they have made life more difficult and more costly for its citizens. Many amenities have been limited, and much discomfort created in ways which seem wholly unnecessary. But even these have not been so serious as other problems, such as the fact that the municipality does not reach far enough geographically to prevent the escape into suburbs outside the city of the most prosperous citizens who then are lost as taxpayers. This problem has grown worse in the past two decades with the creation of a system of magnificent parkways leading outward in every direction and facilitating the daily escape of migrants who work in the city but have their homes in the outskirts. This was already serious because of commuters' services on the rail lines into New

[1] There are a few exceptions to the general rule. Under the Public Works Act of 1933, the administrator was authorized to make direct contracts with municipalities; and certain housing authorities were set up which still have direct relations with the federal government. The same is true of airport construction under the Civil Aeronautics Authority: the Lanham Act, during the war, was a similar exception. Other exceptions are of the informal sort, as when city health authorities confer with the Federal Public Health Service or the Federal Bureau of Investigation with city police. Very recently a controversy concerning civil defence was, however, settled on the general rule that cities may not deal directly with the federal government. There have been proposals for a Federal Bureau of Municipalities which have come to nothing because of state opposition. Teachers of American government are, indeed, accustomed to say that so far as the federal government is concerned, cities do not exist; and it is very nearly accurate. Cities have and do receive federal aid, but with a few exceptions, it comes to them by way of the states.

41. NEW YORK
The R.C.A. building which forms part of the Rockefeller Centre.

42. NEW YORK
Another view of the R.C.A. building, Rockefeller Centre.

[*Photo: Standard Oil Co. (N.J.)*

43. NEW YORK
A pier on the East River, showing oil barges refuelling an ocean-going ship lying on the extreme left.

[*Photo: Standard Oil Co. (N.J.*)]

44. NEW YORK

The harbour. The express highway crosses the railway yards at the foot of West 70th Street.

[*Photo: Standard Oil Co. (N.J.)*

Jersey, upper New York and Connecticut; the parkways, which had their great extensions in the 'thirties, raised the problem to acute proportions.

It is no exaggeration by now to say that New York City exists in a state of chronic bankruptcy. It is not that municipal bills are not paid —although there have been crises when obligations were far greater than resources and further borrowing seemed impossible—but that, facing the fiscal problem, budget makers have had to reduce expenditures until the municipal services reached an almost impossibly low level. Streets have not been properly cleaned and repaired for decades; the school system is miserably maintained; water is chronically short; transportation on the municipally-owned lines is such that daily travel is an ordeal—and similar shortcomings affect every one of the three-hundred odd services the city pretends to perform for its citizens.

And then, of course, there are the slums. No great city in the world, perhaps, has had more incredibly sub-standard housing than New York. The lower East side, East Harlem, West Brooklyn and Queens consisted for decades of filthy and rickety 'railroad' or 'dumb-bell' cold-water flats,[1] without the least pretension to civilized provision for sanitation, in which recently-arrived immigrants swarmed like animals, a disgrace to America and a torment to social workers. The proliferation of these slums seemed endless, and none of the ordinary and minor measures of reform were able to change the pattern. Whole generations grew up knowing no other surroundings, becoming accommodated to them, and finally accepting the slum way of life. This acceptance furnished a kind of inertia and made any corrective effort almost impossible to generate and carry through. Such conditions also create vested interests whose defences are made easy by ignorance of alternatives. Apathy among those whose lives are stunted and circumscribed by such surroundings is one of the chief preventives of reform. And, indeed, reform was delayed in New York until the spreading cause of dilapidation seemed almost beyond control. It was not until the national government, revolutionized by the depression which began in 1929, shook itself and began that movement of regeneration now designated as the New Deal that impulses of change began to reach into the pullulating recesses of the slums. Rebuilding is

---

[1] This type of tenement house construction presented a continuous façade to the street, but toward the rear was narrowed a few feet. There were thus narrow, dark wells in the interior, often almost filled with exterior fire escape structures. These were often used for storage, for drying clothes and other miscellaneous purposes which substantially defeated the purpose of admitting a little light and air.

o

now under way, but the ordeal of the degenerate years has left its mark on generations of New Yorkers.

There began in 1933, gradually gathering momentum, the series of programmes for low-cost housing and public works which has by now begun to change the whole New York scene. It still has far to go; but it is obviously well under way. There are not only enormous new apartment projects in the heart of the old slums; there are also thousands of new parks and playgrounds, highways, schools—all the social amenities necessary to the cultural complex of a great city. But there is still the unsolved fiscal problem; and all the new facilities are badly maintained and many are falling into premature disrepair because of failure to solve the fiscal dilemma of a metropolis which has not attained full organic consciousness, and even if it had, does not control the elements of its organic life.

THE DEVELOPMENT OF NEW YORK

Organized life in New York began in 1626, when Peter Minuit bought Manhattan Island from the local Indians for what is reputed to have been about $24 in trade goods. But the discovery of the harbour dates from a century before this. It was described by Giovanni de Verrazzano in a now authenticated letter to his patron, Francis I of France. Nothing came of this early visit, perhaps because of the preoccupation in those days of European statesmen with the fabled wealth of the orient and their conviction that the American coast was no more than a barrier to be penetrated in reaching the East. Even Hudson's famous visit in 1606 was an incident in the search for a way through the barrier. If a rebellious crew had not forced him to turn south away from the bleak coast of the higher latitudes he might never have seen the river which still bears his name. Even then his exploration of Chesapeake Bay and the Delaware River seemed of more account, for it was there that the Dutch had their more important trading posts. And even in the Hudson, their Albany port was regarded as of greater value than the small settlements at the end of Manhattan Island and on the Brooklyn shore. The real Dutch interest was not in the West; it was in Java; and when in 1664 Charles II presented Long Island to his brother, the Duke of York, and the expeditionary force demanded Manhattan Island too, it was given up without a struggle.

Peter Stuyvesant, who had been governor of the territory since 1647, had been disposed to fight on that occasion. His was an autocratic, a choleric, disposition. But like many possessive executives, he had an urge to build and expand which resulted in at least a beginning of civic consciousness. This had issued in public works of a military sort, but also in improvement of docking facilities. The Dutch were then forced into a wider surrender in 1674, but many patroons had established themselves on the rich lands along the Hudson where they lived in feudal style, and others had become merchants either in Albany or in New York. They were not dispossessed altogether by the succeeding British, and the great Dutch families have persisted as an important element in New York culture to this day.

In the succeeding century the British governed New York in the then colonial fashion. There were thirty successive governors in that century, no one of whom distinguished himself in any way. The settlement around New York bay had by then some 20,000 inhabitants living in a settled way. There was the beginning at least of a local culture—King's College, later to become Columbia University, had been established in 1754— and the merchants and traders lived in feudal style, making such imitation of the governor's elegance as their means permitted.

The town was compressed into the area between the fortifications, which ran across Manhattan Island at Wall Street, and the Battery which was at the island's end, facing down the bay. It was crowded. Streets were narrow; there was no pavement; much of the building was of timber. And when fire broke out in 1776, it destroyed the town. The lesson seems not to have been very well learned, however, for there were almost equally serious conflagrations in 1835 and in 1845. Because of these disasters, only a few landmarks from the early period have survived— such as St. Paul's Chapel and Fraunces' Tavern—and most historic sites can only be imagined from plaques on the skyscrapers which now occupy the locations.

The Revolution ensued, and the State of New York became one of the confederated states of America. There was a time, after the Revolution, when New York seemed likely to be the permanent capital of the new nation. That this was not to be was, however, no grievous disappointment. The city's interest was a good deal less political than commercial. And, in a sense, that has always remained true. Added to commerce and finance, as the nineteenth century ran along, a cultural nucleus also grew. Writers, painters, musicians, and savants gradually gathered there; institutions of learning were strengthened; there was a permanent opera; there were art and science museums. What had been a harbour with its class of merchants, then a national financial centre becoming international, presently became also a cultural capital, first national and then with relations reaching into every part of the world.

From the little settlements at the end of Manhattan Island, and on the opposite shores in Brooklyn and New Jersey, the small revolutionary metropolis gradually stretched out in every direction. It was, in small compass, like the history of the nation. The Americans rolled westward biting off vast stretches of the continent until they reached the Pacific; so the New Yorkers worked out from their centre until they reached the Atlantic at Coney Island and Far Rockaway, spread out into and across the Jersey meadows and up into the Bronx and Queens until, as the national population recoiled into the short grass plains (neglected in the early dash to California), the New Yorkers rediscovered Manhattan. They then began to rework it, at monstrous cost, into a complex machine for the expression of a characteristic kind of metropolitan life. It is fitted out now with utilities, with cultural centres, with skyscraper office buildings and huge apartments. Its bowels are an interlacing network of transportation systems, communication networks, aqueducts and garages, all cut into and through solid rock with enormous labour; its habitations and working places reach toward the sky, fifty, a hundred and more stories. It furnishes a scene to be found nowhere else in the world, an imperial conglomeration, a human ant-heap, in which millions of inhabitants go unconcernedly about their business in the midst of a finely tuned organism which for some purposes functions superbly but for others possesses obvious defects only to be remedied by incredible disturbances of the whole body. But these disturbances are always undertaken, sooner or later, regardless of cost or civil discomfort and presently the machine works a little better, moves more smoothly, and with improved amenity. It is beautiful with a mighty grandeur and ugly with a grinding friction, moving toward some kind of climax not yet clearly visualized.

What was to come, what is yet to come, was never held in view, partly because it was no one's business to conceive the future of the whole, and partly, of course, because of the preoccupation of its citizens with their exploitative activities. When the move up Manhattan first escaped from the Wall Street confines and crept up toward Harlem, the City of New York was all on Manhattan Island. There was another city on the Brooklyn shore; and there were others round about. It was not until the end of the nineteenth century, and after half a century of agitation, political manoeuvres, and a great deal of solid civic persistence, that Greater New York came into being by the consolidation of the five boroughs.

THE GOVERNMENTAL STRUCTURE OF NEW YORK

A political scientist who started out to discover the worst (that is, the most complicated and ineffective) city governments in the world might, for the worst, hesitate between Chicago and London, but for the best he would almost certainly decide on New York. But this excellence is characteristic of the latter-day government, especially that set up by the charter of 1936. It was not always thus. There is, in fact, no more turbid—and sometimes downright disgraceful—municipal history than that of New York during the nineteenth century. Even the charter of 1897, under which unification was completed, had not substantially changed the administrative system. There were revisions, beginning in 1901, some of them important; but when F. H. LaGuardia was elected in 1933 as an independent, a reform mayor, it was obvious that any salutary change would have to wait on comprehensive structural revision. A commission to draft a new charter encountered great difficulties at first; in fact the chairman, who was Alfred E. Smith, resigned, along with other members, and it was abolished. Not until 1935, with T. D. Thatcher as chairman, did a renewed commission begin its task. The work went quickly. The proposed new charter was published in August and its ratification was voted on that same November. It was adopted by a considerable majority.

It was a short charter, requiring an elaborate administrative code for its completion; but that work was also undertaken by the commission and soon pushed to completion. The charter itself was almost entirely confined to the establishment of a governmental structure, the definition of duties and powers of the officers of the government, and the fixing of a time-table for the administrative processes.

The charter of 1936 (which went into effect January 1, 1938) had no preamble. It merely stated at the beginning that 'the City of New York, as now existing shall continue with the boundaries and with the powers, rights and property and subject to the obligations and liabilities which exist. . . .' It then enumerated and defined the duties of the various agencies of government. First came the mayor, who was defined as the 'chief executive officer.' He was to appoint 'the heads of departments, all commissioners, and all other officers not elected by the people.' He was also to have the power of removal; and he was to be a magistrate. His duties were prescribed as follows:

1. To communicate to the council, at least once in each year, a general statement of the finances, government and affairs of the city, with a summary statement of the activities of the agencies of the city.

2. To recommend to the council all such measures as he shall deem expedient.

3. To keep himself informed of the doings of the several agencies of the city and to see to the proper administration of its affairs and the efficient conduct of its business.

4. To be vigilant and active in causing all provisions of law to be executed and enforced.

5. Generally to perform all such duties as may be prescribed for him by law.

These, it will be seen, are the usual duties of a chief executive. The really unique features of the charter were the provisions for the board of estimate and the city planning commission.

There is no difficulty in understanding, at least, the title of the planning commission, although its scope was something hitherto unheard of in American municipal government; but the board of estimate is a very confusing name to describe the body brought into being by the charter. It was, in fact, an upper house of the legislative branch; but it also had assignments which went far beyond those usually belonging to such a body.[1] For this there were historical reasons—there had already been a board of estimate and apportionment going back beyond unification—but there also were practical reasons. Not for nothing had the good citizens who served on the drafting commission watched the course of city corruption. They were determined, even at the cost of being accused of disloyalty to the tradition of democratic representation, to centre responsibility. If anything went wrong with administration, blame could be fixed. It would rest on the mayor, the comptroller, the borough presidents, and the president of the council, who, together, made up the board of estimate.

An old board of aldermen, so usual still in city governments, was changed into the city council of one house; and its powers were greatly reduced, so greatly in fact, that little was left for it to do except to ratify (or object to) the actions of the mayor, the board of estimate, and the planning commission.[2] The charter makers did not hesitate, even, to establish the legislative veto in important matters, in place of the usual affirmative vote. This meant, in practice, that the city's business would go on as determined by other bodies unless a council majority refused permission; and in some matters—those in which log-rolling or vested influences were most likely to interfere—the majority for veto had to be as large as two-thirds or even three-quarters. This may not have been strictly democratic, and it was certainly a serious modification of the

[1] The charter set out, in fact, that the board of estimate, 'shall exercise all the powers vested in the city except as otherwise prescribed by law.'

[2] The council was defined as being vested with 'the legislative power of the city, with the sole power to adopt local laws under the provisions of the city home rule law or otherwise, without requiring the concurrence of any other body or officer except as provided in sections 38, 39 and 40.' But sections 38, 39 and 40 define respectively the duties of the mayor and the board of estimate and provide for referenda. In effect the board of estimate must concur in legislation and the mayor must approve (subject to a two-thirds majority for passage over his veto). Even then it must also pass the board of estimate on which the mayor (or his deputy) sits, with three votes. Since the comptroller and the president of the council each have three votes, their combined votes amount to nine—a majority.

traditional governmental division (legislative, executive, and judicial) with its elaborate interacting checks and balances; but it did seriously hamper the machine politicians, it did centre responsibility, and it did make for efficiency and celerity. These qualities, after the confusions and corruptions to which New York, along with other American cities, had been so subject, were regarded as all-important. There would have been a fearful outcry, on theoretical grounds, if such a proposal had been made for a state or for the national government. Curiously enough, not much was heard of such objection in New York. Perhaps the preceding century of degenerate government, of corruption and inefficiency, had issued in a good result. New Yorkers had had enough. They were in a mood for drastic change. And the old-style politicians, although they squalled loudly, were ignored. But the good citizens, having got the bit between their teeth, went even further. They provided for a planning commission and a department of city planning, which was not only to make a master plan, but was also to be given powers to administer it and to protect its integrity.

The planning commission had seven appointed members and one ex officio member—the chief engineer of the board of estimate. Their terms were to be eight years, which was twice as long as those of the elected officials. The chairman, to be designated by the mayor, was also made head of the department of city planning. There were to be unpaid advisory boards of three members in each borough, appointed by the borough president. This last small appeasement proved to be unnecessary; the advisory boards were useless, partly, perhaps, because they had no staff and no powers. It ought to be said, also, that the commission itself has never fully met the expectations of its conceivers. There arose not only the expected opposition and sabotage from the heads of departments, but constant harrying by the mayor himself (although LaGuardia had made much of the commission in his campaign for the charter), the board of estimate, and the council. This might have been anticipated and provided against, but it was overlooked. After fifteen years, the master plan, which it was the commission's first duty to complete, was not yet adopted; and it appeared that unless changes were made, it might never be adopted.

Besides making the master plan, the commission was given the duty of preparing and presenting an annual capital budget together with a six-year budget. This duty has been capably performed, although experience has shown that the commission ought to have been given responsibility for making a complete financial plan rather than merely one for future capital expenditures, inasmuch as these budgets are often so interdependent as to be practically inseparable. Then, too, the commission was given custody of the city map which involved the granting or with-holding of consent to any changes in it. This power, together with the zoning powers implied in the master plan, gave the commission con-siderable control over private developments in the city—at least any requiring street or utility accommodations or changes in the uses permitted in the various zones. But these duties too have been met in only limited ways. Real estate interests favour zoning so long as they can influence its management; but they immediately bring to bear their considerable pressure for delay and obstruction when the public interest requires that their wills shall be crossed.

The worst handicap has been the necessity each year to ask for an appropriation for staff. The mayor, the board of estimate, and the council soon discovered that they could starve the commission with impunity; so, what with packing and starvation, the commission has remained mostly an example of worthy intention, sabotaged by its natural enemies.[1] It obviously needs commissioners appointed as judges are, during good behaviour, and an automatic allocation of a certain percentage of what-ever the current budget may be. With these protections, which some future charter may provide, the commission may be able to meet the expecta-tions of those who devised it.

The commission has functioned most effectively in performing the duties for which a time-table was laid down in the charter and which therefore could not be escaped. By a certain time each year, it must present to the mayor its capital budget and plan. For that purpose, the planning department gathers departmental data, holds hearings, and constructs a proposal. The commission adopts this, or a modified version, and sends it forward to the mayor. This para-legislative act is a very important one, and the budget is not amended with ease; it may not be amended at all without the commission's consent, except by a three-quarters vote in the council. Of course the mayor may offer any modi-fications to which he is devoted as he transmits it to the council; but these can only be made by the commission itself, when the council requires it to be done. There is a pressure of time throughout. Each step has no more than a few days allotted to it, so that delays which might stop the city's business may not occur.

[1] Further explanation of the commission's difficulties may be found in an article by the author, written while he was still chairman of the commission, 'Implementing the General Interest,' *Public Administration Review*, Vol. I, No. 1. Autumn 1940.

The relation of the planning commission to the mayor and the board of estimate has been disappointing; but it has demonstrated the possibility of a far more orderly and effective undertaking of city enterprises, even though the really independent and public-minded commissioners have, up to now, been very few. Since the commission has been so disappointing and the council has atrophied, the centre of governmental interest has been, since 1936, in the mayor, with the board of estimate sharing the spotlight from time to time. There have been, since that time, exceptionally colourful mayors—notably F. H. LaGuardia, William O'Dwyer and Robert Wagner. Most of the time, indeed, the mayor has been able to dominate the board of estimate and reduce it to a shadow of himself. This has been because it is made up of elective officials who often win together on the same party ballot, though this is more true of the general officials than of the borough presidents. These have usually represented political machines in their boroughs and may or may not have anything in common with the mayor.

Running with the mayor, and usually elected with him, are the comptroller and the president of the city council. These together with borough presidents make up the board. But their votes are weighted. The mayor, the comptroller and the president of the council, have three each; the presidents of Manhattan and Brooklyn two each; and the presidents of Queens, the Bronx and Richmond one each. If the mayor, therefore, has carried with him into office the comptroller and the president of the council, their nine votes will always provide a safe majority over the other seven. And that is the situation which has usually existed.

The board of estimate is in any case a legislative body made up of executives. The mayor does not usually preside at its meetings; a deputy mayor had been provided to cast his vote. The president of the council acts as chairman; also he becomes acting mayor when the mayor himself is absent from the city or is incapacitated. The best feature of the board is that, because executives sit on it, its decisions are smoothly translated into administrative action. It can act quickly; and it must act responsibly. It has, however, stolen most of the power from the planning commission and it frequently proceeds less in accord with considered judgment than in response to political pressures. Its relationships with the planning commission have some of the characteristics of the eternal enmity which exists in Washington between the Presidency and the Senate. The analogy may not be pressed too far; but the board frequently holds hearings, appoints committees for investigation, and in other ways duplicates work already done by the planning commission. The results in these cases are

o*

foregone. For political reasons, or because some interest has not been tenderly enough treated, the commission is reversed. A show is made of fairness in reviewing the commission's findings and specious appeals are often made; but the basis for a new judgment seldom lies in the facts or in concern for the public weal; far more usually it lies in political caprice, or, sometimes, in a mere perverse determination not to allow the planning process to grow in prestige. The best mayors—such as F. H. LaGuardia—are most apt to force their whims or conveniences into law in this way. This possibility of whimsical executive action is perhaps the most pressing single remaining flaw in New York's government.

The almost complete centralization of administration for so vast a city in the board of estimate, and finally in the mayor, has solved some problems and produced a good many new ones. The mayor, in spite of the widespread ramifications of the civil service regulations, is able to build up a formidable political machine. There is nothing necessarily sinister in this; it may be a good machine, made up of those in whom he has confidence. But there are thirty-two executive departments in the city,[1] some carrying on operations of enormous scope. And to maintain the kind of control over these operations which will keep them efficient or even merely honest, has proved to be beyond the capability of any mayor up to the present time. Periodic scandals still break out and ineffective administrators cause one department after another to decline in serviceability. The remedy for this is an extension of the mayor's person into an institution; and some of the necessary institutional devices have been adopted, though evidently not enough.

The deputy mayor, who sits on the board of estimate, has been mentioned. Numerous supervisory duties are delegated to him. There is also a department of investigation;[2] there is a legal department; and there are budget and personnel offices. Besides this, the fiscal affairs of the city are supervised by a comptroller. Because the comptroller is elected, even if on the mayor's party programme, he almost invariably becomes a

---

[1] Not all are called departments; some are boards, commissions, etc.; but all have executive duties.

[2] The department of investigation is peculiar to New York City. It is primarily under the direction of the mayor (but also must undertake investigations for the Council) and has the power to require information concerning the transactions of all city departments, to take evidence and finally to make reports on which the mayor may, if he chooses, act. It has been of notable service in checking the graft which was so prevalent in the older régimes. Concerning its powers, the charter indicates that it 'is authorized and empowered to make any study or investigation which in its opinion may be in the best interests of the city, including but not limited to investigations of the affairs, functions, accounts, methods, personnel or efficiency of any agency.'

political rival and is not to be counted as a faithful subordinate. This, also, is a situation which should be remedied. It has often become an intolerable nuisance to have such division and rivalry at the centre.

In spite of all the devices used to extend the personality of the mayor, he has not been able to free himself sufficiently from routine duties. Very often entirely unexpected issues are thrust upon him for settlement —unexpected in that he has not been able to prepare for them. The world impinges on New York in most insistent ways. In recent years there have been wars, cataclysmic depressions, labour disturbances, and numerous lesser matters, many of them originating elsewhere, which have broken in upon the routine of the city and caused the gravest disturbance. Some sort of emergency adjustment has to be made to meet them; and most of the responsibility for action falls inescapably upon the mayor. He may appoint a committee or a commission to investigate; he may create emergency agencies; but the creation and execution of policy is, politically his. The people, having elected him, expect him to solve all the city's problems or to see that they are solved. And little leeway is granted him for dilatoriness or failure. It is not surprising that routine is sometimes neglected and breaks down, or that new responsibilities are evaded until criticism becomes acute. The mayor of New York must necessarily be among the most harassed and unhappy of men.

Some of these difficulties are expected to be eased by an entirely new device adopted by Mayor Wagner. This is a city administrator who will direct the administration of the municipality and free the mayor from the heavy load of such work which has weighed him down throughout the life of the new charter. This scheme went into effect in 1954. It is obviously an attempt to meet the problems created by centralization just as centralization was an attempt to meet the problems of diffuse responsibility.

The charter of 1936 had the effect of rationalizing the city's government in other ways already prepared for in public opinion. The comptroller became an auditing and conserving officer, having charge as trustee of the sinking funds for which there had formerly been a board. But the duties of collecting revenue and making disbursements were given to a new officer, a city treasurer, who was made head of a department of finance. There was set up also a consolidated department of housing and buildings with authority respecting the planning, construction and inspection of private buildings. The commissioner in charge of this department was to have a deputy in each borough for the convenience of the public. He was, however, to succeed to the formerly scattered duties of co-operating with all the public and private agencies which had long been concerned with the amelioration of housing conditions. In particular he was to advise the planning commission in this field. A new department of public works was established to centralize all city-wide construction of streets, sewers, and public buildings. The charter commission, however, believing that public opinion was not prepared for complete centralization, left with old local improvement boards and the borough president, the planning and construction of local projects. That this was not a logically conceived, but rather a compromise, settlement, can be seen. Also the construction of schools still remained with the department of education, docks and airports with the department of marine and aviation, underground railways with the board of transportation, water supply with the department of water supply, gas, and electricity, and parks with the department of parks.

In its rationalizing effort the commission did not venture to transform all multiple-headed agencies into single-headed ones. There remained, for instance, a board of education, and one of transportation, which last was to have greatly expanded duties when the underground railway system was taken over in 1941. But by 1938 there were no longer boards, but single executives, in charge of such large departments as welfare, law, health, hospitals, correction, police and fire, although advisory boards or boards for the setting of standards and making of regulations were retained.

The duties and powers of these departments and boards are generally those indicated by their titles. Appointments are made by the mayor; but subordinates and deputies are appointed by the designated commissioner. There are sometimes qualifications set out by charter, as for instance that the commissioner of health shall be a physician; but the mayor generally has a free hand in choosing subordinates and is regarded as responsible for their acts.

Mayoral responsibility is complicated by the fact that some departments must comply with state laws and regulations; and some are almost completely governed by them. For example, the commissioner of welfare acts under the state public welfare law, which authorizes and regulates outdoor relief; another instance may be found in the limitations on the department of water supply, gas, and electricity imposed by the board of water supply, which is a city board providing water from distant sources outside the city. In most cases where state or federal funds have been furnished to the municipality, a state agency has been established for their control. This, at worst, amounts to a duplication of duties;

and, at best, seriously limits the autonomy of the executive department concerned.

It will be seen that admirably as the charter conforms to modern notions of efficiency, its framers did not go all the way in establishing a simple and logical structure. Aside from retaining multiple-headed departments, perhaps the most striking concession to tradition was the retention of the borough offices. In its report, the charter commission defended this recommendation by recalling that the borough presidents had been given seats in the board of estimate. This it was said, was in response to a demand for 'separate and direct representation' for each locality. It was admitted that this would probably lead to log-rolling and neglect of the city's general interests. But it was felt that there could be no perfect solution, and most of the incentive to log-rolling had been removed by other devices—notably the planning commission. At any rate the concession was made; and although the borough offices have almost become vestigial, they still exist.

In spite of the obvious exceptions, the intent of the commission to centralize executive power in the mayor was fairly well realized. There is, in most instances, a well defined chain of responsibility. This, together with a clear specification of powers and duties, has considerably reduced the inefficiency and confusion which so long characterized New York city government. It will be noticed, however, that the city's responsibilities are, in accordance with the general American disposition, more confined to regulating private enterprises which furnish services than is true of European practice. The city does own a large part of its transportation system and does furnish water services, although the gas and electricity utilities are privately owned. Its medical services are confined to the indigent; and only a very limited number of low-cost apartments are publicly owned. This mixture of public and private ownership is characteristic of the present transition stage in the United States, and is being modified only with considerable reluctance.

During the depression of the late 'twenties and early 'thirties, the city's productive apparatus came almost to a standstill. Factories were idle and office buildings were deserted. Workers who could not pay rent and were pressed for a family's food were poor taxpayers. But the demands on city services were never so pressing. The resorts to which Mayor LaGuardia was forced were many and sometimes devious. In the end the federal government was induced to come to the rescue. And at one time more than half the expenditures in New York for municipal services (including emergency welfare services) were being paid for (although often indirectly) by the national government. But what manoeuvres were involved in bringing this about, what sleepless nights and endless conferences, can only be known by those who saw something of the process. The city departments had to go on operating; schools had to be kept going; water had to flow; garbage had to be collected; parks had to be maintained; transportation services had to run. And each of these was deeply affected by changes. Revenues were down; but demands were up. The unemployed might sleep in the parks; but that did not lessen the need for park maintenance: children might go to school underfed; but that only made it more necessary that they should be given a school lunch: office buildings might be nearly deserted; but the streets around them and their water and sewer connections must be maintained.

The governmental problem in this is obviously not that of centralizing but of discovering devices which will make the centre effective. Unless these problems can be solved, the old scattered responsibility, corrupt and ineffective as it often was, might still be better. New York has gone a long way towards a solution. Its government is a kind of pilot plant, at least for America, and its failures along with its successes are noted everywhere.

PLANNING AND FISCAL PROBLEMS IN NEW YORK

The physiology of New York fixes many of its problems; it also furnishes the pattern and scene which make the city unique; and it gives structure to the organism which so obviously pulsates with rhythmical busyness and goes about its affairs with monotonous regularity. To a detached observer, hovering above its morning or evening streets, it would have an unmistakable likeness to the habitations of the more highly developed ants. Small creatures emerge from the dormitory areas far-out or near-in, converge on the arteries of transportation and presently erupt into the streets about the downtown centres and find their way to innumerable offices. At noon for an hour or two the streets are crowded again with lunch-time strollers. And from half-past four the subway entrances, the ferries and bridges, carry a returning scramble of home-goers. On all the converging parkways in the morning small beetles hurtle toward the city centre, hustling into parking spaces, discharging a passenger or two and cooling off there till afternoon again. There are eddies in these currents, a little movement against the stream here and there; but these tend to be lost in the insistent pulsations of migration.

The city is bedded in the sea, and the sea encircles many of the city's areas. It reaches about Manhattan and Staten Island (the Borough of

Richmond) and almost about Brooklyn; it extends behind the Palisades on the Jersey side and penetrates far up the Hudson to meet the down-rolling waters from up-state New York. Coming down into the Narrows from the wide ocean to the east, ships turn northward up the bay—Staten Island on one side, Brooklyn on the other—and sight at once the abrupt skyscraper cliffs erected on oldest New York, back of the Battery on lower Manhattan. Getting to their piers, past the statue of Liberty on Bedloes Island and, on the other side, the military installations on Governor's Island, they sail over seven tunnels[1] and if they sail up the East River they slide slowly under old Brooklyn Bridge, then the newer Manhattan and Queensborough bridges. They stop south of the great George Washington Bridge across the Hudson if they come up the west side (the North River to New Yorkers), but it rests, pale and graceful, up ahead against the sky. Upon these crossings, some for rail traffic, some for motors, the traffic rushes, slowing as it concentrates, then speeding to spread into the teeming business districts. There it coagulates and crawls, coming finally to a stop.

Ships go over or under New York's traffic; they also meet it, for the harbour is alive with ferries which the faster tunnel and bridge crossings have made obsolete but have not yet given the *coup de grâce*. The ferries are not all for passengers, of course, for much of the materials by which the city lives—food and fibre, as well as the finished and semi-finished products of a high civilization—come to a rest at dock-side in Jersey. There they may be shifted to barges, the cargoes broken; or they may simply be carried across on car-ferries. There are nearly three million people in Brooklyn and they have no railroad station (a curious lack), and most of their freight comes by water since the one railway tunnel is hopelessly inadequate.

Manhattan Island is, of course, the centre of New York. The old commercial and financial district at its lower end still has the greatest concentration of cliffy office-buildings. There, a hundred skyscrapers reach higher into the air than any but a few structures anywhere else. But even this district is now dwarfed by the mid-town concentration between thirtieth and fiftieth streets. At 34th Street is the Empire State Building, and up at 50th Street, Rockefeller Center slides smoothly

---

[1] Altogether there are thirty-six bridges and tunnels providing crossings between the central city and outlying areas. Some of these are such magnificent structures as the Brooklyn, Queensborough, Triborough, George Washington and Whitestone bridges; some are major tunnels such as the Lincoln, the Queens and the Brooklyn Battery, and some are much shorter but no less necessary crossings of the Harlem and the outer channels between New York, Coney Island, and Far Rockaway.

skyward; and over on the East River, the United Nations has created its modern headquarters.

The great spiring buildings seem to rise from the sea; but actually they are founded on granite. This is fortunately a solid base; but tunnelling through it for underground transport, communications, water supplies and sewer passages is far more difficult than such operations in cities built on less compact substances. Nevertheless, hard as all this drilling and tunnelling has been, the substructure of the city is by now an incredibly riddled complex, at once marvellously efficient and frighteningly vulnerable to interruption. Most New Yorkers never consider the delicate network which functions so regularly beneath and all around them. They take it for granted, roaring through their subways, casually using their telephones, drawing water, turning on and off their centrally-provided warmth. But, actually, it requires a vast army of engineers and maintenance crews to keep this network in operation. Occasionally gas, water or steam erupts through a solid-seeming pavement; less occasionally a service fails; then for the moment the citizen is annoyed by the inconvenience. But ordinarily he lives and works, calmly unaware of the carefully adjusted mechanisms on which he is so dependent.

The cost of maintaining New York's plant and the operating of its services some time ago passed a thousand million dollars a year. Replacements and extensions (never enough) run to half as much again.[1] But all such accountings are confusing and inadequate because of their incompleteness. They are, for one thing, for New York City alone, which covers only about half the area and contains but two-thirds of the inhabitants of the metropolitan area. Then, too, they take no account of expenditures within the metropolis by the state (more than half of New York State's employees are occupied in New York City) and federal governments. Both contribute to the building and operating of housing projects; both have enormous hospitals; both have some specialized schools; and both do a good deal of their work within the city. Any generalized account has to fall back on suggestions of vastness and scope, on descriptions which are admittedly unsatisfactory to the economist or the engineer but which often convey reality better than figures.

This vast municipal mechanism, it is well to recall, is crossed by the 41st parallel of latitude. This northerly location, together with its nearness to the sea, produces a stimulating but not extreme climate. There are frequent storms. There are cold draughts from the north, fresh winds from the west, and occasional damp warm air-masses from the south-east.

---

[1] The capital budget for 1951 totalled $478,761,756.52.

45. NEW YORK

The Triborough Bridge at the point where the roads from Manhattan, The Bronx and Queens converge at Randall's Island.
[*Photo: Standard Oil Co. (N.J.)*]

46. NEW YORK

The skyline of Manhattan Island at dusk.

This changeability produces its effect. New Yorkers are notably restless and energetic; they move rapidly as though they always had somewhere to go in a hurry; they produce freshets of ideas and are apt to want things done without delay. It would be to press the theory of climatic determination too far to attribute these traits altogether to climatic influence: other factors have to be remembered as well. The decades of migration to New York of the more restless folk from elsewhere; the incidence among them of ambitious small enterprisers; and the fact that New York quickly became the headquarters for most of America's great corporations, have each contributed to a unique way of life. New York is a concentration of restless, idea-generating, fast-operating planners and executives together with their staffs. They live tensely; amuse themselves recklessly, and regard themselves, on the whole, as a picked and superior group. Climate and geography between them, however, undoubtedly give colour and character to the organism. The bright, clear, sparkling skies when the west and north-west winds come down from the Appalachians; the furious three-day rain and wind storms out of the north-east; and the close foggy interludes when the Gulf Stream exerts control—all these come in upon the city and give it a characteristic variety. The winds blow through the city canyons, the fogs cling to the skyscraper towers; and everywhere, at the end of the street, is a bay, a river, or a marsh.

It was this restless entity, living to itself, but setting out to manage affairs of vast scope beyond its borders, that New York's planners always had to deal with as they strove for foresight and for a grip on the fulcrums of planning control. Planning was sporadic, piecemeal, and largely futile until the Regional Plan Association was organized and set out to comprehend in one whole the ethos and physical structure of the city. The Association was born partly of a certain conviction of sin. The Equitable Building in lower Broadway, led, in 1916, to the first zoning effort. For that vast structure, surrounded on four sides by narrow streets, covered every inch of its ground space with a many-storied monolithic warren of offices. It imposed its bulk on a whole neighbourhood, thieving light and air from all the abutting blocks and reducing values for an indefinite distance in every direction. It was the sheer necessity for self-protection which led the real-estate interests themselves to demand some control. The first restrictions were farcical, because, of course, everyone wanted everyone else restricted, but himself left free. Into the potentially valuable zones for commercial use were thrown many times the acres needed, even if the most fantastic chamber of commerce dreams should ever come true. But it was a camel's nose within the tent of cut-throat real estate competition. And it had its consequences.

THE NEW YORK REGIONAL PLAN

The very futilities of this first restraint produced the inevitable conviction that zoning was only a means to an end, which itself had not been properly visualized. The Russell Sage Foundation provided the funds; a committee of leading citizens provided the sponsorship; and an able staff produced, a decade later, the Regional Plan of New York and Environs—the first really organic view of the metropolis, and the starting place for all subsequent efforts.[1]

The work done by the Regional Plan Association could hardly have been accomplished by an official body. This was explained in the introduction: 'The facts that the Region lies in three states, and that there are hundreds of public authorities, as well as numerous public utility and trunk railroad corporations . . . indicate that the making of a complete regional plan was a task that could not be undertaken by an official group. . . . Thus the situation is that the various governmental bodies and private corporations that have the powers necessary to legislate for and execute plans have not combined . . . and are not likely to combine . . . while the voluntary body . . . has no power and cannot be given the power to carry out its proposals.'[2] This intimates that, in the American system, no comprehensive operating plan is possible for a region except through the voluntary adherence of the organisms represented. And it remains true. The municipalities and utility corporations are chartered by states; and states are sovereign. In the case of New York City three states are intimately involved; and there are more if other conditions are considered. For a railway or a public utility, operating in many states, may be chartered in only one. That one may be—it often is—Delaware. The requirement that such a corporation submit to direction of an overhead planning authority would be subject to grave limitation. Only strictly police-power regulations may be applied without question; and these are matters for the states, not for cities, unless there has been specific delegation.

---

[1] The date of its copyright is 1929 and 1931. The list of committee members is as follows: Frederick A. Delano, chairman; Robert W. de Forrest; John H. Finley; John M. Glenn; Henry James; George McAvery; Dwight W. Morrow; Frank A. Polk; Frederick B. Pratt; Lawson Purdy. The General Director of Planning was Thomas Adams. In addition to the two large volumes, there were numerous monographs on special features of the plan.

[2] *Op. cit.*, Vol. I, p. 158.

We can see, then, that the New York region is not legally recognized as an entity, however much it may function as one. It is penetrated by and has as functioning members numerous organizations which are subject to only limited control by any governmental unit which might federate with others to form a comprehensive regional organ. This statement perhaps exaggerates. It is possible to do much by indirection —as has been shown by the New York Port Authority and the Bridge and Tunnel Authority. But the Regional Plan Association was quite justified in pointing out at the outset that about all it could do was to analyse and assess the facts and potentialities and suggest the physical means for reaching what seemed to be the needs and aspirations of the region's people. And that is what it did. The plan, when it was published, was a graphic plan, showing physical facilities. This did not mean that a great deal of economic, sociological, and political science talent had not been employed. It had. But it was not a scheme for a governmental institution which could go on devising facilities and seeking better ways to make them operate. It was strictly physical, however flexible, and depended on its graph for conviction. It did suggest joint actions of various kinds; and certain joint efforts were subsequently made, and there are still others to come. But essentially the situation remains now what it was several decades ago. The New York region, because it has no comprehensive government, has no continuing and comprehensive planning.

It has been of immense value, however, to have had, as a starting place and background for all subsequent efforts, the carefully proposed graphic plan of 1929. Most great cities have had such plans made, but none has been technically superior to that of New York. It included, when it was completed:

1. Proposals for extensions of trunk line railways, railway terminals, rapid transit lines, etc., the planning and development of which are usually the function of private corporations, such as railway companies subject to special state laws and some measure of federal, state, and municipal control.[1]

2. Proposals for harbour and waterway improvements, extensions of railway facilities, freight distribution, major bridge and tunnel projects, etc., that primarily lie in the jurisdiction of *ad hoc* authorities such as interstate commissions appointed under special state laws.

3. Proposals for county highways and loop roads, parks and parkways, and, generally, for projects outside New York City, which extend over county areas and are appropriate to be dealt with by counties with the aid of the state on the one hand, and in co-operation

---

[1] An exception to private planning and operation of such utilities exists in the case of the City of New York in connection with a large proportion of its rapid transit system.

with the city, town, borough and village authorities within the county on the other hand.

4. Proposals for arterial and secondary highways, zoning, parks, mapping of land and other features that may be dealt with under the city charter or state enabling acts, by the City of New York.[1]

5. Proposals for:

(*a*) Highways, zoning, parks, mapping of land and other features that may be dealt with under planning and zoning enabling acts of the respective states;

(*b*) Restriction of open areas and development of architectural projects that can only be carried out through co-operation between public authorities and owners of land under present conditions of the law and for which new legislation is needed;

(*c*) Planning of sections such as the Jamaica Bay and Hackensack meadow sections, which requires the passage of special laws and the setting up of special joint authorities.[2]

It was when the planning commission was made part of the city's government that the next long step in planning was taken. The Commission, through its capital budget, has sought, when it could, to press forward the conceptions of the Regional Plan. As has been said, its powers to do this were seriously limited because New York City is far from being coterminous with the New York region. And little progress has been made toward solving the problems posed by this difficulty. Nevertheless as New York has grown, the growth has been notably in the directions indicated by the great original basic study.

MUNICIPAL FINANCE

To discuss adequately New York's fiscal situation and the problems it discloses, would be to consider exhaustively the whole basis of American municipal finance. The problems can only be indicated here. New York reveals them in an exaggerated form: the vast and rapid growth of population; the tendency first to crowd into centres, then, as the techniques of transport and communications are improved, to throw out escape routes; to build irresponsible dormitory and manufacturing satellites along these routes, leaving the centres to decay; the continuing struggle to maintain all the old central facilities for a declining population and, at the same time, to duplicate them for the new fast-growing outlying areas; and further to provide, from shrinking resources, the means to meet inevitable demands for modernization of all the old facilities and the addition of new ones.

Remembering that New York City is by no means identical with the

[1] The area, population and unique character of the City of New York with its five counties requires that these proposals should be classified separately.

[2] These three classes of proposals—(*a*), (*b*) and (*c*)—are those that are appropriate for treatment in the plans of cities, boroughs, villages and towns in New York, although no statutory power exists to deal with (*b*) and (*c*).

metropolitan area, and that data for New York City would have to be increased by about one-third to gain a comprehensive view of the problems involved in taxing and spending for the area, it is still possible to gain some idea of size and scope by inspecting the city's income and outgo. These, of course, balance; for a city is a corporation and it may not have an unbalanced budget. But part of its income may be derived from borrowings; and the extent to which New York City has resorted to borrowing is shown by the size of its debt—which on July 1, 1950 was $3,139,894,613.75.[1] But its borrowing is limited by state law to ten per cent of the average realty valuations in the five-year period immediately preceding any current year. Many exceptions have been permitted to this limitation.

Its power to tax is similarly limited to $2\frac{1}{2}$ per cent of realty valuations. Other sources of revenue have had to be found, and have approximately doubled the income; but the state legislature has been reluctant to grant other taxing powers. At nearly every legislative session more is asked by the city than is granted. In spite of some relief, the limits within which spending must be confined are strict. Revenue seems never to be sufficient; and services, maintenance and capital extensions have to compete for what income there is. How this may affect so important a matter as the provisions of schools was made clear in the introduction to the proposed capital budget of 1951 by the planning commission. It was said there:

> The demand for schools has reached a stage where it is impossible to supply the needs of the children in a reasonable time without the extraordinary measures and greatly augmented funds. . . . It is clear that some aspects of the school problem—certainly its magnitude—have created a real peacetime emergency in this city. . . . Enough is not being done; the long-range plans of the Board of Education are not moving fast enough, and even if they were speeded up considerably, the difficulty of providing sufficient funds to meet the demand for schools would remain. . . .
>
> The school shortage is the result of wholly unforeseeable developments in many fields. Unparalleled prosperity, increases in populations, in marriage and birth rates and employment; the increase in the numbers of families; public and private housing developments which had increased densities in old sections and caused great shifts of population to new areas, and the tremendously increased school population, all help to explain the shortage. Difficulties in relocating tenants so that sites may be cleared is perhaps the greatest single factor in retarding construction.
>
> These developments have not come suddenly and the trends have been apparent for some time, but the city has never been able, financially, to meet the demands of a constantly expanding community. During the depression the city received federal aid and constructed many schools, but it never caught up, either in providing for new neighborhoods or in replacing obsolete structures in old sections. Cessation of building

---

[1] This is gross funded debt. Comptroller's *Report*, July 1950.

during the war and the phenomenal development of new sections since that time have caused the city to fall farther behind.

There was a good deal more to this rather pathetic plea in avoidance. In contained, of course, the usual contradictions. It was said, for instance, that the situation was both unforeseeable and of long standing, although both could not be true. The truth is that the same, or almost the same, account could be written about the arrears in all other facilities.[1]

The situation is not one which was unforeseeable. If the commission had scanned its own reports in former years it would have seen ample notice of the inevitable development of just such a crisis as now exists. In fact the crisis of 1954 is only a deepening of that which already existed in 1940.

The crisis arises from the fact that real-estate valuations control both an important part of the taxes which can be levied and the debt which can be incurred—even when the state legislature makes provision for special taxes of some sort, such as a city sales tax. Realty values are, of course, a function of their earning capacity. If they are allowed to fall because industry and population escapes into outlying areas, or because of wide exemptions, both values and taxes fall. It is, however, unfortunately true that declining population does not result in a proportionately declining cost for the facilities which serve the decimated areas. They, in fact, remain very nearly constant. So there arises the double problem of providing new facilities in the escape areas and of maintaining the old ones in the degenerating areas.

This is a dilemma which can only be met by political unification, by

---

[1] Except, perhaps, parks, playgrounds and highways for ingress and egress from the city. But the relative adequacy in these matters is accounted for by the activities of Mr. Robert Moses, commissioner of parks, chairman of the Bridge and Tunnel Authority, and construction co-ordinator. Mr. Moses, a man of explosive energy, enormous persistence, and ruthless determination, has schemed, coaxed, and driven from his centre in the State Office building (he is commissioner of parks for the State of New York as well), until the city has multiplied many times the number and area of parks and playgrounds (from 1934 to 1947, the number of parks increased from 119 to 492) and is provided with an incomparable system of highways leading out into the rural areas. It has in fact more parks and playgrounds than it can well support; and, of course, the highways, convenient as they are to the motorist, have, by luring taxpayers into the country, done more than anything else to maintain the state of chronic bankruptcy of the city. They themselves have been paid for by separate bond issues. A toll is levied on every passing vehicle. This taxation for a specific and segregated purpose at least provides for their maintenance—and, of course, adequate security and ample interest for the bondholders. Those responsible for other city facilities envy Mr. Moses his sure source of revenues. But they seldom say so; for the Moses temper is notoriously uncertain and his attacks on critics are wholly unrestrained.

checking unplanned escape from the centre and by requiring outlying areas to shoulder their share of costs for maintaining the centre on which they are parasitic. This is not an impossible programme. It requires the overcoming of resistance from real-estate and manufacturing interests which have escaped into areas of low taxation. In New York this could be accomplished for certain areas—like the Borough of Queens (whose population has grown enormously) and even Westchester and Nassau counties. No more than an act of the state legislature would be required. But it happens, as has been noted, that the New York region lies within more than one state. New Jersey is heavily involved and Connecticut, also, to a degree. The situation can be understood when the increase in population in these areas, in the decade 1940–50, is compared with the growth in the city itself.

The counties comprising New York City showed the following percentage increases 1940–50:

New York County (Manhattan)	2·5
Kings County (Brooklyn)	0·7
Bronx County	3·6
Richmond (Staten Island)	9·5
Queens[1]	19·2

But nearby counties in New Jersey have shown, in the same period, the following percentage increases:

Bergen County	33·2
Burlington County	40·1
Hudson County	20·8
Middlesex County	21·9
Monmouth County	38·5
Morris County	30·9
Ocean	48·8

This is the most plagueing difficulty for New York. The problem of extending city limits, and therefore taxing areas, into another state, is, under the American federal system, beset with such possibilities of obstruction as to seem almost impossible ever to bring about.

Yet the metropolitan area which lies within the state of New Jersey cannot wholly abstract itself from the problems of the city. It has so far refused to recognize any responsibility for the decline of realty values in New York and for its problems of obsolete schools and insufficient

---

[1] Queens County has itself been an escape area throughout the decade.

services of other kinds. But certain mitigations have occurred. Federal funds, in time of emergency, flow to places of greatest need. Thus, during the depression, for instance, the city received federal aid based on its needs rather than on population statistics. And federal aid remains as a possible continuing ameliorative possibility.[1] But also certain interstate devices have been set up to meet problems which are recognized as inescapably common. The most important of these is the Port of New York Authority.[2]

THE PORT OF NEW YORK AUTHORITY

The Port of New York Authority has a joint charter from the states of New York and New Jersey. It operates in the field of transportation, terminal facilities, and their subsidiary activities. Like its older prototype in London, it serves to elude many of the difficulties of conflicting jurisdiction; but it is also a way of escaping from the limitations on expenditure imposed by tax and debt restriction. Being a separate corporation, it has its own borrowing powers; and it may require charges for its services. The many facilities it has built and operated have on the whole been well managed and provided at reasonable rates. There has not been great objection to the principle of specific charges, though in theory it would seem an awkward way to provide such services. The excuse of its interstate nature has carried conviction.

A similar role has been played by the Triborough Bridge and Tunnel Authority. This is a merger of the various authorities set up to build and operate such facilities as the Queens and Brooklyn–Battery Tunnels, the Hendrick Hudson Bridge and the Triborough and Bronx–Whitestone bridges. It is hardly possible, now, to find a way of ingress or egress for which specific tolls are not charged. Furthermore it is necessary to use several of these crossings in proceeding from one part of the city to another. The objection that passage about the city or into and out of it ought not to be thus interrupted has often been made and has a certain force. The answer that if it had not been done this way it could not have been done at all, is, however, a true one. The city's chronic bankruptcy would certainly have made these modern improvements impossible to

---

[1] Mostly, again, indirectly; but sizeable funds come directly under the Public Works Act of 1933 (actually Title II of the National Recovery Act), and under the Lanham Act, which provided for defence facilities in preparation for war.

[2] The Authority issues a full *Annual Report*. The information here is from the 29th report of 1949. As to the Bridge and Tunnel Authority the information here is also from the *Report* of 1949.

build or maintain. There is also a certain strength in the argument that a specific charge for such a service is not unreasonable. At any rate the method chosen is in force, the facilities do exist, and until the improvements are paid for, at least, the toll-charges will continue. Between them, the Port of New York Authority and the Triborough Bridge and Tunnel Authority have gone far to bind the metropolitan region together in spite of physical difficulties. The city and its region are far more a whole unit than they would be without the new bridges and tunnels.

Even with this relief the city's fiscal problems are, however, far from being solved. The debt-charge continues at more than $100,000,000 annually.[1] This in itself is an extraordinarily heavy burden. If the debt could be discharged and the sums going to its service saved, many services could be undertaken which are not now even thought of. But the possibility of finding such a sum seems more and more remote. It is, indeed, far more likely that pressure for a higher debt limit will succeed in enlarging the burden.

The city's current revenues include less than $600,000,000 from realty taxes,[2] the traditional source of municipal revenue. Long ago other devices were resorted to, some very doubtful, and once relied on have never been discarded. Such for instance, is the sales tax which is admittedly regressive and undesirable from every point of view, except perhaps sheer practicality. But it does raise about $131,000,000 annually and its loss is something which city administrators no longer even contemplate. There are taxes on business which produce about $80,000,000; and a profit of some $50,000,000 is secured from the water service. Licence fees and other specific charges for services produce smaller amounts. Of more importance is $50,000,000 from state-shared taxes; and contributions of $207,065,188.19 from the state[3] and $42,713,350 from federal aid.

It will be understood that the withdrawal of many facilities from the tax rolls—such as all the public housing—solves some problems but makes others worse. For although these properties are exempt from taxes, the cost of servicing them tends to rise rather than to decline.

The following list of generally classified expenditures (for 1950–51) will give some idea of the cost, even on an admittedly unsatisfactory basis, of the services provided by the city of New York.

[1] City of New York, *Budget for the Fiscal Year*, 1950–51, p. 415 ff.
[2] $565,774,041.22 in 1950–51. *Budget*, p. 420.
[3] Figures estimated for 1950–51. *Budget*, p. 421.

Budget for 1950–51

	Total $
Legislative	518,310.00
General government—City	78,000,774.24
Libraries	6,963,574.50
Education	285,444,484.18
Cultural, scientific, recreation and memorials ..	3,383,778.00
Municipal parks	17,325,635.03
Public safety	171,155,665.46
Sanitation and health	68,929,537.72
Hospitals	80,198,831.50
Social welfare	220,997,521.06
Correction	7,158,193.84
Judicial—City	9,924,634.83
Public service and other enterprises	33,384,543.94
General government—County	3,932,187.50
Judicial—County	11,117,899.59
Debt service (exclusive of education) ..	204,400,993.01
Miscellaneous	30,493,788.96
Total 1950–51 Expense Budget	$1,232,430,353.36[1]

NEW YORK EMERGENT

The people of most great cities in the world are, to a certain extent, cosmopolitan. But none, perhaps, exhibits such an obvious mixture as New York. It is true that London, Moscow, and Paris are the capitals of empires and so draw to themselves the representatives of government, of commercial interests, and of cultural movements from many far and diverse places. New York has not been a capital in this sense until recently —when the United Nations adopted it—but it has long been the home of banking houses and commercial enterprises, especially those which have had far-flung relationships. The embassies of foreign governments, are, of course, in Washington, D.C. But most of them maintain an almost equally important and sometimes larger centre in New York: witness, for instance, the impressive grouping of British, French, Italian, Dutch, and other representatives in Rockefeller Center. But even these, important as they are, would not give New York its mixed culture. That *mélange* was created by the vast tides of immigration which flowed toward the

---

[1] *Budget*, p. viii. New York City is coterminous with five counties whose budgets are included in that of the city. The items are mostly for legal officers—district attorneys, judges and their staffs, etc.

United States during most of the nineteenth and the early decades of the twentieth century. These, as everyone knows, were first English, Scottish and Irish, then in turn Scandinavian, German, Italian, Russian and Eastern European. And then, when restrictions on immigration were established, there still flowed in great numbers from Puerto Rico, the American island possession in the Caribbean, whose Spanish and Negro population already possessed citizenship and could not be shut out.

These peoples from all parts of the world, except Asia, came into the United States largely through the port of New York; and many of them stopped in the city. This was natural. They soon began to form enclaves; and as these grew, additions to them by accretion became easier. New York not only offered as much economic opportunity as could be found elsewhere, but also offered a kind of cultural continuity not to be found in many other places. Immigrants arriving in middle age frequently spent the rest of their lives in the city without learning English, without changing their religious or social customs, and often without bothering to become citizens. New York indeed became for a time a kind of aggregation of communities—Jewish where Yiddish was spoken, German, Italian, Polish or even Armenian. There were, it was said, more Poles than in Warsaw, more Italians than in Rome—and certainly there are now more Puerto Ricans than in San Juan. And since people of these various origins tended to flock together they formed something like cities within the city.

Another distinctive feature of New York's population is the vast numbers of Negroes who have moved in from the South. This was partly because they could find at least some freedom from the racial discrimination they suffered in Southern cities. And the Negroes, although they were Americans, like other incoming peoples, tended to form an enclave in Harlem. This was not altogether their own fault. They were not welcomed in other neighbourhoods. But both for them and for other citizens the concentration there created many difficult problems which are still far from manageable. For the only possible solution to this, as to the other concentrations, is dispersal and assimilation in a new cultural entity made up of many diverse elements, all being in the end Americans and New Yorkers. That is undoubtedly on the way; but the struggles, while the inevitable process works itself out, have been and will be many and sometimes agonizing. There have been dangerous race riots in New York in the recent past; and there will be more. But they are not new. They were happening even in the middle decades of the nineteenth century. Sometimes the antagonism of older arrivals was aimed at the

Irish, sometimes at the Jews, sometimes at the others. And it has not been unusual for those who suffered from discrimination in one generation to treat more recent arrivals with the same discrimination and even outright hostility which they themselves—or their parents—had suffered earlier. The Irish, for instance, once the lowliest of immigrant groups, soon became scornful of the Jews; and all have joined in making life miserable for Negroes and Puerto Ricans.

But sheer size is something. Everyone who becomes a citizen can vote; and anyone who acquires an income can own property and establish himself in business. So gradually the lowly immigrants have come to possess important increments of political and economic power. Many of the largest enterprises in New York are owned by the Jews; and some by the Irish. And politicians nowadays treat all the large distinctive groups with respect. It is no accident that Mayor LaGuardia was a second generation Italian; that Mayor O'Dwyer was born in Ireland, or that all three mayoralty candidates in 1950, including the successful one, were born in Italy.

The gradualness of the absorption of these groups into one another to make an emerging entity has often been noted. Some observers have despaired, as Henry James did in the first decade of the twentieth century.[1] But now, more than forty years later, his picture of New York has become unrecognizable. Only the vestigial remains of those scenes and institutions about which his eloquence wove a delicate tracery of comment still exist. He was obsessed by the Jewish dominance then—not only in their 'ghetto' on the Lower East Side but as he strolled in Central Park or on Fifth Avenue. The Jewish culture has in two generations been transmuted. There is no real ghetto in New York now; and not so much Yiddish is spoken, although there is still a Yiddish theatre and there are numerous synagogues where Hebrew is spoken. But neither is there any longer an old American social 'four hundred,' having immense inherited fortunes, spending conspicuously, and excluding the *nouveau riche*. The modern men of wealth would not have been recognized as such in that older day. Their position is no longer impregnable, what with income and inheritance taxes; and they are no longer a recognizable and exclusive class. Into their ranks Jews, Italians, and Irishmen have now penetrated, and the higher-income groups, like the lower, tend to blur and fuse. There are clubs still where newcomers are not admitted; and there are social circles which hold themselves aloof; but they no longer carry

---

[1] *The American Scene*, New York, 1907.

themselves with the old assurance of superiority. They will not endure for long; and they know it.

Fifth Avenue has been succeeded by Park Avenue, and both by the Westchester and Long Island suburbias. There the old families still fight rearguard actions to keep possession of their 'position.' But all this kind of thing has diminished; it no longer interests others. And it is the incidents of a rising and spreading democracy which hold the interest in recent times. The melting pot is fusing the elements which threw themselves into it as they passed from the processing of Ellis Island; and the emergent is wholly novel. The teeming slum areas, now being rebuilt, are making something new. The Jew and the Italian managed to transform the old-law tenements and the streets which separated them into transatlantic replicas of old-world cities—even more degraded sometimes than the originals. The new housing projects, rising on cleared sites, have no possibility of being thus transformed. But then the inhabitants themselves have lost the backward look which would make them regretful. And they accept life in the new conditions as part of the democratic progress they themselves have helped to create. This is now New York—there in Red Hook, in Stuyvesant Town or Lincoln Houses—it is not Europe and it is not the American South. It is an expression of a new self-consciousness.

This progress in fusing the old and new immigrant groups into a new organic New York ought not to be exaggerated. It can readily be seen to be going on; and there are impressive changes to be noted when accounts of New York life in former decades are read, or when those of us, who are now old, recall what it was once like. The process is continuing, though it is by no means complete. New public housing has been provided for sixty thousand families and accommodation for another sixty thousand families is in sight. Middle-class and upper-income family rehousing has gone on apace. The job is not even half done; but its momentum can be seen and will now carry it through to the end.

Momentum is indeed a potent social force, nowhere better illustrated than in public housing. It seems to the casual New York visitor over several decades, that the transformation now coming into view is a very recent phenomenon. But actually the rehousing effort goes back more than half a century. At first, reform was directed towards improving the private tenement; then it showed itself in philanthropic projects; and it finally emerged, in the 'thirties, as publicly built low-cost apartments. There are honourable names associated with this long effort. One of the documents which stirred the civic conscience was Jacob Riis' *How the*

Other Half Lives; but also—and more practically—a Brooklyn engineer, Alfred T. White, not only wrote about housing reform but engaged in it. He built blocks of houses in Brooklyn in 1876 which in 1950 were still occupied and profitable. White, like A. T. Stewart the merchant, who began the now enormous housing ventures in Garden City, Long Island, was influenced by Robert Owen and Ebenezer Howard. There was passed by the state legislature in 1900, after long agitation, a new tenement house law which outlawed the old railroad and dumb-bell flats. But there are still almost half a million people living in those horrors.

Perhaps the degradation of New York's poor reached its lowest point about 1910. That was the year when Manhattan's population reached its highest point (2,330,000). It was just after this that escape to the outer regions began by way of the new subways and other transit lines and the East River bridges. That escape doomed old Manhattan. For decades decay slowly ate up the old tenements as people escaped from them, creating the conditions which finally made slum-clearance possible. By the 'thirties, when this work began in earnest, it was possible to walk through many east side streets where half or more of the tenements had been closed and boarded-up by the health authorities. It was a region then of incredible squalor; its remaining inhabitants a mass of pitiful poverty-ridden folk. What was more important, perhaps, the slums there no longer 'paid.' So long as they had, there was not much hope of reconstruction on a large scale. But when the properties had fallen one by one into bankers' hands the position was changed. Not only could large tracts be assembled for private development, but also it became possible to overcome the resistance of landlords and real-estate owners to public housing. The depression and the succeeding New Deal completed the conquest. From then on the new policy was dominant and transformation began. By 1950 the Lower East Side, once the ghetto then the half-abandoned slum, was receiving in its rebuilt state the sons and daughters of those who had escaped from Manhattan a generation or two earlier.

So the battle for better housing turned toward victory. But city folk do not live in houses alone; almost as important are the adjuncts of family life, the means for maintaining social cohesion, and the institutions for conserving and improving culture. New York was formerly notably deficient in small parks, playgrounds and open spaces. These might have mitigated, even if only to a degree, the crowding of the slums. Even that lack is being repaired as the slums are cleared and the new housing replaces it. For the new housing at least in the centre of the city, is multistoried and provides a certain amount of open space which can be utilized for community purposes. It is not nearly enough, as such critics as Mr. Lewis Mumford have bitterly insisted. In fact, almost always, the density per acre increases, a function of the inflated land values figured into the per-unit costs of all the dwellings. But still there is some advance; and amenities are somewhat enlarged by the growing freedom with which the prevalent gridiron street-plan is modified to create super-blocks. Through traffic is thus shut off; and the old street-space becomes park or playground.

Then old schools and other community facilities in the neighbourhood of improved housing show themselves in shabby contrast. There is pressure for rebuilding these also which in the long run cannot be resisted. As has been noted, however, the school building programme lags. Such improvements suffer from the perpetual incipient bankruptcy with which the city struggles. There can never be enough of anything of this sort so long as the basic fiscal situation remains unchanged. And that can only come at a later stage with more complete unification and planning for the whole. Transportation has declined in efficiency for the same reason. Subway travel in New York has for many years been an ordeal. The equipment is ancient, shabby, and, in consequence, poorly maintained. Living places have been too distant from work places for reasonable transportation costs. A condition of this sort has either to be met by paying excessively for service or by submitting to inferior service, a dilemma from which no real intention to escape is as yet visualized.

The people of New York have not yet taken hold of their basic problems in any genuinely determined way. In that they do not differ from the people of other American cities. But the materials, and even the institutions, for social direction and control are, except for still broader unification, already in existence. They await only the leadership which will fuse the present conflicting interests into a single purpose. When that leadership is supplied, the objectors and resisters will be subdued and the city will become what is so clearly possible—a vast and beautiful machine for living and for working.

There are, in spite of civic deficiencies, unmatched amenities in New York. The art collections are rich and varied; the institutions of higher learning, led by Columbia and New York Universities and the four city colleges, are unrivalled in the whole nation; most of the books and magazines published in America issue from New York; the theatre has its home along Broadway; there are superb musical organizations; and linked with the several medical schools, some for specialized or post-graduate work not found elsewhere, there are vast hospitals and medical centres.[1]

The New Yorker, conscious of living in and with these cultural enterprises, although his share in them may be very tenuous, has a pride in his city's greatness which gives him enormous satisfaction. If he has certain discomforts, if he lives in crowded quarters, suffers indignity when he travels and moves through ragged parks and half-cleaned streets, thinks these penalties a small price to pay for sharing in the life of his town. The fact is that most citizens are hardly conscious of their disadvantages. Having grown used to them, they seem almost inevitably linked with those values which he holds most precious. New Yorkers certainly cannot be said to be unhappy about their state and are not overly sympathetic to critics who profess to feel sorry for them. These attitudes may very well be reasons for the delay of long-needed reforms.

[1] Although the number of hospital beds is still insufficient. There are some 40,000 but 60,000 are needed.

47. PARIS

The Arc de Triomphe and the Etoile.

[*Photo: French Government Tourist Office*

48. PARIS
The Hotel de Ville, or Town Hall

49. PARIS

A general view, showing the *Cité*, an island in the Seine containing many famous buildings.

[*Photo: Greff*

[Photo: J. Richard

50. PARIS

A closer view of the *Cité*, showing Notre Dame Cathedral on the right.

Paris

BRIAN CHAPMAN

P

BRIAN CHAPMAN

Born 1923. Educated Owen's School, London, and Magdalen College, Oxford. M.A., D.Phil. Sub-Lieut. R.N.V.R. on special service, 1942–45. Student of Nuffield College, 1947–49; doctoral thesis on Italian regional government; Research Assistant, University of Manchester, 1949; Lecturer in Government 1952.

Author: *An Introduction to French Local Government*; articles on Italian and French government and law in *Public Administration*, *World To-day* and *Modern Law Review*.

Paris

PARIS is not only the capital of France but also of Western Europe. It is situated in the centre of the basin of the Seine, linked to the sea by river, to many other centres by canal, and is a great hub of road and rail communications for France and for transcontinental traffic.

THE SIZE AND SIGNIFICANCE OF PARIS

Its importance comes less from its favourable geographical position than from its history. The French kings elevated Paris from being one provincial centre among many to the unrivalled capital of all France. The Revolution confirmed this; the actions of the Paris population determined the course of events and the fate of governments, and the centralization of the government and administration of the country under Napoleon broke the remains of the proud autonomy of the great provincial cities, to the benefit of Paris alone. Today the control of Paris means the control of France, and its political and social significance make it a target for all those hostile to France and to her government.

If Paris is the political key to France, she has also been the political conscience of Europe. In the nineteenth century the exiles of the oppressed nations of Europe came to Paris as the centre of humane ideals, reasoned optimism, and advanced intellectual speculation. Such was her unquestioned supremacy that even now that France has ceased to be the first power in Europe, Paris remains the first city in Europe. She has for long been at the heart of Western fashion, art and literature, and today she is the centre of a world-wide empire and the chosen home of influential international organizations, military, political and cultural.

Paris is not only the intellectual and political capital of France; she is also its greatest industrial and commercial centre. Paris in the widest sense includes the *Ville de Paris*, the Department of the Seine in which technically the *Ville de Paris* is included, and the modern suburbs in the Departments of Seine-et-Oise and Seine-et-Marne. In this essay the term Paris will be used in the wide sense, and the administrative sub-divisions will always be specified. The *Ville de Paris* has a population of 2,869,000 and the Department of the Seine rather over 2 million; these combined figures represent an eighth of the total population of France, and saturation point has now been practically reached. In 1921 the population of the *Ville de Paris* ceased to rise for the first time in over a hundred years, and for the next fifteen years there was an exodus from the *Ville de Paris* to the suburban communes. After 1936 the increase in their population also ceased, and the population of the department as a whole suffered an absolute decline, reflecting a flight to areas well beyond the Department of the Seine. The population of this new metropolitan area, in the Departments of the Seine, Seine-et-Oise and Seine-et-Marne, is difficult to decide but it is well over 5 million people.

From the *Ville de Paris* and the Department of the Seine are collected 50 per cent of all the taxes paid in France. Some 30 per cent of all the industrial and commercial activity of the whole of France is carried on in the metropolitan area. The *Halles aux Vins* and the *Entrepôts* of Bercy and Charenton are the biggest commercial centres for wine in the country, as are the *Halles Centrales* for food. Industrial activity in the Department of the Seine far surpasses that of any other area; nearly a third of all persons concerned with metallurgical industries in all France work there, and two-thirds of all the national automobile and aircraft industries are in this area. Heavy industries, electrical engineering, the chemical industry and paper production, as well as specialized luxury trades such as crystal, glass, ceramics, fabrics, tapestries, precious metals, printing, and book production, all have very important centres in or near Paris.

The population of this immense conurbation is composed of many different elements. There are 308,000 registered foreigners, many of them political refugees, and there are also an unspecified number of non-European citizens of the French Union who can either enter France without a passport or who require only a special permit. Their numbers are difficult to determine, but a reliable estimate puts them at 300,000.

Paris is, then, by far the largest city in France, over four times as large as any other. It is the principal industrial and commercial centre in the country and it is the political and administrative capital of a closely centralized nation. Further, the Parisian population is quite exceptional, having a strong revolutionary tradition, a volatile cosmopolitan element, and strong politically-minded unions and leagues. For all these reasons Paris is the one city in France not administered according to the uniform pattern applied to all the other communes and departments.

The Department of the Seine consists of the *Ville de Paris* and eighty other communes. The boundary between the *Ville de Paris* and these other communes was originally drawn up in the nineteenth century when the city was surrounded by fortifications, and then the *portes* through these fortifications marked the limits of the city. The ramparts have now disappeared and wide boulevards have been laid out in their place, so that today the *Ville de Paris* and the immediately adjacent communes form a continuous built-up area. At times the French legislature has made half-hearted attempts to recognize the change which the outward spread of Paris has brought, and has extended the technical boundaries of the *Ville de Paris* to accommodate parts of the most densely populated adjoining communes. These boundaries are still largely artificial, and correspond to no urban, geographic or demographic reality.

The *Ville de Paris* is the original area contained within the various *portes*, subsequently rationalized by the inclusion of portions of the immediately surrounding communes. The metropolitan area of Paris, however, extends far beyond the administrative limits of the *Ville de Paris* and the Department of the Seine, and dormitory suburbs penetrate deeply into neighbouring departments. Perhaps the principal lesson to be drawn from this chapter is that a well-defined uniform system of local government like that of France, where power and responsibility have been rationally and exhaustively allocated, lessens the problems posed by large conurbations since it eliminates the need to create *ad hoc* authorities. The laws of April 5, 1884, and August 10, 1871, on communal and departmental organization were deliberately framed so as to indicate in advance which authority, communal or departmental, elected representatives or local executive, was competent in any particular class of affairs. Since that time the senior administrative court, the *Conseil d'Etat*, has built up an extensive body of judgments settling the difficult points of conflicting competence. In this way the overflow population and the problems raised can be dealt with within the existing framework and by one of the existing authorities. To understand this point fully it is necessary to glance briefly at the ordinary French system of local government.

THE GENERAL SYSTEM OF LOCAL GOVERNMENT

French local government is based on two territorial units, the commune and the department. The commune is the primordial community; it may be a village in the Haute-Savoie with a dozen inhabitants, or it may be a city like Lyons or Marseilles. Technically the *Ville de Paris* is only the largest commune in France.

In each commune there is an elected municipal council with powers of decision over all matters of purely local concern; and each has a mayor elected by and from the municipal council, who executes its orders and guards the interests of the public and the state within the commune. He is responsible for ensuring public order, morality and hygiene (*salubrité*) in the commune, and he is authorized to issue police ordinances (*arrêtés*) regulating the activities of the citizens. This is the *police municipale*.

The communes are grouped together at a higher level into departments, whose authorities are responsible for providing common services such as highways, education and public assistance, for ensuring that the communes fulfil their duties, and for caring for public welfare and for the security of the population as a whole. The elected body in the department is the departmental council (*conseil général*) which has a decisive voice in all matters where departmental finance is concerned, but the executive, the prefect, is nominated directly by the government and is the immediate hierarchic subordinate of the Minister of the Interior. He has to execute the decisions of the departmental council acting within its powers, but he is also personally charged with the protection of the general interests of the state, with the enforcement of the law, with the tutelage of the communes and with the public security of his area.

At first sight such a system might seem to lack suppleness and adaptability, but an examination of the administration proves the exact opposite. There is no part of France which does not come under both a communal and a departmental authority. At any level, therefore, the precise authority for coping with an influx of population is known in advance; the commune has to provide the basic municipal services, the department the joint services, and in debateable cases the *Conseil d'Etat* decides. The mayor is personally charged with ensuring the welfare and good order of the population within the commune, the prefect in the department. Should the mayor or the municipal council find the task beyond their capacity, the prefect and the department can assist, or even, in case of personal shortcoming, assume control in the general interests of public order. In the last resort the minister can intervene and promote inter-departmental services or order the prefects to take certain steps.

The problems of a large metropolitan area when posed in Paris were never solved by the creation of *ad hoc* authorities. Instead, as the population extended beyond the official limits of the *Ville de Paris*, the suburban communes of the Department of the Seine were automatically called upon to assume, under the co-ordination of the prefect and the departmental council, the responsibility for ensuring the basic municipal services. The department itself undertook the broader issues of common services.

When, eventually, the metropolitan area spilt over into the Departments of Seine-et-Oise and Seine-et-Marne, the prefects of those departments at Versailles and Melun, and the respective communal and departmental authorities, naturally took up the burden.

A study of the administration of Paris shows clearly that a formal system of law and a standardized scheme of local government can have a high degree of adaptability and suppleness, provided that the executive authority be accorded considerable responsibility and independence.

THE PREFECTURE OF POLICE

The executive authorities in French local government are the prefect and the mayor. Paris departs from this normal system in that the executive is reinforced *vis-à-vis* the elected bodies, and the executory power is apportioned differently.

The *Ville de Paris* has no elected mayor, and those powers which the mayor would normally possess are, with one exception, held instead by the Prefect of the Seine. The exception is that a special authority, the Prefect of Police, holds all the *pouvoirs de police* in both the Department of the Seine and in the *Ville de Paris*. He assumes all the police powers possessed by the mayor in the commune and the prefect in the department. All the police forces in the Paris area are under his control, and he has a special residence in Paris, the *Préfecture de Police*.

There are, then, two independent heads of the administrative services in the *Ville de Paris* and the Department of the Seine: the Prefect of the Seine and the Prefect of Police. This division of authority gives rise to several complications.

The Prefect of Police has a triple role. First, he is responsible for protecting the general interests of the state, and for enforcing obedience to the law (*police générale*) throughout the Department of the Seine. For this purpose he is empowered to issue ordinances restricting the liberty of citizens in matters such as the press and public demonstrations which might lead to disturbances, and he has the authority to seize documents or arrest persons on suspicion of anti-state activities.

Second, the Prefect of Police controls the *police municipale* in the *Ville de Paris*. This includes public order, morality and 'salubrity' in the city, and these three terms have a wide sense: they cover the safety of the public highway, control of hotels, theatres, and other public places, and the inspection of all private establishments in which dangerous trades are carried on or where food or drink is prepared for public consumption.

Public order includes the prevention of civil catastrophes, and, in the event of such emergencies, the police authority is responsible for coping with them. Consequently, all floods, fires or famines demand the attention of the Prefect of Police.

In the third place, the Prefect of Police has certain powers of administration throughout the department: prisons and asylums come under his supervision in several respects, and he replaces the mayors of the other communes of the Seine in many matters of *police municipale*; these mayors retain only limited powers in such questions as public health.

To carry out his duties the Prefect of Police has a Secretary-General of the Prefecture of Police and a *directeur du cabinet* as his personal aides, both of whom are also senior members of the prefectoral corps. The police services are grouped under four great general directorates of municipal police, judicial police, information and intelligence (*renseignements généraux*), and personnel. The municipal police is the uniformed branch responsible for maintaining order in the streets and security on the roads. It musters some 18,000 officers and men who are based on twenty district police stations (one per *arrondissement*) in the *Ville de Paris*, and twenty-five district stations throughout the other communes of the department. The judicial police comprises 2,000 plain-clothes men and is responsible for detecting crimes and arresting offenders. Men from this force are attached to each of the district and area stations to deal with purely local matters; at headquarters on the *Quai des Orfèvres* there are specialized sections which investigate serious crimes and those involving more than one district.

The general directorate of information and intelligence collects details of economic, social and political interest to enable the prefect to forestall any threat to public order or the safety of the state. The general directorate of administrative services has eight subordinate sections to deal with particular subjects such as public hygiene, foreigners, factory inspection, and five specialized technical departments for the repression of fraud, chemical tests, food inspection and so on.

The Prefect of Police also has an armed reserve force which he can call out in times of emergency. These are a foot and a horse regiment of the national *Gendarmerie* called the *Garde Républicaine*, comprising some 2,000 officers and men. They are under the War Ministry administratively, but they are deployed under the orders of the Minister of the Interior. They provide a ceremonial guard on state occasions and stand watch on the residences of high state officials and the houses of Parliament.

Thus the security and enforcement services of the state, department and commune are grouped under one man.[1] Whatever affects the interests of the state or harms the private citizen, whether it be an attempt to enter the President's palace, bad meat, or driving to the public danger, the Prefect of Police has adequate means to cope with the situation.

The problem of an almost autonomous police force is posed in its most acute form in Paris, which has known revolutions, and on the control of whose streets the fate of governments and *régimes* has depended. In times of crisis the well-being of France may depend on the efficiency of the Prefect of Police. The force which he controls, the information on secret and political matters which comes to his knowledge, and the dependence of all the other public authorities in Paris on his effective and honest administration, make him a man greatly to be feared and often greatly hated. The Minister of the Interior alone has day-to-day hierarchic control over the Prefect of Police, and he can always dismiss him without explanation if he so desires. The freedom of action and the potential power of the Prefect of Police is always, therefore, dependent on the Minister of the Interior and his strength of character. With a strong minister and an able Prefect of Police, Paris has an extremely efficient centralized police organization. A weak minister can unwittingly turn the Prefect of Police into a political figure of some magnitude, as Chiappe was at the time of the Stavisky scandal: a weak Prefect of Police, on the other hand, can unwittingly create the conditions necessary for a street revolution. Since the time of Chiappe, the Prefects of Police have all been senior prefects of considerable calibre and integrity, held in high esteem as administrators, and there has been little abuse of power for political ends.

THE PREFECTURE OF THE SEINE

The Prefect of the Seine has a more constructive role. He resides in the *Hôtel de Ville* of Paris, which is therefore the headquarters both of the Department and of the *Ville*. He exercises all the powers of a normal departmental prefect together with all those normally held by the mayor, with the exception of the police powers.

As Prefect of the Department of the Seine he is the executive head of all the departmental services. He draws up the department's budget, organizes the sessions of the departmental council, and prepares memoranda on all matters dealt with by it. His powers *vis-à-vis* the departmental council are more extensive than those of the provincial

---

[1] With one exception: the Presidents of the two houses of Parliament have personal authority to requisition military assistance if Parliament is threatened.

P*

prefect because certain extra limitations are imposed by law on this council's freedom of decision. He has the ordinary tutelage control over the decisions of the municipal councils and the actions of the mayors in the other eighty communes of the department. This means that all their budgets and many of their financial transactions require his approval before becoming effective, and in several other fields such as the appointment of higher communal officials and the disposal of communal property his prior consent is needed.

As the administrator of the *Ville de Paris* the Prefect of the Seine controls all the municipal services. He draws up the municipal budget for the *Ville de Paris*, settles the agenda of its council's meetings, and prepares official studies and proposals to serve as a basis for debate. The Prefect of the Seine has, in addition, important powers of decision which normally reside in the municipal council as a body, but which have been withdrawn from the Paris Council by special laws.

The Prefect of the Seine is assisted by two Secretaries-General of the prefecture and a *directeur du cabinet*, all of whom are senior members of the prefectoral corps and appointed directly by the government. Under them come 37,000 permanent employees of the *Ville de Paris*, of whom 16,000 are manual labourers. The senior staff of the prefecture, the *attachés*, are recruited by a national examination held by the Minister of the Interior for the higher personnel of all the prefectures of France. From the *attachés* are chosen the *chefs de division* of the prefecture responsible for the work of a directorate, and they rank immediately next to the secretaries-general in the administrative hierarchy. This is now the only instance of state officials directly involved in municipal administration.[1]

Junior officials in the prefecture and allied municipal services are recruited by an examination held by the *Ville de Paris*, and successful competitors undergo a course of instruction in a special *Ecole d'Administration* set up and run by the city authorities. Specialists, such as engineers, doctors, primary school teachers and architects, are engaged on contract.

The first Secretary-General deals with social and economic affairs and the other with administrative and technical matters. Administratively, municipal and departmental affairs are shared between nine great directorates, some of which are concerned only with the *Ville de Paris*, but most of which deal with matters common to both department and city. These directorates are for administrative control, finance, technical services, personnel, social affairs, commerce and industry, municipal affairs, departmental and general affairs, and architecture and town

[1] From 1938 to 1945 members of the prefectoral corps administered Marseilles.

planning. There are also five inspectorates to examine and supervise the administration of the prefecture, elementary education in the department, libraries, health services and museums. Through the nine directorates the prefecture supervises the work of the communal administrations in the department and has some responsibility for all the public services, except transport, provided in the *Ville de Paris*.

THE ARRONDISSEMENTS

For administrative purposes the *Ville de Paris* is divided into twenty *arrondissements*, ranging in size from the XVIIIth round Porte Clignancourt with some 258,000 inhabitants, to the I[st] *arrondissement*, the business area centred on the Palais Royal and the Place Vendôme, which contains only 36,000 residents. Each *arrondissement* is sub-divided into four *quartiers*, of which the smallest is Quartier Gallon in the IInd, with just over 3,000 inhabitants, and the largest Quartier Clignancourt with over 100,000. The *quartiers* have nowadays a limited social significance, and since they have no elected body or council chamber of their own, they have little importance in municipal government.

The *arrondissement*, on the other hand, is an active area of administration: it has a real centre, the *mairie*, or town hall, which is the focal point of local interest. The *mairie* houses the local administrative services concerned with routine matters such as registering births and deaths, school attendance, various forms of public assistance, and the compilation of statistical information. The *juge de paix*[1] and the *commissariat* of police of the *arrondissement* are also to be found within the *mairie*. Each *arrondissement* has a mayor, assisted by from 3 to 7 deputy-mayors (*adjoints*) according to the size of the *arrondissement*.

The mayor of a Paris *arrondissement* has virtually none of the attributes normally possessed by a French mayor. He is not elected, he has no elected council beside him, he has no regulating powers and scarcely any self-contained authority. Instead, he is nominated directly by the Minister of the Interior from a list drawn up by the Prefect of the Seine, and he is always liable to instant dismissal. There is no elected council in the *arrondissement* to correspond to the municipal council over which the provincial mayor presides, and moreover the municipal councillors of the *Ville de Paris* are forbidden to associate together to form a group to represent a particular *arrondissement*. The only restriction on the

---

[1] The magistrate who deals with trivial civil cases such as small debts. He sits alone, and there are no legal costs.

minister's choice of mayor is that he must *not* be a municipal councillor of Paris. There is no limit to the length of his term of office.

The duties of the mayor are administrative, and principally concerned with ensuring the proper functioning of the services attached to the *mairie*. These services, however, are sub-sections of the central services of the *Ville de Paris*, and their ultimate superior is always in the *Hôtel de Ville*. The *arrondissement* in Paris is a convenient administrative device, and the *mairie* as an administrative centre has some social significance; but it has no political importance and is in no sense an area of self-government.

MUNICIPAL AND DEPARTMENTAL SERVICES

The *Ville de Paris* undertakes the provision of several of the normal public services; cleansing, street lighting, highway maintenance and construction, drainage, public gardens and the Port of Paris are each directed by a special technical division of the prefecture. Where appropriate these services have several branches each dealing with a part of the city.

Certain other services are operated under contract by private individuals and companies, for example the treatment of sewage, organization of municipal stadiums and sports arenas.

The great utility services are undertaken in the *Ville de Paris* by special contracts with the *Gaz de France*, *Electricité de France* and the *Compagnie Générale des Eaux*, which are public corporations running nationalized industries. Transport is run by the *Régie Autonome des Transports Parisiens*, a public company having monopoly powers throughout the Department of the Seine and parts of Seine-et-Oise and Seine-et-Marne. Technical direction rests with the heads of the *Inspection des Ponts et Chaussées* in the department, and his superior, the Minister of Public Works, keeps close administrative and financial control over all other aspects. The *Ville de Paris* is represented on the board of control but it has no decisive voice in its operation. The gas, electricity and water suppliers are, of course, bound by the terms of their contracts with the *Ville*.

In the department the great public utility services of gas, electricity and water are provided by the same public corporations as for the *Ville de Paris*. Groups of communes voluntarily form a joint authority (*syndicat intercommunal*) for each particular service, and the joint board, as a distinct legal personality, makes a special contract with the monopoly corporation concerned to serve associated communes. A special joint board has been created to provide a funeral service (*pompes funèbres*) covering most communes in the department. The other municipal services are organized in each commune (except the *Ville de Paris*) by its own administration, and with these the Prefecture of the Seine has little concern, except for the general powers of tutelage.

Two very important social services, education and public assistance, have a rather complex structure.

In education the system is broadly this. The *Ville de Paris* provides and maintains over 400 junior and primary schools (*écoles primaires*) and some 160 infant schools. In partnership with the state it is partly responsible for 24 modern and technical schools (*collèges*), and it has to pay the salaries of those employed in these schools. The state controls and finances the higher secondary educational institutions, the *lycées*, but the *Ville de Paris* contributes towards their maintenance; amongst these schools are some of high international reputation, for example, the *Lycée Louis-le-Grand*, and the *Lycée Henri IV*. The city also awards scholarships and assistance to worthy students and institutions. In 1950 the combined cost of the educational services to the city was $3\frac{1}{2}$ milliard francs.

The department has to provide teachers' training colleges for its area, and it has set up its own special schools for handicapped children, and craft and professional training centres. It also assists the suburban communes to provide school welfare, dental and medical services. It is required by law to maintain offices for the inspectors and officials of the state educational administration and it pays their expenses. The total charge to the department in 1950 for education was over $2\frac{1}{2}$ milliard francs.

At the head of the teaching services in the educational region of which Paris is a part is the Rector of the University, who is appointed by the government for academic distinction. He supervises higher and secondary education and some aspects of primary education. The head of the administrative services in the department is the *Inspecteur général de l'instruction publique*, an official of the Ministry of Education. The state, therefore, has a strong hold on the educational system, but in practice elected representatives of the city and department are closely associated with the formation of educational policy and with the detailed administration of the various schools. The Rector is advised by a departmental council, presided over by the prefect, composed of departmental councillors, teachers elected by their fellows, and primary school inspectors. Both councils have a permanent committee[1] specially charged with the supervision of education, and these committees appoint councillors to serve on the governing boards of the secondary, technical and special

[1] See below, p. 466.

schools provided by the city or the department. These small governing
boards, the councils' committees, and the councils themselves, provide
a chain, at each stage of which the councillors can influence policy and
enter into the details of current administration.

In Paris there is a unique organization charged with the manifold
duties required from local authorities in providing public assistance. This
organization is the *Assistance Publique de Paris*. The organization governs
over eighty institutions with some 45,000 beds and comprising general,
specialist and children's hospitals, hospices, sanatoria, dispensaries,
orphanages and rest homes; it also has a children's home at Hendaye in
the Basses-Pyrénées and sanatoria in other departments.

The *Assistance Publique* employs about 33,700 administrative, medical,
technical and manual staff, under the control of a director-general
appointed by the President of the Council of Ministers, and a secretary-
general appointed by the Minister of Public Health on the recommendation of the Prefect of the Seine. An advisory committee (*conseil de
surveillance*) of thirty-two members is presided over by the Prefect of the
Seine and comprises the Prefect of Police, fifteen municipal and departmental councillors and members nominated by medical and professional
associations. The senior personnel (inspectors-general, heads of departments, directors of hospitals, etc.) are appointed by the Prefect of the
Seine on the advice of the director-general, and the latter directly appoints
all the junior staff. Administratively and financially the whole organization is subject to the Ministers of Finance, of the Interior and of Public
Health. The Minister of Public Health exercises technical control.

In 1951 the budget of the *Assistance Publique* came to roughly 32 milliard
francs, of which half was contributed by the *Ville de Paris*, and the other
half by fees from patients, bequests, gifts and property, and the tax
on entertainments.[1] The *Ville* is also responsible for the welfare centres
(*bureaux de bienfaisance*) in each *arrondissement* and for the health visitors
and public assistance committees in each *mairie*. Most of the suburban
communes have made arrangements with the *Assistance Publique* for
accepting their patients, but eight suburban communes have set up their
own hospitals and another three hospitals are run by joint inter-communal
boards.

Communes are responsible for the institutions of public assistance, but
it falls to the department to provide a wide variety of allowances, monetary relief and free medical aid to the needy. Allowances to pregnant

---

[1] This is everywhere compulsory and a high proportion must go towards public
assistance.

mothers, milk allowances, financial aid to the aged, infirm, incurable and destitute living at home, grants to large families in straitened circumstances, all are borne by the department; the money is paid from its own revenue, from contributions from the communes, and from state subventions.

In 1950 half the budget of the Department of the Seine was taken up with expenditure on public assistance. Personnel, construction and repairs accounted for some $2\frac{1}{2}$ milliard francs, and over 14 milliard francs was paid out in allowances and relief.

The administration of public assistance is organized in a way that ensures ministerial control. In practice the Prefect of the Seine plays an important part in the efficient working of the system and elected representatives are closely associated with all aspects of the work. Through various committees and boards councillors can intervene at all stages; from the boards of the various hospitals, from the appeal tribunals to which any contested claim for public assistance is referred, and finally from the permanent committees of the two councils, which are charged with the supervision of public assistance.

THE APPOINTED EXECUTIVE

In the *Ville de Paris* the executive authority is shared by two administrative agents of the state, the Prefects of the Seine and of the Police. There are inevitably certain fields where it is difficult clearly to delimit their respective competences. The three most notable are public health, markets and highways.

In all three spheres the police administration is principally concerned with enforcing existing regulations and with ensuring conditions compatible with good order, while the Prefect of the Seine is empowered to make regulations as he thinks desirable for good government. Thus the Prefect of the Seine decides the conditions under which markets shall be conducted, while the Prefect of Police ensures fair dealing and the quality of goods offered for sale. The Prefect of the Seine is concerned with the conditions of pavements, the safety of buildings, street lighting, and so on, while the Prefect of Police ensures the free movement of traffic in the streets. Lastly, while the Prefect of the Seine deals with the broader aspects of public health, the Prefect of Police enforces the law with regard to the demolition of condemned buildings and the accuracy of pharmacists' dispensing. Conflicts can arise, and on the occasions when they are not settled by amicable discussion between the two administrations the matter is referred to the Minister of the Interior for arbitration.

The Prefect of the Seine is always chosen from among the most senior members of the prefectoral corps in service. It is a position of great responsibility and moral eminence. Unlike the Prefect of Police he deals with constructive matters, and the well-being of the metropolis depends to a large extent on his capacity and vision. The effect of the individual personality of the Prefect of the Seine is of the utmost importance; a prefectoral post in France always throws a great strain on the individual prefect, and although surrounded by able subordinates a prefect can never divest himself of personal responsibility. Few ministers of state assume so great a burden as the Prefect of the Seine. Ministerial instability only increases his importance and frequent changes of minister render the position of the prefect more isolated than that of any other public servant; when there is no minister to refer to the prefect must cope with circumstances almost entirely on his own initiative.

THE MUNICIPAL COUNCIL

The unusual concentration of power in the hands of two men, the Prefect of the Seine and the Prefect of Police, means that the elected bodies, the municipal council and the departmental council, have only a limited influence on the government of the city.

The municipal council of the *Ville de Paris* is an elected body of ninety members. Before 1935 a councillor was elected for each *quartier* whatever its population. This completely distorted the suffrage and in 1935 the number of councillors was increased by ten and the *Ville de Paris* was divided into electoral districts according to population. Since 1945 a system of *scrutin de liste* with proportional representation[1] has been adopted, several *arrondissements* being grouped together into large electoral areas. The present municipal council was elected by this method in 1953.

It is elected for a period of six years and is renewed in its entirety. Anyone over twenty-five years old with residence or property qualifications in the city is eligible for office, but the holders of numerous official posts such as prefects and magistrates are ineligible. A municipal councillor is paid a monthly salary of £45 as well as an expense allowance when on official missions.

The administrative authorities have potentially great control over the municipal council: the Prefect of the Seine can suspend its sessions for

[1] The electors vote for a party list, and the number of seats to which each list is entitled is calculated by dividing the total poll by the total number of seats to be filled, and further dividing the resulting figure into the number of votes cast for each list. Any seats not so filled are allotted to the list with the greatest remainder of votes unused.

three months and the Minister of the Interior can extend this suspension to a year. Should the council publish political proclamations or addresses the Prefect of the Seine can instantly suspend all its activities, and the President of the Republic has the power entirely to dissolve the municipal council and not to call new elections for a period of three years. The municipal council has not been suspended since 1887; it has never been dissolved. Finally, the prefect can dismiss individual members from office if they accept a post incompatible with that of councillor, or if they absent themselves from the whole of two sessions without adequate excuse; while the *Conseil d'Etat* can annul a member's mandate if he refuses to perform a duty legally incumbent upon him.

The municipal council has four ordinary sessions a year: in February, May, November and December. In the course of time the practice grew up of running the last two sessions together, and this was formally recognized by two Decree Laws in 1939. Each session must last at least ten days and the last two taken together must not exceed a total of six weeks. The Prefect of the Seine can call extraordinary sessions at any time, and a third of the municipal councillors can also demand a special meeting. This demand can be refused by the prefect, but only after a detailed explanation of his reasons and subject to an appeal to the Minister of the Interior. In practice the prefect prolongs the sessions as long as is necessary to conclude the work on the agenda.

At the beginning of the June session the municipal council elects its own president, four vice-presidents, and the *syndic*.[1] The president has prestige but few. powers. His powers are limited to presiding at the sessions of the council, assuring the proper functioning of the services attached to the council and maintaining order within the council chamber. But he must obtain the consent of the Prefect of the Seine before resorting to force to quell disturbances and the officers required for this purpose come under the direct orders of the Prefect of Police. The president's status therefore depends almost entirely upon the fact that he alone can, with any justification, claim to be the chief elected representative and civic head of the *Ville de Paris*. It is he who receives official visitors from overseas and speaks in the city's name at formal receptions: even this civic eminence was not achieved without a struggle, for it was only in 1897, at the opening of the rue Réaumur, that for the first time he, and not the Prefect of the Seine, represented the *Ville de Paris*.

Most of the important work of the municipal council is done in

[1] The *syndic* is a councillor responsible for the furnishing and equipment of the apartments and chambers used by the council, and for organizing municipal fêtes and receptions.

committees. These bodies have a long history, but only in 1939 were they officially recognized; before that time their meetings, though frequent and fruitful from the point of view of good government, were technically illegal, and even up to 1945 they could only meet within a fortnight of the ordinary sessions of the municipal council.

THE COMMITTEES OF THE MUNICIPAL COUNCIL

There are four different types of committee: permanent, special, mixed and administrative. The six permanent committees deal with finance, administration and police, highways, education and fine arts, public assistance, and public health. These are only general headings, and the matters assigned to any committee often have little relation with one another. The ninety municipal councillors are divided between these committees so that each has fifteen members; each elects its own president, a post of considerable influence in the affairs of the council.

Next there are special committees charged with a single important function; elaborating the rules, procedure and privileges of the municipal council, aid to ex-servicemen and child protection, commerce and industry, and so on. Such committees can be appointed *ad hoc*, but most of them are in effect permanent.

The other two types of committees are of a very special nature, as both have a mixed membership in which the municipal councillors are sometimes in a minority. There are mixed committees set up to deal with matters of interest to both the municipal council and the Departmental Council of the Seine, to which both assemblies send representatives; reconstruction, transport, labour and unemployment, and social services, are all dealt with by this sort of committee, but it should be emphasized that they are concerned with long-term problems and do not take a very active part in day-to-day affairs. There are finally administrative committees on which members of the municipal council sit side by side with members of the municipal administration. There are some ninety such bodies, some of which include in addition members of the departmental council. They are grouped under the six permanent committees and the members of each administrative committee are nominated by and from the members of the permanent committee. They ensure close contact between administrators and elected members and enable the city's representatives to enter into the smallest details of municipal administration. They range in importance from the *conseil d'administration* of the *Régie Autonome des Transports Parisiens*, through the *conseil de surveillance* of public assistance to the jury of the *Grand Prix Littéraire* awarded by the city of Paris, and the boards of the schools and hospitals.

The permanent committees fulfil very important functions. Each is a miniature municipal council reflecting the political composition of the council, and in them all future policy is brought to the test of discussion and compromise. Although the municipal council is legally forbidden to delegate any of its powers of decision, the pressure of work is so great that without the assistance of these committees it could never cope with a fraction of the problems set before it.

In practice some 80 per cent of the decisions taken by the committees are accepted without debate by the municipal council meeting in full session, the practice being that unless a councillor puts his name down explicitly against a motion when it appears on the agenda for the day that motion is given a formal vote of approval.

Over and above these four types of committee there is the *comité du budget*, composed of all the members of the municipal council meeting in private session under the chairmanship of a specially elected president. Its work is prepared by the *commission du budget et du personnel*, which comprises the president of the *comité du budget*, the *rapporteurs généraux*[1] of the budget and of personnel, the president of the municipal council and various other senior members, principally the presidents of the permanent committees and the *syndic*. The *commission du budget et du personnel* is informed by the prefect of the contents of the budget before the budgetary sessions of the municipal council. Its members examine his proposals and elaborate any new suggestions which seem desirable, and the prefect and the administration can be invited to give advice, provide further information and explain details. The *commission* presents its report to the *comite du budget*, making its own proposals and counter-suggestions. At this stage the budget is regarded as a whole, but in the subsequent public sessions the details of the budget are examined and voted one by one.

THE POWERS OF THE MUNICIPAL COUNCIL

There is no branch of municipal life which the municipal council cannot examine through its own representatives, and the annual examination of the city's budget gives rise to widely ranging discussions on

---

[1] The councillor charged with formulating and presenting to the whole council the conclusions of the committee.

broad lines of policy as well as to penetrating questions on past administration. However, although the municipal council is virtually uninhibited in its role of guardian of municipal interests and subjects all the work of the administration to critical scrutiny, its powers of decision are circumscribed and it can rarely force its will on the administration.

Decisions of the municipal council are only automatically self-sufficient and binding on the prefect when they are concerned with the way in which the municipal property is administered, with the organization of public social services, and with fixing the part the *Ville de Paris* should contribute to work undertaken by the state which is of interest to the city, for example the building of a new *collège*. Any other decisions it takes are strictly subject to the subsequent approval of the Prefect of the Seine or another higher administrative authority. Until 1939, the Municipal Council of Paris, like all other municipal councils, was entitled to discuss all matters of communal interest and to take conditional decisions on them; since the decree laws of April 30 and June 13, 1939 (liberalized by an *Ordonnance* of April 13, 1945), the Municipal Council of Paris is entitled to discuss only the most important matters of communal concern, and these decree laws drew up an exhaustive list of those matters on which the municipal council could take conditional decisions. Most of these concern expenditure: the budget, loans, the grant of funds to other bodies, the control of communal property and the organization of municipal services, new construction and highway development, whether or not the *Ville de Paris* should start legal proceedings, and whether new markets or fairs should be opened. All this comes within its competence provided that the expenditure involved is above a certain limit; for example, if the sum involved in the legal action is more than a million francs. Beneath this limit the administration is free to act as it pleases, subject only to the *a posteriori* control of the municipal council exercised by oral or written questions addressed to the prefect, or by demands for information in the council chamber.

The system of conditional decisions means that virtually any project approved by the municipal council can be vetoed by an administrative authority; most frequently by the Prefect of the Seine and the Minister of the Interior, but sometimes by the Ministers of Finance and the *Conseil d'Etat*. But the administrative authorities have a purely negative power, and in no circumstances can they substitute their own project when it deals with one of the matters enumerated as being within the competence of the municipal council. In sum, while on the one hand no important decisions of municipal concern can be taken without the consent of the elected body, on the other hand the municipal council cannot conduct affairs without the approval of the administration.

There are further niceties to the balance of power between the municipal council and the administration. There are certain items of expenditure declared obligatory by law for which every French local authority must provide in its annual budget. The list of such items is extensive, including the police forces and the normal administrative services, the upkeep of public buildings and the servicing of the municipal debt. If the municipal council fails or refuses to meet this expenditure, the prefect can write the required sums into the budget on his own authority, either by transferring credits already allowed for non-obligatory services or, when this does not suffice, by requesting a higher administrative authority to raise fresh taxation within the commune. This puts the Prefect of the Seine in a relatively strong position in face of an unco-operative municipal council. On the other hand the elected body can resist the prefect over the question of the so-called non-obligatory services, many of which (e.g. the construction and repair of highways) are vital to the good order of a city like Paris. But in the event of such resistance the electors of the municipal council would be the first to suffer, and there is little doubt as to the superior position of the administration in a show of strength. But the politics of the municipal council's relations with the Prefect of the Seine are subtle and peculiar, and it will be best to return to these later after dealing with the other elected body, the departmental council of the Department of the Seine.

THE DEPARTMENTAL COUNCIL OF THE SEINE

The departmental council is a body of one hundred and fifty members. The department, excluding the *Ville de Paris*, is divided into five large electoral areas each of which elects a number of councillors according to population, by *scrutin de liste* with proportional representation. Sixty departmental councillors are elected in this way. The ninety municipal councillors of the *Ville de Paris* complete the departmental council; they sit as of right without further election, and they receive the same salary as the other departmental councillors (£40 per month) in addition to their salary as municipal councillors. Because of its special composition the Departmental Council of the Seine (unlike other departmental councils), is re-elected in its entirety every six years at the same time as the municipal council, and there are no by-elections.

The attributes of the departmental council closely resemble those of the municipal council. The two assemblies hold their ordinary sessions during the same periods, are subject to similar standing orders, have similar internal organizations and the same system of six permanent committees. The departmental council has its own *comité du budget* of the whole assembly and a corresponding *commission du budget et du personnel* comprising the senior officers of the assembly. It also has special committees and sends representatives to the various mixed committees appointed in common with the municipal council. Many of the administrative committees also have a delegate from the departmental council.

Before 1939 the departmental council was empowered to discuss and to take conditional decisions upon all matters of departmental interest. The restrictive decree laws of 1939 limited the competence of the departmental council to roughly the same subjects on the departmental level as those to be treated on a municipal level by the municipal council of the *Ville de Paris*. Certain public services like public assistance and education, which are organized throughout France on a departmental basis, swell the list for the departmental council. Similar minimum financial limits operate for the department, and the prefect can take a personal decision on all matters not specified. The budget for the department is not voted by individual items but by chapters, and in executing the budget the Prefect of the Seine is entitled to transfer credits from one item to another provided they fall within the same chapter and provided he does not exceed the total expenditure under any one chapter. A provincial departmental council elects each year a *commission départementale* to act on its behalf between full sessions and to keep watch on the activities of the prefect. The Departmental Council of the Seine elects no *commission départementale*, and the prefects are therefore more free from continual supervision by the elected bodies. This appreciably modifies the balance between the elected council and the executive, and renders the Departmental Council of the Seine the largest and yet in some ways the weakest of its kind.

FINANCE

Undoubtedly the most important single function of both councils is to discuss and vote the budgets of the *Ville de Paris* and the Department of the Seine.

In 1950 the budget of the *Ville de Paris* was 42 milliard francs while that of the Department of the Seine was 33 milliard. To these figures should be added the budgets of the eighty communes in the department, which in 1950 came to 14 milliard francs. The exceptional size of these budgets is in part due to the fact that the *Ville de Paris* is over four times the size of any other French city, and that no other department has a fraction of the urban population of the Department of the Seine; but it is also augmented by the special requirements of such a concentration of people. Problems of social welfare, education, public order, public health and housing are particularly acute, and a highly developed system of public assistance is necessary to public order. Lastly, the pride of a great city demands that labour and care be lavished on its public monuments, architectural treasures and the city's museums and art galleries in which are embodied its civic eminence and history.

Since about 1936 all local authorities in France have been pre-occupied with rising prices and the threat of open inflation. In 1938 the city spent some 13 per cent of its total budget on wages and salaries; since the war this has risen to nearly 32 per cent. There has been a similar increase in the department's budget. In face of the increasing cost of staff and of materials, the budgets have been balanced by reducing the work done on public buildings and on the maintenance of public services. This policy of economy has been easier to follow in the department than in the *Ville de Paris*. The city is liable for the lighting, streets, traffic control, public buildings, and so on, but the department is only responsible for departmental highways and for work beyond the scope of the communes, such as the development of port installations at Gennevilliers and the building or repair of mental hospitals.

Before examining the budgets in detail it is important to mention two points. First, several services which affect the daily welfare of the citizens are provided by other public bodies with separate accounts. The transport system (*Régie Autonome des Transports Parisiens*), the *Crédit Municipal* and the *Assistance Publique de Paris* come under this category. The transport budget came to 36 milliard francs in 1950, the *Assistance Publique* to 32 milliard francs, and the *Crédit Municipal* to 4½ milliard. There are various conventions between these organizations and the local authorities whereby the latter either undertake to make good any deficits incurred or to make annual contributions towards running expenses.[1]

The second point is that local authorities are required by law to present their budgets in a standard form, and no provision is made for separate accounts to be kept for particular aspects of municipal or departmental activity. There is, therefore, no 'education' or 'highway' budget; instead all employees—doctors, labourers, teachers, engineers—are grouped

---

[1] See item 8 of budget below.

together under item 3, personnel. And similarly with the other divisions.[1]

TABLE I

Expenditure 1950

in million francs, roughly £000s

	Ville de Paris	Department
1. Debt service	1,816	1,324
2. Councils	58	162
3. Wages and salaries	12,984	6,252
4. Equipment and operating cost of services provided by local authority	7,842	2,739
5. Repair of property, construction work and acquisition of property	1,502	1,955
6. Police	4,275	402
7. Public assistance and social services (direct aid)	1,174	14,145
8. Subventions to *Assistance Publique* and *Crédit Municipal*	8,484	—
9. Contributions towards deficits of public corporations (gas, transport, etc.)	949	3,143
10. Grants to other bodies (allowances to cultural institutions, scholarships, fêtes, etc.)	638	849
11. Miscellaneous expenditure (legal expenses, tax collection costs, unforeseen expenditure, etc.)	1,776	834
12. Deficit from previous year	760	1,467
Total in million francs	42,258	33,270

The *Ville de Paris* pays the salaries of prefectoral employees except those concerned with purely departmental matters. It receives contributions from the department and the state to counterbalance this outlay; in addition, the most senior officials—the members of the prefectoral corps, the *trésoriers payeurs généraux*, the inspectors-general of education, public works, and members of the state engineering corps—are paid directly by the state and do not appear in these budgets. This item also includes the salaries of all those employed in communal or departmental utility services.

A substantial part of the city's budget is absorbed by the operating costs of the utility services provided directly by the municipality (item 4); these include street lighting, the heating of schools and offices, markets, abattoirs, street cleaning and drains, the last two items alone accounting

---

[1] The figures given for education on p. 461 were arrived at after a detailed analysis of the sums allowed in different chapters of the budget on salaries, repairs, administration, etc.; they are at best a rough calculation.

for nearly 3½ milliard francs. The department provides social rather than utility services, and the most considerable elements in its expenditure here are on hospital equipment, ambulances, special schools, orphanages and workhouses.

As was explained earlier, the department is responsible for financial aid and medical assistance to those in need, and its expenditure under this head is by far the most important item in its budget (item 7). It has to provide outdoor relief to the old, the infirm, the incurable and the blind, allowances to pregnant and nursing mothers, the protection of children in moral danger and the provision of free medical assistance to the poor. Poverty in capitals is exacerbated by anonymity, and expenditure on public assistance in the Department of the Seine is greater both absolutely and proportionately than elsewhere in the country. The city has to bear the cost of caring 'for certain categories of sick and helpless persons; the most important items are the hospital expenses of tuberculosis sufferers, school fees for abnormal children, and contributions to other communes and departments for caring for the indigent whose registered address is Paris.

The Prefecture of Police has its own budget, to which the *Ville de Paris*, the suburban communes and the department are all required to contribute (item 6). The state contributes up to four-fifths of the total cost, though this does not take into account the cost of maintaining police establishments and the buildings of the Prefecture of Police. This burden is borne by the Department of the Seine, and this expenditure is included in item 5, along with the cost of the other offices and property which the department is called upon to keep in repair.

It is evident that a very high proportion of both budgets is used for providing necessities, and whether or not such matters are legally declared to be 'obligatory' it is unlikely that they will be made the ground for serious debate in the elected assemblies. The part of the two councils in elaborating the budgets is principally to decide on the order of priority of the work to be done, to weigh whether it is desirable to undertake large public works by adding to the departmental or municipal debt, and to apportion the limited revenue available between competing minor demands.

Separate capital budgets are kept for both the *Ville de Paris* and for the department, in respect of large scale works such as the construction of new squares and gardens, additions to hospital and school buildings, and the rehabilitation of the drainage system. Capital expenditure for the *Ville de Paris* in 1950 was some 11 milliard francs, and for the department

$2\frac{3}{4}$ milliard. These works are financed by the floating of new loans, by borrowing from banks and by subventions from the state.

The councils have considerable powers in deciding how revenue is to be raised and in apportioning the burden between the various kinds of taxpayer.

TABLE II

Revenue 1950[1]

In million francs, roughly £000s.

	Ville de Paris	Department
1. Additional centimes	3,262	15,850
2. Local taxes	28,807	—
3. Rents and dues from property	1,382	88
4. Revenue from services organized directly by local authority	4,328	1,412
5. Payments for concessions and monopolies granted by the *Ville*	1,428	—
6. General administrative charges	344	322
7. Payments and grants for public assistance	—	8,333
8. Interest and capital repayments on loans and grants	829	646
9. Grants, subventions and contributions	1,669	3,296
10. Carried forward from previous year	7,176	8,559
11. Remaining to be found	1,680	7,334
Total in million francs	50,905	45,840

The principal source of revenue for the department and an important item for the *Ville de Paris* are the additional centimes. This tax, which bears upon property and rents, is extremely complicated and is raised according to an archaic system, whereby for every franc paid to the state for four national taxes[2] an additional centime was paid to the local authorities. The state stopped collecting these taxes for its own use in 1917, but the amount it raised in that year has been fictitiously maintained as a basis for calculating the additional centime, and within the maximum limits laid down each year by the national finance law, the local authorities are empowered to raise 'additional centimes.'

In addition to these resources, the *Ville de Paris* has a large source of income not open to the department in the special communal taxes. They can be roughly divided into four classes. In the first there are the dues levied by the city for providing services; for example, the highway taxes,

[1] The excess of revenue over expenditure shown in Table I on p. 472 is due to the budgetary deficit incurred in previous years. See *post*, p. 476.

[2] Levied on built-on land, un-built-on land, personal estate, and business premises.

the rates for drainage, refuse collection and road cleaning. A second group of taxes can be added at the discretion of the local authority to those already collected by the state, and these bear principally upon rents, furnished rooms and business licences. Third, there come a number of optional taxes which the local authority can levy upon a strange variety of possessions: they range from balconies to domestic servants, from the consumption of gas to draught animals. Finally, there comes an extremely important group of taxes which must be raised by the commune: the most prominent are a tax on bars, cafés, etc., a tax on entertainments (most of the proceeds of which go towards public assistance), stamp duties, and a percentage tax on all business transactions. This last is the most important local tax, and it accounted for over half the revenue which the city received under this item.

The *Ville de Paris* also received a substantial income from the public utility and commercial services which it provides; the most profitable of these are the charges levied in the abattoirs, dues from the Port of Paris, the municipal funeral services and municipal motor services. The department has far fewer interests in this field, since most communes prefer to provide their own services either directly, or through joint boards, or by granting concessions. The only important service undertaken by the department is sewage disposal, and this brings in over two-thirds of its revenue under this item (4).

The *Ville* also receives revenue from conceding to public or private companies the right to operate or to use public property; these two sources are kept distinct. One item (5) shows the rent paid for the use of communal property by public companies, principally the *Electricité de France*, and the other item (6) the charges made for the use of communal property by private persons—licences for stalls in the markets and outdoor fairs, municipal sports stadiums, and outdoor advertising.

The department spent over 14 milliard francs on public assistance. It received back from various sources over 8 milliard francs in contributions. Over 5 milliard francs came from the state as a grant-in-aid to which the department is legally entitled. The remainder came from payments made by communes and private individuals in departmental hospitals, institutions and sanatoria.

The state and the communes also make substantial grants and contributions to work of common interest undertaken by the department (item 9). The state makes grants for a variety of matters; for administrative costs, for social hygiene, for firemen's barracks, for the equipment of special schools and for war damage. The communes contribute principally towards the salaries of teachers in departmental schools and toward public works of particular interest to them, such as port extensions at Gennevilliers and departmental sanatoria.

The *Ville de Paris* receives only half the amount of contributions that the department does. These come from the state, from the Department of the Seine and from private companies and individuals. The state's grants amount to 900 million francs; principally for education, road repairs (this is peculiar to Paris), public health and housing. The department contributed some 400 million francs towards the cost of personnel and made allowances for construction work and repairs of common interest, and for receptions and fêtes in which both authorities took part.

Both authorities find it impossible to keep strictly within the limits of the financial year. For the *Ville* the deficit incurred in 1949 was over 8 milliard francs, and it consequently had to budget for a surplus of income over expenditure of the same amount in 1950. There was, however, over 7 milliard francs revenue for the year 1949 to set against this (item 10), so that in fact the *Ville* had to find only an additional milliard francs. The department inherited a deficit of 12 milliard francs from 1949, against which could be set $8\frac{1}{2}$ milliard revenue carried forward from the same year.

This time lag is partly due to continually rising costs forcing up the estimates; in part it is a result of close ministerial control of the budgets of the city and department.

The Prefect of the Seine prepares the budgets of the *Ville de Paris* and of the Department of the Seine, and he presents them to the councils with explanatory memoranda. After the municipal council and the departmental council have discussed the draft budgets and made any amendments to them, they are voted upon and sent for approval to the Ministers of the Interior and of Finance. Within the next few months these two authorities make such changes as they consider necessary (numerous since the war) and eventually they are approved by ministerial ordinances. The Prefect of the Seine is responsible for executing the provisions of the budgets, the payments being made through the offices of the *trésorier payeur général* of the department (who also acts as the *receveur municipal* of the *Ville de Paris*, and is an official of the Ministry of Finance). The accounts of the city and of the department, prepared by the Prefect of the Seine, are submitted to the municipal council and the departmental council respectively, which can demand explanations. Finally, they are approved by the *Cour des Comptes*, the supreme financial tribunal.

All local finance in France, and especially that of the *Ville de Paris*, is in an extremely confused state, since many of the factors on which the budget is based are unknown until half the financial year is passed. Furthermore, these factors are often subject to bureaucratic delays or political considerations, or both.

In the *Ville de Paris* it has been the policy of the Right majority to stabilize the existing level of taxation; the Prefect of the Seine has, however, repeatedly had to call upon the elected assemblies to vote new taxation in order to balance the budget, and when all else fails it falls to him to act as their advocate before the central ministries to gain state aid in order to achieve this end. The councils themselves feel that such assistance is a right, since the city and the department provide services and amenities which are utilized by all citizens of the country who never make the smallest contribution to their maintenance.

CENTRAL CONTROL

The elected authorities of Paris are in many respects subject to the authority of the prefect; he in turn is often unable to act without the prior consent of the ministries. In part this is due to the prefect's personal status, since his appointment and dismissal are in the hands of the government, and he is liable to removal without notice or cause. In addition, most statutes relating to local government or the provision of national services explicitly provide a special system for the *Ville de Paris* and the Department of the Seine, whereby final decisions on matters of importance, the appointment of higher officials, and the control of expenditure, are subject to close ministerial control. The Minister of Public Health in matters of public assistance, the Minister of Public Works for the transport system of Paris, the Minister of Reconstruction for the planning of the Department of the Seine, all possess powers of control and decision over Paris unparalleled in any other commune or department. Further, the Minister of the Interior may intervene in the administration of the Prefecture of Police and the Prefecture of the Seine, and the Minister of Finance has very extensive powers of financial tutelage not only over the legality of the financial operations of the department and city, but also over financial policy. These two ministers exercise some supervision over all departments and great cities, but in practice this is greatly intensified in the case of Paris.

Councillors of both assemblies complain that such close central control tends to make them into servants of the state rather than representatives of the electors. They further hold that the administration of the city and department is in fact harmed by this extreme form of tutelage. To illustrate these grievances, they can point to the fact that the Parisian pays nearly ten times as much as the average Frenchman for his police forces, yet the Minister of the Interior uses these forces at his discretion to protect national interests, and can transfer them to duties which are of no direct service to the Parisian. Again, the legal provision that all financial measures require the approval of the Ministers of the Interior and of Finance subjects the budgets and loans voted by the councils to intolerable delays. The budgets voted in December of each year are seldom passed by the Ministry of Finance before the following July, so that for half the financial year the city does not know its exact position and cannot usefully debate long-term projects. Finally, many of the services which affect the material welfare of the Parisian population are so strictly controlled that the city's representatives are unable effectively to make their voices heard. Most of the higher officials in the hospital and welfare organizations are nominated not by the city but by the Minister of Public Health; the transport organization is so constituted that councillors of the city and of the department are always in a minority *vis-à-vis* the technical officers and ministerial representatives.

POLITICS

If the law were rigorously applied the elected assemblies of the *Ville de Paris* and the Department of the Seine would be little more than consultative bodies; in fact what amounts to binding conventions regulate the relations between the councils and the executive. There have, in the past, been very heavy-handed prefects; for instance, between 1883 and 1893 Prefect Andrieux provoked such antagonism that the municipal council refused to vote the police budget, and it had to be written in by the Government. His successor, Prefect Lépine, restored amicable relations and since that time the prefects have, with very few exceptions, set out to court harmony whenever possible.

A dispute between the councils and the prefects can either be resolved by reliance on legal powers (in which case the prefect will win) or by compromise and mutual concessions. In the committees conflicts are frequently resolved without the publicity of the council chamber, and in most cases every effort is made to establish agreement inside the committees rather than provoke public dissension. Against the overwhelming legal superiority of the prefect, the principal weapon of the elected assemblies is the threat of public opinion; and there are occasions when they will choose to use it and focus public interest on a burning issue. Moreover, there is always the possibility that they will succeed in involving one or more of the national party machines. For the government of Paris is frequently a matter of high politics, and it is not wholly comprehensible except in terms of national party politics.

Paris has always been a city of extremes, and since the turn of the century there has been a decline in the central parties in favour of the extreme left and right. Before the Second World War the various semi-fascist leagues found their greatest support in Paris, and after 1947 it was the Communists and the R.P.F. who were most favoured by the electorate. The system of proportional representation introduced after the liberation permits the M.R.P. and the Socialists to retain some influence. The R.P.F. and the Communists are strongly centralized, and municipal politics tend to reflect the directives of their party headquarters. The M.R.P. and the Socialists are less closely organized. Each of these parties has party federations in every department, which send representatives to the central party committees, and the relative strength of the federation of the Seine on the central committee largely determines the interest which the party will take in Parisian affairs. But in any event no political party which seeks urban support, as a party must do if it is to be of first rate importance in the National Assembly, can afford to ignore Paris and its problems. The Paris electorate is too large and the political influence of the city too great ever to be neglected.

Consequently no party can monopolize any social problem of Paris to its own advantage: all the main municipal issues such as transport, housing, public health and public assistance, are fully recognized by all political groups within the councils. Programmes naturally vary a little: the right wing parties tend to voice the demands of the bourgeoisie and the small independent business man, while the Communists, M.R.P. and Socialists consult the needs of wage-earners, casual labour and unemployed. The R.P.F. complicate this picture by including in their programme the 'municipalization' of many commercial and industrial services, and by basing their party organizations on the place of work rather than the place of residence—a point borrowed from the Communists. The R.P.F. have, however, consistently evinced a typical right-wing desire to balance the budget at a steady level of expenditure by economies in other fields rather than by increased taxation. This is the principal dividing line between the parties of the right and those of the left and centre, which are virtually agreed that any new service must be accommodated by higher taxation.

In the 1947 elections the R.P.F. had an absolute majority, but after

1950 internal dissensions broke the unity of the party. The seceding members went either to the Independents or the R.G.R., and the 1953 election resulted in a right-wing coalition majority of Independents, R.G.R., and R.P.F. There is a good deal of political bickering in the public sessions of the council, but sensible and serious administrative work is done inside the committees, and an immense amount of ground is covered each year.

In the departmental council there is no clear majority, the centre, right and Communists having roughly equal representation. The right have tacitly allied themselves to the centre, and the centre parties can always hold them to ransom. The tendency to demagogy among the Communists is less marked in this body than in the municipal council; effective participation in local government has a softening effect on a party which often courts favour by seeking to exploit distress.

The difference in political atmosphere between the two assemblies is to some extent due to a difference in the type of member which they attract. Few municipal councillors are proposed by their parties for election unless they are first and foremost active politicians: this is not true of departmental councillors, several of whom are local personalities attached to a particular party, but principally interested in social questions and in active participation in local government. The political aspect of the departmental council is therefore less marked than that of the municipal council.

In a system of government which lays such stress on the part of the administration, no member of the elected bodies is ever directly responsible for any act of municipal government. There is no reason why any councillor should ever propose or accept a proposition distasteful to his own clientèle. The administrative authorities alone bear a direct responsibility for curbing expenditure, restraining demands and actually getting things done. Consequently they tend to become the butt of all parties, and on some issues involving social services or increased expenditure on public welfare the political contest in the assemblies assumes the character of a struggle between elected members and the administration rather than one between rival parties.

The position of the prefect is then a very delicate one, as it is his duty both to represent the interests of the city to the ministries and to justify the ministers' decisions to the councils. Were he faced with Utopian parties on the one hand and a rigid authoritarian government on the other, he would find it extremely difficult to promote positive measures of municipal government. But normally he can rely upon the high

Paris

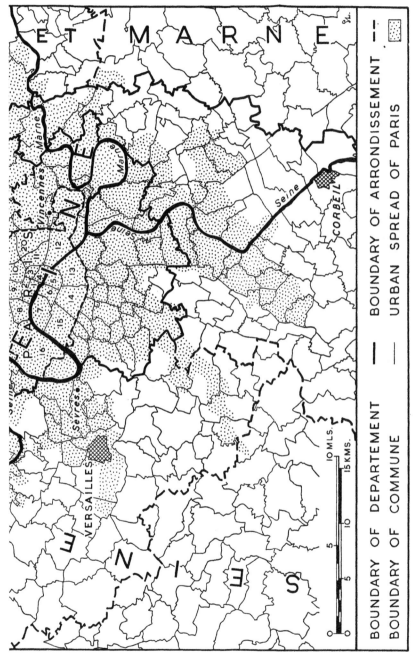

THE GREATER PARIS REGION. The stippled portions show how the urban population has spread from the Ville de Paris (in the centre of the map) to the Départements of the Seine, Seine-et-Oise, and Seine-et-Marne. It will be seen that the Ville de Paris is within the Département of the Seine, which is completely encircled by the Département of Seine-et-Oise.

BOUNDARY OF DEPARTEMENT — — BOUNDARY OF ARRONDISSEMENT — —

BOUNDARY OF COMMUNE — URBAN SPREAD OF PARIS

prestige he enjoys with the ministers as a very senior administrator, and upon the trust he inspires in the councils as the conscientious and steadfast supporter of the city's interests. Political wrangling gives way before the practical needs of municipal government.

PROBLEMS AND PLANNING

The social problems of the Paris region are those which arise in all great cities. The provision of services, the maintenance of order, the demands of public health and housing, the difficulties posed by distance between residence and employment, and the curious social climate engendered by close and inescapable proximity, all affect the life of Paris as they do that of London, New York, Calcutta or Shanghai.

The Frenchman, however, does not see these problems in quite the same perspective as the Anglo-Saxon. He gives more weight to curbing the annoyances of daily life than to the cure of radical social evils, to the protection of existing interests rather than to the creation of new social demands which may conflict with them. One must assess the administration of Paris by this standard, not by one of good government in the abstract.

The key to the modern problems of Paris lies in the person of Baron Haussmann, Prefect of the Seine under the Second Empire. Not that he intentionally created problems, but the sweeping demolitions and reconstruction which he carried out to combat the difficulties which faced him as a nineteenth-century administrator were so extensive and of so definite a character that they have profoundly influenced all subsequent work on social and urban development.[1]

Haussmann's two principal aims were to embellish the capital city, and to eliminate as far as possible the threat of insurrection. The construction of broad imposing boulevards seemed to serve both these ends; they set off the ornate style of public buildings and private apartment houses of the period, and they were also far more difficult to barricade than the slums which they replaced. However, when Haussmann elevated the teeming centre of the old city to a bourgeois stronghold, the population of the former slums were driven to the north and east; this dispersal of the working population to the city limits crystallized the geographical and social pattern of Paris. The high-class area of the centre has attracted the best hotels, the luxury stores and the principal offices; and the bourgeoisie, now dispossessed of their residences, have moved westwards

[1] B. Chapman: *Baron Haussmann and the Planning of Paris.* Town Planning Review, October 1953, pp. 177-93.

Q

to the areas bordering on the Bois de Boulogne, to the zones in that area left free by the demolition of the old fortifications, and to the parts contiguous to the fashionable centre. The lower middle class and the workers have consolidated their hold on the north and east, and with the growth of the automobile and aircraft industries they have spread to previously undeveloped outskirts in the south and south-east.

Such well-defined social geography poses a serious moral problem and leads to localized and therefore more concentrated social evils. Overcrowding, lack of sanitation, juvenile delinquency, infant mortality, tuberculosis and stunted growth, all have their highest incidence in the north (XVIII, XIX, XX *arrondissements*), the south-east (XIII *arrondissement*) and the south (XIV *arrondissement*). There are also black spots around the central *Halles* and the Gare d'Austerlitz, but these are inward tentacles from the outlying areas rather than faded patches of a prosperous past.

This geographical division of bourgeois and working-class partly accounts for the peculiar social climate of Paris. The hostility of the working-class and the reciprocal fear of the bourgeoisie find their natural reflection in the tendency for Parisian politics to fly to extremes: the centre parties are numerically weak everywhere, the R.P.F. meet negligible opposition in the better residential districts (I, IX, VIII, XVI, XVII *arrondissements*), and the working-class vote oscillates between the Communists and the R.P.F.

It is the enforced proximity of antipathetical social groups within the framework of a large city which drives those groups apart and foments an exaggerated degree of social animosity. This is reflected in the polarization of political voting and also in the ever-present threat of street disturbances. Paris is a city with a revolutionary tradition. Haussmann took drastic steps towards eliminating open warfare in the streets, but the development of modern trade union organizations today offers a far more effective method of besieging the city, merely by a large-scale refusal to work. Famine cannot be defeated by machine-guns; the police problem would be insuperable, were it not for that same volatile individualistic quality in the Parisian which is responsible for the original tension: he will not lend himself easily to organized demonstrations without personal provocation or a direct threat to his personal liberties. The Paris police forces could not subdue the combined forces of the potential militants of all groups, but such a position has never arisen unless the government has acted in such a way as to arouse the concerted hostility of antagonistic groups of individualists. If the whole population of Paris rises against the government it is not a riot but a revolution, which might perhaps be quelled by the military but certainly not by the police.

Baron Haussmann solved neither the political nor the police problem of Paris, and he bequeathed to the future administrator two new problems to which the Parisian attaches considerable importance. First, the movement of the working population to the outskirts required an extensive transport system to get them to and from work in the city. These transport services have been a continual source of trouble. Since 1918 (except for a brief period in 1940–43 when virtually no services were provided), the Metro system has never been able to balance its budget, and the bus services have had a deficit for the last thirty years. Policy has been partly responsible for this, since the Government has always hesitated to raise tariffs on political and social grounds, and there is now little scope for economy, since out of a total budget of 35 milliard francs wages take 21 milliard and 5 milliard are spent on the maintenance of existing equipment. The state contribution towards meeting the deficit is limited by law to 15 per cent of the total receipts, and the remainder has to be borne on the budgets of the *Ville de Paris* and the Department of the Seine. Periodic strikes for higher wages, inadequate bus services, and the organized protests and demands of the Parisian traveller echoing in the councils, all testify to the continuance of the transport problem.

In the second place, although Haussmann's magnificent boulevards provide east-west routes through Paris and fine radial boulevards in the centre and the modern quarters, in the old parts and in the industrial quarters the roads are quite inadequate for the present traffic. The problem of traffic movement is regarded by the Parisians and their elected representatives as extremely acute, and they perhaps underrate the inestimable advantages of the existing boulevards on which high speeds may be maintained. Experiments are being made with an elaborate system of one-way streets in the most crowded quarters and experience suggests that this is the best solution. The Prefect of Police, however, estimates that an expenditure of some 14 milliard francs will be necessary before the problem of circulation will be solved.

Compared with other cities Paris is in many respects already a model of town-planning: consequently it is not surprising that the plans that exist are concerned with severely practical issues of highway development, industrial zoning and space for housing.

Paris is planned as a region, comprising the departments of the Seine, Seine-et-Oise, Seine-et-Marne and five cantons of the Oise. The communes included in this region are required to present plans conforming to the regional plan. These communal plans are drawn up by a qualified architect, approved by the municipal authorities, scrutinized by the Inspector-General of Town Planning (an official of the Ministry of Reconstruction), and ratified by prefectoral ordinance. In the Department of the Seine the plans are prepared by the *Direction des Services d'Architecture et d'Urbanisme*, a directorate of the Prefecture of the Seine.

All these plans, as well as the regional plan, have to be submitted for approval to a special advisory body, the *Comité d'aménagement de la région parisienne*, composed of senior ministerial officials, representatives of local authorities, members nominated by associations and specially appointed experts.

The regional plan for greater Paris was drawn up in 1934–35, and it marked the main routes, both existing and projected, and indicated a green belt round Paris to be preserved. This is still the nominal masterplan for the development of the greater Paris region. At the same time the first steps were taken towards the creation of a satellite town at Orly, to be built by private enterprise with the assistance of the city of Paris. Gradually, however, this plan fell into abeyance, as the site chosen proved unsuited for large-scale building, insufficient private contractors came forward, and financial circumstances at that time had so reduced the resources of the city that it could not meet the cost of the road, transport and public utility services it had contracted to provide.

Eight different plans have been drawn up for the *Ville de Paris* since the war, but all these have, with two exceptions, been concerned with charting existing resources and marking areas to be preserved from interference. They show, for instance, roads, public open spaces, private grounds which are to be preserved as open spaces and areas in which the characteristic style of architecture is to be maintained in all new building. One exception to this is the policy of the administration as far as possible to zone industrial activity; all new factory buildings require a permit from the administration, even when on the site of a previous industrial plant. The other exception to this 'preservative' policy is a project for creating compulsory associations of private landowners in order to promote the rational use of large contiguous properties. This project has not yet received the approval of the legislature, but in any event the extent and the importance of the areas so planned is negligible in relation to the problem of urban development in Paris as a whole.

It is fair to say that the principal preoccupation of the architectural administration is to make the best use of the very small available space left for building, and of any houses condemned on grounds of public health. Interest in planning is secondary to problems which are felt to be more pressing: the need for more schools, for new thoroughfares, and for the demolition of dangerous and insanitary buildings. The French custom is that public authorities should carry out projects by assisting private enterprise rather than by direct intervention, and their powers of expropriation and coercion are always tempered by a high regard for private interests. The extensive development of new areas beyond the built-up region of Paris is regarded as impractical, since this would only aggravate the existing problems of transport and circulation.

On financial grounds alone the Government will certainly refuse to consent to any major development schemes for some time to come.

CONCLUSION

The problems of Paris are those of a great metropolitan region, the centre of national commerce and industry, the political capital and focal point of the French nation and empire. Its rapid growth during the nineteenth century, like that of so many cities, posed the basic problems of providing and administering services for a growing population over an ever-wider area. Elsewhere this process has raised the question whether these services should continue to be administered by the original city authority or whether they should be shared with minor local authorities. In France the solution was self-evident: the increasing population was automatically absorbed by existing local authorities with adequate legal powers, and these authorities formed part of a national unified system of administration within which collaboration could effectively be maintained. The Prefect of the Seine's position as administrative head of the department ensured co-ordination between the *Ville de Paris* and the other communes within that area, and financial assistance where necessary through the departmental budget; and interdepartmental matters could be fruitfully discussed at the highest level between the respective prefects. Above them stood the appropriate ministers, who could in the last resort decide for Paris as a whole, in relation to the complete perspective of French politics and administration.

Almost any modern government presupposes some division of responsibility between elected representatives and executive authority. For the Anglo-Saxon, municipal administration is the natural field for self-government, and he tends only to grant executive powers to administrative authorities on grounds of efficiency. French law presupposes a monopoly of executive power in the hands of the state, and if it allows powers to elected authorities these are concessions. The fate of Paris entails in too many ways the destiny of the state for any appreciable decentralization to elected bodies to be envisaged. There is, however, scope for a fruitful deconcentration of executive authority from the central ministries to the Prefecture of the Seine. But it is on grounds of political control rather than on those of organizational efficiency that the government of Paris must be considered. As Baron Haussmann said: '*La capitale appartient au Gouvernement.*'

Rio de Janeiro

JOSÉ ARTHUR RIOS

JOSÉ ARTHUR RIOS

Born 1921. B.L., Law School, State of Rio de Janeiro,
1943; M.A. Sociology, Louisiana State University, U.S.A.,
1947.

Taught Sociology at Vanderbilt University, 1948; Rural
Sociology at School of Arts and Letters, University of
Brazil, and Human Relations at the Centre for Selection of
Personnel, Brazilian Army.

Is now a Lawyer in Rio de Janeiro, and Consultant in
Sociology and Education in the Public Health Service, Rio
de Janeiro.

[*Photo: B.O.A.C.*]

51. RIO DE JANEIRO
Sugarloaf Mountain and a modern hotel.

52. RIO DE JANEIRO

A general view of the magnificent harbour and scenery surrounding the city.

[Photo: B.O.A.C.

Rio de Janeiro

THE GEOGRAPHY OF RIO AND ITS EFFECT ON LOCAL GOVERNMENT

RIO DE JANEIRO was built on the left margin of Guanabara Bay, between its quiet waters and the first elevations of the Orgaos mountains. From the very beginning, the hill and the beach were basic elements, not only in its geographical constitution but also in its social structure. The geological cleavages which originated the vast bay cut through the mountain range and depressed the soil, which was then isolated from the ocean. Later, the ground crumbled and this big hole was enlarged. An old river which had been stopped by sand mounds changed the edge of the coast into a swamp. This swamp is another geographical constant in the city's history. In fact, a good part of the history of Rio is devoted to efforts to dry out the ground on which it stands.

Anyone who comes to Rio from the sea, after having passed through the natural gateway formed by the Sugar Loaf on the left, and the Pico on the right, sees the city massed before his eyes on a series of panels which seem to open up as the ship proceeds. On the opposite side of the bay lies Niteroi, the capital of the neighbouring but independent state of Rio de Janeiro. At first, the city of Rio belonged to this state. But in 1834, it was separated from state territory and transformed into the so-called Neutral District.

The city owes its beauty to its wonderful setting between mountain and sea. But man has paid a heavy price for this beauty. He has to fight against these two elements for his survival. The city grew through the reclamation of considerable portions of the bay and the draining of the swamps. The first inhabitants established themselves on the hill-tops and, little by little, came down to the seashore and the dry ground which was constantly extended by the labour of generations. On the tops of those hills, which became the residential areas for the poorest class of the population, are still found some of the oldest buildings, baroque churches and ancient fortresses.

The city of São Sebastião do Rio de Janeiro, which since the republic has become the Federal District, is divided into an urban and a rural zone. The system of incorporated townships and cities, which is a

Q*

characteristic of America, does not obtain in Brazil. The Brazilian county (*municipio*) covers not only the rural but also the urban area within its limits. The Federal District follows this rule. Its area is about 1,111 square kilometres, but if we add the nearby Bay islands the total area is 1,167 square kilometres. The urban section, which is roughly bounded by the Gavea district in the south and by the Engenho Novo district in the north, covers about 133 square kilometres. The remaining area is suburban or rural.

It is very difficult to establish a sharp division between the urban and the rural zones in the district, and still more difficult to differentiate between the urban and the suburban areas. By a convention which grew out of the local usage 'suburban areas' are all areas which are served by local railways. No administrative division corresponds to the term. But, undoubtedly, the suburb is a very definite social area in Rio, with peculiar customs, styles of living and traditions. The upper-class elements of the population, situated along the Marquês de São Clemente (Botafogo-Gâvea) and the mixed sections of Copacabana, Ipanema and Leblon, remain aloof from these suburban areas. However, the expansion of the suburbs along the railways is one of the most important facts in the city's history.

The extension of the district is only one of the problems its administration has had to face. The provision of certain services is made difficult by the great differences in altitude at relatively close points of the city. Rio is both a seaport and a hilly city. In some places, at a point 20 kilometres from the sea the ground rises only 30 metres above sea level. Nevertheless, there are altitudes of 700 metres within a distance of forty minutes' tram journey from the city centre. Even this is not the highest altitude within the city limits, and the Andaraí and Tijuca peaks rise to more than 1,000 metres. The topography is quite irregular. The mountains come to the sea now in abrupt walls, now in small hills, spreading out like the fingers of a hand. These mountains are covered by remains of the primeval forest. All the splendour of a tropical forest can be found in a few minutes' ride from the city.

Situated at 23 degrees south latitude, Rio is the greatest city of European civilization in the tropics. After fighting against the swamps, the forest, the mountain, and the climate, Rio, in the words of a French geographer, is 'a superb human triumph.' In fact, here man developed new methods of combating the forces of nature. The swamps were drained by channels, some of them running underground, others at ground surface. In summer, abundant rains fall, sometimes with a precipitation of 200 millimetres daily. Certain districts under the slope of the mountains are periodically flooded. In order to avoid the scourge of flood, Rio has built a system of storm sewers. Unfortunately this system lacks continuity and its capacity is still inadequate for the heavy rains of the summer season.

Another problem for the administration of Rio, provided by its natural setting, has been the lagoons which extended behind the sand mounds at the ocean shore. The imprisoned waters spread epidemic fevers and prevented city expansion. Channels had to be cut, and dykes were built to open the stagnant waters to the movement of the tides. A typical example of this achievement is the Rodrigo de Freitas lagoon, around which today are built some of the finest homes and the racecourse of the city.

In other places, the sea broke everything. Breakwaters had to be built all along the beaches and avenues which connected the northern and the southern sections of the city. Great stone blocks were placed against the banks to protect them against the big waves which assail the coast in the winter. In some sections large-scale land reclamation was needed, and whole districts, such as Urca, Calabouço, and the main airfield, were conquered from the waters. The nearest islands were connected to the mainland by a system of bridges and serve as arsenals, piers, reception centres for immigrants, and store-houses for oil and minerals.

An even more serious problem deriving from Rio's setting is that of communications and transportation. The city engineers have had to fight the mountain as an ever-present enemy. Sometimes they took radical measures against it and levelled it down. Sometimes, the mountain was simply cut through. Cutting, like tunnelling, is very expensive and requires highly specialized equipment. More often than not the nature of the rock does not allow the first, and the municipal finances will not stand the second. Thus, the lack of adequate roads and transport in Rio is the cause of many of its social and administrative problems. It is responsible for the high percentage of tuberculosis in the city. The long journeys to work and the crowding of vehicles impose a terrible strain on a great part of the population who work in the central section. This condition has encouraged also the development of satellite cities, with their own facilities and community characteristics.

It is important to emphasize the peculiar disposition of the city. Nothing could provide a better introduction to its problems. Rio does not present the radial distribution characteristic of many European and American cities. As its expansion was barred by the mountain and the swamps, Rio was forced to extend along the sea coast, stopped only by the hills.

Little by little, the latter were overcome. Growth thus proceeded by jumps which scattered the city districts over a large area, but left them practically isolated. The main transportation axis still runs along the coast which it leaves only at narrow gorges. The Avenida Rio Branco, the main street, connects the coastal avenues to the northern section and the suburbs. Its overcrowding has compelled the government to open up another main route, the Avenida Getulio Vargas, which serves a similar purpose. Along both, the business area is concentrated. Here is the banking and commercial section, and the administrative centre of the city. This expansion of the city is illustrated by the fact that Leblon, a high-class residential district in the southern section, is 30 kilometres from the Penha suburb.

The irregular landscape also explains a curious social fact. The tops of the hills are cooler and healthier during the summer season, but owing to the high altitudes are not supplied with water, electricity and other services. Land values are therefore quite low in those places, and they are eagerly sought by the poorer classes. On these hills, there has grown up a peculiar type of communal agglomeration known as *favela*, derived from the name of one of the hills and which has become a general designation for a hill slum.

If communication between sections of the city is difficult, it is much harder to connect the capital with the interior of the country. Geography alone can never explain the position of Rio as a capital. But it justifies the attempts which have been made to transfer the capital to another place in the interior. Even though it is the most important railway and airway centre of the country, all its systems of transportation prove more and more insufficient to bring food and the other necessities of life to the population of 2 millions.

This difficulty of communication explains why the relocation of the capital has been a *leit motiv* in Brazilian constitutions since the founding of the republic. Many studies have been made and the argument has been reduced to two or three sites on the plateau of Goiás. The present Federal District would either become another state, or would revert to the State of Rio de Janeiro, from which it has been taken. However, it is not easy to promote such a change overnight. The army of officials has no illusions about the discomforts of a new capital. The issue remains unsettled.

THE POPULATION AND ITS PROBLEMS

Geography has already told us much about Rio. Demography will tell more. The composition of Rio's population is far more homogeneous than one usually finds in capitals. The percentage of foreigners is not high. In 1950, there were only 195,881 foreign-born persons in a total population of 2,377,451. Portuguese and Italians, the two largest groups, make up more than half the total alien population. Ethnically, there is a large dominance of the white element, for 71 per cent of the population is classed as 'white.' But this is a sociological rather than an anthropological term, for among these whites are found all the nuances of dark skins.

Rio has a high level of literacy compared with the rest of Brazil. The 1950 census revealed only 11 per cent illiterate in the capital, while the national average was 60 per cent. Nevertheless, the educational position even in the federal district is far from satisfactory. Among the 1,200,000 inhabitants who know how to read and write, only 16 per cent have completed the elementary school course; only 6 per cent the secondary course; and not more than 2 per cent have achieved a higher education.

One of the peculiar features of the population is that 19,000 people practise agriculture. Agricultural activities account for less than 1 per cent of the total population, but it is not usual to find such a group among the population of a great city. In 1940 there were 7,994 agricultural enterprises. These are mostly responsible for furnishing Rio with its daily supply of vegetables, milk, chickens and eggs. The organization of this agricultural production is, as we shall see later, one of the most serious problems of the Rio government.

The municipal services have to adjust themselves not only to a considerable mass of people, but also to a population which is growing very rapidly. This requires a constant reshaping of the services. In 1672, Rio had only 275,000 people. In 1890, its population was 522,600. In 1900, there were in Rio 700,000 people. In 1920, 1,157,863. In 1940, Rio had a population of 1,764,141 people. In 1950, it numbered 2,377,451, an increase of 36 per cent. Thus in the last thirty years Rio has doubled its population.

The main cause of this growth is internal migration from the country. Rio, like São Paulo the second Brazilian city, is the great centre of attraction for the maladjusted and rural proletariat of Brazil. Each new highway, each improvement in the transportation system, and every technological advance intensifies the movement. The Brazilian countryman is constantly seeking the greatest centres of the country. To study this migration would lead far from the present subject. It is enough to say that the countryman finds in the cities greater opportunities and a chance to escape from the oppressive social structure, based on the latifundia system, which obtains in the hinterland. In the cities, far as they still are from providing democratic conditions, immigrants from the country can find better facilities for health, comfort and education. The 1950 census found in the Rio population 1,153,991 (or 49 per cent) 'displaced persons,' that is, people from other states of the federation.

Migrants make up the largest part of the *favelas*.[1] An enquiry recently made into these shacks showed that most of their occupants came from the States of Rio and Minas Gerais. The large infant population of the *favelas* and the high percentages of illiteracy and vagrancy are enough to reveal the various problems they bring to the city government. As these migrants arrive, largely by family groups, the schools and hospitals are insufficient to accommodate them. They bring to the city problems of health, education, and housing which overtax its resources. They increase the popular pressure on the city services and lengthen the queues at the administrative offices and public departments. It is among these newcomers that the political bosses of the district find their agents and recruit their bodyguards.

EVOLUTION OF THE GOVERNMENT OF RIO DE JANEIRO

The building of the city of Rio answered the strategic needs of Portuguese colonization in Brazil. In the sixteenth century, French pirates following the steps of Villegaignon and still nursing his dream of an antipodean France, came for shelter to Guanabara Bay and maintained good relations with the Indians. The Portuguese sank their ships and hunted them on the sea, but in 1560 the Governor-General Mem de Sá determined to attack them in their headquarters. The action, in which both Portuguese ships and Indian canoes took part, was a success. However, Mem de Sá made the mistake of leaving without consolidating his victory. The Frenchmen came back, making necessary a second expedition led by the Governor's nephew, Estácio de Sá, who not only expelled the pirates but also laid the foundation of the city.

Estácio de Sá also determined that the city should be removed to a safer place not so close to, but not too far from, the entrance to the bay. He chose as the best place for the city a hill facing the mouth of the Guanabara. He built there a sort of citadel, and churches, the cathedral, the customs house, and the Crown stores.

In order to understand the city's government and its evolution, it is necessary to bear in mind that the Portuguese brought to Brazil a century-old form of political life, the basic unit of which was the county

---

[1] See *ante*, p. 492, for an explanation of the term *favela*.—Ed.

(*municipio*). This had been the traditional system of local government in the Iberian Peninsula since the Romans. At the beginning of colonization, however, only villages could be founded in Brazil, because the establishment of cities would necessarily have involved the creation of independent *municipia*, which according to the Portuguese law, owned the lands within their limits. As all the land in Brazil had been given to the old Brotherhood of Christ, and as only its chief, the Portuguese king, could administer it, this was not feasible. Yet there arose the need for new bishoprics in the new territory, and the bishops, as titled princes and noblemen, could not live in villages. Therefore, when the first bishoprics were created in Brazil, the king, as Grand Master of the Brotherhood of Christ, emancipated the land and dedicated it to the service of the Catholic faith, converting into cities the villages which had been chosen for the bishops.

Consequently, the *municipio* has been the first cell of Brazilian governmental organization since the early years of the country. It had to face, here as well as in Portugal, the expansion of the state which little by little infringed its privileges.

In 1603, when the Philippine Ordinances were promulgated, the system underwent deep changes. 'From then on,' as one historian states, 'the *municipio* began to play a purely administrative role, and relinquished judicial functions, which were restricted to judgments by the magistrate, who was also the president of the chamber (as the alderman's council came to be called), in cases of defamation, petty thefts and fiscal suits.'

The chamber, composed of aldermen selected from the local gentry, was continually struggling for independence with the governor-general of the colony. Indeed, the history of Rio, during its first two centuries, is full of protests and mutinies against the governors. The disputes were nearly always about the heavy taxes and the numerous prohibitions which aimed at maintaining the Crown's monopolies against the interests of the colony.

In the eighteenth century, the discovery of gold and silver mines gave tremendous importance to the city. It became the natural outlet for these precious metals. Its population increased fourfold. The eighteenth century brought wealth and splendour to Rio. It is the century of governors with a sense of urbanism, and they enriched the city with public works and important buildings. The raising of the colony to the category of a vice-kingdom, after 1763, increased the political and economic role of Rio de Janeiro, but was a death blow to its independence. Directly appointed by the king, the viceroys possessed so much power that a contemporary publicist called it 'monstrous.' The elected senate of the city, which was created in 1787, was unable to resist this encroachment of the vice-regal power, and the senate began to lose its former autonomy.

The second revolution in the history of Rio was the coming of Don João and the Portuguese court to Brazil in 1808. They fled from the French army of Junot which invaded Portugal and were brought to Brazil under the protection of the British fleet. After staying a few days in the northern city of Bahia, Don João determined to establish the capital of his kingdom in Rio, which became, in this way, the capital of the Portuguese empire.

When Don João returned to Portugal, and Dom Pedro I proclaimed the country's independence in 1822, the position of Rio de Janeiro as capital was firmly established. In 1834, the additional act to the first imperial constitution, which laid the foundations of the province's autonomy, also created the neutral county (*municipio neutro*) in the city of Rio de Janeiro, dismembering it from the Provincia of Rio de Janeiro.

During the empire, the city progressed little. Rio is, indeed, an achievement of the republic which was proclaimed in 1889. When the neutral county of the empire was transformed into the present Federal District, many services which were monopolized by the central government were transferred to the district administration. Only after the city attained its new status was it possible for men like the mayor Pereira Passos and the sanitarian Oswaldo Cruz to achieve their revolutionary improvements.

The republican constitution, promulgated in 1891, created the Federal District and gave power to the national Congress to pass laws to organize it. Paradoxical as it may seem, in a régime which was based on municipal independence, the capital's *municipio*, the most important in the country, was not able to determine its own form of government. The constitution did, nonetheless, place the city's administration entirely in the hands of the municipal authorities and vested in them the power to vote their own budget, a privilege never enjoyed during the empire.

On September 20, 1892, the national Congress issued the first organic law[1] of the Federal District. It provided for the city of Rio to be governed by an authority divided into legislative and executive branches. The first was the municipal council, a body of twenty-one aldermen (*vereadores*) one for each of the city's districts. The executive was, and still is, represented by a prefect (*prefeito*) appointed by the President of the Republic on the approval of the Federal Senate. Any veto by the prefect to a resolution of the city council must be submitted to the Federal Senate.

---

[1] Law No. 85.

This organic law also transferred to the government of the Federal District many services, including public health and cleansing, child welfare, primary education, drainage and fire protection.

A great effort was made, when the draft of the organic law was under consideration in the national Congress, to provide for a mayor to be elected by the city council or by direct election, in place of the prefect, but this did not succeed.

The organization described above prevailed until 1930. The revolution which ended the so-called liberal era in Brazilian history changed the city government. The provisional government dissolved the city council and put in place of the prefect a federal deputy (*interventor*). When the constitution of 1934 was promulgated, the legislative assembly was urged to give the district municipal autonomy. Accordingly, it made the mayor as well as the city council directly elected by the people. Nevertheless, when the constitution began to be applied, there was some fear of carrying it out to the full extent and the mayor was eventually elected by the aldermen.

After 1937, both the executive and the legislative powers for the whole of Brazil were invested in the President, who enacted decrees, assisted by purely bureaucratic bodies acting as advisers. The constitution of 1937 kept the office of mayor, but its holder was henceforth to be appointed by the President without consulting any other governmental body. The constitution also established a deliberative council, but it never met.

The *mot d'ordre* was to centralize as much as possible and to rationalize every service. A certain 'technicism' developed which recruited a large bureaucracy. In a few years, it was necessary to regulate minutely the privileges and duties of the officials, and the President issued a law regulating the civil service of the Federal District. This statute, the benefits of which are undeniable, had also the effect of fostering continual claims for promotion and advancement. As the dictator had to govern with the support, among other forces, of this bureaucracy, he could not avoid granting it favours. In order to pay the tremendous cost of these successive readjustments, banknote issues were increased, and to this we certainly owe the scourge of inflation which was later aggravated by war conditions.

When Vargas fell on October 29, 1945, a new assembly met which promulgated a democratic constitution, re-established the chamber and the senate and, after ten years, called for fresh elections. The new régime, however, could not erase overnight the effects of the dictatorship. All its clumsy and demagogic legislation remained, making the good intentions of the new government completely ineffective. Such is now the case with the Federal District.

THE PRESENT GOVERNMENT OF RIO

The Federal District is now governed by a mayor and by the city council. They represent, respectively, the executive and the legislative power. The council, but not the mayor, is directly elected by the people. The congressmen of 1946, like their predecessors of 1891, hesitated in giving the district full autonomy. Therefore the mayor is appointed by the President of the republic, who can also dismiss him *ad nutum*. Even though the constitution of 1946 requires the choice of the mayor to be approved by the Federal Senate, the head of the executive is inevitably the President's man, and owes no allegiance to the council or the electorate. This is the great difference which still sets Rio apart from the Brazilian counties. Rio alone cannot elect its mayor.

To say that the mayor exercises the executive power and the council the legislative power does not give a true picture of their respective functions. The mayor's authority goes beyond the mere execution of laws. He can promulgate laws himself; he enacts decrees, regulations, and instructions. Furthermore, he can block the action of the council by exerting his veto against the council's bills. There has been permanent friction over these powers in the Federal District. When a veto occurs, the senate is called up to settle the question. When tension arises between the mayor and the city council, a deadlock frequently results, and the work of the senate is handicapped by the intervention of the parties who wish to see the veto cancelled or maintained. The meetings of the senate committees are disturbed by frequent visits of the deputies sent by interested groups. The situation has become so intolerable that the senate is considering placing the final decision on the mayor's veto with the city council itself. This would do much to enhance the autonomy of the district.

The authority of the mayor now covers a large field. He directs and supervises the municipal public services, borrows money in anticipation of revenue when authorized by the city council, and makes agreements with creditors or debtors of the district. Whenever the law deems it necessary to the common weal he can requisition private property, providing the owner is paid compensation. He can appoint, promote, dismiss, retire, and re-employ public servants. He is authorized to collect taxes, fees, contributions, legal costs, and rents due to the district, and to apply them in accordance with the law. He is responsible for the conservation and management of the public property of the district and can alienate or change it provided the legal requirements are fulfilled. Another of his duties is to promote planning and to submit schemes to the council, indicating the resources available for their execution. The inter-relationships between the district and the Union or the other states are in large part his charge. As legal representative of the district, he can make agreements and conventions *ad referendum* of the council. He represents the Federal District in the courts, through the medium of municipal lawyers and public attorneys.

The mayor has to report to the council on his government and to render accounts to them. He may also be required to submit in writing any information which the council requests. In the management of the district he is assisted by one secretary and five secretaries-general. They attend to the affairs of the several departments, issue instructions, recommend the appointment, promotion and dismissal of the personnel in their departments, present an annual report on the activities of their services, attend the council whenever required, and lastly, draft the decrees affecting their departments.

The council is composed of fifty aldermen who are chosen by direct vote of the district's citizens. The conditions of candidature are few. The candidate must be a natural born Brazilian, in full possession of his political rights, and more than twenty-one years of age. The holders of certain public offices in the Union and the district cannot become candidates until from three to six months after termination of appointment. In a country where the family group plays an important role, some prohibition is to be expected excluding the mayor's relatives from holding office. The law stipulates that until six months after he leaves his office no relative, either consanguineous or by marriage up to the third degree, can become a city alderman. The same rule applies to the relatives of the President of the Union, ministers of the states, and the secretaries of the district.

The Federal District Organic Law of January 15, 1948, forbids many activities to the aldermen in order to prevent corruption. While holding office, they cannot make contracts with the district or the Union. They cannot accept any charge, commission, or salaried position. Moreover, they cannot hold any office of supervision, management, or direction in companies holding a concession for local public services if they are subsidized either by the district or by the Union. There are many other provisions intended to promote honest administration.

On the other hand, the person of an alderman is inviolable. He cannot be prosecuted for a criminal offence or imprisoned unless the council has given permission. When bail cannot be granted, the offence has to be reviewed by the council and it may decide either to authorize or prohibit prosecution.

The municipal legislature sits for four years. The council meets on the first of April and continues without interruption for seven months. It can be convened for special sittings provided a request comes from four-fifths of its members or from the mayor. The council's decisions are taken by a majority of those present, provided the required quorum of twenty-six members exists. For action on a new tax or on expenditure, an absolute majority of the full membership is necessary.

The council has authority to provide for its own internal organization, for police, and to assign duties to its secretaries. It also fixes the salary of the mayor and the aldermen for the next period in the last year of its sitting, any alteration at another time being strictly forbidden. In constituting the various committees the council in practice tries to give proportional representation to the political parties according to their relative strengths.

POWERS OF THE MUNICIPAL GOVERNMENT

Thus far, only the structure of the municipal government has been analysed. The scope of its powers, or the matters on which it can legislate, have not been considered. Broadly speaking, the district has the right to do anything which is not explicitly or implicitly forbidden to it by the Constitution or by any federal law. Thus, the municipal government can (1) organize its administrative services; (2) provide for the necessities of its government and administration, resorting, when necessary, to the Union; (3) organize its civil service provided it respects the principles established by the Constitution; (4) issue laws to supplement or complement federal legislation; (5) impose taxes other than those which are within the sole competence of the Union; (6) collect fines, fees, and any revenues which may result from the exercise of its powers or from the utilization of its property and services; (7) borrow money; (8) perform any public service which is not reserved to the Union and, finally, (9) establish agricultural settlements in order to cultivate its public land.

Besides these separate powers, the district government shares certain responsibilities with the Union. It is expected to preserve the Constitution and uphold the laws, to care for the public health and welfare, to protect natural beauties and historical or artistic monuments, to foster land development, and to assist in financing public education and welfare.

But the federal law-makers have also added to this list specific duties concerning the organization of services which bring greater comfort to the city's population, to improve public health and welfare by assisting mothers, infants, the aged and the crippled. Lastly, the municipal constitution enjoins the municipal government 'to ensure in the best way the moral and material conditions on which depend the development of individual energies, the utilization of capacities, and the improvement of culture.'

These purposes should be implicit in any government worthy of its name. However, the legal formalism which still dominates Brazilian culture demands their explicit enumeration.

FINANCE

One aspect of the city administration of Rio which deserves particular analysis is the budget. The Federal Constitution of 1945 prescribes the general features of the municipal budget. Accordingly, the budget must be unified, and all income must be included in the general revenue. On the other hand, every item of expense to maintain the public services must appear as an expenditure. The budget Bill must include a special article dealing with any item for which borrowing is required.

The constitution further establishes that the expenditure schedule must be divided into two parts: part I, which cannot be altered except by a subsequent law; and part II, which may be modified by a rigorous procedure. The transposition of sums, borrowing of money, or borrowing in anticipation of revenue without the council's permission, are forbidden. The opening of extraordinary credits or long-term loans can only be made in the event of civil commotion, war, or public calamity. The opening of special short-term or revenue-anticipation credits can only be made after the first three months of the financial term, and supplementary loans are allowed only after the first six months of the budget year.

A financial board of review is charged with the supervision of district finances. It is composed of seven councillors appointed for life by the mayor. They are chosen among Brazilians over thirty-five years of age who possess outstanding financial capacity and legal experience. The board controls the accounting of all public servants responsible for money and property belonging to the district or entrusted to it. The board registers all transactions of the municipal administration which involve any financial obligations, such as deeds, contracts, orders granting aid, decisions concerning the retirement or dismissal of civil servants, and any executive instrument involving expenditure. It follows closely the execution of the budget and the application of the budgetary and extra-budgetary borrowing. It also verifies the precautions taken by officials who are responsible for public moneys. Contracts which involve revenue, borrowing, and the issue of property titles are also examined. Only after this examination can they be registered. Finally, the board reviews the mayor's final report and has thirty days in which to publish its findings.

It is evident that the board is one of the most important organs of the municipal administration. To refuse registration to an administrative transaction or instrument, owing either to insufficient funds or to mis-application of funds, is equivalent to forbidding the execution of that transaction.

When the mayor has prepared the draft of the budget, he sends it, with his message, to the city council in order to receive its approval. It is very rarely that the budget is returned without changes which modify its whole framework. The standards prescribed by the Constitution can seldom be followed to the last item. Financial pressures break the unity of the budget through the inevitable 'extraordinary accounts and credits.' On the other hand, as it is annual, the budget must be adjusted to economic and financial plans which usually go beyond the yearly period. Budgets in Rio are drafts which have the value of simple sets of rules, and are not rigid programmes. The mayor himself regards his budgets as mere forecasts of the annual expenditure and revenues.

The analysis of the budget reveals many things concerning the adminis-tration of the district. In the first place, taxes which are collected by the Federal District consist of levies, (1) on property in general, (2) on the transmission of property *causa mortis*, (3) on the transmission of property *inter vivos* and its incorporation in any society's funds, (4) on sales and consignments by merchants and producers (the first operation of the small producer being the only exception), (5) on the export of locally produced merchandise to foreign countries up to a maximum of 5 per cent *ad valorem*, (6) on industries and professions, (7) stamp duties on title deeds, contracts and other legal instruments, (8) on licences, and (9) on public amusements. Besides these taxes, the district also collects, (10) contribu-tions for public works improvements which increase the value of private property, (11) fees, (12) fines of whatever nature, and (13) any other revenues which derive from the utilization of the district property and services.

The revenue collected in 1948 amounted to Cr$1,781,094,000. The importance of the district in the country's finances is illustrated by the fact that only one state, the State of São Paulo, achieved a larger collection. A classified table of revenues is shown below:

Estimated Revenue for 1949
Ordinary revenue

Taxes	Cr$ 1,404,500,000
Fees	Cr$ 125,950,000
Estate revenue	Cr$ 27,280,000
Business revenue	Cr$ 110,000,000
Other revenue	Cr$ 5,960,000
Extraordinary revenue	Cr$ 158,900,000
Total	Cr$ 1,832,590,000

The extraordinary revenue resulted from taxes on public utility undertakings (e.g. the electricity supply and tramway company), from the sale of municipal buildings and urban land, from borrowings, and from the payment of indemnities, fines, and from the highway fund.

The most productive tax is that on sales and commissions, which it was expected would yield in 1949, Cr$742,000,000. The tax on houses is next with Cr$195,000,000. In the third place, the tax on the transmission of property *inter vivos* was expected to produce Cr$160,000,000. Under the same heading are included the tax on land, on licences for peddlers, on the exhibition of advertisements, on the operation of vehicles, on the letting of shops, and the tax on public amusements, industries, and professions, as well as on the transmission of property *causa mortis*.

Fees cover the charges for licences, registration, and the revenue from municipal services, etc.

The revenue on property or estates includes expenses on land titles, renting of municipal buildings, tenures, interests, and dividends. The business revenue includes the rates on water consumption, drainage charges, and taxes on production. Under the heading 'other revenue,' the municipal budget usually includes receipts from cemeteries, markets, and slaughter-houses.

In recent years, some of these sources of revenue have increased, whereas others have diminished. The revenue resulting from taxes, for instance, increased in 1949 by 11·9 per cent over 1948. Income from fees also increased by 10·4 per cent. Rent from property, however, decreased by 33·1 per cent. Among the general divisions of the budget there was also an increase in 1949 in the collection of the taxes on houses (11·4 per cent), and on the transmission of property *causa mortis* (50·0 per cent). Some of these advances are due to improvement in enforcement and collection, but others have economic causes.

It has been felt in certain circles in Rio that a rise in the rates would necessarily bring an increase in the total revenue. The mayor, however, in his annual message to the council in 1949 took a different point of view. 'It is well known,' he stated, 'that the budget of the Federal District depends mostly on the taxes levied on sales, commissions, on houses, and on transmission *inter vivos*, all of which are subordinated to economic conditions independent of the administrative action of the mayor. These activities result from causes which the local public powers cannot reach. The worsening of the commercial situation, resulting from the weakening of internal production and from additional restrictions imposed on imports, the fixing of rents by a federal law, and the withdrawing of estate operations force us to be pessimistic about the immediate tapping of greater sources of revenue, notwithstanding improved methods of collection. The resources of the taxpayers do not keep pace with the needs of the local administration and the increase of municipal debt is a sure symptom of this disparity.'

This means that the exhaustion of the taxpayers' capacity is becoming more and more apparent, while the taxes continue to increase. The extent and the seriousness of the problem is better revealed by municipal expenditure. The expenditures were calculated, in 1949, to reach Cr$1,832,419,857. This sum represents an increase of 7·7 per cent over outgoings in 1948.

Expenditure for 1949 was distributed among the different departments of the municipal government as follows:

City Council	Cr$	32,730,400
Mayor	Cr$	11,345,000
Municipal attorneys	Cr$	259,400
Secretariat for Administration[1]	Cr$	767,324,641
Secretariat for Agriculture, Industry and Commerce ..	Cr$	55,343,940
Secretariat for Education and Culture	Cr$	116,018,696
Secretariat for Finance	Cr$	171,146,284
Secretariat for Transportation and Public Works ..	Cr$	191,088,500
Secretariat for Interior and Security	Cr$	401,706,196
Financial Board of Review	Cr$	424,700
Supervision of transportation	Cr$	41,768,400
Total	Cr$	1,832,419,857

[1] The Secretary-General for Administration controls the whole system of the executive branch of the city government. He controls the municipal civil service and his authority is second only to that of the mayor.

The distortion of this budget, which favours the bureaucratic departments more than other bodies, becomes apparent when expenses for personnel are revealed. In 1949, personnel expenditure amounted to Cr$1,057,251,180,000. This represents 57·7 per cent of the total budget. We notice, however, that these figures have been increasing tremendously. In 1948, such expenditure required only 55·5 per cent of the total municipal outlay.

The mayor is the first to acknowledge the urgency of this problem. In his 1949 budget message, he declared: 'I take it for granted that the sums at my disposal are not enough for the achievement of a programme which might correspond to the needs of our community. However, I have to subordinate my plans to the amount of the budget which goes for personnel. The payment of municipal personnel weighs heavily on the budget of the city government, and restrains the planning of public works which might be of greater utility.' Further, the mayor has stated in more urgent words: 'We regret that the considerable amount of money devoted to the payment of personnel does not allow the city government to start public works to promote the public welfare.' The mayor, however, reveals only half the truth. He is himself responsible for having appointed 8,000 officials to different departments of the government. It is estimated that the district government now employs around 53,000 people.

This problem is not merely administrative, but results from deep social maladjustments in Brazil. The growth of the bureaucracy is not confined to the municipal government. The federal public service has seen an increase in its officials which has not been accompanied by an increase in services or efficiency. The payment of this tremendous army of bureaucrats is absorbing more and more of the country's resources and explains the difference in the standard of living between the capital and the country as a whole. It has been said that Brazil is trying a new and advanced form of society in which unemployment is eliminated by placing every citizen on the state pay-rolls.

THE POLITICS OF THE CITY OF RIO DE JANEIRO

Therefore, the real state of the government of Rio de Janeiro is not revealed only by the formal structure of powers. The blueprints of the government do not tell the whole truth, and the administration of the city works within limiting social, economic, and political conditions.

One of the greatest difficulties of Rio government arises from the political relationship of the mayor to the city council. The mayor has a privileged position, being appointed by the President of the Union. As he can only be dismissed by the President, he feels strong enough to govern without reference to the city council. Under these conditions, a conflict may easily arise. The mayor vetoes many of the council's decisions, which then must be submitted to the Federal Senate. Municipal administration becomes handicapped, and the impasse even threatens national affairs.

Administration and politics in the Federal District cannot be understood apart from the national pattern. It is necessary to bear in mind that the country has recently undergone a period of dictatorship, followed by a political régime closely akin to it, namely the Presidentialism established by the constitution of 1945. In fact, Brazil has not known since 1937 a truly democratic régime. Presidentialism, even though it did not create nepotism and other political plagues, has kept them alive. Municipal, as well as federal, administration has become crippled by these maladies.

Sometimes the mayor has political ambitions of his own which aim beyond the office of mayor. He may wish to be a senator, a congressman, or perhaps president. He develops, therefore, an administrative and political programme designed to secure the support of the masses in the district. To begin with, he usually appoints many people to public offices in the municipal government. These appointees, paid by the public treasury, become his political bosses. In a country such as Brazil, where public opinion is not organized or educated, a faithful bureaucracy is a tremendous political asset. It can be employed both as a means of disseminating propaganda and as a machine for coercing the voters. An ambitious mayor may also try to win the masses by the old method of offering them bread and circuses. In such circumstances administrative ability may go unrecognized and unrewarded. Indeed, administrative ability may even be penalized if it appears to challenge the authority or the popularity of the mayor.

A foreign reader may ask why public opinion does not react against this type of situation. It is difficult to explain without going into details of the relation between the political power system and the Brazilian social structure. The country is still going through a transition from the caste to the class system. Patriarchism still exerts a tremendous influence on politics. The state has not overcome the institution of the family which absorbs many of its functions and permeates public relations with its peculiar styles and ways of action. Power, therefore, is not only obtained through elections or impartial choice, but through an elaborate system of personal connections which at the same time works as a system of social classification. Consequently, Brazil presents problems which do not exist in European countries. At present, for instance, it is difficult for public opinion to obtain the dismissal of an astute prefect who understands how to manipulate the sentiments of the masses.

It is, moreover, highly doubtful whether the public would want to dismiss him. One must not confuse the opinion of a very small *élite*, which may be expressed through one or two newspapers, with mass opinion. As the latter is still a product of the class structure, it has not achieved an organic unity and is unable to influence public affairs. The country is now going through a slow and painful process of political evolution. We see the rising of the masses, who through the vote are trying to secure better social conditions. They do not vote for candidates who offer them democracy, but for those who promise them relief from oppressive conditions. This undoubtedly represents the stage of demagogy. Be this as it may, it is necessary to take account of this fact to explain the Brazilian political situation.

Rio de Janeiro has always been a stronghold for the opposition. When the candidates of the government have won elsewhere, they have been defeated in the district. That is why it is vital not to abolish Rio's political rights and to reduce its citizens to a position of political inferiority like those of Washington, D.C. The economic and political importance of the city would not now allow the government to do such a thing. The country would also suffer from such a measure because the district represents one of the most cultured and developed units of the Federation. On the other hand, the government must often feel uneasy about the district. It is hard to govern when one is sitting on a powder barrel. Fear of anarchy and political upheaval is always present for our statesmen.

However, even this opposition is now becoming meaningless. In the elections of October 3, 1950, the government, as well as the traditional opposition, were defeated by a third force which represented a movement for the return of the former dictator Getulio Vargas under the name of the Brazilian Labour Party. In the district, the candidates of this party received the most votes, not because they advanced ideas of social reform, but only because they had the political support of the Vargas machine.

The democratic parties, like the National Democratic Union (U.D.N.) or the Social Progressive Party (P.S.D.), failed in appealing to the middle

class which they were supposed to represent. The very narrowness of
this middle class in Brazil explains the lack of balance in the political
and economic situation. The great parties have to compromise between
the vested interest groups and the masses. Those who achieve this difficult
feat are likely to win.

The most alarming aspect of the political equation in Brazil is the lack
of popular interest in democratic institutions. More and more, the people
look forward not to perfecting these institutions, but to some Messianic
leader who promises higher salaries and social equality. This Utopian wish
is fed by the lavish standard of living of a bourgeoisie which has forgotten
habits of thrift and indulges in foolish speculation and excessive luxury.
In addition to feeding their resentment, this national spectacle sets the
pattern for the lower classes who also want motor cars, radios, and
refrigerators before anything else.

In the district, the problems resulting from the expansion of this urban
proletariat are very vivid. These problems reflect a situation which is
shared by the whole country. When Vargas fell, some people thought
that the city council would act as a restraint on the mayor's ambitions.
It would express the popular ideals and would block any encroachments
on local freedom by the central government. These hopes began to
vanish when in 1946 the council lost the power to override the vetoes
of the mayor. As the council had lost authority, the aldermen felt less
responsible. At that time, some of them even had the courage to resign
from office.

Partly as a consequence of this loss of power to override the mayoral
veto, the aldermen grew less conscious of the public welfare and more
aware of their private interests. Many of them were recruited from among
radio speakers, popular singers and stage artists, and had no training
whatsoever for a political career. The council voted increases in their
own salaries and also tried to employ every friend in the council's
secretariat itself. Functions were created only for this purpose of rewarding
supporters. Supporters began to look upon public offices as a fair reward
for their efforts. Sometimes aldermen have no connection at all with
the constituency which elects them. They negotiate with the local boss
for his support and the latter provides blocks of votes. Many of these
bosses have been famous in the political history of the district.

On the other hand, the voters expect from their candidate not only
local improvements, such as the paving of a street, the extension of light,
water, and gas services, but also personal favours. In many different ways
jobbery and personal influence vitiates administration.

PRESENT PROBLEMS AND FUTURE PROSPECTS OF
RIO DE JANEIRO

It is evident, therefore, that the continuing political crisis of the district is a result of the lack of political education. The fact that democracy in Brazil came from the top downward, was first a conquest of the *élite* and then came to the people, is all-important to an understanding of Brazilian political problems. Only after a period of thorough education, of daily practice in the democratic process, will our people be able to discriminate between candidates and hold them responsible.

The independence of the district, that is, its capacity to elect its own representatives, has been a favourite slogan with politicians for two generations. The idea was first uttered in 1856, during the Empire, when congressman Candido Borges tried to sketch the government of the so-called neutral county. According to this project, the central government would choose from among the elected aldermen one who would become mayor. In 1862 another project gave the aldermen authority to elect the mayor. Alfonso Celso, in a book published in 1883, advised the imperial government to give full autonomy to the city. It is important to recall that even Ruy Barboza, one of our greatest democratic leaders, hesitated when called upon to give his opinion concerning the independence of the district. His first utterance on the subject was a refusal: 'Everything makes us believe that the simultaneous existence of a federal and a municipal government in Rio de Janeiro is a threat to the basic necessities of order and the elementary principles of the federation.' He even advised that the capital city should have no representation in the Congress and that all political functions should be taken from it. Later he abandoned this first position and advocated full independence for the district.

However, the problem is not easy. The experience of autonomy from 1930 to 1937 provides little guidance. The mayor became an idol of the masses and his power even threatened the authority of the President. Perhaps the time was not ripe for such an experiment. The President, it will be remembered, was Getulio Vargas and, at that time, he was already planning to destroy the democratic régime and establish his dictatorship. The autonomy of the district was necessarily a handicap to his schemes, as the autonomy of the states was also. The result, therefore, is not conclusive.

The present trend is towards greater municipal autonomy. The Federal Senate is perfectly aware of the necessity of giving the council responsibility for voting on the mayor's vetoes. The work of the senate has been constantly blocked by pressure groups which send delegations to the senators who have to consider these vetoes. Moreover, the new government has announced its decision to fight for the direct election of the mayor. Of course, all these good intentions are conditioned by a third factor which has already been mentioned, namely the extent to which the council will become conscious of its duties and the people aware of their role in a democracy.

Thus, the city government does not seem to be equipped to solve the greatest problems of Rio. The first of these is transportation. From what has been said about the city's topography, it is clear that Rio badly needs an adequate system of transportation and the tunnels which might bring the several sections of the city closer to one another. In the present situation, the trend is towards the creation of satellite towns, each section striving for full independence. It is understandable that the outlying sections want to get rid of the 'centro,' that is, the central section which comprises the Avenida Rio Branco and covers the commercial and administrative area. Only the building of tunnels or the establishment of the 'metro' can change this trend towards suburban decentralization.

No one has stated the effects of the transportation problem in Rio better than Dr. J. Fernando Carneiro, who put it as follows: 'For the working masses of the Federal District the transportation problem is a tragic one. They spend an average of six hours a day solving it: three hours from the suburbs to their jobs, and three hours from the office, the factory, or the shop, back to their homes. They arrive home late. They leave early in the morning, bringing with them in small cans a lunch which was cooked the night before. They leave at 4 or 5 o'clock in the morning, and are back at 8.30 or 9 o'clock in the evening. Many of them can only see their children awake on Sundays. It is utterly impossible for them to have lunch at home. They have to choose between the restaurants and the meals they bring in the cans. They spend six hours a day in queues, in crowded trains, or standing in buses. Transportation exhausts them much more than their jobs. I am thoroughly convinced that better transportation for the workers would have a definite effect in reducing physical exhaustion, tuberculosis, and mortality; and it would also allow them to frequent night classes, meetings and lectures.'

The problem of water supply is also very serious. It is astonishing that the city does not suffer from epidemics. The water shortage leaves many sections of Rio completely dry for periods which vary between a week and a month. The tremendous growth of the city in part explains the problem, but to explain is not to solve. Water shortage is not, however, entirely a local problem. It has a definite connection with the intensive cutting of wood for fuel and the resulting erosion in the watersheds surrounding Rio. The water level is sinking everywhere.

Moreover, it becomes more and more difficult to feed Rio. Milk supply is particularly inadequate. The city now consumes 300,000 litres of milk daily. Years ago, there were stables right in the city and milk was distributed direct from the cow to the customer. These stables, however, were responsible for many diseases. There were sick cows, and milk handling was very insanitary. The stables were then eliminated, and milk was distributed from rural areas through dairies or by delivery. As during the Vargas régime it became a fashion for the government to interfere with everything, milk distribution followed the rule. There was a milk committee, and milk became a state monopoly. The committee was authorized to distribute milk and to invest the collective profits as it pleased. The money was devoted to luxurious buildings, and soon the committee was so heavily in debt that it had to be dissolved, its functions being given to the co-operative of milk producers.

The position of the consumer, however, is no better. Milk is still bad. There is no adequate grading. The two present types are not fully satisfactory and the largest part of the population drinks the inferior one. It has been shown that much of the gastro-enteritis among the city's children is due to infected milk. Handling in the dairies is also far from clean. The imposing array of officials which supervises this service does not prevent the mixing of water with the milk or the use of unclean equipment. A doctor summarized the problem as follows: 'High profits for some; fraud by many; lack of responsibility in the retail trade . . . here are some of the factors which explain the low consumption of milk among us.' The same authority recommends the British method of the regulation of milk supply, but this is impossible in the prevailing Brazilian social and economic conditions.

Meat supply is another food problem of serious dimensions. The secretary-general of agriculture supervises the supply of meat. Since the war, there has never been a sufficient supply. The problem is not confined by any means to the district. Its solution demands close co-operation between the municipal and federal authorities. It is evident that the municipal government is completely unable to control the meat supply from producer to the customer. However, it could at least influence transportation and distribution. The customer would benefit from the distribution of previously cut and classified meat. At present, meat is still processed by butchers in shops which are unclean and ill-smelling.

The Federal District could be supplied with 400 tons of meat three times a week. In order to achieve this supply, however, it would be absolutely necessary to rebuild the most important cold storage warehouse in the city, so that its capacity could be increased from 5,000 tons to 30,000 tons. It is necessary to observe, nevertheless, that under the present conditions of excessive valuation of cattle, excessive meat exports, increase in home consumption, and decrease in the weight of cattle due to the shortage of grazing fields, the problem of meat supply in Rio is only one of the aspects of a general crisis in Brazilian agriculture, and therefore can be solved only by co-operation and planning by both the municipal and federal authorities.

The city also lacks a well organized green belt. Strange as it may seem, Rio also has its agrarian problem. Its rural population does not enjoy full ownership of the land, and 70 per cent of it consists of share owners and tenants. Tenants and share croppers pay high rents for the land and have almost no security on it. They can be thrown out at any moment by an unscrupulous landowner.

The city's rural surroundings are controlled by landowners who are not farmers. They have bought land as a speculation. The owners are mostly banks and estate agencies which have been waiting for an increase of values in order to realize an easy profit. Meanwhile, they keep the land in an uncultivated state, or ask a rent such as the poor tenant cannot afford. Another type of landowner is the so-called 'asphalt agriculturist,' who owns a small farm only for pleasure. He keeps a manager on the farm and visits it on week-ends, living mostly in a comfortable apartment in Rio. These two kinds of landowners, however, are better equipped as far as political and social connections are concerned than is the poor tenant. In consequence, when they learned that the municipal bank was going to make money available on certain terms, they employed their influence to secure at least part of it. Thus, money which should have gone to small farmers and co-operatives went into their pockets. The plan of rural credit in Rio was a complete failure.

This account of the problems and prospects of Rio does not present a happy picture. Those who come to Rio for a week or two will probably be unaware of its problems. Rio is still a beautiful city. In Copacabana and the fashionable sections, life is very pleasant. Some of the problems dealt with above are mere symptoms of growth which every metropolis has at some time experienced. However, in Rio they are aggravated by the general social and economic condition of the country. In some ways Rio is the product of a vast maladjustment. Its growth is unparalleled by any other city in the nation. The tendency has been to enslave the whole country to this abnormal growth, like a tumour which drains all the energies of the body. Its high-life is fed upon the misery and backwardness of the rural population. Its army of civil servants lives on taxes levied throughout the country. Its *élite* is made up of persons from the other states who did not find there opportunities equal to their capacities.

A plan is now under consideration to move the Brazilian capital. It is not a new idea. The Congress of 1891 included it among the articles of the constitution. Since then, the matter has been mentioned in every federal charter. Studies have been made to choose another place. A site has been more or less settled in the central plateau of Goiás, hundreds of miles from the coast. Much evidence has been adduced to show that Rio de Janeiro is the worst place for the national capital. Climate, strategic position, marginal situation, have been listed among its handicaps. The last government intensified the studies and made efforts to foster this change. Relocation cannot be expected, however, for the next ten years. It is not an easy task to take civil servants away from the easy life of Copacabana beach and the Avenida Rio Branco, and throw them into a pioneer city cut out of the jungle. These studies are a healthy symptom, nevertheless, and should be continued. At least they shed some light on the city organization itself, and cause the people of Rio to realize what they are sometimes prone to forget, that behind their backs a country still exists.

R

Rome

GIUSEPPE CHIARELLI

Degree in Jurisprudence, University of Rome, 1926; Assistant at Istituto Giuridico, University of Perugia, 1927; Acting Professor of Administrative Law and the Law of Labour Relations, University of Camerino, 1928; subsequently Professor of Public Law and the Law of Labour Relations in the Faculty of Political Science, University of Perugia, and also Director of Istituto Giuridico. Professor of Public Law in the Faculty of Economics and Commerce, University of Rome since 1942, and Dean of the Faculty, from 1952.

Has a legal practice in Rome and is particularly concerned with the Supreme Court and the Council of State. Has taken part in the reform of the civil code.

Author of: *La personalitá delle Associazioni Professionali*, (Padova, 1931); *Gli Organi di elaborazione del diritto del Lavoro* (1938); *Il pensiero giuridico italiano e i problemi attuali del diritto pubblico* (1943); *Istituzioni di Diritto Pubblico* (1944); also notes and articles in principal Italian legal periodicals and newspapers.

Rome

ON the left bank of the Tiber, at that time the only navigable river in the Italian peninsula, rose Ancient Rome, the city with a unique destiny. The coasts of Etruria and Latium, flat and lacking all natural harbours, offered no practical means of access to the interior of the region other than that of the Tiber and its tributaries. And on the Palatine, round the foot of which swirled the river, the primitive city arose, then spread over the seven hills and became the political and commercial focus of central Italy. Under Etruscan and Greek influence it grew in riches, and, having evolved a highly-developed legal structure, dominated first Italy and then the whole Mediterranean basin.

Favoured by its geographical position, Rome thus found itself at the centre of the ancient world and hence of Christendom. And when, after long and varied vicissitudes, modern Europe began to take shape, based on the principle of nationality, and Italy became a simple unified state, Rome, because of its geographical location and its past history, was the inevitable choice as capital.

Rome is situated almost half-way down the long narrow Italian peninsula, in the region of Lazio, which touches Tuscany and Umbria in the north, the Abruzzo and Molise in the east, the Campania in the south, and the Tirrhenian Sea on the west. Across these regions, the ancient consular roads and the modern Italian railway system converge on Rome.

THE HISTORY OF ROME

The history of Rome, a city so closely linked to the history of the whole Western world, is too well known to need more than a brief summary here. It may be useful to remember that classical historians now agree in placing the foundation of the city in the middle of the eighth century B.C. A tradition accepted since Imperial times and vouched for by Varrone, fixed the official date of the foundation at 753 B.C. On one day in that year, the circumference of *Roma Quadrata* was traced on the Palatine Hill; later the city was extended beyond the Cloaca Massima to the Quirinal Hill, including the populations of the ancient villages, united by the need to defend themselves from their enemies, especially from the Etruscans who lived on the right bank of the Tiber.

There was certainly a period of monarchic rule, of which no historical details are known, and this was followed, in the sixth century B.C., by the republican era. In the third century B.C., after the Samnite and Pyrrhic wars (280–272), Rome dominated the whole Italian peninsula, the peoples of which were in part under her direct power, in part federated with her. In the following century, with the conquest of Macedonia, of Greece, and of Pergamo, with the fall of Carthage in the Third Punic War (146 B.C.) and the expansion into Northern Italy and into Spain, Rome dominated the entire civilized world.

The events of these centuries left their mark on the face of the city, which alternated between building up its defences and embellishing its own interior. The so-called Servian walls were probably constructed after the invasion of the Gauls (390 B.C.), when Camillo rebuilt the monuments of the burnt-out city. During Silla's dictatorship (82–80 B.C.), the city was first systematically laid out and beautified with marble brought from Greece and the East; but Silla let loose civil war and so, via the triumvirate and the rule of the Caesars, the Imperial Age began.

Augustus (63 B.C.–A.D. 14) rebuilt and embellished Rome, as celebrated by Horace in the *Carmen Seculare*. To him was attributed the phrase: 'I found a city of bricks and I leave a city of marble.'

There followed the stormy vicissitudes of the Empire, the famous fire which broke out under Nero (A.D. 64) and which involved a rebuilding of the city, largely undertaken by Domitian (A.D. 51–96). Aurelian (A.D. 214–279) surrounded it with walls, which enclosed ten hills, as a defence against the barbarians. Finally, the period of Diocletian (Emperor from A.D. 284 to 305) and Constantine (Emperor from A.D. 311 to 337) marked the end of the ascendancy of the city under the Empire. Then the decline set in.

When Diocletian established the tetrarchy (A.D. 291), or government by four sovereigns (two Augusti and two Caesars), Rome still remained the moral capital of the Empire, but the seats of government were fixed in cities nearer the boundaries (Treviri, Milan, Sirmio and Nicodemia), and Diocletian himself, who as *senior Augustus* had supreme authority, set up his government in the last-named of these cities. Constantine later proclaimed Byzantium to be the capital of the Empire (A.D. 331), with the result that Rome and Italy became of secondary importance and were the more easily conquered by the barbarians from the north. The Emperor Theodosius, shortly before his death (A.D. 395), divided the Empire into an Eastern and a Western Empire, and under Honorius, Ravenna was chosen as the capital in the West. In A.D. 410, Rome was captured and sacked by Alaric, and there followed the long series of barbaric invasions, sieges, conquests, devastations.

The Empire of the West having broken up, Rome became a dukedom dependent on Byzantium, and fell to the position of a provincial city. But at the same time, the influence of the Bishop of Rome was becoming stronger.

Leo I (A.D. 400–461), the Pope who affirmed the primacy of the Bishop of Rome over Christendom, had already been the defender of Rome against Attila; and thus the destiny of Rome became united with the destiny of the papacy. At the end of the sixth century began the resurrection of Rome, which started with its independence from Byzantium. With the gift by the Longobard king Liutprand of the city of Sutri to the Pope (A.D. 727), began the temporal dominion of the church. The king of the Francs, Pepin, then gave the dukedom of Rome to the papacy, thus freeing it from the Empire of the East.

It is well known that on Christmas Day 800, the Pope crowned the Emperor Charlemagne in St. Peter's, so setting up the Holy Roman Empire. But in the succeeding century interminable struggles developed between supporters and adversaries of the former and of the German Empire, and Rome fell into feudal anarchy. During these struggles ancient monuments were destroyed or transformed into towers and fortified. At the head of the administration of the city in this period there was sometimes a senator, at others a patrician.

In 1143 the people re-established the senate, and the commune arose, with the title of Roman Republic, the council of which had its seat on the ruins of the Campidoglio. But the communal autonomy was of brief duration, and ended shortly after the condemnation to death of Arnold of Brescia (1155), the friar who had upheld it against the papacy. During the papacy of Innocent III (1198–1216) a single senator was nominated as head of the city, under the Pope. In 1266, the office of senator was conferred by the Pope on Charles D'Anjou.

During the time the papacy was at Avignon (1309–1377), there occurred the brief adventure of Cola di Rienzo, who was proclaimed tribune of the people in 1347 and then killed by the people. Infatuated with classical culture, he tried in vain to raise up again the destiny of Rome.

After the return of the popes from Avignon, the troubles continued and there raged the schism of the West. In 1398 Pope Boniface IX abolished the Roman republic.

The schism having ended with the election of Martin V at the Council of Constance in 1477, the new Pope returned to Rome to find it almost deserted and uninhabited. The Roman forum had become pasture ground for cows, from which came the name 'Campo Vaccino' (Cows' Field). But the popes who succeeded Martin V undertook the reconstruction of Rome, which was reborn as a Renaissance city and knew the splendours of the century of Leo X.

In 1527, when Rome was sacked by the Landsknechts mercenaries of Charles V, the buildings of the city were spared; but the population, which during the papacy of Leo X had risen to some 60,000 inhabitants, subsequently fell to 54,000, and then, as the tragic result of the sack of Rome, to 33,000.

After the Council of Trent (1545–63), Rome, as centre of the counter-reformation and movement for restoring the power of the papacy, acquired the appearance which it retains to this day, with its characteristic baroque architecture.

In the Napoleonic period, Rome was occupied for the first time in 1798 by French troops. Pope Pius VI was arrested and the city was proclaimed a republic. This, however, lasted only a short time because of the success of the anti-French coalition. In 1809 Napoleon incorporated Rome in his empire and the Napoleonic government lasted until 1814; but the foreign domination was unpopular with the people.

After the restoration, the pontifical government continued in Rome until 1870, except for the glorious interlude of the Roman republic of 1849, under a triumvirate including Giuseppe Mazzini. On March 27, 1861, the Parliament of the new Kingdom of Italy, sitting in Turin, proclaimed Rome capital of Italy, but it was only on September 20, 1870, that Rome was occupied by troops of the kingdom, to which it was annexed by the plebiscite of October 9, 1870. Thus ended the temporal power of the Pope, whose position and independence were regulated by a special law (*Legge delle Guarentigie*) which he himself, however, did not recognize. The relations of the Italian state with the pontiff were later improved by the Lateran pacts of 1929.

At the time when Rome became part of the kingdom of Italy, it was administered by a senator and three conservators. With the annexation, elective bodies were placed in charge of administration in conformity with the constitutional régime of the state.

This brief historical survey suggests certain general considerations. Although legend has always exalted the humble origins of Rome, and the poets of the Augustan Age used to contrast the Greater Rome of their days with the *parva urbs* of Romulus,[1] in actual fact Rome from the

[1] Cf. *Properzius:* Elegies, IV, 1.

53 · ROME

The Piazza del Popolo, showing some of the main arteries.

very first showed herself to be the political centre of a community subordinate to her power, unlike the other European capitals, which were originally borough towns, gradually developing in size and authority, while around them, more or less synchronizing with their growth, the unitary national state was being formed. Rome represented in the ancient world the most important form of city state, enlarging its dominion to the outermost confines of contemporary civilization. And when, with the end of the republic and the disappearance of the ancient liberties, the original political structure was transformed, Rome still remained the sumptuous centre of an immense political organism, the Empire.

With the fall of the Empire and the disappearance of the Western Empire, the political function of Rome decayed and she became first a provincial city and then the capital of a small territorial state. Yet at the same time she was the seat of one great institution, the Catholic Church, which from this centre radiated a world-wide influence, thus impressing on her the distinguishing signs of her new function in the world.

It is only within recent times, during the second half of the last century, that Rome became the capital of a unified national state, that is to say, of a state in which were united regions which possessed certain national characteristics in common but which had for many centuries subsisted under different forms of government and had undergone a different economic and social development.

These events of ancient history, the survival through the centuries of her function as a centre of radically different political organizations, and her only recent assumption of the position of capital city of a unitary national state—all go to explain the particular political, administrative and social position of present-day Rome, and are at the root of the problems she now has to face.

THE ECONOMIC AND DEMOGRAPHIC BASIS OF ROME

From an economic point of view Rome cannot be called an industrial city. Civil servants form the largest group of its population, and administration rather than production is characteristic of the city's life. Among the business activities carried on in Rome, medium and small concerns predominate, covering a vast and varied range of production. Commerce plays a greater part than industry; the building industry is important, and also the various subsidiary branches connected with it, such as cement works and brick works, as well as food production, printing, etc. Special

R*

mention may be made of the Government printing works known as the *Poligrafico dello Stato*, the most important concern of its kind in Italy, which also does some unofficial printing as well.[1]

Many companies have their headquarters in Rome, but carry on their industrial undertakings in other parts of the country, especially in the north of Italy. A samples fair takes place in Rome each year, organized by a special corporation, with the aim of extending the capital's economic activities and making it a co-ordinating centre for the economy of both Northern and Southern Italy. It must also be remembered that Rome is the most important centre of consumption in the whole country.

When Rome became part of the Italian state in 1870, it had a population of 226,022 inhabitants, a figure which had grown by 70,000 since the beginning of the century. On December 31, 1871, when Rome had been set up as the capital of the kingdom, the population had risen to 244,484. Since then there has been a steady increase, with a sharp rise in recent decades, as shown by the following figures. In 1914, the population was approximately 580,000; in April 1931, it had risen to over one million, and by December 31, 1940, to 1,368,440. By 1949 the population had risen to 1,649,684.[2] According to the census figures of November 1951, the population of Rome was 1,695,477, and the population as at April 30, 1953, had increased to 1,735,354.

---

[1] According to provisional figures (not yet available in full detail) provided by the latest census (of November 5, 1951), business and industrial concerns within the Commune of Rome were distributed as follows:

	No. of concerns	No. of employees
Industry	16,668	137,173
Transport and communications	1,187	47,573
Trade, banking, insurance, and various other services	28,306	106,809
	46,161	291,555

These figures do not include concerns whose domicile in Rome is purely legal—in other words, the many factories and companies which merely maintain the headquarters of their administration in Rome but which carry on their actual business activities elsewhere, retaining no factory or shop or other place of business in the capital. The number of firms and companies registered with the Rome Chamber of Commerce is very much higher (c. 90,000).

[2] For the decade 1940–49, the following are the official figures:

1940	1,368,442	1945	1,495,867
1941	1,415,890	1946	1,550,817
1942	1,471,971	1947	1,591,163
1943	1,501,116	1948	1,626,635
1944	1,497,673	1949	1,649,684

The increase was due to two main factors: the natural excess of births over deaths (e.g. in the six months January–June 1950, this excess was 6,916), and above all to immigration into the city (in the same six-months' period the municipal authorities registered 18,946 immigrants from other communes or from foreign countries, as against 9,879 emigrants). Official data for the period January 1, 1939–December 31, 1949, show a natural increase, by excess of births over deaths, of 160,819, and an increase through immigration of 240,481. But the effective immigration has been estimated at about 400,000, so that during these eleven years, the total real increase in population must have been at least half a million.

In order to emphasize the striking nature of these facts, we may note that during the same period, the city of Milan had a natural increase of population of 24,825, and an immigration of 76,443, while the population of Naples grew by 135,674, almost entirely as a result of the net excess of births over deaths. This sharp rise in the population of Rome occurred in the Second World War and immediate post-war years, when building was almost at a standstill. The housing crisis has therefore become particularly acute in Rome: we shall discuss later the efforts being made to deal with this problem.

The whole territory of the commune covers 150,760.54 hectares, which is about five times that of the Commune of Milan and is indeed greater than that of entire provinces, such as those of Naples, La Spezia, Pistoia.

In 1871, the city, contained within its ancient walls, was divided into 14 *rioni* or districts, as it had been in the days of the Emperor Augustus. Inevitably the outer districts expanded both in population and in buildings, till in 1911 the area outside the walls was divided into *quartieri*.

At present the communal territory is divided into 21 *rioni*, covering 1,568 hectares, 18 *quartieri* covering 10,354 hectares, fifteen suburbs, covering 8,770 hectares. The rest of the communal territory consists of the *Agro Romano*, which extends over 129,888.54 hectares and includes the Lido of Ostia.

This latter, which was formerly called the Lido of Rome, is tending to develop from a purely holiday resort into a residential area with a stable population. Before the last war, the winter population of Ostia Lido was only 5,000 persons; today, as a result of the housing crisis, that figure has been trebled, and the town planning project for the Lido provides for some 135,000 regular inhabitants.

By the Lateran treaties of 1929, Italy recognized the absolute and exclusive jurisdiction and sovereignty of the Holy See over the Vatican, together with all its adjuncts, properties and endowments. Thus was created the Vatican City. It was, however, agreed that St. Peter's Square, although forming part of Vatican City, should normally remain open to the public and be subject to the police jurisdiction of the Italian state. Vatican City covers an area of 0·44 square kilometres, and according to the census of December 1932, it had a population of 1,025 persons, 853 of whom were Italian nationals.

THE GOVERNMENT AND ADMINISTRATION OF ROME

The administration of the city of Rome today is practically identical with that of the other Italian communes. The principle of uniformity in local government has existed in Italy since the time when, following the unification of the nation, the whole system of public administration was re-designed. The law of 1865 dealing with communes and provinces established a uniform type of administration for all the communes including the capital city, which at that time was Florence.

The autonomous statutory powers of the medieval communes had disappeared some centuries previously and, even in the states which were united into the Kingdom of Italy, the communes were administered under state laws. One of the first tasks of the new kingdom was to give a uniform administrative pattern to the whole country, thus replacing the laws of the now vanished states. The French system was adopted as a model, with the one important difference that no special provisions were laid down for the capital city. Italy thus became one of the few countries in which not only was there no distinction between urban and rural, or large and small, communes, but also none between provincial capitals and smaller towns, while the capital city itself was treated exactly like any other commune.

When the capital was transferred to Rome in 1871, it was soon realized that the metropolis had particular problems of its own—not only the problems appertaining to every capital city which is the seat of the main organs of a nation and of all foreign representations to it, the centre of the national administration, but also problems peculiar to the city of Rome. In fact, the city was singularly ill-prepared to play its new role, especially as regards building and housing accommodation. It needed considerable adaptation, and this was rendered exceptionally complicated by its historical character and its wealth of ancient monuments. Furthermore, in addition to its new function as capital of the Kingdom of Italy, Rome remained the seat of the pontiff of the Roman Catholic Church, of the Roman Curia, and of the diplomatic representation to the Holy See.

Politicians and experts argued, therefore, the need for special legislation to give Rome a type of administration that would meet her requirements. But all that could be obtained over a number of years were certain economic provisions regulating national financial help for the city's expenses.

Fascism did give Rome a special administrative status. A law of October 1925 created a governor of Rome, the functions and powers belonging to this position being reaffirmed by the Law on Communes and Provinces of March 1934. The governor was nominated by royal decree, on the proposal of the Ministry of the Interior, following a cabinet decision. At that period, the heads of all other communes (called *podestà*) were also nominated by royal decree, but the governor of Rome was distinguished from the other local leaders by being regarded as a state official, forming part of the hierarchy of the state. And whereas the term of office of the *podestà* was four years, there was no limit to the appointment of the governor of Rome. In addition to the governor, there were two deputy-governors, and a *consulta* (a purely consultative body) comprising twelve members. During eighteen years there were five governors.

After the fall of Fascism, the governorship of Rome was abolished by decree in November 1944, and the elective system reinstated for all municipal authorities. The same decree laid down certain special provisions for the administration of Rome. In regard to any matters not specified in this decree, the 1915 law covering communes and provinces was restored partially after the end of Fascism, and all other laws concerning local government were operative.

The government of Rome is thus in a transitional stage, regulations being contained in a number of different legal texts, some of which are only partially operative. All these provisions need to be co-ordinated in a single comprehensive law, and at the same time a general administrative reform considered. The increase of population in recent years, with the resulting social problems, combined with the expansion of governmental and business activity now concentrated in the city, make it even more essential that Rome should be treated as a special case and that special legislation be designed to meet her needs. Draft legislation to this effect is now being prepared.[1]

---

[1] At the present time, the fundamental laws which govern the administration of the Commune of Rome are the following: (*a*) Legislative Decree, No. 426, of November 17, 1944, for the suppression of the Governorship of Rome and the organization of the administration of the Commune of Rome; (*b*) Communal and Provincial Law of February 4, 1915, No. 148, and subsequent modifications; (*c*) Communal and Provincial Law of March 3, 1934, No. 1265, in those parts which are still in force; (*d*) Law No. 84

At present the organization of the Commune of Rome is as follows: all provisions regarding communes in Italy and their relations with the central government are based on the principle of autarchy, which is part of the more general principle of local autonomy sanctioned by Article 5 of the constitution. By autarchy is meant, in general, the self-governing powers of the municipal corporations. This concept, therefore, covers (*a*) a collection of administrative powers attaching to the autarchic corporations; (*b*) organs of self-government, usually elective, which exercise these powers; and (*c*) various economic means for securing the interests of the community and carrying out all necessary public services. The autarchic corporations are subject to control by the state in such measure and degree as is laid down by law.

The organs of self-government consist of (i) the communal (or municipal) council; (ii) the municipal *giunta*; (iii) the mayor.

THE COMMUNAL COUNCIL

(i) The Communal Council is elected by universal suffrage, voting being by party lists. The system of voting, in communes having over 10,000 inhabitants, is that known as 'proportional representation with a majority premium.'[1] The electors are all persons over twenty-one years

of February 24, 1951, on communal elections. There are also the regulations made by the Commune itself for implementing the above mentioned laws and for its own internal organization, among which should be noted: (*a*) Regulation on the Communal Council of Rome, approved after discussion with the said Council, March 27, 1950; (*b*) General Regulation, 1932, on the personnel of the city offices and utilities.

[1] This system was introduced by the Electoral Law of February 24, 1951, which amended the former system of voting by pure proportional representation. Under the existing system, several lists of candidates can be presented for election; in communes with a population of over 500,000, the lists must be presented by at least 500 electors. Lists can be 'related' or 'linked' to one another, in order to obtain a single 'electoral figure' in the allocation of seats; such grouping of lists must be announced on the occasion when the lists are formally presented. The 'electoral figure' is secured by the number of votes gained by each list or group of lists.

To the list or group of lists which secures the highest number of votes, or 'electoral figure,' is assigned two-thirds of the vacant seats; in the case of Rome, this means—in round figures—fifty-three seats at the present time (June 1953). The remaining seats are distributed among the minority lists or groups of lists, on a proportional basis (d'Hondt method). If the majority list or group of lists gains more than two-thirds of the total valid votes cast, seats are allocated to all the lists or groups of lists by the above-mentioned proportional method.

When voting for a list, the elector can indicate his preference for certain candidates on the list. Where, as in Rome, there are eighty councillors to be elected, the elector can indicate five preferences. The number of preference votes given to a candidate, added to the 'electoral figure' of his list, compose the 'individual figure' for each candidate. The of age, of Italian nationality, and domiciled within the municipality, who are not disqualified owing to physical or legal incapacity. Any citizen whose name has been entered on the electoral register of any commune is eligible for election to the council of his own or any other commune (except for a very few situations of 'incompatibility').

The council is elected for four years. The office of councillor is unpaid. The council is the basic organ of communal government, institutionally vested with deliberative functions. The Municipal Council of Rome (which in this resembles all other cities with a population of more than half a million) is composed of eighty members.

The main subjects falling within the purview of the council are: (*a*) general regulation of the municipal offices and salaries of the employees; (*b*) personnel questions (appointment, dismissals, and suspension of employees); (*c*) property and capital (purchase, sale, investments); (*d*) use of municipal properties; (*e*) legal matters; (*f*) public works; (*g*) rates, taxes, excise; (*h*) markets and fairs; (*i*) budget; (*j*) regulations concerning public hygiene, municipal and county police, building, etc.

In addition the council exercises supervisory powers over all institutions set up for the benefit of the citizens and over the budgets of all administrative authorities that receive subsidies from the commune.

It elects from among its own members the *giunta* and the mayor.

seats gained by each list are therefore assigned to the individual candidates on that list on the basis of their 'individual figures.'

The aim of this complicated system, which attempts to combine the systems of majority vote and proportional representation, is, on the one hand, to ensure the formation of a homogeneous and stable majority, in a position to implement its own programme and administer the public interests, while at the same time providing, on the other hand, the necessary stimulus and control through the minority groups.

This system was put into operation in the 1952 elections and produced the following results:

	Seats		Seats
Democrazia Cristiana	39	Partito Socialista Democratico (Social Democrats)	4
Partito Communista	8	Partito Republicano	3
Movimento Sociale Italiano (Neo-Fascist)	8	Partito Monarchico	3
Partito Liberale	7	Left Independents	3
Partito Socialista	4	Fronte Economico	1

The Democrazia Cristiana, Partito Liberale Italiana, Partito Socialista Democratico and the Partito Republicano were linked together, while a single list was also presented by the parties of the left (Partito Communista, Partito Socialista and Left Independents).

The administration of the city is thus principally in the hands of the Christian Democrats, who are also in control of the central government.

In the present Council there are 74 men and 6 women.

THE EXECUTIVE COMMITTEE

(ii) The Municipal *giunta* is composed of the mayor and the 'assessors,' who comprise twelve full members and two substitute members, in communes of more than 500,000 inhabitants.

The *giunta* is in the nature of a committee of the council and represents the latter in the intervals between meetings of the council. The *giunta* (*a*) has preparatory functions before a council meeting and executive functions following such a meeting, (*b*) assists the mayor in the discharge of his duties as head of the commune and government official, and (*c*) has its own deliberative functions in matters of minor importance. It fixes the dates for meetings of the council, submits drafts of the municipal budget and by-laws, and supervises the regular functioning of municipal services. At the suggestion of the mayor, it can appoint and dismiss municipal employees, draw up the tax tables, fix the charges of certain public services (e.g. taxis and carriages for hire), and discuss certain matters to be brought into court.

In addition, the *giunta* can exercise the functions of the council by delegation, within the limits allowed by law, and in case of emergency it can assume on its own responsibility all the powers of the council, which body must later ratify any action taken. It submits an annual report on its activities to the council.

Each assessor, in addition to taking part in the collective work of the *giunta*, is in charge of one particular branch of the municipal administration.

THE MAYOR

(iii) The mayor is the head of the commune and represents it on all official occasions, but his activity is subordinated to the decisions of the council and the *giunta*. He has also certain functions as a Government official.

As head of the commune his functions are:

(*a*) preparation for and direction of the work of the council and the *giunta*; (*b*) execution of their decisions and of those matters that by law are attributed to him; (*c*) supervision of municipal offices and institutions; (*d*) legal representation of the city in respect of third parties; (*e*) certain powers of certification and registration. The mayor summons and presides over the meetings of council and *giunta*, proposes the items for discussion, distributes the items on which the *giunta* must make decisions among the various members of that body, and sees that decisions are

54. ROME
The University City.

55 . ROME
The Forum.

[Photo: B.O.A.C.

56. ROME

The river Tiber winding through the city.

[*Photo: B.O.A.C.*]

57 · ROME

St. Peter's, showing the basilica. On the left are the Vatican palace and gardens.

implemented; he supervises all municipal offices and institutions (if necessary, he can suspend from duty any municipal employee); he signs all contracts, represents the commune in all legal matters and in court, is present at public auction sales of communal property, issues certificates, attests signatures, etc.

As an official of the central government, he is responsible for publishing the laws of the country and Government proclamations within the commune; he is responsible for maintaining public order, safety and public health; he keeps the civil registers and population registers, and can celebrate a civil marriage.[1] He can also take action, in case of urgent need, in matters of local police, housing and public health.

The mayor can be removed from office in the following instances: (i) on the proposal of the municipal council, if at least one-third of the councillors or the prefect of the province have submitted a request which has been approved by two-thirds of the council; (ii) if he suffers any legal penalties that restrict his personal liberty for more than one month, or if such penalties should lead to a condition of ineligibility; (iii) if he is removed by the Government, by decree of the head of the state, for grave reasons of public order, or when, having been formally reminded of his legal obligations, he persists in non-compliance. Parliament must be informed of any such removal from office of the mayor. As may be expected, the Government uses this power sparingly and only in cases of extreme gravity. One such celebrated occasion was the removal, by the Di Rudini government, of Prince Torlonia from the office of Mayor of Rome, because he had paid an official visit to the cardinal vicar of the city, at a time when the conflict between the Italian state and the Holy See was still acute. This action, which reflected the particular bitterness of the conflict at that time, now purely an historical memory, was undoubtedly an arbitrary exercise of power, since the cardinal vicar formally represents the Bishop of Rome and not the Holy See.

The mayor and the *assessori* receive an honorarium, determined by the Ministry of the Interior but levied on the municipal budget.

The communal council has set up the same number of standing committees as there are *assessori*; each committee has eight members,

[1] The Concordat of 1929 between the Holy See and the Italian state, and the law of that same year on religious cults in Italy, provides that a marriage celebrated according to canonical law before a Roman Catholic priest or an authorized minister of any other religious cult recognized by the state, takes effect in civil law. The mayor, therefore, as a state official, celebrates only civil marriages between persons who do not intend to contract a religious marriage; and he is also responsible for the correct registration in the state records of marriages celebrated before a minister of religion.

elected by the council and in proportion to the representation of the political parties on the council, with either the mayor or one of the *assessori* acting as chairman. The committees examine all proposals submitted to them by mayor, *giunta* or council, and report to the council. These committees have only recently been set up and have not yet begun to function; they should be able to lighten and speed-up the work of the council.

The seat of the representative bodies of the commune is the *Campidoglio*. On the site of the ancient Acropolis (*Arx Capitolina*), this has remained through the centuries the seat of the senate of the city, and the meetings of the communal council are still held in the senatorial palace.

The paid officials of the city consist of the secretary-general and the employees. The secretary-general assists the mayor, the council and the *giunta* in their deliberations; he is the head of all the personnel employed by the commune, and has supervisory powers over all the communal offices. The employees (administrative, technical, sanitary and teaching personnel) are divided into three groups (A, B and C) in accordance with their qualifications and duties, and are classified in twelve grades.

The appointment of the secretary-general and of the permanent employees is made as a result of public competition.

At the present time, the permanent and temporary employees of the commune number about 15,000, which is almost double the figure of 1932.

THE EXECUTIVE DEPARTMENTS

The municipal offices, all controlled by the secretary-general, are organized in thirteen departments. Each department is concerned with one particular public service and one branch of the administration is normally assigned to each assessor. The departments are then organized into divisions and sections. The present scheme is as follows: (i) staff; (ii) land, buildings, municipal property, etc.; (iii) rates and taxes; (iv) register of births, deaths, marriages, issue of identity cards and other certificates, electoral lists, military service lists; (v) public works; (vi) technological services and municipal control of services; (vii) city police, district sub-offices; (viii) sanitation and public health; (ix) schools, libraries, public assistance and welfare; (x) historical monuments and fine arts; (xi) food inspection, food controls, public markets; (xii) equipment and supplies; (xiii) Rome Lido and *Agro Romano*.

In addition there are the following offices: (*a*) the mayor's office; (*b*) the general secretariat which ensures technical co-ordination between the various offices; (*c*) the accountant-general's office; (*d*) the solicitor's office, dealing with all legal matters.

Certain offices that do not fall under any of these heads deal with special services (e.g. public cleansing, public gardens, special office for the outlying suburbs).

In the territory of the Commune of Rome there are fifteen municipal sub-offices, called *delegazioni*. They are designed, mainly for the convenience of the public, to deal with simple activities such as issuing certificates, in the various areas where they are situated. There is a project now under discussion by which the activities of these offices will be largely increased, in order to obtain a greater decentralization of municipal functions. Of the fifteen *delegazioni* there are at present twelve urban and three rural; the new project would raise these figures to eighteen and six respectively.

MUNICIPAL TRADING SERVICES

According to Italian law, communes may acquire the plant and equipment, and assume control of public utility services, in addition to those which by law they are obliged to provide. Normally, the taking over of such services is intended to guarantee their efficient functioning in response to the public needs: in economic terms, to achieve a social or public utility (by offering the service at a lower price, extending its range and distribution, etc.) which would not be effected, or only in small measure, if the service were wholly left to private initiative. Only a few services specified by law can be taken over by the commune to the total exclusion of all private competition (slaughter-houses, markets, street hoardings and publicity). The others are undertaken in competition with private concerns, except when in hard fact a monopoly has developed. Public services taken over by the commune can be managed directly by the communal administration, if they are relatively unimportant or are not of an industrial character; otherwise they can be carried on as commercial ventures or as municipal business. A municipal business *Azienda municipalizzata* is separate from the general municipal administration. It has its own executive organs (a general manager and a governing board nominated by the commune) and its own budget, and it may take all necessary action to achieve its purposes. The city council has certain supervisory powers over a municipal business. According to the law, the profits accruing can be divided, the one part between municipality, manager and staff, and the other part to form a sinking fund, according to the rules governing the particular business. Alternatively, after deducting the amounts required for maintenance, extension and improvement of services and for the lowering of the tariffs, the profits may go to the municipality. The city budget meets any losses that may be incurred.

In the city of Rome, the following municipal businesses are administered on these lines: the water and electricity company (A.C.E.A.), and the tram and bus transport company (A.T.A.C.). In practice these companies in recent years have registered an annual deficit.

There are also municipal undertakings running trading services on a commercial basis, such as the central milk station, the funeral services, the zoo, the poster publicity organization, etc.

Finally, there are corporations having legal personality for the administration of municipal property, such as the Opera House (*Ente Autonomo del Teatro dell'Opera*) and there are private companies carrying out public services and financially controlled by the municipality, e.g. the STEFER Company, which runs transport services between Rome and the *Castelli Romani* hill towns; a majority of the shares in this company are owned by the city of Rome and the mayor is chairman of the company.

THE SOCIAL AND PUBLIC SERVICES

The main public services run directly or indirectly by the Commune of Rome are the following:

(*a*) *Police.* These consist of the *Vigili Urbani* and the *Polizia Metropolitana*.

(*b*) *City Cleansing.* This is administered directly by the municipality in the central zone of the city and given out on contract in the suburban areas.

(*c*) *Food Supply.* The sale of produce is carried out by means of the general markets and the district retail markets. The former, for the sale in bulk of fish, meat, poultry, fruit, etc., are organized on a sale on commission basis. Moreover, as in all communes with a population of more than 200,000, there is an *Ente Communale di Consumo* which acts as a non-profit making wholesaler. This is an autonomous body directed by a commission nominated by the city council.

(*d*) *Public Health and Sanitation.* Health services for the poor, who are specially registered for this purpose, are run by dividing the area into districts (*condotti*) to each one of which a medical officer is attached. Public sanitary conditions are controlled by the public health office, through its health inspectors, all-night emergency service, chemical and bacteriological laboratories, etc. For the purposes of the anti-tuberculosis campaign, the municipality participates in the provincial anti-tuberculosis authority and has one tuberculosis institute, four dispensaries, one seaside hospital and permanent residential preventive centres. An anti-malarial prophylactic service is attached to all the suburban medical services, as well as to those in the *Agro Romano*, and in addition there is an anti-malaria centre.

(*e*) *Transport*. For transport purposes, the city is divided into three zones: a central zone, surrounded by a circular tram route, and served by diametrical omnibus lines; an outside circular zone, with radial tram, bus and trolley-bus lines; and a suburban zone, served by special bus or extended tram lines. In addition, there are suburban bus, tram, and rail services which link Rome with the surrounding countryside. These transport services are run by the organizations mentioned above.

(*f*) *Other Public Utility Services*. Electric light and power is provided by a municipal company (A.C.E.A.) and by a private company (*Società Romana di Elettricità*). Recently the commune has acquired rights in the *Società Imprese Centro Italia*, which is setting up plants for the utilization of the water of the Tiber for the production of electrical energy. In this way a large part of the electricity produced by the power stations so formed will be used by Rome. Drinking water is provided by the above-mentioned A.C.E.A. and by the Acqua Marcia Company, the latter holding a concession from the municipality. With the construction of the Peschiera aqueduct (undertaken during the period 1936–44 and not yet completed), the supply of drinking water in the city has increased by 1,500 litres per second and will finally be increased by 4,000 litres per second. Gas is supplied by the *Società Italiana Gas* (formerly the Anglo-Romano Company) on a concessionary basis. In January 1949 the city council decided to revoke the concession, but the issue is now in litigation before the Council of State (*Consiglio di Stato*).

(*g*) *Education*. By law, the commune is obliged to make certain contributions, at its own expense, to the state schools: that is, it must supply the premises for elementary and secondary schools, maintain them, and provide them with heating, light and water. It must contribute to the state university and the *Patronato Scolastico* which is an organization for the assistance of elementary school pupils. It can, beyond this, contribute to teaching and educational institutions and can run its own schools at its own expense.[1] The city of Rome also runs open-air schools

[1] In Italy, teaching is undertaken by the state as a public service, but private individuals have the right to open schools and undertake teaching (Art. 33 of the Constitution). In other words, the state undertakes teaching not under conditions of monopoly but of free

and a number of technical and professional schools. These include evening classes in arts and crafts, in science, public health and maternity welfare; training schools for gardeners, etc.

(h) *Public Assistance and Welfare*. In every commune, public assistance to individuals and families in necessitous conditions is entrusted to an appropriate body, called the *Ente Communale di Assistenza* (municipal assistance board). This board has its own property, and the assistance and services it renders are provided by the income on this property, by a special surtax, and by gifts from the province, commune and certain public and private organizations. It is administered by a committee, nominated, in the case of Rome, by the Minister of the Interior, and presided over by the mayor.

(i) *Hospital Service*. Hospital services in Rome are provided by the *Ospedali Riuniti* (joint hospitals board), which form an administration of their own, comprising nine hospitals, containing about 10,000 beds. This is under the Ministry of the Interior and is controlled by a special commission. It enjoys financial support from the state and from the commune. The charges for poor patients are paid by the state, in contrast to the other communes, where the commune itself has to pay these expenses. Reorganization of the *Ospedali Riuniti* is at the present time being considered. The health services of Rome are also carried out by institutions of social insurance, apart, of course, from nursing homes and private doctors.

FINANCING THE CITY

One aspect of municipal 'autarchy' is financial autonomy. It is shown most clearly in the power of the commune to hold property and to levy and collect taxes. The commune controls its own income and expenditure, in conformity with the law and under the supervision of the state, which, in the case of the city of Rome, is directly exercised by the Minister of the Interior, acting in agreement with the Minister of the Treasury and the Minister of Finance.[1] The state can subsidize municipal expenditure

competition, so that there are private schools as well as state schools (*scuole governative*). The private schools can also be set up and run by public organizations such as the commune, and can be recognized by the state. The recognition implies that the school conforms to the general regulations established by law, and the teaching follows the syllabuses in those regulations; the pupils are treated in the same way as pupils in state schools, and the course of study they follow has legal validity. The state examinations are prescribed on leaving school.

[1] In Italy, the Ministry of Finance and the Treasury are separate. However, the distinction between their respective spheres of action does not follow any rigid logical rule. The Ministry of Finance deals with the administration of taxes, state property and state

and, through extraordinary financial provisions, meet possible budget deficits. Such intervention by the state is particularly necessary in the Commune of Rome, owing to the fact that, in fulfilling its function as a capital city, it incurs expenditure that is, in reality, national rather than local in character.

Municipal income consists of: (*a*) income from property belonging to the commune; (*b*) rates and taxes; (*c*) profits from fee-paying public services; (*d*) grants and subventions, both regular and extraordinary, from the state or from other bodies.

The Commune of Rome has recently decided to set up forty special commissions (*Consulte Tributarie*), nominated by the communal council, to collaborate with the municipal offices in the preparation of lists of taxpayers and in working out their taxable income and property.

Municipal expenditure is either obligatory or optional. The former category covers all outlays on communal property, general overheads, the cost of the local police force, public health and sanitation, as well as certain public services the costs of which are partially borne by the municipality, such as the maintenance of law and order, public works, education, agriculture, public assistance and welfare. All other expenditure is optional. It must, however, always be intended for a public utility purpose or have a public service character, subject to specific controls and limitations exercised, in the case of Rome, by the Minister of the Interior in agreement with the Minister of the Treasury, if the local taxes exceed certain specified limits.

The municipal financial estimates are passed by the city council in the autumn preceding the year to which they apply; they must then be approved by the Minister of the Interior, in agreement with the Ministers of Finance and Treasury. This is peculiar to the Commune of Rome, since in the case of all other communes the budget goes for approval to the provincial administrative *giunta*.

The city treasurer levies all rates and taxes, and pays all expenses; the accounts must be ready for inspection three months after the close of the period to which they refer. They are first submitted to the auditors nominated by the city council and subsequently examined by the council itself, in conjunction with the auditors' report, and that of the *giunta*. The council's report is then made public, in order that any individual

monopolies, and customs duties; the Treasury is responsible for national income and expenditure, the administration of the public debt, the supply of materials to Government offices, war pensions, etc. The Treasury was set up in 1877, and in 1922 the two ministries were combined into one. But they were separated again in 1944.

ratepayer may make observations or raise objections, and finally the rectified balance is approved by the Minister of the Interior. Appeal against the decision of the Minister lies with the *Corte dei Conti*, which is the ultimate controlling body in Italy in financial matters.

Municipal property consists of both landed property and real estate. The former is normally reserved for public use and is inalienable (streets, public spaces, markets, etc.); the latter is subject to the customary legal controls on property ownership.

Real-estate owned by the municipality of Rome consists of (a) buildings, divided into about 3,000 units, each one of which is let on lease or on concession, and (b) communal lands, extending over about nine million square metres. A portion of these lands is utilized for communal services and therefore brings in no return.

Rates and taxes form the major part of the municipal income, as is the case in all communes. The Commune of Rome levies taxes today on rent receipts, on industry, trade, professions, on carriages and taxis, on domestic servants and dogs, on pianofortes and billiard tables, on 'Espresso'-type of coffee-making machines,[1] etc., in addition to certain sumptuary taxes, a hearth-tax, and general consumption taxes. There are also taxes on public services (e.g. school fees, stamp duty on official documents and certificates, payments for the occupation of public spaces, etc.) and advertising taxes (on posters, billboards, etc.). The city also receives a portion of certain state taxes, such as the entertainments tax, and levies certain surcharges over and above the state taxes on land and buildings and receipts from landed property.

All these rates and taxes, superimposed on state and provincial taxes, create an extremely heavy tax burden for the individual citizen of Rome.

At present, the municipal rates and taxes are levied at the maximum rates permitted by law; in other words, the city is using its taxing powers to their fullest extent. Nevertheless, the city has to meet such heavy expenses that periodic grants-in-aid from the state are necessary.

Finally, by the law No. 103 of February 28, 1953, the Commune of Rome has been granted an annual contribution of three milliard lire for the years 1952, 1953 and 1954, as state recompense for the burdens which are borne by the commune because it is the seat of the capital of the Republic. Moreover, loans to the Commune of Rome, amounting to a total of 55 milliard lire (that is, 11 milliard a year, from 1953 onwards) have been authorized to finance certain public works.

---

[1] Coffee-making machines of this type are widely used in Italy in bars, cafés and restaurants.—Ed.

One can get a more concrete idea of the actual financial situation of the Commune of Rome from this brief statement of the provisional balance for 1953:

TABLE I

Income and Expenditure, 1953

	Italian lire
EFFECTIVE INCOME	
Capital receipts	1,000,000,000
Income from various sources	2,079,780,508
Government and private subsidies ..	743,652,300
Rates and taxes	18,318,000,000
	22,141,432,808
MOVEMENT OF CAPITAL	
Sale of properties	600,372,000
Credit balances, legacies, etc... ..	48,311,664
Passive loans	681,000,000
	23,471,116,472

EFFECTIVE OUTGOINGS		
Upkeep of property	6,312,840,000	
General expenses	12,713,354,000	
Local police and hygiene	7,314,696,000	
Public security and justice	454,935,000	
Public works	6,205,886,073	
Education	853,775,000	
Fine arts and monuments	638,950,000	
Religion	100,000	
Welfare and assistance	1,034,668,000	
Unforeseen expenditure	100,000,000	
Reserve fund	200,000,000	
	35,829,204,073	
MOVEMENT OF CAPITAL		
Purchase of properties, improvements, etc.	851,462,000	
Active loans	—	
Settlement of debts	469,929,871	
	37,150,595,944	
DEFICIT		13,679,479,472

In 1949 the deficit was around 6 milliard lire, in 1950 almost 7 milliard, and in 1951 almost 10 milliard.

RELATIONS WITH THE CENTRAL GOVERNMENT

The central government exercises control over the Commune of Rome through the prefect of the province and the Minister of the Interior.

The prefect is the representative of the executive power of the province.[1] The decisions of the communal council and of the *giunta*, for which special approval is not required, must be sent to the prefect within eight days. They become operative when they have been sent in this way to the prefect and posted at the town hall for a period of fifteen days. If the prefect considers the decision contrary to law, he must annul it within twenty days of its receipt. The prefect cannot, on the other hand, judge whether such a decision is opportune. His control is therefore solely one of legality. If the prefect annuls a decision, appeal may be made to the Minister of the Interior. Against the decision made on such an appeal, and also against decisions which have become operative, interested parties may have recourse (on grounds only of legality) to the Council of State[2] or to the President of the Republic.

The Minister of the Interior exercises control over matters which have wide economic importance. Decisions on such matters must have the approval of the Minister, and this also covers approval of the desirability of such a decision. This is the point at which the organization of the Commune of Rome differs from that of other communes, in that, while such control for Rome is vested in the Minister of the Interior, for other communes it is vested in the provincial administrative *giunta*.

Decisions subject to the approval of the Minister are those concerning expenses which will affect the budget for more than five years; the alienation of property, bonds, shares, etc., valued at more than five million lire; law suits and transactions to a value of more than 2,500,000 lire; leases for more than twelve years when the sum involved is more than 2,500,000 lire a year; direct assumption of public services; town

---

[1] The territory of the Republic is divided into 92 provinces. The province of Rome includes 113 communes, and covers an area of 5,337 kms. The population at January 1, 1950, was 2,044,336 inhabitants.

[2] The Council of State is the supreme body for consultation in judicial administrative affairs and has judicial functions on matters concerning public administration (Art. 100 of the Constitution). It is composed of six sections; the first three have consultative functions and give advice to the Government on questions of a juridical or administrative nature. The other three sections have duties of administrative jurisdiction. They are competent to decide what remedy lies against the public authorities for persons or concerns, whose interests have been infringed by actions of the administration. The members of the Council of State are nominated either from high grade administrative civil servants or by competition. They have a position of complete independence *vis-à-vis* the Government.

planning; regulations concerning the use of communal property, or concerning hygiene, building, local police, communal institutions, organization of the municipal departments; conditions of employment and legal treatment of personnel.

Finally, the Government may send its commissioners to take over temporarily the communal administration if, through exceptional circumstances, this cannot function through the ordinary organs of local government.[1]

PLANNING THE CITY

In Italy until 1942, planning still came under the law of 1865 concerning expropriations of land for public purposes. This, like the corresponding French legislation, provided for (*a*) so-called town-planning schemes concerned with the reconstruction of derelict and sub-standard buildings for reasons of public health, or demolitions for purposes of road construction; and (*b*) development schemes, laying down the conditions governing new building. The execution of this law frequently necessitated supplementary legislation to meet the particular needs of certain cities, and for Rome such special laws were passed in 1932 and 1935.

In 1942, the present law on town planning was enacted which gave the commune power to draw up town-planning schemes. These schemes may be either general or limited in character, the former covering the whole territorial area of the commune and regulating the network of communications, the division of the area into residential and non-residential zones, the allocation of land for public spaces, housing, public works, schools, community buildings (town hall, etc.); the latter referring only to specific zones which are being planned or re-planned in detail. Both types of schemes must be accompanied by detailed financial estimates. Projects for detailed local re-planning in Rome are examined by a special consultative committee of experts.

General and local plans must be available for public inspection in order that citizens may criticize or raise objections. The plans must then be submitted to the Public Works Advisory Council and finally approved by the head of the state on the advice of the Minister of Public Works.

Municipal regulations control certain aspects of building activity: building licences, height of buildings, sanitary requirements, etc. The mayor, assisted by the Municipal Building Commission, ensures the execution of these provisions.

---

[1] This general regulation is naturally rarely brought into force, particularly as far as the administration of the large cities is concerned.

The chequered history of Rome has left innumerable traces on the face of the city. Periods when its development was planned and controlled (as in ancient times, under Silla, Caesar, who produced the law *de urbe augenda*, and Augustus) alternated with periods when these problems were wholly neglected and, as Livy wrote,[1] Rome resembled more a town temporarily occupied by a military outpost than a city regularly laid out for its residents' use and pleasure.

In contradistinction to other historical cities, such as Athens for instance, in which the modern city has been able to develop in a zone entirely separate from the ancient city with its historic monuments, the centre of activity in Rome has always remained inside the ancient city walls and the problems of modern town planning have been intertwined with the need to preserve and display the wealth of historical monuments and ruins, the evidence of so many successive epochs and vicissitudes.

From the moment Rome became the capital of the kingdom, these problems were recognized as acute. The reader of Stendhal's *Promenades* will be familiar with the picturesque but archaic condition of Rome during the last decades of the papal government. But in 1870, Rome was already in process of transformation. Shortly before the fall of the papal government in 1867, a brilliant Belgian architect and prelate, De Merode, had drawn up a town-planning scheme for Rome based on modern principles, and a beginning had been made towards executing it. A few days after the conquest of Rome, studies were begun on a master-plan which was completed in 1873, re-drawn in 1883, and again in 1909.

But although Augustus Hare was certainly exaggerating when he wrote that 'the years between 1870 and 1903 have done more for the destruction of the artistic beauty of Rome than all the invasions of the Goths and Vandals,'[2] it is undoubtedly true that after 1870 the city developed in a haphazard and wretched fashion that disfigured the grandeur that was Rome. The new was clumsily grafted on to the old by an unco-ordinated architecture, in most cases conspicuous for its bad taste.

The town planning problem has become steadily more complex with the increase in population and in traffic. After some years of study, a new master-plan was produced in 1931, in large part the work of the architect Marcello Piacentini. It was later supplemented with several detailed plans for separate districts, some of which diverged from the master-plan. Since this plan is only legally binding until 1955, a new master-plan for Rome is now once more under study, and will come

[1] Livy, V, 56, 2.
[2] Quoted by Reed, 'Rome, The Third Sack,' in the *Architectural Review*, February 1950.

into force on January 1, 1956. The plan at present under consideration
is an overall one, which will gradually be brought into force using
detailed plans for the various districts. It will include all the communal
territory, that is the city, the suburbs and the *Agro Romano* as far as the
limits of the neighbouring communes. An inter-regional and inter-
communal plan is also being considered so that a harmonious solution
of the various urban problems may be reached.

Town planning in Rome has various aspects: the problem of the
expansion of the city as a whole, the problem of its central zone where
traffic congestion is most acute, the aesthetic and historical problem, and
the social problem.

Since before Fascist days, the city has expanded towards the sea. In 1919
a Board of Industrial and Maritime Development of Rome (*Ente S.M.I.R.*)
was set up, but suppressed in 1923, and its activities were handed over to
the commune. The railway between Rome and Ostia was then
constructed.

In 1939, when Rome was chosen as the seat of a world exhibition to
be held in 1942, an attempt was made to combine work for the exhibition
with desired municipal expansion, and a site for the exhibition was
selected between Rome and the sea coast. A group of large buildings
was constructed, or at least begun, as the nucleus of the exhibition, with
the intention of later transforming them into museums and public
buildings to serve as the centre of an entirely new residential district.
But the war effectively prevented the realization of this large project.
So there remain the grandiose buildings erected before the war and the
roads of communication laid out, and hence this zone offers admirable
possibilities for building expansion. The buildings designed for the
universal exhibition have already been fitted out again to house various
exhibitions which have been organized in Rome. Meanwhile in the same
district, many blocks of flats are being built.

The 'Universal Exhibition of Rome' (E.U.R.), which has been established as a public corporation, possesses an area of 420 hectares which
has already been almost entirely planned with large streets and squares,
retaining, however, all the trees that are in the zone, and in particular
safeguarding a eucalyptus grove which covers 11 hectares.

The building plots are on sale at reasonable prices, but the purchasers
must respect the rigid conditions laid down by the E.U.R. regarding
the height and size of buildings, the care and replacement of trees,
maintenance of gardens, etc.

This district is therefore designed to become a sort of garden city and model quarter, in which it is favoured by its position. The Via Cristoforo Colombo unites it on one side with the centre of Rome and on the other with the Lido at Castelfusano, joining also with the new Via Pontina, which will be the fastest road to Naples. Other roads join this district directly with the big arterial road to the North.

This district will, of course, be capable of full development only on completion of the underground railway which will connect it with the main railway station and the centre of Rome by a rapid and cheap means of transport.

The problem of the city centre and its traffic congestion is today the most serious aspect of town planning in Rome. Traffic has increased enormously in recent years, owing to tourist visits, the concentration of the expanding central administration, and the activities of the Roman Catholic Church, together with the geographical position of Rome in the centre of a long, narrow peninsula. A revealing indication of the increase of traffic is the fact that every month about 1,400 new vehicle registrations are effected in Rome. Some 60,000 motor cars, 15,000 lorries, 22,000 vans, 25,000 motor-scooters and motor-bicycles, 1,000 public traction vehicles and 300 horse carriages, comprise the average traffic circulation of Rome.

This traffic piles up in the narrow central streets that cannot be widened, largely owing to their historical character. The creation of an 'office zone' outside the city centre would be an obvious remedy, but a long term one. The 1909 town plan provided for an outer circular boulevard to link all the residential districts on the periphery, and this idea reappeared in the 1931 plan. Today this idea has been superseded by projects for two wide streets *de dégagement*, designed to connect with the main roads leading into Rome from both north and south. They should relieve the centre of the city of all transit traffic but whether, in fact, this relief will be substantial is open to doubt, as Rome offers many attractions to anyone who is travelling the length of the country as a place in which to stop and meet people.

Of these roads, which are designated 'circular by-pass' (*Raccordo annulare*), the one connecting the Via Aurelia (the Tyrrhenian coast road —Rome, Pisa, Genoa) with the Via Appia (the principal road from Rome to the south, Campania and Apulia as far as Brindisi), crossing the Ostia *autostrada* by means of a fly-over; and the one joining the Via Tiburtina (Rome—Tivoli—Abruzzi) and the Salaria (Rome, Abruzzi, Marche), have already been completed. It is hoped that the complete circular by-pass will soon be finished, connecting in the north-east the Aurelia with the other northern routes, the Cassia and the Flaminia, continuing to join the Salaria, and in the south-west continuing eastwards to join the Appia with the Casilina (the road to Cassino), the Prenestina and the Tiburtina.

Construction of an underground railway was begun before the war, to connect the main railway termini with the exhibition site, passing by the Colosseum and St. Paul's Outside the Walls. But, owing to the fact that the subsoil of Rome has, at a depth of about six metres, a strata of considerable archaeological remains, and for other reasons, construction of such a railway is extremely difficult and costly; it is not likely that a second line will be attempted.

Underground parking spaces for motor cars are greatly needed. The long rows of cars and coaches parked in the most picturesque streets and squares impede the flow of traffic and have completely ruined the physiognomy of the city. Another partial solution of the traffic problem would be to ensure that only a uniform kind of vehicle should be allowed through the narrow streets of the centre. The one-way traffic rule has already been enforced in some of them.

The problem of historic Rome and its monuments has been studied with especial care since the end of the First World War. A municipal commission was set up in 1918 to supervise the modernization of the 'Renaissance area,' that is, the zone of papal Rome which was developed during that period of history, and in 1920 a similar commission was set up for the Capitoline Hill and its surroundings. The two commissions reached quite different conclusions, the earlier commission wishing to preserve the general harmony and the original aspect of the buildings, with a minimum of demolition, while the second commission proposed to free the *Campidoglio* (Town Hall) by clearing away all the old buildings which had grown up around it. Between 1928 and 1940 the *Campidoglio* was, in fact, isolated, and in 1932 the Via dei Fori Imperiali was cut through from the *Campidoglio* to the Colosseum, including among its flanking views all the most important remains of Imperial Rome. It would be a good thing if some of the neglected features of mediaeval Rome could now be adequately displayed.

From the aesthetic point of view, it should be noted that insufficient attention has been given to the problem of green spaces within the city. Already in the nineteenth century, building speculators had destroyed some of the most beautiful gardens of Rome, and D'Annunzio deplored that the cypresses which once solemnly shaded the head of Goethe had now been torn up by the roots. At the same time, the restoration and isolation of historical monuments was inspired by the concept of creating large archaeological zones set amidst extensive gardens and wide tree-lined avenues. A good example of the practical application of this concept was the creation of the *Passeggiata Archeologica*, within which are the remains of Caracalla's Baths. But at present the few trees still left in Rome are being destroyed—we are not, of course, referring to the public parks, such as the Villa Borghese, the Pincio, the Janiculum, Villa Sciarra, and the parks surrounding certain former private residences—and in their places rise tier upon tier of new ferro-concrete buildings. The gardens belonging to private villas thus acquire a high business value and their owners find it more profitable to sell them, at remarkable prices, for building purposes.

From the social point of view, the main town-planning problems of Rome are (1) the housing shortage, and (2) the need to control the development of the outlying suburbs. The housing shortage has become progressively worse in recent years, owing to increasing population and wartime cessation of building. As a result, it is exceedingly difficult to relax the system of blocked rents, the ban on evictions, and other restrictions still operative throughout Italy, although recently there have been some relaxations in the network of controls.

The municipality of Rome raised a loan of some 5,000 million lire, of which sum four-fifths were allocated to the construction of small-sized flats of a popular type. But this is a problem that the municipality cannot tackle for itself; the Institute for People's Housing, the Communal Public Assistance Board, the National Insurance Institute (I.N.A.) and U.N.R.R.A. must all co-operate with the Government in finding a solution.

The following data show the increased building which has been carried out in recent years: (The figures refer to habitable rooms):

	Building permits	*Licenze di Abitabilita*[1]
1951	79,489	31,325
1952	94,286	37,422
1953 (first quarter)	38,422	12,892

An increasing city population and consequent overcrowding of dwellings has led to a spill-over of population, mostly unemployed, into suburbs in the surrounding countryside, in some of which the public services are as yet quite inadequate, the general living conditions somewhat primitive, and where poverty, squalor and their attendant social evils flourish. There are over one hundred of these suburbs,

[1] This *licenza* is a certificate that the house is habitable.

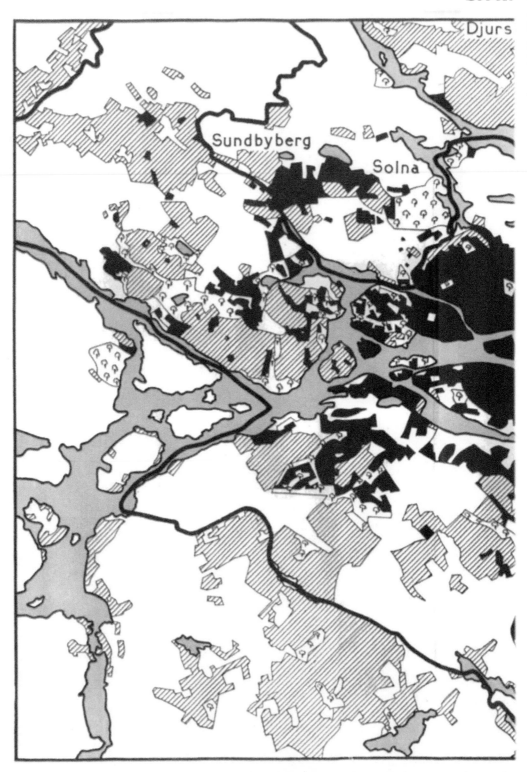

Djurs

Sundbyberg

Solna

STOCKHOLM AND ITS WATERCOURSES. The black portions are the densely-
suburban settlements consisting of houses built for individual fami

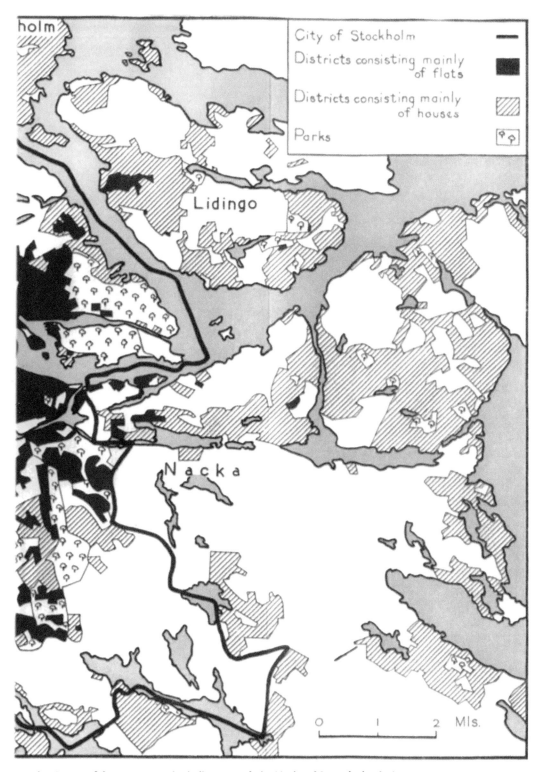

City of Stockholm

Districts consisting mainly of flats

Districts consisting mainly of houses

Parks

Lidingo

Nacka

0   1   2   Mls.

populated parts of the city occupied wholly or mainly by blocks of flats. The hatched portions represent
ilies. The map clearly shows the extent of urban sprawl beyond the city boundary.

some of them (e.g. Primavalle) containing over 50,000 persons, which in some cases have developed in a normal and semi-controlled fashion. The idea of building outlying suburbs for the poorer classes is already discredited, as it has produced the worst possible results from all points of view, in particular the economic and social.

So, beginning from an entirely opposite idea, an attempt is now being made to encourage the construction of a new district near the Via Prenestina, to be called the Quartiere di Villa dei Gordiani, which will include various classes of people in the same district. The commune has taken the responsibility for building 2,000 apartments, two schools, a playing field, a market and a public park. Then there will be put up low-priced houses for the type of person who is able to pay a minimum rent for them, while certain areas will be reserved for private building of blocks of flats for middle class professional and business people.

Besides controlling building activity, the commune must provide immediately an extension of the public services to the new districts. Under the pressure of an extraordinary development in building, the commune is faced with the necessity of constructing roads and drains, extending public lighting, water supply and public transport, and of providing schools, out-patient clinics, municipal *delegazioni* and many other services. In the present state of building development in Rome, the commune is faced with an expenditure of more than 50 milliard lire to meet these tasks; this sum would absorb its financial resources available for public works for four years.

CONCLUSIONS

We have now seen that in Rome today there is a town-planning problem, a social problem, and an administrative problem. Town planning itself comprises three problems—the expansion of the city, the rearrangement of its traffic system, and new housing for its citizens. The social problem centres on the crying need to clear up the suburban area, lest a belt of squalor grow around the city. And, inevitably, all these problems hinge on the underlying problem of finding adequate financial means to effect the desired reforms. This is a permanent, and not merely a post-war difficulty. The municipality of Rome could never manage on its own to meet the needs of the city—needs which, to a great extent, derive from its function as capital of the state and as a business and administrative centre. Even a measure of administrative decentralization would be unlikely to reduce to any appreciable degree the close network of public relations and offices centring on Rome.

S

It is for these reasons that special legislation and special powers are needed for Rome. All political parties represented in the municipality are in agreement on this point. Such legislation is now being drafted and it is to be hoped that it will soon be ready, for the problems outlined above are becoming more acute as time passes and they are problems of urgent social necessity, not at all of political grandeur.

The main elements in this new legislation must be: first, to ensure the municipality of Rome the means necessary to carry out its functions—functions which are, in many cases, undertaken in the national and not in the local interest, and to the cost of which all citizens should contribute. Second, to work out a system of controls within the municipal administration for the firm yet rapid execution of its plans. It is clear that, given the Italian administrative system, Rome cannot be withdrawn entirely from the general system of controls applying to all communes, all the less since she hopes for grants-in-aid from the national treasury. But these controls must be effected by simple, speedy and direct methods that will strengthen and not weaken the practical executive powers inherent in communal autonomy.

Stockholm

GUNNAR HECKSCHER

PER HOLM

GUNNAR HECKSCHER

Born 1909; D.Phil., Uppsala University, 1934; Lecturer, University of Uppsala, 1934–41; University of Stockholm, 1941–48; Dean, Stockholm School of Social Work and Municipal Administration since 1945; Professor of Political Science, University of Stockholm, since 1948; Visiting Professor, University of Chicago, 1948; Member, Swedish Social Science Research Council, 1951–52; member of several Swedish Royal Commissions; Vice-President, International Political Science Association.

Publications: *Konung och statsråd i 1809 års författning* (1933); *Parlamentarism och demokrati i England* (1937); *Brittiska imperiet* (1939); *Svensk konservatism före representationsreformen*, I–II (1939–42); *Staten och organisationerna* (1947, 2nd ed. 1952); *Svensk statsförvaltning i arbete* (1952); various articles.

PER HOLM

Born 1915; educated University of Stockholm, Pol. mag. (M.Soc.), 1940; Statistician, Statistical Bureau of the City of Stockholm, 1940–43; Research Officer, 'Svenska Riksbyggen,' 1943–45; Assistant Research Officer, Department for Social Affairs, 1945–48; Research Officer, 'Gustavsbergs Fabriker' since 1948; Secretary, Royal Commission on Industrial Localization; Editor of 'Plan' since 1947.

Author: *Swedish Housing* (1949) (with A. Hald and G. Johansson); various articles.

Stockholm

STOCKHOLM is a comparatively old city. It dates from the Middle Ages, when it first grew up as a centre for trade in the Baltic region. Although Stockholm was never a Hanseatic city, the German influence was very much in evidence, and one of the later medieval charters of the city expressly provides that not more than half of the city councillors must be Germans. Indeed the city owes its original position to the Hanseatic cities, rather than to the economic development of Sweden itself. But its geographical position, on islands controlling the single passage between the Baltic and Lake Mälaren in the heart of Sweden, also gave it considerable strategic importance. This fact also explains why the Swedish state never allowed it to come under complete foreign control.

THE RISE AND SIGNIFICANCE OF STOCKHOLM

From an early date, Stockholm became the capital of Sweden. With the growing power of national Swedish government from the beginning of the sixteenth century, this function became more and more important. For the better part of the year, the king held his court in Stockholm, exercising from his palace in the city power over the rest of Sweden. Central administrative agencies, which came into existence from the same time, were invariably situated in the capital. As, during the seventeenth century, Sweden developed into a great European power, Stockholm became the political as well as the economic centre of a Swedish realm including Finland, Estonia, Latvia, and Pomerania, together with Sweden proper. The burghers of Stockholm thrived, and particularly in the eighteenth century—when the Swedish realm was already dissolving—their city was one of the very few places in Sweden where the guild system was a reality and not only a paper structure.

Stockholm today retains some of the characteristics of these times. There is still the old city of the islands, with streets and buildings dating back at least to the fifteenth century. Government departments still occupy quarters on the brink of the old city. But the business centre has shifted somewhat to the north, and, of course, the city has outgrown by leaps and bounds its earlier limits as new residential quarters have come into existence in all directions from the old city. However, Stockholm remains one administrative unit, even though a rather considerable number of the people working there actually live in adjoining cities, such as Sundbyberg, Solna, Nacka, Lidingö, and Djursholm, which have practically no administrative ties with Stockholm.

The economic structure of the population differs from that of the country as a whole. The following figures, showing the distribution by percentages in different occupational groups (in 1945) should illustrate this:

	Sweden	All cities	Stockholm
Agriculture, etc.	29·7	3·0	0·7
Industry	39·7	48·9	40·0
Commerce and transportation ..	20·9	32·1	39·4
Public administration	9·7	16·0	19·9

Thus, Stockholm is not predominantly an industrial city. Its industries are by no means unimportant—there are electrical and mechanical industries such as L. M. Ericsson, which are situated in the capital—but a considerable part of the industrial population is occupied in building and in service industries. Its dominant position in the country (Stockholm is more than twice the size of the next city, Gothenburg) is not due to its direct industrial importance, but rather to the fact that a considerable part of the economic life of Sweden is administered from there. Many industrial firms with productive units elsewhere have their head offices in Stockholm, partly because it is convenient, partly because the rate of taxation is comparatively low in the capital.

But Stockholm is also an important centre of commerce and transportation. Banking in Sweden has concentrated rapidly in the past fifty years, and all the major banks have their headquarters in Stockholm. Its port is second only to that of Gothenburg in the amount of in- and out-going tonnage, although it is true that the greater part of it is bound for other parts of Sweden. Stockholm is the most important road and rail centre of the country. It is also the administrative centre, a fact which is becoming more and more important in economic life.

Higher culture in Sweden used to be concentrated in the two university towns, Uppsala and Lund. More than a hundred years ago, however, a medical school was founded in the capital, where opportunities for research and training in this field were greater than in smaller cities. It was followed in the latter half of the nineteenth century by the beginnings of a university, meant to provide facilities for research and general education but not to give degrees. This university now comprises faculties of law, science, and arts, but not of divinity; and the medical college is still independent. In addition, new independent schools of engineering, dentistry, veterinary, surgery, commerce, social work, etc., have grown up. There is close co-operation between all of these, and administratively they are beginning to merge. So far, however, the characteristic fact about the university of Stockholm is that it does not exist; but there are many more university students in Stockholm than anywhere else in the country.

LOCAL GOVERNMENT FUNCTIONS

Local self-government is very strong in Sweden, and is jealously guarded, a tendency which is the more emphasized because members of the national Parliament generally have begun their political careers in municipal and county councils. The capital is no exception and it differs from other cities in one respect only, namely, that it is both a city and a county, so that its city council has the additional functions carried out elsewhere by county councils.

Functions are of two types. The law contains a general clause, empowering municipal bodies to 'manage their affairs, in so far as such management is not confided to any other body.' In addition to this, legislation empowers or requires them to assume a number of specified duties, and the latter have become increasingly dominant. Thus, while the city has of its own free will decided to take an interest in housing, to provide gas and electricity, and to finance partly the university of Stockholm and certain schools of the academic type, it is on the basis of national legislation that it administers primary schools, the police force, public relief, hospitals, and child welfare. Consequently, its freedom of action is far more limited in the latter fields. It *must* provide schools up to a certain age, but it *may* (and does) require children to go to school for a longer period than that which is obligatory elsewhere. It *must* provide relief securing a certain standard, but it *may* (and does) raise the standard above that level. The decision of the child welfare board to take a maladjusted child away from its parents is subject to the approval of the Governor (*överståthållare*) and, in the last resort, of the supreme administrative court. The same is true of decisions by other boards of a similar character. In regard to the management of hospitals, control is exercised by the Royal Medical Board (*medicinalstyrelsen*), which has to assure conformity with national hospital legislation.

The functions of the city government of Stockholm can be most easily shown by the accounts in Table I giving expenditure and revenue in respect of the principal municipal services.

In the expenditures of the city, the dominating items are hospitals, streets, and schools. Stockholm differs in this respect from smaller

TABLE I

STOCKHOLM

Current Revenue and Expenditure Account, 1950

	Revenue (Kronor)	Expenditure (Kronor)
City council, etc.	157,624	4,452,049
Board of finance	1,361,914	3,392,187
Revenue authority	1,250,075	12,025,622
Municipal board of judicial administration[1]	1,776,788	9,908,512
Real-estate property board	57,870,183	62,671,583
Police board	3,191,659	22,428,085
Building control board	505,866	2,197,946
Zoning board	113,460	1,552,823
Streets and roads board	69,554,906	109,488,345
Cemetery board	1,765,750	3,717,753
Sports and hobby grounds board	1,000,982	5,595,310
Water works	13,536,586	10,404,311
Gas works	44,385,150	44,125,928
Electricity works	69,333,474	45,571,398
Port authority	21,151,916	21,842,146
Market hall board	2,736,869	2,484,823
Public relief board	9,529,305	39,510,140
Child welfare board	5,056,603	23,017,608
Unemployment relief board	1,145,558	3,194,497
Primary schools board	27,646,610	64,990,097
Vocational schools board	2,293,800	7,204,341
Library board	163,603	3,247,042
Hospitals' board	30,234,614	103,642,350
Public health board	176,606	2,624,233
Dentistry board	205,194	4,425,282
Old age pensions (municipal grant)	—	19,112,141
Other current income and expenditure (other than taxes and interest on capital)	67,748,203	99,216,813
Interest on capital	51,508,663	—
Taxes	253,394,049	1,942,161[2]
Total	738,796,010	733,985,526

municipalities, where public relief and child welfare are relatively more important than in Stockholm. This does not mean, however, that the capital maintains a lower standard in the latter fields, but rather that

[1] Part of cost of the city court of justice has to be paid by the city itself.   [2] Refunds.

her economic strength is such that they are a smaller burden than elsewhere.

Some of the city's functions, such as police, primary education and poor relief, are strictly regulated by national statutes and ordinances, leaving the city little opportunity to influence either the amount of money to be spent or the principles to follow. On other points, such as management of real-estate properties, gas and electricity distribution, the city acts with the same independence as would a private company working in the same field. For certain purposes, finally, the city receives grants-in-aid necessitating a certain amount of state control; this is to a certain extent true of housing. For a city with the financial strength of Stockholm, however, such considerations do not weigh too heavily, and if forced to break loose from state control, e.g. in housing, the city could certainly afford to do altogether without state grants.

Swedish cities are governed in accordance with the Local Government Act of 1953. This law, however, does not apply to the capital, which has a separate charter dating from 1930. The provisions are similar, but there are certain differences necessitated partly by the need for a more highly-developed administrative structure, partly by the fact that Stockholm is not part of any county but forms a county unit by itself under its own Lord Lieutenant (*överståthållare*). This means, also, that city authorities are responsible not only for the duties applying in all cities, but also for the duties elsewhere delegated to the county council (*landsting*) and its committees and officers.

THE GOVERNMENT OF STOCKHOLM

The national government exercises no direct positive power over the city administration and little general control. The city charter as well as certain other ordinances regulating city government being acts of Parliament, they could be altered by legislation, but so far no such legislation has been passed except on the demand of city authorities or in order to assure conformity with general legislation. A certain amount of control is exercised over the floating of loans, but the financial strength of Stockholm makes this less important than in other cities.

The 'parliament' of Stockholm is the city council (*stadsfullmäktige*) with 100 members, elected for four years. The last election took place in September 1950, and the composition of the council in 1952 was 43 Labour, 35 Liberals, 17 Conservatives, and 5 Communists. The council meets at least once a month in the town hall (*stadshuset*), and its debates are apt to become rather long—a meeting of four or five hours is not

s*

unusual. The budget is passed at the end of each calendar year, and at the same time the council decides the tax rate for the coming year. These decisions are not subject to any administrative control by the Lord Lieutenant, who may only disallow an illegal decision, but not put his own decision in its place.

Political parties play an all-important part in the city government, far more in Stockholm than in smaller cities and boroughs. Most problems coming before the city council are solved without any open difference of opinion, while others cut across party lines. But in crucial questions, the decision rests with the political parties, who have developed a number of strong leaders, such as Z. Höglund (Labour) and John Bergvall (Liberal).

It is interesting to note, however, that personal leadership is much more in evidence than municipal party programmes. Municipal elections, which take place on the same day all over the country, are fought on national issues, largely because county councils and some city councils (such as that of Stockholm) have the duty of electing members to the Upper House of Parliament. Although the party organizations in Stockholm all publish specific election programmes, these tend to be conventional in character, and it would be difficult to give tangible instances of real municipal policies. If Labour in the capital has been in the past more radical than elsewhere in the country, this has been evident less in the type of policy pursued than in the means employed: for instance, its co-operation with Communists—which, incidentally, seems to belong to the past.

The city council, from among its own members, elects the board of aldermen (*stadskollegiet*) of 12 members. It is the duty of this board to prepare all business for the city council, as well as to exercise general control over the administration of the city. The president and vice-presidents of the council may attend its meetings, but they cannot be elected members of the board and are not entitled to vote. The board is the central authority of the city and wields great actual power—greater than appears from the charter; for the leaders of the parties in the council all sit on it, and its proposals, therefore, are generally, if not invariably, accepted by the council. On the other hand, although the board meets much more frequently than the council, its members are unpaid (like the city councillors) and consequently are not able to participate in the actual administration of affairs.

The administrative heads are the city directors (*borgarråd*), at present eight in number. These officers are elected for a period of four years at the first meeting of a new city council. The election is on party lines, and the directors may or may not be members of the council. They draw salaries which are among the highest in the Swedish public service (45,000 Swedish kronor, which exceeds the salaries of Ministers of the Crown by about 20 per cent). As a rule, these posts are divided between the parties more or less according to their strength on the council, but there is no legal obligation to that effect, and 'coalitions' occasionally result in a seeming lack of balance. A Labour-Communist majority governed the city from 1946 to 1950; from that year, Liberals and Conservatives have co-operated. At present there are 3 Labour, 4 Liberal and 1 Conservative directors. Each of them is in charge of one department (*rotel*) of city affairs: finance, personnel, schools, hospitals, public welfare, housing, streets and planning, and public utilities. The directors are obliged to participate in the meetings of the council, whether they are members or not, and of the board of aldermen, of which they must not be members. In the board, each director has to report on matters belonging to his department. The directors also form a board of their own (*borgarrådsberedningen*), which meets under the chairmanship of the finance director to prepare business for the board of aldermen and to discuss more or less informally all matters of importance to the general administration of the city. While the city council corresponds rather closely to the corresponding institution in other Swedish cities, and the board of aldermen also finds its counterparts elsewhere, the city directors are peculiar to Stockholm. Their offices were created just after the First World War, and there is still some discussion as to whether such a combination of political and full-time administrative offices is really warranted in a city, indispensable as it may be in a national government. Also, opinions differ as to whether the present partition of offices between the parties is the best solution, or whether it should give way to some sort of responsible cabinet government, all the directors being drawn from the majority in the council. So far no party has been prepared to take the responsibility for a pure majority rule (which would in any case have to be based on a coalition), and as a matter of fact the directors have quite frequently been able to co-operate across party lines and to solve political differences by mutual consent at an early stage of proceedings.

The city directors, however, do not exercise supreme power within their departments. Actual administrative decisions of greater importance are taken not by them, but by separate boards elected independently by the city council. The directors generally act as chairmen of the boards belonging to their respective departments, but cannot dictate their decisions. Thus, the finance director presides not only over his fellow directors, but also over the board of finance (*drätselnämnden*); the school director over the primary schools board (*folkskolestyrelsen*), the vocational schools board (*yrkes- och lärlingsskolestyrelsen*), and the library board; the public welfare director is chairman of the poor law board (*fattigvårds-nämnden*), the child welfare board (*barnavårdsnämnden*), the unemployment relief board (*arbetslöshetsnämnden*), etc. Each of these separate boards consists of members of the different parties, elected by the city council for four years (or in some cases for a shorter period) on a basis of more or less proportional representation. Board members may also be councillors, but a great proportion of them consists of other citizens nominated by the respective parties; and membership is virtually unpaid. Consequently, while the board of aldermen is generally able to gain acceptance of its recommendation by the city council, this is not equally true of the separate administrative boards; but on the other hand, the latter exercise more direct administrative power, and take decisions which neither the board of aldermen nor the city council can influence either directly or indirectly, except through the election of members and the passing of appropriations. And the paid officials of the city are generally employed by the separate boards, not by any central municipal authority.

The doctrine of separation of powers applies to the relationship between the city council on the one hand and the board of aldermen, the city directors, and the independent boards on the other. Thus, matters of policy should be decided by the council, administrative matters by the other authorities. This means, in practice, that the city council decides in some detail all questions of a budgetary character, but on the basis of recommendations from the various administrative boards. Thus, each independent administrative board also acts to some extent as a committee of the council, and the same is even more true of the board of aldermen. If a motion is made by a councillor, it is referred to the competent administrative board, which reports back to the council. Its report, together with the original motion is considered by the board of city directors and the board of aldermen, where the actual political decisions are frequently made. The same procedure is observed where preliminary investigation through a select committee has taken place; on the other hand, all communication between the independent boards and the city council goes through the board of aldermen.

In administration, on the other hand, the decisive power lies with the competent board, and the council must not interfere directly. The city director who is the chairman of the board may have and almost invariably has power delegated to him by the board, but in certain respects he will remain a mere chairman, particularly if he belongs to a party which does not command a majority. Appointments of personnel are invariably made by the board itself. In certain fields, where no specialized board is competent, administrative decisions are taken by the board of aldermen.

This structure of innumerable boards may, at least at first, seem rather confusing. Actually, it exists not only in Stockholm, but to an almost equal extent in other Swedish municipalities, where the absence of a board of city directors makes for even less administrative co-ordination. At the bottom of the system lies a profound distrust of municipal bureaucracy and the desire to engage a considerable number of citizens in the administration of local affairs. Also, it can hardly be denied that the system has served to create a practical, matter-of-fact attitude to public business and to mitigate to some extent the bitterness of party strife: when members of different political parties have perforce to co-operate for the attainment of tangible aims, abstract principles lose some of their hold over their minds, and it becomes easier to reach agreement even on points outside their immediate administrative duties. It is true that this latter result has been less evident in Stockholm than in smaller municipalities. In a larger city, more questions of principle arise, and even the separate boards are not able to participate as much as in a city of 20,000 to 30,000 inhabitants in the day-to-day work in their respective fields. But this does not prove that party strife might not have been even more bitter under a more unified system, where elected representatives must be almost completely excluded from deliberation over practical administrative problems.

SPECIAL FEATURES OF STOCKHOLM

Another difference between Stockholm and the smaller municipalities is in fact more important. In the latter, elected representatives are able to maintain, at least to some extent, personal contact with public opinion among their electors. In Stockholm, this is hardly the case. Indeed, it may well be asked whether the government on a unitary basis of a city of 700,000 inhabitants can at all be called local self-government. Not only the city council and the board of aldermen, but also the separate administrative boards are almost as centralized authorities as Parliament and the national government. The child welfare board knows little or nothing about the individual children for whom they are responsible, the primary school board knows neither the children nor the teachers personally, members of the board for the administration of building regulations (*byggnadsnämnden*) are not able to study the plans for each individual house, and the housing board (*fastighetsnämnden*) has never met the tenants of city-owned dwellings. The whole system is planned to avoid the evils of municipal bureaucracy, and yet the power of deciding individual cases falls into the hands of municipal bureaucrats.

Can this be avoided? City authorities are only just beginning to take cognizance of the problem. Plans are being developed for the creation of advisory councils for different parts of the city (*stadsdelsråd*), but interest in these plans is not wholehearted, for there is reasonable doubt whether members of such councils, having no real power, will prove willing to put sufficient work and interest into their duties. And so far, wholesale decentralization of city government has not been seriously advocated by any political party or any public figure of importance.

On the other hand, relations between the city of Stockholm and adjoining municipalities also present difficulties. Greater Stockholm includes, in addition to the capital itself, five separate cities and six or seven boroughs (*köpingar*) and townships (*landskommuner*), which even belong to a different county. You cross a street or a bridge, and find yourself in a different community. When a citizen of Stockholm goes to see a football match, he has to pass into another city. And co-operation with these separate entities is difficult: taken together, they do not even approach the population of Stockholm proper, and consequently they deprecate the creation of any joint authority, in which their individuality would tend to disappear. With the growing density of population in the Stockholm region, this problem becomes even more acute, but hardly more easy of solution. Its importance, however, is clarified only by a study of the planning problems of Greater Stockholm.

STOCKHOLM CITY PLANNING

When Stockholm was founded in the thirteenth century upon a group of islands which controlled the straits linking the Baltic Sea with Lake Mälaren, it was a small city judged by international standards. And it remained so for a long time. At the close of the seventeenth century, its population was about 45,000. Industrialization, together with urbanization, came relatively late to Sweden, but proceeded then at a very swift pace. The growth of the city into a metropolis has occurred entirely during the last one hundred years, as Table II shows:

TABLE II

| Year | Population | | Greater Stockholm's per cent of the country's total |
	Stockholm proper	Greater Stockholm	
1850	93,000	100,000	2·9
1900	300,000	320,000	6·2
1950	740,000	980,000	14·1

Stockholm's population has increased tenfold between 1850 and 1950. At the present time Stockholm, together with its suburbs is in the one million class, ranking in size with such cities as Manchester in England and San Francisco in the United States.

The map facing page 544 shows Stockholm's location. Originally, the city was confined to a strategically located island (between Lake Mälaren and the Baltic Sea) called Stadsholmen. Gradually, settlement extended to the mainland north and south of Stadsholmen, even though the terrain here is relatively ragged and partly ill-suited for building construction. Today, the area of Stockholm proper is about 18,000 hectare (45,000 acres) and that of Greater Stockholm about 250,000 hectare (625,000 acres).

Within and around the city there are many watercourses. From an aesthetic point of view, these are great assets; but they create serious problems for construction activity and especially for management of traffic.

The presence of these watercourses means that Stockholm cannot expand concentrically. A consequence is that areas to be exploited for construction must extend much further from the city centre than would be necessary for a city of similar size expanding under normal conditions. For one thing, good connections between different parts of the city require expensive bridges. Likewise, traffic into the city must pass through various 'bottlenecks'; their number is limited by Stockholm's peculiar geographic layout.

The above table may also suggest the age structure of buildings in Stockholm. In simplified, schematic form, three distinct phases can be discerned in the settlement of the city. First, the buildings located on the tiny island (Stadsholmen) in the city centre, dating back to the Middle Ages. Second, the 'City of Stone,' mostly built between 1850 and 1920, in accordance with a 'chess-board' design. This period was characterized by square or rectangular blocks, supporting stone houses four to six floors in height. Third, the garden suburbs of 1920–50, featuring modern apartments, open blocks and large park areas.

A short historical summary will give the reader a background for discussion of planning problems. It should be emphasized that the pattern of past settlement strongly influences the planning of today. It provides a framework for new construction, a framework which must be progressively adjusted in accordance with the new demands which an improved technology, a higher standard of living, and a new philosophy of life make upon it.

The central part of Stockholm's Old Town is characterized by a medieval layout, with its irregular network of streets suited to the terrain. It still bears the marks of Stockholm's old role as a fortified town. The closely built houses overlook streets whose widths vary generally between two and five yards. The royal palace, parliamentary buildings, and many government offices are located in the Old Town. Elsewhere the buildings have largely retained their medieval character, thus making the Old Town somewhat unique in northern Europe. At the same time a number of important commercial thoroughfares can be found here. This part of Stockholm has created difficult problems for current town planning. Its buildings, so valuable from a cultural and historical point of view, must be retained unaltered so long as it is at all possible to do so. Nevertheless, a large number of the houses are unsuitable for living quarters. In terms of traffic management, big problems arise because the huge stream of traffic between north and south must pass through this part of the city, and do so without impairing its quaint character.

During the latter part of the nineteenth century, stone houses four to six stories high flourished in the swiftly growing areas both north and south of the Old Town. The town plans in the City of Stone are largely Renaissance plans which were designed as early as the middle of the seventeenth century. Their chief features are (*a*) the rectangular network of streets and (*b*) the width of the streets (broad for their time)—ranging from eight to ten yards and up to twelve yards for main streets.

When Stockholm really expanded at the end of the nineteenth century, similar 'chess-board' plans were superimposed on the Renaissance plans. The only difference was the somewhat greater width of the streets, twelve to eighteen yards.

During all of this period, methods of construction followed those in vogue in the large cities of the Continent. Whole blocks were encircled by houses, and eventually houses were erected even on the inner courtyards. In several blocks, the density of population was very high; as many as 1,000–2,000 persons per hectare (400–800 per acre) was quite usual at the turn of the century. As a rule, the well-to-do occupied the larger and more elegant quarters which fronted towards the watercourses

58. STOCKHOLM

Fishing by this method is a common sight in the heart of the city.

59. STOCKHOLM
The Town Hall, a modern work of genius.

60. STOCKHOLM

The Old Town with its narrow streets. In the background is the Cathedral.

61. STOCKHOLM

A view of the Old Town taken from the tower of the Town Hall. Modern sections of the city can be seen on the mainland in the distance.

and the centrally-located thoroughfares. Low income groups lived in those flats which faced away from the street and towards the courtyards.

However, the housing of this period did not involve a distribution of social classes to different parts of the city. Labourers, white-collar workers and executives mingled on the same streets, and lived in the same blocks throughout the whole city. Only the locations of houses and floors separated them. As a result, Stockholm never acquired the pronounced slum areas or working-class districts which often characterize the chief cities of other countries.

After the turn of the century, Stockholm began to grow outside of the central city. A reaction against the City of Stone set in. Suburban railways were built, garden suburbs appeared. The change in town planning and construction, gradual and tentative at the outset of this period, became quite radical in time. The community, which since the beginning of the century had made sporadic attempts to influence developments, step by step expanded its influence by initiating active land, town-planning, and housing policies.

Town planning has now undergone a complete change. One can detect attempts to put radical architectural ideas of international scope into Swedish practice. However, the confrontation of these ideas with reality, has led to results which must have been quite unsatisfactory for the creators of the dream.

MUNICIPAL HOUSING ACTIVITY

The men who were most advanced in questions of housing policy and town planning aimed at creating better living quarters and a better social milieu for the broad masses, for those labourers and white-collar workers who were badly paid and poorly housed. This new outlook received wide, active support from different political parties; the only differences between parties resulted from disputes concerning the means to be employed in attaining the higher standard of building construction. General agreement prevailed that the community in various ways must undertake to raise the dwelling standard.

The housing constructed during this period is quite unlike that of previous periods. A new type of apartment-house dwelling, the 'three-story shallow house,' became the standard. Municipalities have approved this type of house because it is supposed to confer certain social advantages. The reason for building it three stories high is to avoid installing expensive elevators, and because it gives families with children 'direct contact with the earth.' Small children can reach the playground outside all by themselves, and mother can easily keep watch over them from her window. These houses are a little over nine yards wide, and each floor has two apartments, one on each side of the stairway. Each apartment covers the whole floor front to rear. Windows face in two directions, usually east and west, so that sunshine penetrates all rooms at some time during the day. All apartments are easily ventilated. And each one is as good as the next.

Considering Stockholm's historical development in terms of building construction, the reader will now be better able to understand the various policies currently in force. The following account will attempt to describe local government policy under four heads: (*a*) land policy, (*b*) planning of new residential areas, (*c*) housing, and (*d*) traffic management. Developments in these fields will then be summarized in relation to the general plan for Stockholm's future development.

LAND OWNERSHIP POLICY

Feverish building activity in Stockholm towards the end of the nineteenth century resulted in steeply rising land prices and intensive private speculation in real estate. The authorities thereupon decided to insure the city's expansion in the interests of the many by buying up tracts of land. Above all, they were interested in land necessary for the future development of the city. Thus, as far back as 1904, the city purchased houses and grounds on its outskirts. Such purchases have continued to the present day. By 1950, the community was the biggest landowner in Stockholm. Up to now the city has acquired 9,000 hectare (22,500 acres) for residential purposes and a much larger area for recreational and other purposes.

Ownership of land in Stockholm is divided as follows: the city owns one-third, and private interests own the rest. However, much of the city's holdings consist of streets and sidewalks, parks, and grounds for public buildings. Actual residential land in the business districts is for the most part privately owned. This circumstance involves considerable difficulties for the execution of plans calling for the rehabilitation of inner Stockholm.

A large portion of land acquired by the community is under consideration for future development. Moreover, the city of Stockholm has also purchased large tracts of land in various peripheral areas, which are intended solely for athletic and outdoor recreational purposes. Among these may be noted Tyresta and Tyresö, with a combined area of 2,000 hectare (5,000 acres). These areas still retain wild, unspoiled, natural beauty containing, among other things, five lakes and a stretch of salt-water beach six miles long.

Out in the Stockholm archipelago, the city has purchased large areas and there built vacation resorts, piers for small boats, and bathing facilities. These acquisitions give all citizens some chance during the summer to make use of the archipelago for sailing, fishing, and bathing.

When the city began building garden suburbs in 1907, the authorities did not sell the land outright; instead they granted leasehold rights, which gave the tenant use of the land but not title thereto. Nowadays, leasehold rights are usually granted for a 60-year period; the fee for land use is estimated at 4 per cent of the assessed value of land. In 1948 the total area of land granted under these conditions was 969 hectare (2,400 acres). Almost all new construction occurs with leasehold rights.

SUBURBAN DEVELOPMENT

During post-war years, strong criticism has been directed against the newly-built suburban residential areas. They have been attacked as failures, both from the social and aesthetic points of view. According to critics, the outstanding contribution of closely-built flats has been the creation of a monotonous and gloomy urban environment. People should live in homes of their own, they say, not in rented flats. The absence of industry and other business (with recreational facilities of their own) has gradually converted the suburb into a dormitory. At this point, Swedish discussion on the subject bears resemblance to the current international debate on 'neighbourhood units.'

Similar ideas for community betterment had gained currency among Sweden's bolder and more forward-looking town planners, who sought to apply them to Swedish conditions and problems. Thus, when designing new residential areas in Stockholm, planners have followed the system of dividing them into neighbourhood units. The size of each unit is dictated primarily by technical considerations. The reasoning along these lines has been as follows.

Within an area of this type, residents have easy and quick access to various local establishments, such as shops, cinemas, schools, and playing fields. Maximum distance to be covered on foot (from means of transportation to dwelling) has been calculated at 500 yards for apartment-house areas and 900 yards for one-family home areas. Considering the hilly terrain of peripheral Stockholm and this new type of residential development planned for it, authorities estimate that a population of between 10,000 and 15,000 can be accommodated in the area.

Certain rules are followed in locating different types of dwellings. In general, concentrated blocks of flats are built on areas close to the city centre and within easy reach of near-suburb railway and tram stations. One-family houses, much less densely spaced, dot the immediately adjoining areas. Houses and street networks are placed in such a way that all pedestrian traffic is removed a safe distance from street and highway traffic. For example, children on their way to school or to the nearest tram station can walk the distance without encountering any heavy vehicle traffic. Similarly, playgrounds and games fields remain free of cross-cutting arterials or roadways.

Planners use two rough measures of quality of residential areas: population density and 'floor space index.' Density of population measures the number of inhabitants per hectare (or per acre) of land while 'floor space index' measures the relation between total residential surface and total land area.

In those areas described herein, population density is 80 inhabitants per hectare—32 to the acre—in one-family home areas, and 200–300 inhabitants per hectare—80–120 to the acre—in apartment-house areas. Corresponding floor space indices are 0·2 and 0·5–0·7 per hectare.

HOUSING STANDARDS

Town planning in 1950 clearly deviates from the ideas which held sway in the early 1940's. Instead of the three-story shallow house which dominated then, a more differentiated pattern of building construction is the order of today.

The elegant designs on an architect's drawing board immediately capture an observer's fancy. So do the natural surroundings of a residential area. Less obvious, though equally important in listing the components of a 'dwelling standard,' are the size and interior equipment of each individual dwelling. A couple of facts stand out in this connection: (1) the smallness of the individual dwelling and (2) the high rents. Crowding inside of four walls is the usual consequence. However, this does not necessarily mean that the average dwelling standard is low. The newer apartments can display the blessings of modern technology. At the same time, loans and subsidies from state and municipality have reduced rents for the mass of the people. During the 1940's a speedy alleviation of the crowding problem coincided with a relatively drastic reduction of construction costs. It is a paradox that general standards have risen despite the pressing housing shortage, which arose during the war years and still remains.

The table below reveals that, in 1939, 50 per cent of all dwellings consisted of not more than one room and kitchen. Even up to the outbreak of the Second World War, this type accounted for half the new construction. A radical change set in during the 1940's. House building began to allot a lesser share to the small apartments. In 1947, 40 per cent of production consisted of apartments with three or more rooms.

TABLE III

The size of dwellings

Type of dwelling	Existent 1947	Production 1939	1947
One room and kitchen	49	47	17
Two rooms and kitchen ..	28	31	43
Three or more rooms and kitchen	23	22	40
	100	100	100

To speak of one 'dwelling standard' is perhaps misleading, since older and newer parts of the city offer striking contrasts. Those apartments built since 1930 boast a uniformly high technical standard. In this connection it is meaningless to distinguish between dwellings for different social classes or levels of income; dwellings for labourers, white-collar workers, and persons of high incomes are approximately of the same standard. With occasional exceptions, no dwellings are built exclusively for special social groups. Production uniformity is largely the result of minimum requirements set up by the state for the granting of loans.

All newly-built apartments thus feature central heating, water closet and bathroom, as well as modern kitchens equipped with stainless steel dish racks, ingeniously designed cabinets (pre-fabricated), electric ranges and refrigerators. Kitchen planning and equipment mostly follow the model of an 'ideal kitchen,' which the state-subsidized Home Research Institute designed after intensive scientific study. In almost every large new block of flats—or in an area serving several smaller houses of similar type—can be found a laundry room, with mechanical equipment available without charge to all residents.

The average dwelling standard is high chiefly because 50 per cent of all dwellings were built after 1930, and are, therefore, less than twenty years old. Moreover, even houses existing before that time have been renovated to a large extent. The following figures show the number of dwellings with central heating and bath for the years 1915 and 1950.

TABLE IV

Percentage of dwellings with	1915	1950
Central heating	10	90
Bath	9	70

Many dwellings are of course far below the standard in various respects. Their number has increased during latter years because of the ever more limited scope of clearance and reconstruction activities; thus, current deterioration has not been remedied.

Stockholm has been called 'the city of small families.' A low birth rate, a heavy influx of unmarried persons and the unattached Stockholmer's penchant for acquiring a dwelling of his own, have resulted in a large number of households consisting of one or two persons. In 1945, nearly 50 per cent of the city's 23,000 households fell in this category. Only 23,000 households consisted of five or more persons (about 10 per cent). Families with three or more children were therefore few in number. This circumstance accounts for the heavy demand for small dwellings, and for the stubborn persistence of contractors in building them right up to the end of the 1930's.

Population density in Stockholm is not remarkably high when judged by living conditions in Sweden. The number of people per dwelling has steadily declined from 4·5 in 1900 to about 2·7 in 1950. However, far too many families with children continue to live in the older one-room-and-kitchen apartments. In Sweden, a dwelling is usually considered overcrowded and a family 'cramped,' when there are more than two persons per room (kitchen not included). A family with three children would therefore be classified as cramped if it lived in a two-room apartment, and a family with five children if it lived in a three-room apartment, and so on. According to this criterion 25 per cent of all Stockholmers, and 43 per cent of all children under fifteen years of age, live in crowded dwellings (1945).

THE COST OF HOUSING

To make international price comparison of dwelling costs in different countries is facilitated if not only rents (calculated according to current rates of foreign exchange) are reckoned, but also dwelling costs in relation to the incomes of a representative social group. A comparison, to be

TABLE V

Rents in Stockholm, 1950

Type of dwelling	Equipment	Floor Space		Yearly rent 1950 excluding fuel costs		including fuel costs		Rent in dollars (incl. fuel) per sq. foot	Rent (incl. fuel) as % of an ind. worker's average income 1950[1]
		sq. metres	sq. feet	Sw. kronor	$	Sw. kronor	$		
1 room and kitchen									
Built before 1921	Without central heating and bath	33	355	670	129	900	173	0·49	13 %
Built 1950	Central heating, bath, modern kitchen equipment, refrigerator, laundry machine	38	409	770	146	1,000	192	0·47	14 %
2 rooms and kitchen									
Built before 1921	Without central heating and bath	53	570	955	183	1,200	231	0·41	17 %
Built 1921–1931	With central heating and bath	63	678	1,500	289	1,850	356	0·53	26 %
Built 1950	With central heating, bath, modern kitchen equipment, refrigerator and laundry machine	55	592	1,500	289	1,875	361	0·61	27 %
4 rooms and kitchen									
Built 1921–1945	With central heating and bath	103	1,109	2,360	453	2,880	554	0·50	41 %
Built 1950		95	1,023	2,500	481	3,100	596	0·58	44 %

[1] = 7,000 Sw. kronor.

valuable, must also reveal understanding of the products to be compared. Table V lists the costs of dwellings of varying types and standards within Stockholm. These costs are also presented in terms of an industrial worker's income (average income during 1950 is calculated at 7,000 kronor).

Since 1939, rents in Stockholm have remained practically unchanged despite the rise in construction costs (up to 60 per cent, according to various indices) and the pronounced shortage of dwellings, which has been chronic since 1945.

Rents have lagged far behind the other price indices. During the same period, for example, the wages of industrial workers rose at least 70 per cent. A larger mass of people have thus been enabled to improve their dwelling standard. Before the war, an industrial worker setting aside 20 per cent of his income could not possibly afford to rent a modern apartment with more than one room and kitchen, or an older apartment with more than two rooms and kitchen. Today, the same portion of his income can command a modest three-room apartment.

POPULATION PROSPECTS

During the post-war years, the population in the Stockholm region has increased by leaps and bounds. Current predictions do not rule out the possibility that the population will number between 1,200,000 and 1,500,000 by 1970. The city administration reacts to this prospect with mixed feelings. A big problem is, of course, whether such an expansion will be advantageous for the people already living in Stockholm. Doubts are expressed not only about maintaining a decent human way of life, but also about the ability to manage an economy of metropolitan dimensions. These two problems are inextricably intermingled. In order to improve conditions of civilized living, the city must rearrange its layout for maximum efficiency. This implies a heightening of technical standards which will be very expensive to maintain. The big question is: will further growth of the city confer economic advantages to compensate for additional expense to the community?

Doubts flourish particularly in those areas outside of Stockholm most immediately affected by the growth of the central city. People in the country, having seen more and more of the region's youth migrating to Stockholm, fear threatening stagnation. Farm and forest regions in Sweden are sparsely populated; many responsible people therefore favour encouraging the growth of small and medium-sized cities in other parts of Sweden.

62. STOCKHOLM

A summer's night scene in the Old Town

[*Photo: Refot. A-B, Stockholm*

63. STOCKHOLM

Looking north from the centre of the city across Lake Maeler. In the foreground is a series of highway ramps leading up to the Southern Heights.

[Photo: *Refot A-B, Stockholm*

Nevertheless, planners reckon with a 400,000 population increase within the next twenty years, a rise from the present figure of 970,000 to 1·3–1·4 million. This calls for a frontal attack on several problems. To begin with, additional construction in the 'Inner City'—including the Old Town and the City of Stone—is an impossible idea. On the contrary, the city plans to extend its holdings for public purposes and to reduce the number of dwellings. In achieving this aim, the city relies on a 'rehabilitation campaign,' to be completed by 1970. Many buildings will be torn down; a higher dwelling standard should be the result. Removal of dwellings is expected to force 140,000 persons to move out of Stockholm. Add these to the 400,000 expected via channels of normal increase, and the difficulties mount. Housing for about 540,000 people must be found in dwellings outside of Stockholm during the next thirty years.

FLATS OR HOUSES?

Those unfamiliar with Swedish conditions have probably posed this question more than once: why build so few one-family homes and so many blocks of flats in Stockholm? A related question, widely discussed in Sweden, is: aren't these apartment-house areas too monotonous, ugly and unpleasant to live in?

Among the factors which influence dwelling production in favour of multi-family houses, the high cost of construction has certainly been decisive. To this must be added a historical tradition: in the past Sweden's ties to the Continent, especially to Germany, were very strong. The result was Sweden's uncritical adoption of continental methods in the building of cities during the nineteenth century.

Also, for a long time the Swedish standard of living had been very low. A labouring man in those days usually had a large family to support. Out of his meagre wages, he could not afford to pay rent for more than a single room; at best, he might rent a flat with one room and kitchen. Dwelling units of such limited size clearly ruled out all thoughts of private house building. Moreover, the terrain of inner Stockholm proved to be hardly suitable for house building. The arrival of the twentieth century with its more abundant resources promised improvement, but by that time the prejudices of architects and resident families were too deeply rooted to permit a normal development. True enough, economic expansion did encourage a rise in the dwelling standard. However, this improvement was not in the direction of more living space, but rather towards the installation of superior equipment. Thus, central heating, bathrooms, and refrigerators took precedence over larger dwellings. In this connection, it should be mentioned that one-family houses, both detached and in rows, are almost never rented. No Swedish law forbids renting, but Swedish custom does. Therefore, the only alternatives available to a Swede are to rent an apartment in a multi-family dwelling or to buy a one-family house.

The housing market was characterized by these conditions until the 1940's. We have shown how recent production is more and more emphasizing larger residential types which readily coincide with the requirements of detached dwellings or terrace houses. The big question remains: do Stockholmers really want to live in private homes? If so, under what conditions is the city prepared to fulfill this wish?

In 1946, an investigating committee prepared an exhaustive report which summarized the opinions of the public on the subject. A large number of families were personally interviewed (non-family households were excluded). Families were asked whether or not they were satisfied with the type of dwelling in which they then lived. The answers revealed some interesting attitudes. Thirteen per cent of the families investigated lived in private homes; 51 per cent declared that they would like to move into homes of their own. Of those living in private homes, 95 per cent said they were satisfied. For people living in suburban blocks of flats, the percentage of satisfaction was very low—21 per cent. Indeed, only 7 per cent of all respondents gave suburban flats as their first choice. Strange then, that 90 per cent of all dwellings now under construction are of just this type! Nevertheless, this much is clear: the Stockholmer's attitude towards his dwelling closely approximates the attitude held by an American or Englishman.

Municipal authorities have not as yet committed themselves publicly. However, in order to meet the demand for one-family houses, present production would have to be quadrupled.

Despite the fact that Stockholm is a relatively small metropolis, the general plan for the city recommends a limitation to future concentration of business within the central zone. Quite intelligible economic and social considerations stand behind the idea.

DECENTRALIZING THE CITY

Were concentration of business within the city centre to continue, the resultant congestion of traffic arteries would require drastic and expensive solutions. Decentralization of business would reduce the amount of traffic (i.e. in the bottlenecks of the Old Town), and, above all, would lessen the risk of traffic jams during rush hours. It is primarily these traffic jams which account for added construction difficulties and, of course, higher costs.

However, planning does not go so far as to call for erection of special 'satellite towns' (that is, independent towns located some distance away from the metropolis). As far as possible, business and commercial establishments are urged to follow the example of industry in locating outside of Stockholm. If such drastic relocation is not feasible, every attempt is to be made to move them out of the city centre, at the very least. Estimates say that half of all gainfully employed workers will in the future both live and work in the suburbs.

A new tube is under construction, and will soon be added to the existing transportation network. If all goes as planned, no more than 30 minutes will be needed to negotiate the distance from suburb to business centre, and no more than 45 minutes from door to door. Maximum capacity of transportation facilities is based partly upon an estimated number of trips per annum for every inhabitant, partly upon the rush hour load of main lines.

In 1950 there were five automobiles for every 100 inhabitants, but this ratio is expected to rise to 10 per 100 by 1970. To this must be added the heavy volume of bicycle traffic in Sweden.

We have seen that the heart of the city, lower Norrmalm, was built mostly in accordance with plans which are now three hundred years old. Rebuilding of this part of the city has long been considered, but municipal authorities did not come forward with a plan until recently. The chief reason for the delay is the difficulty in reconciling old Stockholm's aesthetic values with the modern era's demand for efficiency. In most cases the 'materialists' have scored convincing victories over the 'aesthetes,' but the latter have not given way easily. Every new bridge or traffic innovation has had to overcome the bitterest opposition.

The new city plan leaves the Old Town with its historical landmarks (like the royal palace and government administration offices) undisturbed. However, radical measures are planned for the business district; large modern streets will penetrate existing structures to become arterial highways for through traffic.

Some residential areas in the City of Stone will be gradually cleared or renovated; it is expected that this project will be completed by the year 2,000. Although rehabilitation plans are already drawn up for large areas, only a small proportion of these have been carried out. An inevitable consequence of rehabilitation will be a much lower rate of land utilization. Density of population in affected central areas is expected to be 700 inhabitants per hectare, which corresponds to a floor space index of about 1·5.

Rehabilitation is practicable only if the city buys up property in sub-standard areas. In the past few years, the city has gone in for systematic purchase of houses, but, as yet, it owns only a small part of the total to be demolished. The rest will apparently have to be expropriated. Officially, nothing has been said about the costs of such an operation, but they will probably be high since the city must pay those prices prevailing in the open market.

Sydney

F. A. BLAND

M.P. since 1951 (in Federal Parliament, Canberra) for constituency of Warringah; Chairman, Joint Committee on Public Accounts. M.A., LL.B. (Sydney); Emeritus Professor of Public Administration, University of Sydney (occupied Chair from 1935 to 1948). Assistant Director of Tutorial Classes, University of Sydney, 1917–35. Visiting Professor of Government, New York University, 1929–30. Fellow of the Senate of the University of Sydney since 1944. Chairman, Local Government Examining Committee, New South Wales; and formerly Chairman, Public Service Examinations Committee, New South Wales. Editor (1937–48) of *Public Administration*, the Journal of the Australian Regional Groups of the Institute of Public Administration.

Author of *Shadows and Realities of Government* (1923); *Planning the Modern State* (2nd ed., 1945); and *Budget Control* (4th ed., 1946); Editor of *Government in Australia: selected readings* (2nd ed., 1944), and *Changing the Constitution* (1950); contributor to learned journals.

Sydney

THE founding of Sydney, the capital of the state of New South Wales in the Commonwealth of Australia, was a direct result of the successful revolt against British rule by the American colonies in 1776. Let us trace the connection.

Medieval geographers had been confident that a large southern continent existed, and their beliefs were fortified by the explorations of Portuguese and Spanish seamen commencing with the sixteenth century. In the seventeenth century, the Dutch explorers sketched in parts of the north, west, and south coast of Australia, while Abel Tasman found Tasmania and New Zealand in 1642.

SYDNEY'S HISTORY: 'A THOUSAND SAIL'

Our story begins with the British scientific expedition to the South Pacific led by Captain James Cook, and including Joseph Banks, which was instructed to determine the nature and extent of the southern continent. In 1770, after sailing round New Zealand, Cook sighted the east coast of Australia on April 20th and charted the coast up to the Torres Strait.

Transportation had long been a statutory punishment in England, and when it was impossible to continue the transportation of criminals to North America, an alternative outlet had to be found. The matter became more urgent when space in the prison hulks on the Thames proved inadequate, and penal settlements in Africa failed. Sir Joseph Banks pressed the claims of Botany Bay, which he had explored during Cook's voyage, and during the next decade Banks's suggestions were expanded to cover a possible home for North American loyalists, displaced by the revolt.

After much hesitation, the Pitt ministry issued an Order in Council on December 6, 1786, authorizing the establishment of a penal settlement on 'the Eastern Coast of New South Wales,' and selected Captain Arthur Phillip, a man of considerable resource, initiative, and human kindness, as the first governor.

Phillip acted with expedition in preparing for the adventure, and on May 13, 1787, sailed from England in 'the first Fleet,' which consisted of eleven small ships totalling less than 4,000 tons, and which transported 586 male convicts, as well as 695 officials, marines, and crew. The voyage occupied eight months, and ended in Botany Bay on January 18, 1788, when Phillip had the satisfaction of knowing that he had only lost 32 by deaths on the way.

Botany Bay proved barren and inhospitable, as it is today, and Phillip sent a party to explore another inlet that had been noted a few miles to the north. The reports being satisfactory, Phillip moved north and on January 26, 1788, unfurled the British flag on the shore of Port Jackson, which he enthusiastically described as 'the finest harbour in the world in which a thousand sail of the line may ride in perfect security.'

Phillip's enterprise not only secured for Sydney a most picturesque setting, but he enhanced the beauty of the natural surroundings of the city, whose greatness he envisaged, by setting apart areas that are now the famous Botanical Gardens, the Domain, and Hyde Park. Other large parks within the city boundary, later dedicated, are Moore and Centennial Parks, similar as to area and purpose to Kensington Gardens, Hyde Park, and Green Park, London. Outside the city are the National Park (34,000 acres) to the south, and the Kuringai Chase to the north covering 38,600 acres.

During the four years of Phillip's administration, over 4,000 convicts arrived, most of whom were ill-fitted for the pioneering work of housing and feeding the populace, and for winning from a reluctant soil and sparsely watered territory a livelihood for an ever growing company.

For sixty years Sydney was used as a penal settlement, although its future as a free colony was strongly pressed upon the 'Home' government. The conditions attaching to a penal settlement plainly mark the differences between the foundation of Sydney and the colonization of New England in the United States by the Pilgrim Fathers and their successors, and show clearly why the free institutions of New England could have little scope in Sydney.

Governors for the first three decades (until Macquarie) combined in their persons all the civil functions of law-maker, judge, and administrator, as well as police powers and defence. After Macquarie, judicial and political reforms in 1823 and 1828 provided for a hierarchy of criminal and civil courts, and for a nominated executive council to advise the governors on legislative and administrative matters.

Experience gained by participation in the conduct of public affairs, and colonial echoes of the agitation that resulted in the passing of the Municipal Corporations Act in England in 1835, prompted Governors

64. SYDNEY

A general view of the magnificent harbour and its bridge. In the foreground are quays and warehouses.

65. SYDNEY

A closer view of the Harbour Bridge seen from Lavendar Bay.

[Photo: B.O.A.C.

66. SYDNEY

William Street, a main artery leading to King's Cross, a residential suburb seen in the background.

[*Photo: Australian News*

67. WELLINGTON
The city viewed from the East.

[Photo: National Publicity Studio. Wellington

Bourke and Gipps in the late 'thirties to urge the colonists to take some responsibility for the administration of Sydney. Beginnings had been made with those services inseparable from the working of any organized community—markets, roads, and police.

Markets established in 1806 passed first to magistrates and then, in 1839, to elected commissioners. Roads, constructed by the governors, were maintained by tolls (1810) and in 1833 were committed to trusts which might levy rates for construction and maintenance. A police force, constituted by Macquarie in 1810, was maintained from a Colonial Police Fund credited with the proceeds of three-fourths of the revenue from customs and excise. After 1820, the fund was known as the Colonial Fund, and it was charged, in addition to the cost of police, with 'ornamenting and improving the town of Sydney, and in constructing and repairing quays, wharfs, streets, and roads within the limits thereof.' In 1833, three magistrates were appointed to regulate the police in the town and port of Sydney.

INCORPORATION OF SYDNEY: THE CITY'S PEOPLE

Such was the extent of the collaboration in local administration by appointed officials when, with the approval of the colonial secretary, Governor Gipps secured the enactment of 6 Vic. No. 3—'An Act to declare the town of Sydney to be a City, and to incorporate the citizens thereof' (July 20, 1842). The area of the city was approximately four square miles.[1]

The council was elected in November 1842 by the males of the city of 21 years of age who had resided in or within seven miles of the city for the past twelve months, and who possessed property of the annual value of £25.

The qualification for a councillor was the possession of real or personal property either in his own right or that of his wife of £1,000, or being rateable on an assessed annual value of £50.

The structure and functions of the council followed closely the model of the English Act of 1835, and provided for elected councillors, co-opted aldermen, assessors and auditors, and the annual election of a mayor by and from the council. Functions covered matters relating to public

[1] Adelaide, the capital of South Australia, has the distinction of being the first municipality of the Commonwealth to be incorporated, having been established two years prior to Sydney. Melbourne (1842), Perth (Western Australia, 1851), Hobart (Tasmania, 1857) and Brisbane (then part of New South Wales, but now capital of Queensland, 1858) follow in the order named.

T

health and sanitation, lighting and maintenance of streets, markets, water supply, licensing of vehicles, and partial control of police.

The duties of the new corporation were not matched by adequate resources, and it had difficulties in obtaining a competent and experienced administrative staff. Complaints by dissatisfied ratepayers led to re-criminations between the council and members of the Government, and finally to the suspension of the council between the years 1853–57, during which time the city was administered by three appointed commissioners.

When the charter of the corporation was restored in 1857, the composition of the council was limited to sixteen aldermen elected from eight wards for two years, and a mayor elected by and from their number, while the police and licensing powers were withdrawn. The franchise was exercised by enrolled adult males who were ratepayers.

The corporation was dissolved again on December 31, 1927, by the Sydney Corporation (Commissioners) Act because of malpractices in letting contracts and in administration. Three commissioners were again appointed until June 30, 1930, and given the task of 'cleansing' the administration. Under another amending Act (1929), the city was divided into five wards, each electing three aldermen, and the new council took office on July 1, 1930.[1]

Although the area of the city of Sydney was at the time of incorporation slightly more than four square miles, additions occurred in 1870 and 1908 which enlarged the area to approximately five square miles. The suburban area was then 180 square miles, but now it is approximately 240 square miles.

The population of the city and its development are indicated by the following figures:

Year	Sydney	Suburbs	Total, Metropolitan	Total for State
1911	112,921	516,582	629,503	1,646,734
1941	83,720	1,253,330	1,337,050	2,812,321
1947	95,925	1,388,079	1,484,004	2,984,838
1952	207,540	1,413,500	1,621,040	3,421,768

The city proper is mainly inhabited by hotel dwellers, caretakers, occupiers of service flats and some few professional men and women. On the immediate outskirts of the city are the wharves, docks, ware-

[1] The city of Melbourne has had the distinction of maintaining almost intact the origina form and character of its governmental structure until 1938 when a government bent upon democratizing its structure abolished the body of aldermen and many of the old forms and ceremonies, and left the city with a council of 33 members (the largest in Australia) elected from eleven wards, three for each ward.

houses, and some industries, and in these districts the population is almost entirely made up of industrial workers. Here the population shows a progressive decrease, a decline that will continue as the demolition of slum areas proceeds and as transport and housing accommodation in the suburbs improve.

Similar features are present in the industrial suburbs abutting the city boundaries, where there is a large number of sub-standard houses about a hundred years old, and where factories have tended to replace poor-class dwellings. Eight of these industrial municipalities were added to the city by the Local Government (Areas) Act, 1948, which provided for the amalgamation of smaller municipalities into larger areas, and the addition of these areas to the city will tend to strengthen the control of policy by Socialists in the council.[1]

The addition of these areas to the city has increased its size to 11 square miles and brought its population up to 207,540 (1952), which is still only about one-seventh of the total metropolitan population.

SYDNEY IN THE COMMONWEALTH

As capital of the 'Mother State,' Sydney enjoys a political pre-eminence in the Commonwealth that is only now being challenged by the Commonwealth's garden city capital at Canberra. It is the seat of the New South Wales Parliament, and the headquarters of all the state and some of the Commonwealth departments. The Parliament consists of the Governor, appointed by the Queen on the nomination of the Secretary of State for Commonwealth Relations, with the approval of the state cabinet, and two houses of the legislature, viz. the Legislative Council and the Legislative Assembly. The Legislative Council is elected by members of the council and the assembly, and consists of 60 members elected for twelve years in groups of fifteen every four years. The Legislative Assembly consists of 94 members elected for three years by adult suffrage. All enactments are 'by the Queen's Most Excellent Majesty, by and with the advice of the Legislative Council and the Legislative Assembly in Parliament assembled.'

Sydney is also the registry for the Supreme Court of the State of New South Wales. The court was created by the Charter of Justice in

---

[1] This shift in the Socialist influence in the council is produced by the re-districting of the city into ten wards, each returning three aldermen, instead of the arrangement of five wards in the old city, each returning four aldermen. As a result of amalgamation and the re-districting, the Socialist industrial suburbs will thus cause stronger Socialist representation in the governing body of the city.

1825, and consists of a chief justice, and a number of puisne judges. The Supreme Court exercises both a civil and a criminal jurisdiction, the former being distributed into various branches, including a jurisdiction at common law, in equity, in probate, and in divorce. The common law jurisdiction is exercised in part at Sydney, and in part at circuit courts held throughout the country. There are also a number of other courts including a vice-admiralty court, land appeal court, courts of general and quarter sessions, district courts, courts of petty sessions presided over by stipendiary magistrates in the city, or by police magistrates in country towns.

Sydney is one of the capital cities for sittings of the High Court of Australia. The fact that the High Court and the Supreme Court have their registries in Sydney makes that city the headquarters of the legal profession which in New South Wales is separated into solicitors and barristers, and only the latter may appear in certain jurisdictions. There are therefore no barristers practising in country towns other than when on circuit.

The headquarters of the medical and other professions is situated in Sydney. As far as medicine is concerned, the teaching hospitals associated with the University are all in Sydney, and save in Newcastle, a city of about 140,000 over a hundred miles distant from Sydney, few, if any, medical specialists could be consulted outside Sydney.

The University of Sydney was the one university in the state of New South Wales until 1948, when there was launched the University of Technology. One country college affiliated with the University of Sydney and conducting courses up to degree standard in arts and science has been established at Armidale—the New England University College—but attempts to found other colleges have been unsuccessful. The college was given autonomy in 1954.

The concentration of political, judicial, and educational facilities in Sydney results in attracting all the enterprising and ambitious youth of the country towns to the metropolis. The dominant position of Sydney is also enhanced by the centralized character of the administration of the state. There is little effective local government, measured by standards obtaining in England and the United States, and hence an administrative career is thought to be crowned when a public servant returns to Sydney after an involuntary term of office in country towns to gain experience.

These things naturally suggest that Sydney is also the unrivalled economic capital of the state. The fetish for centralization of administration results in all the railway lines of the state fanning out from Sydney so that the bulk of the overseas imports are distributed through Sydney, and exports, such as wool and wheat, are brought back by the railways to the waiting ships. Only recently has there been any attempt to use Newcastle, the only other important seaport, for the shipment of wool and wheat overseas. Consequently, in Sydney will be found the head offices of the banks, the insurance companies, the brokers, the wool companies, and the shipping lines. Railway mileage has grown from 3,761 in 1911 to 6,113 in 1953, while capital expenditure increased from £61 millions to £213 millions. In 1952–53, 271,698,000 passenger journeys were made and over 17,877,000 tons of goods carried. The volume of shipping through Sydney is revealed by the table below:

Year	Cargo discharged (tons, 000's)	Cargo shipped (tons, 000's)	No. Vessels	Tonnage, vessels (tons, 000's)
1930	4,125	2,034	5,877	8,143
1940	5,143	2,860	6,530	9,984
1950	5,658	2,637	3,927	8,149
1952	6,977	1,804	3,938	8,261
1953	5,204	2,022	4,163	8,540

THE MACHINERY OF THE MUNICIPAL GOVERNMENT

The government of the city is entrusted to a council elected for three years,[1] the members of which are unpaid.[2] The franchise extends to every person who is qualified for enrolment either as owner of property or as ratepaying lessee. The division of the city into wards has been abolished, and the principle of proportional representation applies to the election of aldermen. Voting has recently been made compulsory, thus bringing the municipal practice into line with that which has existed in parliamentary elections for about three decades.[3]

The lord mayor is directly elected by the citizens and holds office for three years. He is an alderman by virtue of his office. He receives an allowance, not to exceed £5,000, fixed by the council for each year

[1] Brisbane, like Sydney, has a lord mayor and aldermen; Melbourne and Perth have a lord mayor and councillors; Adelaide and Hobart have a lord mayor, aldermen and councillors, but aldermen are not co-opted as they were in Melbourne until 1938. In Hobart and Perth, the tenure of the lord mayor is two years, in Brisbane three years, and elsewhere it is one year. In Hobart, Brisbane, Adelaide and Sydney, the lord mayor is elected by the ratepayers, in the other cities election is by and from the council.

[2] In Brisbane, aldermen are paid £750 per year for their services, while the lord mayor's allowance there is £2,000 per year.

[3] Substantial changes have been introduced recently by the Local Government (Amendment) Act, 1953. Prior to this the lord mayor was indirectly elected by the council, and the city was divided into wards. An Act of 1952 abolished plural voting.

of the mayoral term. On a certificate issued by the Minister of Local Government he can receive such additional sum not exceeding £2,500 as the minister may consider reasonable, having regard to any special circumstances which may arise.

The work of the council is organized on the committee system, the more important of which are (*a*) the general purposes committee (comprising the whole council and dealing with staff matters and anything not relegated to other committees), (*b*) finance committee, (*c*) works committee, and (*d*) health committee. These committees have modelled their methods upon the British committee system, but they have no experience of the British system of co-option.

The city has no jurisdiction over any of the suburban municipalities: it is merely one council in a large number operating throughout the metropolitan area. There is nothing akin to the London metropolitan boroughs' joint standing committee for securing voluntary co-ordination.

THE CITY'S FINANCE

The government of Sydney is financed from several sources, the most important of which is the rate on real property. Until 1908, rating was on the annual rental value of realty, but in that year the council was empowered to change over to rating on unimproved capital value, a system introduced for general use by the Local Government Act, 1906. The city had its own types and limits of rates and operated under these conditions until the general rating system was applied to the city on January 1, 1949. Rating limits were removed in 1953.

The following table shows the amount in the £ of the general rate, the unimproved capital value, and the yield, for each year since 1949.

TABLE I

GENERAL RATE LEVIED

	Amount in £	Unimproved Capital Value of Rateable Land	Yield
1949	6d.	£68,269,814	£1,699,159
1950	6d.	£69,155,177	£1,725,968
1951	7½d.	£70,556,374	£2,219,294
1952	9¼d.	£71,602,659	£2,794,411
1953	8d.	£104,180,423	£3,472,682
1954	7d.	£106,000,000 (approx.)	£3,115,103 (est.)

In 1953 the improved capital value of rateable land in the city was £289,451,509, and the assessed annual value was £17,059,403. The unimproved capital value was £104,180,423 and the produce of a general rate of 8d. in the £ was £3,472,682.

Other revenue is secured from municipal rents, licences, and dues. No grants-in-aid or shared revenues are realized from the state or the Commonwealth. The table on page 584 reveals the revenues and expenditures of the city of Sydney for the fiscal year ending December 31, 1952.

RELATIONSHIPS WITH THE STATE

Under the federal constitution of the Commonwealth of Australia local government is one of the residual powers retained by the states. It follows, therefore, that the national parliament has no authority to legislate for the organization and regulation of local government within a state. The general pattern of the legal relationships between the state governments and their cities is determined by the acts creating the municipalities. From its incorporation in 1842, and the restoration of 1857, until 1948, the city of Sydney was dealt with in special state acts supplementing or varying the original charter. When a tentative system of general local government was begun in New South Wales in 1858, and consolidated in 1906 and 1919, the city remained outside the over-all plan. By the Local Government (Areas) Act of 1948, however, the special position of Sydney was swept away, and the city was brought into the general local government scheme for the state.[1]

Under the general system of local government, the city of Sydney is subject to the direction and control of the State Department of Local Government. In many ways, this department is similar to the British Ministry of Housing and Local Government. As one of sixteen ministries, headed by a member of the cabinet, the department has been charged since 1906 with the oversight of the activities of all bodies engaged in local government.

The Department of Local Government is the primary focal point for all administrative relationships between the city and the state. The department makes regulations for the organization of local government. It establishes rules covering the procedure at council meetings, the qualifications and status of local government officers, the maintenance of accounts, and many other specific activities of the municipalities. City councils are thus dependent upon the department for their detailed powers of operation under the Local Government Act, and cannot make their own ordinances, regulations, or by-laws. The department demands

---

[1] The difference between the treatment of Sydney and other municipalities at the hands of the state has always been a matter of degree, except that in the case of Sydney the control by the Department of Local Government has been less rigid.

TABLE II

REVENUE AND EXPENDITURE ACCOUNT, 1952

	£	£
Rates and extra charges	2,797,555	
Rents, dues, fees, etc.	1,193,322	
	3,990,877	
Sales of assets and repayments by home purchasers	8,466	
		3,999,323
Less: applicable to segregated funds— Loan repayment reserve—		
Interest on investments	148,446	
Sales of residue lands	24,262	
Insurance fund—		
Interest on investments	14,215	
		186,923
.		£3,812,400
General expenditure		4,671,157
Less: Charged to loans	268,166	
Charged to insurance fund	23,987	
Capital expenditure and depreciation written off	950,496	
Loan discount and expenses written off	68,531	
		1,311,180
		3,359,977
Contributions to loan repayment reserve ..		101,263
Insurance fund premiums		24,916
Assets purchased	545,182	
Less: charged to loans	210,834	
		334,348
Loans and advances repaid	1,063,765	
Less: met from reserve	1,057,452	
		6,313
Previous year's expenditure in anticipation of loans transferred against available funds ..		24
		£3,826,841
Balance being excess of expenditure over revenue for year carried to available funds account		£14,441[1]

[1 *Footnote opposite*

that its approval be obtained to all fixed loans, including over-drafts, the fixing of wards and ridings in municipalities and shires, and to the establishment of local councils. It examines all the annual statements of accounts of councils, and makes test inspections to see that the provisions of the ordinances dealing with accounts are observed. It polices the integrity of local government officers, and, if necessary, may serve notice on any officer to show cause why his certificate of qualification should not be cancelled. In addition to reviewing all state legislation relative to local government, the Department of Local Government provides an advisory service to the localities, supplying without cost technical advice and model plans for engineering works.

After the passing of the Government of the Australian Colonies Act, 1850, all the states had to find some way of compelling the localities to look after their domestic matters, and most of the early local government acts prescribed a range of functions which has seriously shrunk with the passing of the years. No local authority in existence today possesses powers as comprehensive as those of its predecessor. Powers have either been resumed by the state, or have been committeed to *ad hoc* bodies.[2]

AD HOC AUTHORITIES

Roads and bridges, essential in any community, became the duty of a state department in the 1850s, after the failure of the Road Trusts. Following a return to local government at the beginning of the twentieth century, main and developmental roads have become the responsibility

[1] Current data are not available at the time of this writing. While later figures will modify the absolute amounts shown here, the relation of the elements of the revenue and expenditure patterns will not be altered substantially.

In addition to the above accounts, the city has outstanding loans amounting to £8,308,715 for capital works. To meet these loans there is a sinking fund which is invested in Commonwealth stock and in City of Sydney debentures, amounting to £4,509,852. Total assets of the General Fund in 1952 were £16,783,769.

[2] The refusal of the early colonists to back up the governors and the 'Home' authorities in their efforts to establish a comprehensive system of local government in New South Wales is well known. What has not been equally realized is that a hundred years later that failure makes the working of popular government very difficult, because there is no strong local opinion to counteract the forces of bureaucratic centralization. Vast distances, sparse settlements, immense tracts of unoccupied Crown lands, were admittedly unfavourable to the development of local government because, without very generous financial assistance from the state, no services of municipal government could be supported by the locality from rates, fees, licences, and other normal forms of revenue alone. And behind all the proposals for local effort, there has been the recollection that for sixty years or more, the state has provided roads, schools, harbours, water supply, and other services from its own funds. Why should the individual pay rates as well as taxes?

T*

of a Main Roads Commission which is sustained by government grants, municipal contributions, licence fees, endowments, and revenues that were persistently refused to local authorities. The commission is appointed by and is under the control of a state minister.

Water supply and sewerage in Sydney was first the job of the governors, and then of the city council from 1842. Insufficient and inefficient water services were partly responsible for the suspension of the city charter in 1853, but these services went back to the city in 1857. For thirty years there was a constant failure of supply to keep up with demand, and, finally, in 1888 an *ad hoc* authority was created which has since controlled water, sewerage, and drainage for the whole metropolitan area. Repeatedly reconstituted, that authority has always preserved a link with local government in that some of its members have been elected by the city council, and others by the suburban municipalities. It is as nearly 'independent' as a statutory authority can be with the Government appointing the full-time executive president, as well as a part-time vice-president.

Abattoirs were a city function, but complaints of inadequate facilities offered an opportunity for the undertaking to be handed over to an independent statutory body whose personnel are appointed by the Government.

The fish markets have gone through the same cycle, and there are serious suggestions for taking the fruit, vegetable, and other commodity markets away from the city council and handing them over to a government appointed commission.

Road transport and tramways are also controlled by a government appointed commission.

Fire protection is in the hands of another statutory corporation and, like the Water and Sewerage Board, has maintained direct connection with local government in that city and suburban municipalities elect representatives. The president of the board is a full-time executive officer appointed by the Government.

Sydney's city council was authorized by statute to undertake the supply of electric light and power in 1896, but it was not until 1904 that the service commenced. The demand constantly pressed upon supplies, and after the First World War large extensions of the plant provided opportunities for corrupt dealings with tenders and contracts. This was the immediate reason for the suspension of the council in 1927.

After the restoration of the council, difficulties again arose in connection with the city's electricity undertaking. Two suburban councils, through which the city's transmission lines ran, threatened to refuse a renewal of the city's franchise and to establish their own plants in competition with the city. The Government decided that the difficulty would be solved by creating an *ad hoc* authority empowered to operate anywhere within the metropolitan area, and in 1935 there was created the Sydney County Council to generate and distribute electric light and power.

SYDNEY COUNTY COUNCIL

This county council illustrates the search in New South Wales for a local government body capable of doing things that are beyond the competence and outside the area of individual municipal or shire councils. A council has been used to clear navigable rivers of obstructive marine growths; control noxious weeds; and to provide water supplies and electricity. In Sydney, it was used to generate and supply electric light and power to the whole metropolitan area. The council is elected by and from members of constituent municipal and shire councils, and may only do what the constituent councils authorize it by delegation.

In Sydney the county council was limited to five members, two representing the city and three the suburban councils. By the Areas Act of 1948, the council was reconstituted to comprise nine members elected from four constituencies. The city electorate returns three members, and the suburban municipalities are divided into three constituencies, each returning two members. The council elects its chairman from amongst its members.

The organization of the Sydney County Council is an adaptation of the American city manager system and, accordingly, the legislation makes a distinction between policy and administration. The council appoints the general manager, the chief engineer, and the secretary, but all other staff appointments and the whole field of personnel management are controlled by the general manager. The council decides financial policy, including the fixing of rates and charges, and general policy, but the manager is by legislation made responsible for day-to-day administration. One important administrative duty is calling tenders and the right to recommend acceptance or rejection to the council.

While there never can be a clear line of demarcation between policy and administration, experience has shown that large municipal undertakings, employing hundreds of trades unionists and allocating very valuable contracts, offer temptations that prove irresistible to popular representatives unused to the responsibilities of office. It would be too naïve to suggest that expert administrators are immune from such temptations, but the history of the modern civil service of any country is a story of methods, practices, and safeguards to limit, if not to eliminate, individual or party patronage in appointments and corrupt practices in handling contracts and supplies. There is, therefore, no doubt that the removal of appointments from the municipal representatives and committing that power to an expert general manager has been signally successful in ensuring the integrity and efficiency of the staff of the Sydney County Council.

Since the end of the First World War the immense growth of secondary industries throughout the city and the suburban areas, the deliberately restricted output of coal by the miners, the delays in installing new power plant and equipment, apart from the difficulties of acquiring it, and widespread industrial disputes, have contributed to repeated failures in supplies of power and light. These black-outs have been seized upon by the state Socialist government as showing the incompetence of the county council and as a justification for a change. In 1950, therefore, Sydney again witnessed the performance of a drama in public administration, the production of which is revived whenever the party political prizes seem adequate and assured. Legislation was passed in that year depriving the Sydney County Council of the right to generate power, and transferring the generating plants of the county council, the state railways, the Southern Electricity Supply system (owned by the state government) and the privately owned Balmain plant (if negotiations for purchase succeed) to another *ad hoc* state authority—the Electricity Commission of New South Wales—all the members of which are appointed by the Government.[1]

Notwithstanding that local government has been responsible for generating the bulk of the present supplies in New South Wales, municipal councils have no representation upon the new commission. The militant unionists, however, who have carried on a vigorous agitation for nationalization, have one of their officials on the commission.

THE DISTRIBUTION OF FUNCTIONS

Before leaving this section, it is well to repeat that the allocation of functions between agencies at the several levels of government in Australia —federal, state, and local—is seldom decided in terms of principle. Thus, local authorities in New South Wales, in common with those of

---

[1] The council retains the right to distribute electricity, however, purchased from the commission.

other Australian states, have no police duties, do not conduct or support any schools or other educational establishments, and do not maintain hospitals, institutions for the sick, delinquent, or aged. The city of Sydney does not have a municipal museum or art gallery, and does not provide services of water, sewerage, gas, electricity, fire protection, harbour facilities, nor any of the social services. And though it may seem incredible to many readers, the city does not have the right to regulate parking of vehicles in the streets. Sydney does provide secondary street construction and maintenance, street cleansing, garbage collection, a municipal library, golf links, bowling greens, and tennis courts, and maintains fruit, poultry, and vegetable markets, cold storage works, supervised playgrounds, and a public bath on the shores of the harbour. The city has acquired certain land and erected buildings on it which are leased to tenants. It also leases land and buildings to business and commercial tenants as a result of acquisitions for street widening or slum clearance. Sydney contributes funds to the support of a symphony orchestra, also subsidized by the state.

It is clear that little consideration has been given to the level of government to which a service or function should be allocated, or what are the likely social, economic, or political consequences of this assignment. Had there been an understanding of the importance of local participation as a training for democratic citizenship, more patience might have been shown for the sometimes clumsy efforts to supply local services. More care, too, might have been taken to link local government directly with the *ad hoc* authorities that sprang up. Allocation has followed the most advocacy and the most aggressive demands, and in such an atmosphere, local government is cast in Cinderella's role.[1]

THE MUNICIPAL CIVIL AND POLITICAL SERVICE

When the general scheme of local government was reorganized in New South Wales in 1906, provision was made for testing the competence of candidates for appointment as town clerk, engineer, or auditor. From

---

[1] This attitude partly explains the constant recourse in Australia to the statutory corporation. For seventy years, the corporation has been used to overcome party political interference in administration, as a means of ensuring some continuation in public policy, and of releasing the rate of development of an enterprise, e.g. water conservation or railway extension, from the restrictions of an annual budget. Since the Second World War, however, Socialist governments have departed from the generally accepted policy of corporation independence and have legislated to provide that the corporations shall act under ministerial direction. The effect of this is to reproduce in the commissions themselves the features which formerly characterized departmental operation and led to the creation of the commissions. And the end is not yet.

that date, only those with a certificate of qualification could be appointed to the offices named. These tests did not apply to junior recruits, nor did the system remove the power of appointment from the councils. This system of 'certification,' however, was not applied to the Sydney city council, whose staffs of all grades were appointed by the council. Here, the higher administrative and professional departmental heads have all been men of recognized attainments, while the existence of closely organized trade unions covering all types employed by the council, together with internal appeals committees, has prevented the emergence of practices of nepotism. Shortages of manpower have operated to make appointments to jobs such as street cleansing, garbage collection and destruction less attractive than they were when unemployment was rife. In consequence, there is little cause for concern about the general integrity and competence of the employees of the city council. With the application of the general provisions affecting local government to the city, future administrative officers of the city will have to possess a certificate of qualification to be eligible for appointment.

The number of employees in the municipal services has declined during recent years with the removal of services and functions from the control of the city. For the year 1952, the civil staff of Sydney was as follows:

Classification				Number employed	Wages paid
Permanent salaried officers	..	..		594	£523,728
Permanent wages staff ..	..	..		2,205	£1,629,101
Total	..	..	..	..	2,799	£2,152,829

As far as the elected representatives are concerned, there is a prevailing nostalgia for the political giants of other days. To the press of a hundred years ago, commenting upon the character and capacity of the men offering for the newly created city council, the candidates for municipal office were anything but giants. Nevertheless, there seems to be agreement, quite apart from the effects upon representation of the intrusion of party politics into municipal administration, that there is some deterioration in quality as compared with the material available before the First World War. Councils are small, and members tend to make their jobs onerous by attempting to deal with matters that could much better be left to the administrative staff. Burdens are not shared, as in England, by co-opting citizens to committees of the council although the restricted character of the work of the city council makes outside advice and assistance less necessary than it is in England. The work is unpaid, although there have been suggestions that aldermen should receive some remuneration.

It is possible that the municipal representative is suffering handicaps similar to those affecting parliamentarians. The old days of broad sweeps and wide generalizations have gone, and there are fewer men willing to attempt the mastery of the problems of government, now so complex and so confused with economic issues. In addition, there are fewer who can neglect their professions or trades, for these too have become more exacting and more complicated; it is less easy to take either politics or business in one's stride than in the more leisurely days before the motor car and the cinema. Both politics and business are today tending to be full-time jobs, and political honours are tending to be sought principally by those who can make a better living from their pursuit than from business. It is suggested that virtue has gone out of the system when one is cynically reminded that people do not go into politics for their health.

Whatever be the conclusion reached about the capacity of the modern representative, it is safe to say that political scandals are infrequently revealed in Australasia and that criticism can be levelled less against integrity than ability.

THE FUTURE PROSPECT

For the past fifty years, there have been a series of efforts to reorganize the government of the metropolitan area of Sydney. There have been two Royal Commissions, some Parliamentary investigations, some government bills that failed to pass the legislature, and many private plans.[1]

Generally speaking, proposals have ranged over the whole field from unitary to federal forms of administration. Unitary schemes were favoured before the First World War when the prestige of the city was greater than that of the suburbs. The most notable federal scheme was that proposed in 1931, which envisaged the creation of a Greater Sydney Council to control an area of over a thousand square miles (including the county of Cumberland and part of the county of Camden), the amalgamation of suburban municipalities into thirteen groups, the enlargement of the actual city, and the reabsorption into the Greater Sydney Council of all the *ad hoc* bodies that managed services affecting the metropolitan area.

It is doubted whether such an expansive scheme would now find any more acceptance than it did twenty years ago, and official thought has

---

[1] The city of Melbourne, like Sydney, has been agitated for the past sixty years over the question of metropolitan government, and there have been no fewer than twenty plans proposed there, without result, during that period.

veered to less ambitious plans than the revival of the idea of an all purposes council for the metropolis.[1]

By the Local Government (Areas) Act, 1948, effect was given to the recommendations of the Royal Commission of 1946 for amalgamating the city and the areas adjacent to it, and for the re-grouping of thirty-six suburban areas into sixteen new municipalities. Seventeen other municipalities and five shires in the metropolitan area were not altered. Since nothing was done to enlarge the functions of city and suburban councils, the only immediate effect had been to reduce overhead administrative costs while removing that administration farther from the citizen.

Perhaps the most promising of all attempts to deal with the metropolitan area was the passing of the Local Government (Town and Country Planning) Amendment Act, 1945, which constituted the Cumberland County District and authorized the election of a county council from and by the councils of the cities, municipalities, and shires within Cumberland County.[2] The county council operates over an area of more than 1,500 square miles, and is charged with the responsibility of preparing a master plan for that area. The council's master plan has been drawn in collaboration with the individual councils in the area and covers the whole land use for the future development of the metropolis. It has been submitted to the minister for consideration by Parliament. These developments are probably as far as it is possible to go. It is too late to attempt to bring back into a metropolitan council all the *ad hoc* bodies that have hived off and set up independent establishments. Water and sewerage, transport, fire protection, abattoirs, electric light and power, main roads, hospitals and public health, maritime services (controlling the harbour, as there is nothing in Sydney like the Port of London Authority), milk inspection, and housing, are each provided by special

[1] The city of Brisbane in the state of Queensland broke with the prevailing indecision about metropolitan government in 1925, when it introduced a system unifying an area of 385 square miles by abolishing all existing municipalities, shires, and other authorities. It also gave the Greater Brisbane Council power to absorb utilities, such as street transport and electricity generation, which were owned and operated by private companies. In addition, the state also abandoned the prevailing British practice of specifying the powers to be exercised by the city and, instead, gave it a general grant of power to do anything it deemed desirable for its good order and government. The council is required to observe prescribed forms when planning new activities, and the state government reserves the right to veto any such plan which is elaborated in an ordinance promulgated by the Greater Brisbane Council. Save in one matter affecting police powers, the state has not found it necessary to intervene to restrain the city council, but that result may be due to the excellent system of consultation which has been built up between the state and the council.

[2] It is pointed out that there is no connection between the Cumberland County Council, an *ad hoc* body for planning, and the Sydney County Council referred to earlier.

bodies. And current staffing preference runs to state nominees, rather than representatives of local government. It would now require a city council of super-men to bring about any effective co-ordination of their activities.

Even more important from the city council point of view is the task of persuading the state either to hand back or to transfer to the council services such as the regulation and control of theatres and halls, the regulation of weights and measures, the licensing of vehicles, the control of traffic, the regulation of lifts and scaffolding, the provision of baby clinics, and so on—functions more appropriate to the city than to state government departments. The redistribution of functions is not likely to be solved by the master plan of the Cumberland County Council, which will, however, throw into relief the difficulties involved in supplying water and sewerage, light and power, food supplies, arterial roads, as well as transport by road, rail, and ferry. The congestion in the inner city streets, where governmental, commercial, business, and professional offices are situated, is already such that motor traffic is often reduced to a stand-still in peak hours.

The difficulties might be a little less intractable were the state government more willing to give the city the power and the resources to control its own domestic duties.

There is much to be said for throwing more responsibility upon local authorities, for sociological as well as for political reasons. But if this is to be done, steps must be taken to ensure that patronage does not affect the integrity and efficiency of personnel. Is it desirable, also, to enhance the status of the town clerk, and adapt the American city manager scheme, even if that does involve reversing something of the traditional relations between the elected representative and the appointed official? Affecting every consideration is the problem of finance. Inadequate resources have from the beginning resulted in the progressive contraction of the scope of the work of local authorities, and that condition still embarrasses them. Finance, functions, and areas seem likely to continue to vex the relationships between state and local authorities in Australia.

Wellington

RALPH H. BROOKES

RALPH H. BROOKES

Born 1924. Educated at Sidcot School, Birmingham University, and the London School of Economics and Political Science. Graduated B.Sc.(Econ.) in 1949. Lecturer in Political Science at Victoria University College, Wellington, N.Z. since 1950. Sub-Lieut., R.N.V.R. during Second World War.

Wellington

IT is doubtful whether Wellington is entitled to a place in a book which deals with the world's great cities. The claim could not be based on its population—in terms of population New Zealand itself, with just over two million inhabitants, presents a problem in local government. Nor could it be based on the size of the urban area; for geographical reasons Wellington will never form part of a conurbation. Indeed, this fact perhaps constitutes its main claim to inclusion. In a volume devoted to the great wens of the world there should be at least one example to illustrate the physiology rather than the pathology of local government. This should be taken to imply not that Wellington poses no problems —some of them will be developed in this chapter—but merely that those problems are, at least in theory, capable of solution. They may yet be sufficiently similar, in kind if not in degree, to those of Megalopolis to be of interest to students of municipal inturgescence.

In 1826 Captain James Herd, exploring an inlet on the northern side of Cook Strait (the channel which separates the North from the South Island of New Zealand), found himself in a large and almost land-locked harbour encircled by high hills, the steeply sloping sides of which came down in most places almost to the water's edge. This harbour he named 'Port Nicholson,' and his subsequent description of it in the *Nautical Almanac* ('Here all the navies of Europe might ride in perfect security') must have impressed the New Zealand Company in London, for it was at Port Nicholson that in 1840 the Company's first settlement was founded. After a few months as 'Britannia' the settlement was given its present name of Wellington, in honour of the Iron Duke. Its first government, a 'Council of Colonists' with taxing powers and authority to maintain law and order, was forcibly dissolved after a few months by the first Colonial Governor, British sovereignty having by then been proclaimed.

The Wellington settlers were soon agitating for the establishment of a system of local government within the colony, and in 1842 a Municipal

Corporations Ordinance was passed (based on the United Kingdom Act of 1835) under which Wellington was made a borough. The ordinance was subsequently disallowed by the British Government. The New Zealand Constitution Act of 1846 provided for the setting up of municipal corporations, but this act never came into operation, and it was replaced by the Constitution Act of 1852 which divided the colony into provinces, each with its own subordinate government. The Wellington provincial government set up first a Town Board (1863–66) and then a Board of Works (1866–70) for the Wellington urban area; these were rating authorities aided by a subsidy, and their functions were in the main confined to the care of streets (including construction, maintenance, cleaning, drainage and lighting). In 1867 the central government passed a Municipal Corporations Act, a permissive statute which provided a legal framework within which urban areas could establish local authorities with fairly wide powers. Under this act Wellington became a municipality in 1870, and in 1876, when the provinces were abolished, a mandatory Municipal Corporations Act was passed; since that time this statute, with subsequent amendment, has been the basis of the urban local government of the Dominion, the most recent consolidating act being that of 1933.

The 1876 act provided for the establishment of boroughs in urban areas containing not less than 250 resident householders, and two years later an amending act made provisions for the amalgamation of boroughs on petition from their councils or inhabitants. Both of these provisions were important in the period 1881–1921. Under the former the growing suburban areas around Wellington separated themselves from the surrounding administrative county (Melrose in 1888, Onslow in 1890, Karori in 1891, Miramar in 1904): under the latter these suburban boroughs, drawn closer to the city partly by the introduction of the internal combustion engine, partly by the growth of the tramways (municipally-owned from 1900, electrified in 1904, with suburban extensions opened particularly in the years 1905–11), partly by recognition of the advantages of providing common services more economically over the whole metropolitan area, and partly perhaps by a desire on the part of the suburbs to enjoy a share of the revenue from the city's high rateable value—these suburban boroughs were amalgamated with the city (Melrose in 1903, Onslow in 1919, Karori in 1920, Miramar in 1921) and Wellington took its present shape. As a result of the amalgamations and boundary adjustments, Wellington increased in size from 3,620 acres to nearly 16,300 acres. Johnsonville

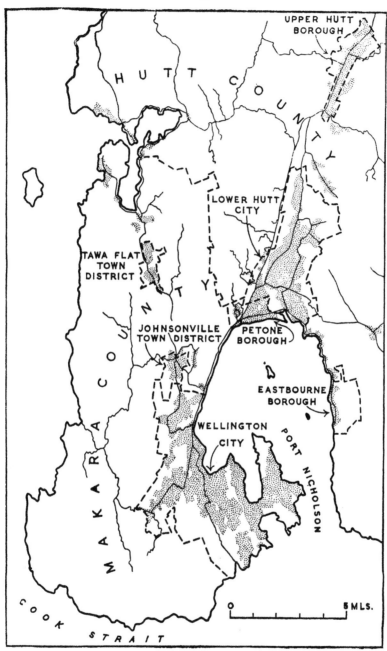

WELLINGTON AND SURROUNDING DISTRICTS. The broken lines denote local authority boundaries (prior to 1953, when Johnsonville Town District was amalgamated with Wellington City). The thin black lines denote main roads. The stippled areas show the extent of urban settlements.

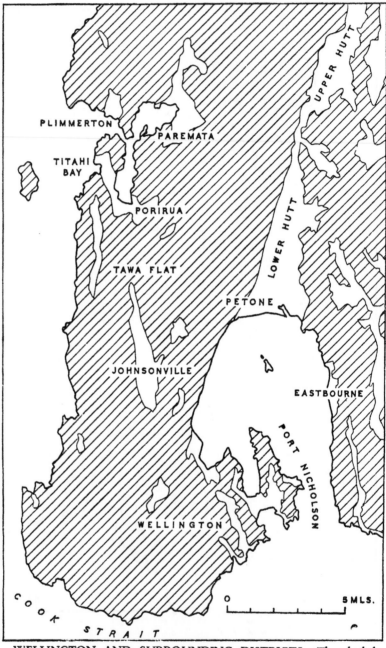

WELLINGTON AND SURROUNDING DISTRICTS. The shaded portion indicates areas of more than 10 per cent land-slope. Johnsonville was amalgamated with Wellington in 1953.

Town District was amalgamated in 1953, thereby adding a further 842 acres to the city.

The geographical relationship of Wellington City to the other local authorities in the district is represented on the maps on pages 599 and 600. A map showing land-slope has been included in preference to one showing contours, as giving a better impression of areas capable of urban development, and it may be gathered from this that there are only two such areas close to Wellington, one in the Hutt Valley and the other in a rather more sloping valley north of the city, running from Tawa Flat to Porirua. Wellington City is linked to the Hutt Valley only by a very narrow coastal strip along which run the road and railway; similarly there is as yet only one road over the hill which separates Wellington from the Porirua basin, whilst one double-track railway passes through the hill. To complete the triangle, there is one road from Paremata (at the seaward end of Porirua harbour) to the Hutt Valley. The census figures for these three districts will give an indication of their recent growth. (See Table I, page 602.)

It will be seen that Wellington has for the time being ceased to grow, that there has been a marked increase during the past fifteen years in the Hutt Valley, and that there are signs of a rapid post-war increase in the Porirua basin, where a town of 20,000 people is expected to develop in the next decade or so. These population trends and their implications will be discussed in the final section of this chapter.

Wellington is important in three respects—as a port, as a commercial centre, and as the capital city. Its excellent harbour and central situation have made it second only to Auckland in aggregate tonnage handled, and a key point for transhipment in a country which, like Britain, depends on external trade; all traffic between the two islands passes through it, most overseas ships call there, and its natural advantages such as a small tide rise and little need for dredging, plus its wharf facilities and a floating dock acquired in 1931, have enhanced its importance. This central situation was also the principal factor inducing the transfer from Auckland to Wellington of the central government in 1865. When modern techniques provided the means for remote control through large-scale organization, both public and private, it was inevitable that Wellington, as both a governmental and a trading centre, should spawn what Mumford has called a 'tentacular bureaucracy.' In addition to the government departments, which employ in Wellington about 15,500 men and women (including post office and railway employees, but not teachers), the city is also a headquarters for many commercial organizations—banks, insurance companies, shipping lines, import and export firms, and those auxiliaries of a pastoral economy, the 'stock and

TABLE I

	1926*	1936	1945	1951
A. WELLINGTON				
Wellington City	98,661	115,705	123,771	120,072
Johnsonville Town District	1,188	1,740	2,474	3,588
	99,849	117,445	126,245	123,660
B. HUTT VALLEY				
Lower Hutt City	7,847	15,960	31,254	44,474
Petone Borough	9,216	10,933	10,877	10,851
Upper Hutt Borough	2,840	3,871	5,494	7,449
	19,903	30,764	47,625	62,774
C. PORIRUA BASIN				
Tawa Flat†	226	415	598	—
Linden†	—	—	273	—
Tawa Flat Town District (includes Linden)	—	—	—	2,459
Porirua†	1,842	2,372	2,051	2,616
Titahi Bay†	60	216	526	1,166
Paremata‡	181	309	444	510
Plimmerton‡	332	463	963	1,152
	2,641	3,775	4,855	7,903
D. OTHER AREAS				
Eastbourne Borough	1,843	2,279	2,561	2,750
Hutt County	6,681	8,705	11,524	17,522
Makara County	3,758	4,305	5,005 (4,133)§	5,778§
TOTAL URBAN POPULATION IN THE WELLINGTON DISTRICT	129,014	159,357	188,091	208,292

* The 1926 figures do not include the Maori population, save in the final total.
† In Makara County.　　　　　　　　　　　　　　　　　　‡ In Hutt County.
§ Makara County in 1951 no longer includes Tawa Flat, now a Town District. The 1945 figure for a similar area has been given for comparison.

station' agencies. There is little large-scale industry in the city by overseas standards. The freezing works which prepare carcases of meat for export are important, and there are a few large engineering firms, but generally production is carried on by smaller units in such lines as the processing of food and tobacco, and the manufacture of paint and plastics, furniture and electric lamps, felt and textiles, woollens and men's clothing, footwear and hosiery. Petone is now predominantly an industrial area, with a similar range of activities (including freezing works), and there has been some spread of industry in Lower Hutt, notably in motor engineering. With a few exceptions—ship repair, woollen manufacture, carpentry and joinery—these industries are a fairly recent development, spurred on by autarchic remedies for the depression of the 'thirties and by the enforced self-reliance of an isolated community in wartime. These industrial beginnings (small, of course, in comparison with manufacture overseas) suggest that the indictment of Wellington as a parasitic growth on the New Zealand economy is not altogether justified: it is truer, I think, to say that its main contribution is in the provision not of goods but of services, always remembering that the inhabitants of any metropolitan area tend, to an increasing extent, to live by taking in one another's washing.

THE MACHINERY OF CITY GOVERNMENT

Like the municipalities of Great Britain, Wellington is a corporation, i.e. a legal entity distinct from the organs of the central government. Unlike many of those municipalities, Wellington derives this status from a statute, not from a charter, and the powers possessed by 'the Mayor, Councillors, and Citizens of the City of Wellington' are those conferred by statute. The grant of general powers in such acts as the Municipal Corporations Act, 1933, and the Town Planning Act, 1926, is supplemented by special grants in local acts such as the Wellington City Milk-Supply Act, 1919. Powers granted to the corporation are in most cases to be exercised by the council, though the council may delegate to its committees any of its functions except borrowing money, levying a rate, passing a by-law, making a contract, or instituting an action. In practice the council will usually reserve the right of decision, though committee recommendations will normally be adopted without much debate unless they involve the passing of a by-law or the raising or expenditure of money, or impinge on the activities of another committee.

In outline the structure of city government in New Zealand is so similar to that in England (e.g. in the characteristic division between the elected amateur and the appointed professional, with responsibility concentrated in the former, and control exercised largely through a number of functional committees) that exposition can almost be confined to listing the differences between the two systems.

In the first place, the mayor is directly elected and not chosen by the city council. He is the chairman of the council, and may also be a chairman of one or more of its committees—the present mayor, for example, is chairman of the Works Committee. In addition to these 'efficient' functions, he has also 'dignified' or formal duties, such as welcoming distinguished visitors to the city. He is *ex officio* a Justice of the Peace. Though not servants of the central government, the mayors of New Zealand were called upon to organize emergency supply services at the time of the recent waterfront dispute, a task which in Wellington was both onerous and of vital importance.

Secondly, there are no aldermen, and both mayor and councillors are elected triennially *en bloc*. The continuity of personnel which is achieved in Britain through an aldermanate holding office for six years seems nevertheless to have been secured in Wellington. In none of the last six council elections have less than ten of the sixteen sitting councillors (including the mayor) been returned, the average being between eleven and twelve. Allowing for the fact that a few sitting members will usually not be seeking re-election, these figures suggest that the electoral mortality rate is very low.

Thirdly, the city council is smaller than its English counterpart; in Wellington there are only fifteen councillors, in addition to the mayor who is a councillor *ex officio*. (The small council is a feature noted by Sir John Maud in his study of Johannesburg: perhaps it is generally characteristic of city government in the Dominions. In New Zealand, councils are limited by statute to not more than twenty-one members, exclusive of the mayor.) Service on the council is a part-time activity: the mayor receives an honorarium of £750 per annum, but councillors are unpaid, and there is no regular grant for expenses. Council meetings are held in the evening and meetings of the Finance Committee (a committee of the whole) in the late afternoon, each once a month, though special meetings may be arranged by resolution, or on application by the mayor or at least five councillors. Only the meetings (usually monthly) of the other committees, therefore, fall within working hours. This means that councillors can engage in full-time occupations other than their council service, particularly if these are such as to permit them to take an occasional half-day off for a committee meeting.

The council is largely composed of business and professional men (often lawyers), with trade union officials numerous in the Labour ranks.

Retired people are usually only a small fraction of the membership. The only important legal requirement for candidates is that they should be on the District Electors' Roll, though the usual prohibitions regarding aliens, undischarged bankrupts, convicted prisoners serving sentence, and persons of unsound mind, also apply. Formerly, paid servants of the council were also debarred from becoming council members, but this restriction was removed in 1944.

The implication of all this is that the conduct of day-to-day business must be left largely in the hands of the professional administrators. Committees are small, too small to allow the growth of a system of sub-committees; most are of seven members (a few of five or six), and the statutory provision permitting co-option has—unfortunately, I think—not been utilized, so that each councillor necessarily has several committee assignments, which must reduce his opportunity for acquiring expert knowledge of any one field. Half the members of the present council serve on five or more committees. The small committee is helpful in favouring reasoned discussion rather than rhetoric, and in facilitating speedy despatch of business, but the monthly meeting and the relatively inexpert committeeman make it more effective as a brake on the paid official than as a steering-wheel to direct policy or as an engine to propel reforms.

THE ELECTORAL SYSTEM AND THE PARTIES

These organizational impediments do not in practice prevent the council from carrying out an electorally-approved programme, for no such programme actually exists. In local politics as in local government, the state of affairs in New Zealand is not dissimilar from that which recently obtained in many English towns. Labour has been contesting municipal elections in Wellington ever since it parted company with the Liberals early in the century, but it has only once elected a mayor (in 1912—he subsequently left the party), and has never controlled the council. The anti-Labour forces have ostensibly refused to accept the party challenge, arguing (like the 'Independents' in Britain) that party politics and local government do not mix, an argument which has considerable appeal in New Zealand, where the politician is viewed with even more contempt than in most other democratic countries. But despite these protestations an anti-Labour machine has had to be constructed for electoral purposes, the need being even more urgent than in England, for since 1905 there has been no ward system in Wellington, all councillors being elected 'at large.' The 'ticket' system is a natural development: if the elector has to vote for fifteen out of, say, thirty-five candidates (the number has varied between thirty-two and forty-four in the last six elections), there is little likelihood of his being able to distinguish between most of them except by means of a party label; equally, the party must unite behind the candidates on its slate, for a split vote may mean the loss not of one seat (as with single-member constituencies), but of all. It may of course happen that a party secures no representation even if its members solidly 'vote the ticket,' for the opposition may be equally united, and if only two parties are in the field it is a case of 'the winner takes all'; in three of the last six elections Labour has failed to win a seat on the council for this reason. What is perhaps more remarkable is that the minority party should ever secure any representation, and the other three elections are therefore worth investigating.

In 1935 the six successful Labour candidates were respectively first, second, third, fifth, sixth and thirteenth in the poll. The late Mr. Peter Fraser (who subsequently became Prime Minister) headed this list with over 28,000 votes compared with 21,500 for the successful Citizens' candidate in the mayoral election. If we may assume, on an analogy with American gubernatorial election figures compared with those for state legislatures, that at least as many people vote in mayoral as in council elections, it seems clear that there was a good deal of split voting in 1935, and the 1938 figures, though less striking, tell a similar story. It is significant that the three leading Labour candidates in 1935 (also the top two in 1938) were Members of Parliament for Wellington constituencies. If voting tended to be for persons rather than for parties, men thus in the public eye might be expected to top the poll. But in 1941, 1944, and 1947 even these Labour M.P.s (now, admittedly, a little older) failed to win seats on the council, and there was a perceptible gap between the voting figures for the fifteen successful Citizens' candidates and those for the unsuccessful remainder: party lines were hardening. The uniformly poor showing of Independent and Communist candidates, who (except in 1950) have in all cases been at the bottom of the poll, is further evidence of increased ticket voting. The 1950 figures do not invalidate this conclusion—but here some explanation of local party organization is necessary.

Labour party candidates, for local as for national elections, are selected by the local Labour Representation Committee, which in Wellington is a body of representatives from constituency party branches and affiliated unions in the Wellington area. There is no similar link between the national party organization and the Citizens' candidates, nor is there a permanent Citizens' party organization—indeed, the very title of 'party' would be repudiated. Instead, applications are invited, from members of the public interested in standing for election, by a Citizens' committee representing the Wellington Ratepayers' and Citizens' Association and the Greater Wellington Electors' Association; this committee makes a preliminary selection of candidates, the final selection being made by the full Council of the Ratepayers' and the Electors' Associations. This machinery broke down in 1950, when four sitting members of the council (three of whom were chairmen of committees) were not renominated by the Citizens' committee. The supporters of these members argued that the associations no longer effectively represented the various ratepayers' and electors' organizations, that the councillors concerned were being victimized for exercising their right of independent judgment in council affairs, and that in any case no one knew the identity of the Citizens' committee (soon dubbed the 'Secret Six') which had made the decision. The Labour Party took up the cry of 'Tammany'; the four members decided to stand as Independent Citizens' candidates, one of them squeezed into fifteenth place in the poll, and the split in the ticket permitted Labour to regain six seats on the council. One would not be justified in deducing from this instance any return to the split voting of the 'twenties and 'thirties.

The figures for the 1953 election, which again returned 9 Citizens' and 6 Labour councillors and a Citizens' mayor, present a further problem. The split in the Citizens' vote may have been partly due, as in 1950, to controversial nominations: their leading vote-getter of 1950 was refused a place on the ticket in 1953 and finished sixteenth in the poll as an Independent, while the presence on the ballot of six other Independents (who polled much better than the nine Communist candidates) may be in part an after-effect of the 1950 split. It is significant, however, that among the sitting councillors defeated were the chairmen of the Transport and By-laws committees, who seem to have paid the price for the Council's recent lack of success in coping with transport and traffic problems (which are discussed below); the chief protagonist of a new traffic policy—another Citizens' councillor—rose to third place in the poll. This suggests some return to split voting, but of a different and potentially more valuable type; it also provides a neat illustration of the irrelevance of the party label to important policy questions.

If a system existed for Parliamentary elections which normally gave the majority party all the seats and the minority party none, there would be an immediate demand for fairer representation, to be secured either by a form of proportional representation or by the use of single-member constituencies. Nothing illustrates better the artificiality of the party division in Wellington's local politics than the absence of such a demand. There is no support for the adoption of proportional representation for local body elections, even though the fragmentation of parties which would make it objectionable in national politics would seem to accord with New Zealand views on the place of party in local government. There are, on the other hand, keen advocates of a ward system, but the arguments commonly put forward do not rest on the need for more equitable party representation; they are based in part on the short-ballot principle—that the 'bed-sheet' ballot, involving a large number of choices between candidates largely unknown to the voter, encourages electoral apathy—but to a greater extent they are expressive of suburban sectionalism, and give force to the counter-argument that councillors should be elected to promote the interests of the city as a whole, and not merely of particular wards. The demands of these localities are in any case brought to the notice of councillors through pressure groups such as the various suburban ratepayers' associations and progressive associations which are active in Wellington.

The lack of any effective demand for fair representation of parties on the council suggests that even party voters see local body affairs as a matter of administration rather than of policy, and this conclusion is borne out by the nature of election campaigns, by the lack of voter-interest, and by the absence of party lines in the council.

Election campaigns rarely raise issues of principle. An occasional policy item may be put forward—the provision of a municipally-owned produce market, for example—but in general one is reminded of Bryce's remark about 'two bottles, both empty, but bearing different labels'; one party favours economy and efficiency, the other efficiency and economy, everyone claims to be the best man for the job and questions the fitness for office of his opponent. Party controversy in New Zealand is marred by over-indulgence in the argument *ad hominem*: in Wellington politics one hears little else.

Election meetings are poorly attended; of the 109,880 people on the District Electors' Roll, only forty-three attended the opening meeting of the Citizens' campaign in 1950, and the twenty candidates and dignitaries on the platform at Labour's opening meeting faced an audience of but twice their own number. The percentage of electors exercising their right to vote is not small by overseas standards—60 per cent in the mayoral election of 1935 and between 50 per cent and 60 per cent for the four

68. WELLINGTON
The main business centre.

[Photo: National Publicity Studio, Wellington

69. WELLINGTON

The Hutt Valley, looking across the harbour from the northern suburbs of the city.

[*Photo: National Publicity Studio, Wellington*]

70. WELLINGTON

A view across the Porirua basin, site of a proposed new town.

[*Photo: Evening Post, Wellington*]

71. ZÜRICH

Fraumünster Church, with the Town Hall on the right and old guildhouses on the left.

[*Photo: Swiss National Tourist Office*]

succeeding elections; the figure of 30·4 per cent for 1950 exaggerates the drop, as the Electors' Roll was then badly in need of revision to exclude the names of persons no longer in the district—but it should be remembered that anything less than 90 per cent participation in a New Zealand Parliamentary election is regarded as low. At this point it should be explained that the franchise is practically universal: anyone who either (*a*) owns freehold property in the city of not less than £25 capital value, or (*b*) pays rates on property within the city, or (*c*) has resided in New Zealand for twelve months, and in the city for the last three months, is required by law to be on the District Electors' Roll (though I have heard of no prosecutions for failure to enrol) and may vote at mayoral and conciliar elections. Only persons registered under the freehold or rating qualifications may vote at rate or loan polls (i.e. polls required for the approval of a change in the basis of rating, or for the raising of a loan by the local authority).

Although elections are fought on a party basis, and voting is on party lines, divisions in the council—and particularly in its committees, where most of the work is done—do not reflect ardent partisanship. If indistinct policies and campaigns fought on personalities remind one of American elections, the blurring of party divisions in the legislative process is reminiscent of the situation in the Congress and its committees. As in the Congress, the parties are united on matters relating to personnel: the allocation of councillors to the various committees, though formally decided by the council, is largely settled by caucuses and party leaders, and the majority party will normally take care to secure a majority on each committee, and will appoint one of its own members as chairman. The key to party unity in the United States is the need to control patronage, but this is clearly not the case in Wellington. Why, then, in the absence of division on matters of political principle, should a party system have developed?

At least two partial explanations can be given. In the first place, service on a council, and even the unsuccessful contesting of local body elections, is valuable experience for those who aspire to a career in national politics. As the Labour Party is more openly 'political' than the Citizens' organization, the correspondence between Labour candidatures for local and Parliamentary elections is closer than that between Citizens' and National candidatures, but the connection is visible in each case. Secondly, although one must exercise care in reading wider implications into the results of municipal elections, it is evident that many votes are cast not on local but on national issues, that is, for or against the central govern-

U

ment of the day. Nor is this as illogical as it seems. If national issues are important and exciting whilst local issues are uninteresting or non-existent; if it really makes very little difference who is on the council; and if the voter knows that the results of local elections will be treated (at least by the party which tends to benefit) as indicative of a trend of national opinion, then it makes sense for him to regard the election of mayor and councillors as the by-product of what is, in effect, a referendum on the performance of the central government.

Can such a system of government be called democratic? If one defines 'democracy' in terms of the *provision of opportunities* for popular partici-pation in the process of political decision, then the wide franchise, the lack of restriction on candidature, the freedom to form parties, the opportunity to bring influence to bear on the council through pressure groups—all these features would justify an affirmative answer. If, on the other hand, one's definition involves the *existence* of such participation—a grass-roots conception, implying the approval by a majority of citizens of a policy to be implemented by elected representatives—then a negative answer must be given. Elections are held largely on irrelevant issues, there is no popularly-approved policy, and the organization of the council is such that the main share in determining the direction of policy must lie with the professional administrator.

Is the system, then, efficient? This question can be answered only by studying the functions performed, and the services provided, by the city government.

THE SCOPE OF CITY GOVERNMENT

The fact which must be remembered in considering the scope of municipal government in New Zealand is the abolition of the provinces in 1876. At that time the functions of the provincial governments had to be divided between central and local authorities. The central adminis-tration had been in existence for over thirty years, and for the last twenty had been controlled by a responsible cabinet; the war against the Maoris had been successfully concluded, and an extensive programme of public works was being undertaken. The municipalities, on the other hand, were few in number, restricted in scope, lacking in trained personnel, and had no tradition of successful local administration behind them. It is not surprising that several functions which elsewhere fell within the sphere of local government—notably those of education, housing and police—in New Zealand passed to the central government, nor that more central supervision of local authorities was possible than in countries where those authorities had enjoyed substantial independence in the past.

The activities of a city government may be classified in numerous ways; there is much to be said for a functional classification—e.g. Sir John Maud's distinction between activities promoting wealth, health and happiness respectively—but in order to give a picture not only of the services provided but also of the organization which provides them, I have preferred to adopt a division corresponding more closely to the internal structure of Wellington's city government. This still leaves a choice between the structure embodied in administrative departments and that of council committees. Both present some quaint juxtapositions: the former groups—under the town clerk—the abattoirs manager, the sexton, and the curator of the zoological collection; the latter, evidently under the influence of an alliterative principle of classification, includes the hotch-potch of 'Legislation, Leaseholds and Libraries.' As the accounts are presented to show expenditure under committee headings, I have adopted the latter scheme of division.

It will of course be impossible, in an essay of this length, to do more than mention cursorily a few of the significant aspects of the activities controlled by the council. Of its twelve main committees, three—the staff, finance and estimates committees—deal with 'staff' (as opposed to 'line') functions. The remaining nine fall into three groups: two of them (milk, and transport and electricity) control trading departments; four (works, by-laws, housing and town planning, and reserves) will be taken together in considering town planning; and there are three (libraries, airport, and public health) which provide miscellaneous services.

THE STAFF COMMITTEE

This committee is responsible for staff matters generally, and particularly for the classification and regulation of the senior clerical and administrative employees. In February 1951 there were 2,285 council employees, of whom 729 were in the clerical and administrative category, and 1,556 were tradesmen and 'outside' staff. Current figures are probably higher, but there are at present over 400 vacancies for permanent positions on the staff, most of them for 'outside' employees. Remuneration and working conditions of most of the council's servants are governed by awards under the Industrial Conciliation and Arbitration Act. These awards apply to members of industrial unions; there are several such unions covering different groups of council employees—labourers, drivers, carpenters, painters, milk roundsmen, transport workers, electricians, abattoir workers and so on—and employees (including clerical workers) earning less than £650 per annum who do not fall into any of these groups are members of the Wellington Local Bodies' Officers' Industrial Union of Workers. Those above this salary bar not covered by an award are entitled to be members of the Wellington Municipal Officers' Association, an organization which is not a union for the purposes of the Industrial Conciliation and Arbitration Act, and their salaries and working conditions are governed by an agreement with the council (the classification of these salaries and the framing of staff regulations regarding working conditions being functions of the staff committee), save that the salaries of heads and deputy heads of departments are fixed by direct negotiation between those officers and the council. The last few years have seen a prolonged jurisdictional dispute between the Local Bodies' Officers' Union and the Municipal Officers' Association, the former arguing that the latter association is in effect a 'company union' controlled by departmental heads and designed to evade the statutory provisions regarding compulsory unionism, the association producing evidence to show that its members are almost unanimously satisfied with their agreement and do not wish to be unionized. So far, the result of the dispute has been the raising of the salary bar from £375 to £650, with a consequent reduction in the membership of the association, now numbering about three hundred.

Save for the statutory requirements regarding the qualifications to be held by such officers as the city engineer and the health inspectors, the council makes its own rules regarding recruitment. Formerly an educational standard equivalent to matriculation was demanded of applicants for clerical posts, but in the present situation of labour shortage this has been discarded, though a higher salary on commencement is paid to entrants with such a qualification. I know of no post-entry training schemes other than one whereby the best of the horticultural staff are given assistance enabling them to go to Kew Gardens in England for a course, but officers are rewarded by bonuses of from £10 to £30 for taking a degree (in a subject relevant to their employment) or a diploma such as those awarded by the Local Bodies' Administrative Officers' Institute. One feels that more could be done in this direction.

The standard superannuation scheme (optional for employees) which was introduced in 1947 facilitates transfers of officers between local authorities and also between the local government service and the civil service. Wellington City, with a wider range of functions than other local authorities in New Zealand, is a useful training ground for local government officers, who can readily secure promotion by transfer. In consequence, the city loses many of its best officers to other local bodies or to the government. It would not be fair to infer from this that the council is a bad employer. In spite of a black record in the depression years due to the employment of 'relief labour' on public works in the city (the ratepayer thereby profiting at the expense of the government and the worker), present conditions of work are for the most part good, though one has the impression that improvement might well have been made voluntarily in some cases where the corporation has required prodding through arbitration proceedings.

THE FINANCE AND ESTIMATES COMMITTEES

The budgetary process. It has already been explained that the Finance Committee is a committee of the whole council; it is, in effect, the council when concerned with financial business or the acquisition or sale of property. The Estimates Committee consists of the mayor and the chairmen of committees in charge of 'spending departments.' Towards the end of the financial year each 'spending department' prepares estimates of expenditure for the following year; these estimates are revised (usually upwards) by the committee which supervises the department concerned, and the estimates from all these committees are then collated by the city treasurer. He inserts in the resulting budget the estimates for non-attributable expenditure (i.e. expenditure not within the purview of any 'spending committee,' such as interest and sinking fund charges, salaries for the staff of the town clerk's and city treasurer's departments, and so on), and figures for estimated receipts, with an indication of the rate which would be required to bring estimated expenditure and revenue into line. This budget is considered in detail and revised (usually downwards) by the Estimates Committee, each member of which will, as a committee chairman, have a thorough knowledge of one branch of the proposed expenditure. The streamlined budget then passes to the Finance Committee, which almost invariably accepts the recommendations of the Estimates Committee, and approves the suggested rate. The modified departmental estimates are then returned to the committees concerned, which make the appropriate changes in their departmental programmes for the year.

Revenue and central control. Three points should be noted regarding revenue. In the first place, rating in New Zealand may, at the choice of the council, be on the annual or the capital value; or (if approved by a majority in a poll of all ratepayers) on the unimproved value, of property. Like most other boroughs, Wellington uses unimproved value as a basis, the effect being to shift the rate burden from residential property in the suburbs to the valuable land sites in the centre of the city. Valuation is a central government function, one-third of the cost incurred in a district valuation being met by the local authority.

TABLE II

Revenue for year ending March 31, 1950

	£	Per cent of total receipts
Rates	893,683	76·7
Licences, fines and fees	120,594	10·4
Rents	67,511	5·8
Sale of water by meter	42,138	3·6
Petrol tax	31,773	2·7
Sundries	9,025	0·8
	1,164,724	100·0

Secondly, the figures given in Table II relate to the general account only. Trading department accounts are kept separately, and as a general rule their profits are put to reserve and not to the relief of rates. Thirdly, those figures are of interest principally in showing how much of the cost of local government is borne by the ratepayer, and how little is met from grants-in-aid. Only the revenue from the petrol tax can be regarded as a government subsidy, this being a grant from the Main Highways Board for the construction and maintenance of streets forming continuations of main highways. It must be remembered, however, that in New Zealand education is a function of the central government, whilst many public health functions are performed by hospital boards which derive over four-fifths of their revenue from the central government; if local authorities in England had few responsibilities in the fields of education and health, they too might be almost self-supporting.

The case for the grant-in-aid in England has rested not only on the financial need of local authorities, but also on its value in permitting the central government to require from the recipients certain standards of performance. It should not be supposed, however, that the absence of such grants means that the financially independent local bodies of New Zealand are free from central government control. Their autonomy is limited, though the sanctions are judicial and administrative rather than financial. General responsibility for local government is vested in the Department of Internal Affairs which, for example, prepares legislation affecting local authorities, may (on application) confirm their by-laws, and administers some two dozen acts relating to local government, of which the Local Elections and Polls Act is one of the most important. The powers conferred on a local authority are granted by statute, and actions exceeding the scope of those powers can be declared *ultra vires* by the courts; conversely, a writ of *mandamus* may be sought to compel a local authority to perform its statutory duties. There are statutory maxima for rate charges both for general purposes and for the provision of special services such as water supply, lighting, libraries and drainage. Legality both of rating and of expenditure is secured by audit; all revenue and expenditure accounts, also the profit and loss accounts of trading departments, must be kept in a form approved by, and are subject to the scrutiny of, the Audit Office (under the Controller and Auditor General), each unlawful expenditure being surcharged against the members of a peccant local authority. Shortage of staff in the Audit Office limits this review to a cash or 'legality' audit. Constant co-operation between the office and local government servants, plus a continuous internal audit on the part of the local body, make the penal clause in practice a dead letter. The borrowing powers of a local authority, like its rating powers, are circumscribed by statute. All loan proposals require the approval both of a majority of ratepayers (save when statutory exemption has been granted) and of the Local Government Loans Board, a central government agency which scrutinizes closely the soundness of the proposed scheme, the financial ability of the local body to shoulder an increased burden of debt, and the adequacy of the provisions regarding amortization.

The Audit Office, the Loans Board and the Department of Internal Affairs are instances of administrative organs exercising control, distinguished only by their importance from a host of similar supervisory agencies. For example, abattoir works and fees are controlled by the Department of Agriculture, statistical returns by the Census and Statistics Office, cemeteries by the Health Department, superannuation by the National Provident Fund, traffic control by the Transport Department, town planning by the Town Planning Board, by the Lands and Survey Department and by at least two divisions of the Ministry of Works, which also controls electricity distribution, roads and streets, tramways, water supply, and public works generally. In these circumstances it is fortunate that relations between ministers and councillors, and even more important, between civil servants and local government officers, seem for the most part to be harmonious and based on a spirit of co-operation.

The direction of expenditures. An indication of the relative importance of the various activities of the city government can be gauged from Table III.

TABLE III

Expenditure for year ended March 31, 1950

	£	£	Per cent of total expenditure
Reserves Committee		97,458	8·6
Works Committee—			
Street works	88,213		
Water works	144,817		
Sewerage and storm water ..	31,809		
Street cleaning	33,416		
Refuse collection and disposal	61,275		
Other services	35,749		
		395,279	35·0
Libraries Committee		48,180	4·3
By-laws Committee—			
Traffic control		32,044	2·8
Public Health Committee			
Cemeteries	16,592		
Sanitary inspection	11,686		
		28,278	2·5
Airport Committee		3,734	0·3
Non-attributable—			
Interest, sinking fund, etc. ..		206,564	18·3
Contribution to outside bodies		146,704	13·0
Salaries		74,609	6·6
Appropriations		15,647	1·4
Superannuation		11,192	1·0
Miscellaneous payments ..		70,520	6·2
		1,130,209	100·0

It should be noted first that these figures relate to the general and not to the trading departments of the corporation; and secondly, that they cover expenditure both on current and capital accounts. Capital expenditure is abnormally low at present due to the shortage of labour and materials, and the figure for public works can be taken to represent the cost of repairs and maintenance of existing installations with but few extensions or additions. The heading 'Contribution to outside bodies' covers not only sundry grants to many educational and benevolent organizations, but also the levies paid to *ad hoc* authorities such as the Hospital and Charitable Aid Board, the Fire Board (covering Wellington and Johnsonville), the Regional Planning Council, and the City and Suburban Water Board which administers the watershed and (in co-operation with the government) is developing new sources of water-supply for the city and other local authorities in the district. Another important *ad hoc* authority which should be mentioned at this point is the Harbour Board, consisting of representatives elected from the very wide area which is served by the port of Wellington.

LEGISLATION, LEASEHOLDS AND LIBRARIES COMMITTEE

It is often argued by students who are anxious to revivify local government that the apathy which afflicts the local electorate could be overcome by an extension of municipal activities into the sphere of culture—by changing the emphasis, that is to say, from drains to drama and from abattoirs to art. Whatever the experience of European towns, it is difficult to see any solution of Wellington's problems in this direction.

The cultural activities at present sponsored by the council fall mainly within the scope of the libraries department. In addition to the central library, which contains good reference and lending sections housed in a fine modern building, there are eight branch libraries in the suburbs served by a circulating stock from headquarters, four other suburbs are catered for by a mobile library service, and there is a special school libraries section. The library runs its own printing and bookbinding department. It is less creditable to note that only in 1951 did the lending library service become a free one. Besides these activities, the libraries department has branched out in more novel directions. There is a growing collection of gramophone records, which may be played in special rooms fitted with electrical gramophones. An arts prints collection has recently been started, and these prints—mounted, framed and glazed by the library staff—may be borrowed on payment of a small weekly fee. There is a library film group which organizes lunch-time programmes of documentary films weekly during the winter.

It cannot fairly be said, therefore, that the council and its servants have been lacking in imagination, but it is doubtful how much more can be done along these lines. A municipal symphony orchestra is out of the question: it has required the backing of the state to launch and keep going the National Orchestra as an adjunct of the New Zealand Broadcasting Service, and there is no demand, even were the financial and

U*

artistic resources available, for another such organization. There is a string orchestra in Wellington at present endeavouring to secure sufficient support to put itself on a full-time professional basis, and the council could be of assistance here, but the probable electoral repercussions are too powerful a deterrent; it has been said that the only sorts of music which Plato would have allowed in his Republic would have been those typified in our day by Sousa and Moody-and-Sankey, and in this respect New Zealanders are thoroughgoing, albeit unwitting, Platonists. A 'sound-shell' for a silver band in the botanical gardens is one thing, a subsidy for the devotees of Bach or Vivaldi quite another. In fields other than music the situation is similar. New Zealand needs a National Theatre with its own company of players, and this will come in due course. One cannot see the citizens of Wellington paying the 4s. per head per year (in terms of pre-war values) which was the cost to the inhabitants of Freiburg of maintaining their municipal theatre. A Little Theatre is planned, to form part of the projected civic centre; its construction doubtless lies well in the future, and it may be that Wellington will then choose to emulate Birmingham and Liverpool in establishing a full-time repertory company of actors.[1] The city is fortunate in that it contains the National Art Gallery and Dominion Museum. This is a national and not a civic responsibility, but the city council has a representative on the board of trustees, and makes an annual contribution towards the cost of the undertaking.

In conclusion, I would suggest that the advocates of 'civics through culture' are putting the cart before a somewhat refractory horse. It is untrue that the people get the government they deserve—the citizens of Wellington would receive a nasty shock if they were governed according to their deserts—but, even given an enlightened council and intelligent officers, there is a limit to the number of steps which a local authority can take in advance of public opinion, particularly when loans for developments (other than those which are mandatory under statute) require the approval of a majority of ratepayers. If issues more exciting than garbage collection and sewage disposal are needed to capture the voter's imagination, they are not to be found (at least in New Zealand) by engaging in a *Kulturkampf*: that way lies electoral suicide.

[1] Early in 1953 a cinema, offered for sale, was purchased by the council, in co-operation with the amateur dramatic groups of the city, for conversion into a civic theatre. At about the same time was formed New Zealand's only full-time professional company of players. I shall be gratified if these and subsequent events prove my judgment to have been mistaken.

PUBLIC HEALTH AND SANITATION, CEMETERY AND ABATTOIR COMMITTEE

The public health functions of this committee are much narrower than those of its counterparts in Great Britain. Since the passing of the Hospitals and Charitable Institutions Act in 1885 the provision of hospitals, sanatoria, maternity homes, ambulance services, district nursing services and 'outdoor relief' (the last is now almost obsolete since the introduction of social security) has been the responsibility of Hospital and Charitable Aid Boards. These are *ad hoc* authorities, directly elected every three years on the local government franchise, administering areas large enough to warrant separate hospital facilities, i.e. normally covering a group of contiguous local authorities. The Wellington Hospital Board thus serves the inhabitants of Wellington and Lower Hutt cities, Petone, Eastbourne and Upper Hutt boroughs, Johnsonville and Tawa Flat town districts, and Makara and Hutt counties. Many other activities in the field of public health (e.g. the school medical and dental services) are carried on by the central government. Drainage, refuse collection and disposal, water supply and housing conditions, fall within the province of other council committees, so that practically all that remains to the Public Health Committee (apart from its responsibility for cemeteries, crematorium and abattoir) is food control, the removal of condemned buildings, and the suppression of nuisances. There is no medical officer of health under the council, nor is there a separate health department, the sanitary inspectors coming under the control of the city engineer.

AIRPORT COMMITTEE

The insignificance of the item under this heading in the list of expenditures is misleading. Until 1947 Wellington City owned in Rongotai Airport (situated on the isthmus which links the easternmost peninsula to the rest of the city) the busiest aerodrome in the Dominion. In that year the airport was restricted to handling light aircraft, as its runways were considered too short by modern standards, and most planes were routed to Paraparaumu, some thirty miles north of the city. Since then there has been a tremendous increase in air traffic; the air freight service across Cook Strait is the busiest in the world, New Zealand is second only to Australia in the use of air travel per head of the population. There is a powerful demand for the return of these services to a city terminal, and a plan has been approved for re-designing the Rongotai field by an operation which will involve slicing the top off a hill and using the spoil to reclaim land from the sea. When completed, the airport will be a valuable asset, only a few minutes' journey from the city centre, with passenger and freight facilities underground to suit the use of jet aircraft. These facilities will be under the control of the council, whilst the operational part of the service will be a government responsibility. The Government is to provide two-thirds and the city one-third of the capital outlay (estimated at £3 million) and profits and losses will be shared in this proportion.

TRADING ACTIVITIES

Two points need to be made about the trading departments whose activities are summarized in Table IV. The first concerns that familiar feature, the deficit on transport undertakings. High costs due to increasing wage-rates and raw materials prices, plus the necessarily high rate of expenditure on maintenance of tram tracks (aggravated by wartime neglect), have swelled total expenditure, whilst fares have been raised until there has been a sharp falling-off in the number of passengers carried,

TABLE IV

Summary of trading accounts for year ended March 31, 1950

	Working expenses	Capital charges	Total income	Profit
Tramways and omnibus	660,981	110,245	674,491	−96,735
Electricity supply ..	440,624	137,961	597,976	19,391
Milk supply	1,084,722	46,942	1,134,211	2,547
Abattoir	47,044	5,933	53,178	201

particularly for short distances. The increased amount of private motoring has also contributed to this decline. To reduce the maintenance bill it has been decided to replace trams by trolley-buses, which are already operating on some routes, but this will involve a large capital outlay, not only on new vehicles but also in re-making roads to carry heavier traffic. Fortunately for the ratepayer there is still a buffer of reserves accumulated from past profits which may tide the transport services over the inflationary period.

Perhaps the most interesting of the corporation's trading activities is the municipal milk supply. The city assumed control in 1919, and itself took over the distribution in 1922, partly to improve the quality of the milk, partly to secure hygienic handling, and partly to ensure a delivery even to those houses situated on the topmost pinnacles of Wellington's hills. The undertaking is an undoubted success, providing excellent milk at a reasonable price; an incidental result has been the elimination of bovine tuberculosis in the metropolis. The city government may be conservative: by no stretch of the imagination could it be called reactionary.

AGENCIES OF PLANNING

The land-slope map and the primacy of the Works Committee in the table of expenditure may help to bring home to the overseas reader the amazing topography of the Wellington area. One feels inclined to say that almost any function of local government—drainage or water supply, housing or traffic control, milk supply or refuse collection—presents the city council with a problem not of administration or of finance but of engineering: as a former mayor of Wellington remarked, 'God made Auckland, but the engineer made Wellington.'

In many respects, it must be granted, the city engineer's department has done a good job. Extensive reclamations from the harbour have provided the land for the city's main commercial area. Not only have tunnels, bridges, cuttings and retaining walls given access to the most remote of 'goat sections' ('Only goats buy them: only goats can reach 'em'), but some intelligent landscaping has been done in filling gullies with refuse and with the spoil from excavations, and making them into lawns. The reserves department too, benefiting from the foresight of early Wellington settlers who set aside large areas for public recreation, has planted about a million trees which give variety against the background of the mountains, beautiful yet bare, which surround the harbour. In this setting, where (as Professor Wood has said) the views would be breathtaking were they not at the end of every street, one cannot help but call to mind the line, 'Where every prospect pleases . . .'—but it would be unfair to complete the quotation. Not vile: inheritor, rather, of a frontier settlement and a frontier outlook, the Wellingtonian has been ambitious but shortsighted. The engineer has not yet listened to the architect, the geographer, the economist or the town planner.

If the Works and Reserves Committees have, on the whole, done well, the same cannot be said of the By-laws Committee (in charge of traffic control[1]) and the Housing and Town Planning Committee. John Cox, writing of the central area of the city, mentions 'the confusion of traffic, the shortage of parking space, the crowded footpaths, the danger to the pedestrian, the crudeness of the public eating houses, the squalor of the

---

[1] After the 1953 election this function was transferred to the Transport and Electricity Committee.

pubs, the shortage of space for organized games, the almost complete absence of any provision for indoor recreation'—but the list could be continued indefinitely. The key to the whole problem is the limited amount of flat land and the wasteful use of much of it, the one- or two-storied detached houses (increasingly being converted for industrial purposes) combining the evils of low density and congestion. The suburbs are less congested, but (with a few exceptions) are no more aesthetically satisfying. Mumford might have had them in mind when he wrote: 'Every avenue was designed in accordance with the most inexorable of all laws—the law of chance. The street picture lost all consistence: it was a jumble of competing styles . . . ordered to no common visual end.' Here the single-story wooden house, iron roofed, in its own plot of ground, predominates. Wellington was never a Coketown, nor did it see the long terraces of 'by-law houses.' Each house went up independently of its neighbours, and the result is an architectural hotch-potch.

What legal powers has the council to deal with these problems? The Town Planning Act of 1926 (as amended) provides that all boroughs must prepare town-planning schemes which after provisional approval by the council would be submitted to the Town Planning Board (a central government agency of miscellaneous composition). If the board gave its provisional approval, the scheme would be published, and protests from occupiers of rateable property (with the council's comments) would be considered by the board, which might set up a committee to hold a public hearing. Final approval might then be given to the scheme (amended if necessary) which would thereafter be enforced by the council in regard to new works, public or private. The council would be entitled to claim from an owner half of any increment in the value of rateable property which resulted from such a scheme, and these payments would form a betterment fund, to be used for the payment of compensation claims. Compensation would be payable in full, save in regard to restrictions relating to density of building, to the height or design of buildings, or to zoning.

Town-planning schemes were to be submitted by January 1937, but the depression and the war led to postponements, so that of the local authorities in the Wellington district, only Petone has reached the stage of a public hearing. It is no secret that Wellington City has been stalling on the issue, though whether this is because of defects in the act (an amending act is expected shortly) or because of lack of enthusiasm for town planning generally, is not clear. The fact that no energetic action seems to have been taken under the Housing Improvement Act, 1945, which provided that overcrowded, degraded or insanitary districts might be declared 'reclamation areas' and acquired and redeveloped by the local authority (subject to the approval of the Town Planning Board), suggests that lack of enthusiasm may be the answer; a part of central Wellington has been declared a reclamation area, but no redevelopment is yet visible. It must be remembered, however, that post-war shortages of men and materials, plus the clause in the act requiring alternative accommodation to be made available for persons displaced by demolitions, have set the council a difficult problem. Until a town-planning scheme has been brought into operation, local authorities may exercise some control through an 'interim development' clause in the Town Planning Act; proposed works which contravene town-planning principles or which are incompatible with the council's draft scheme may be forbidden, subject to appeal to the Town Planning Board.

The lack of drive which characterizes the activities of the city government in the field of town planning is also the principal feature of its housing policy (if policy it can be called). Local authorities in New Zealand have never been housing authorities on the scale of their English counterparts, and since the government entered the housing business in 1937, the New Zealand equivalent of the 'council house' so widely found in Britain has been the 'state house.' Municipalities have long had power, however, to erect workers' dwellings, and the central government will make funds available to them for this purpose at a low rate of interest (subject to the approval of the Local Government Loans Board and the Minister of Finance). By 1950 the corporation had built ninety-three houses, of which fifty-eight were let to its employees and the remainder had been sold. Comment is unnecessary: it is sufficient to draw attention to the absence of the Housing Committee from the list of spending committees in the expenditure accounts, and to the fall in the city's population recorded in the 1951 census.

This brings us to the problems of overspill and the extension of the urban area, of regional and extra-urban planning. The legal framework is, broadly speaking, that counties may be required by Order-in-Council, or may voluntarily choose, to prepare extra-urban planning schemes, which are subject to the same procedure for approval as town-planning schemes; both Hutt and Makara Counties are preparing plans at the present time. Under the Land Sub-Division in Counties Act, 1946, any proposal to sub-divide land for sale or building requires the approval of the Minister of Lands, who may refuse this, after consultation with the relevant local authority, on planning or other grounds, subject to appeal to an *ad hoc* administrative tribunal. If an extra-urban planning scheme is in force, the minister's decision must be compatible with it.

The Town Planning Amendment Act, 1929, made provision for regional planning. A group of local authorities might, with the approval of the Town Planning Board, set up a regional planning council, which after making a survey of the natural resources of the area would produce a regional planning scheme. This would have no statutory force, but would serve as a guide to the constituent planning authorities in preparing their own schemes. The Wellington Regional Planning Council (representing all the local bodies shown on the map on page 599) is still at the stage of surveying natural resources, but it also assists the local authorities in their preparations.

How has this machinery functioned? The important points are those associated with the development of the Hutt Valley and the Porirua Basin. The Hutt Valley expansion, based on the construction of a large number of state houses, was in large measure controlled by the housing division of the Ministry of Works. Space was reserved for community centres and shopping facilities to serve each neighbourhood, but insufficient provision was made in the early stages for industry, and though the position has been remedied to some extent, there is still a large daily movement of workers from the Hutt Valley to Wellington, causing congestion of transport services and a waste of millions of man-hours per year. It is intended that these 'costs of costiveness' shall be avoided in the Porirua development, and that the end product shall be a satellite town and not a dormitory suburb, but there are already disconcerting creaks from the planning machinery. The weakness is not in the legislation but in its administration, and that principally at the top level; the Town Planning Board, the housing and town planning divisions of the Ministry of Works, the Lands and Survey Department, the Main Highways Board, the Railways Department, the Local Government Commission and numerous other agencies each take decisions which will affect the project, and they do not always see eye to eye. A stronger planning division and more effective co-ordination are essential. This should not be taken to imply that the regional and local authorities are above reproach.

But apart from the mistakes which may be made, the very conception of expansion northward may be at fault. The untidy, single-storied sprawl which is extending from Tawa Flat will cover large areas of good agricultural land, and will involve a heavy outlay on new roads, drains and water mains, when the population of Wellington City shows signs

72. ZÜRICH

A portion of the old town on the bank of the Limmat, with St. Peter's church in the centre.

[*Photo: Swiss National Tourist Office*]

73. ZÜRICH

The river Limmat seen from the Quaybridge.

Photo: Swiss National Tourist Office

74. ZÜRICH

In the foreground is the cantonal Parliament building. On the right are old guildhouses, and behind them rise the spires of the cathedral.

of falling. The regional planning problem, in fact, is due to the failure of Wellington's own town planning, i.e. to the preservation of low densities where topography demands high ones. Blocks of flats, intelligently placed in areas within easy reach of the city centre, could provide more people with more sunlight, better views, more open spaces and better designed homes, with no less privacy than is at present enjoyed by the inhabitants of the small and close-packed dwellings in those parts. It will of course be argued that New Zealanders dislike flats, but with the present housing shortage there can be no question that such accommodation would be popular, and in any case one feels that the costs of the alternative have never received adequate consideration.

CONCLUSION

Earlier in this chapter the question was posed whether the city government is efficient. The answer seems to be that in the services which it provides, and in its engineering aspects, it is efficient; but in the services which it fails to provide (such as housing), and in its planning aspects, it has fallen down lamentably. Its sins are those of omission rather than of commission. Jerningham Wakefield remarked after the election of the abortive borough council of 1842 that 'the gentry secured a very good council.' One may perhaps say of Wellington's subsequent governments that the ratepayers have secured very good councils: their short-sighted efficiency is characteristic of government by the businessman.

In its legal powers (having regard to its functions) and in its territorial extent (having regard to the lie of the land), the government of Wellington City requires no extensive reform. If the present pattern of development continues, one can envisage a structure of three mainly urban authorities (for Wellington, the Hutt Valley, and the area at present covered by Makara County) and one mainly rural one to the north (for most of the Hutt County area) which should be able efficiently to administer local services, with *ad hoc* authorities (as at present) for the control of hospitals and water supply and for regional planning. A regional fire board might well be added.

The two outstanding problems are those of town planning and electoral apathy, and like a pair of simultaneous equations each of them may provide the solution for the other. It was argued earlier that local government could not be revitalized by adventures into the field of culture: I believe, however, that it could be transformed over the issue of town planning. Electoral apathy cannot be overcome by leader-column preaching (hard as the city's editors try), nor will efficient planning come from spontaneous combustion within the electorate. What is needed in each case is imaginative leadership, leadership which is obviously not provided at present by the party organizations within the city. This may be due to lack of effort, to lack of imagination or to lack of nerve—it is not, I think, due to any deficiency in town planning as an election issue. A town planning exhibition prepared by the Architectural Centre in Wellington in 1947 attracted 20,000 visitors; more recently, a group of young enthusiasts won a majority on the Upper Hutt Borough Council by campaigning on a town planning programme. Such a campaign, to be successful, would of course have to be tailored to suit the New Zealand elector, with the emphasis not on possible social or cultural or aesthetic improvements, but on the economic waste involved in unplanned development—but this is merely to say that sound strategy must be supplemented by sound tactics. Given both, the City of the Strait might yet be worthy of its natural surroundings: it could have no higher ambition.

Zürich

MAX IMBODEN

Educated at Geneva, Berne and Zürich. LL.D. Zürich, 1940. Qualified as Solicitor, 1942. Privatdozent, University of Zürich, 1944. Legal Adviser to City of Zürich, 1946. Extraordinary Professor of Administrative and Tax Law, University of Zürich, 1949–53. Member of Tax Commission of Zürich. Professor of Constitutional and Administrative Law, University of Basel, since 1953.

Publications: *Federal Law breaks Cantonal Law* (1940); *The Void Act of State* (1944); 'The Consequences of the Guarantee given to Private Property by Constitutional Law' in *Schweizerische Juristenzeitung* (1944); *Protection from the Arbitrariness of the State* (1945); 'The Organization of Swiss Municipalities' in *Zentralblatt für Staats—und Gemeindeverwaltung* (1945); *Experiences in the Sphere of Administrative Jurisdiction in the Cantons and in the Federation* (1947); 'Immediate Democracy and Public Finance' in *Festgabe für Eugen Grossmann* (1949); *Rechtsstaat and Administrative Organization* (1951); '*Système et organisation des Finances de la Suisse*' in *Archiv für Abgabevecht*, 1951; '*Gemeindeautonomie und Rechtsstaat*' in *Festgabe für Z. Giacometti*, 1953; *Das Gesetz als Garantie Rechtsstaatlicher Verwaltung*, 1954.

Zürich

WHEN compared to other great municipalities, Zürich, with its population of 410,000, may appear to be a city of only medium proportions. But for Swiss conditions, Zürich has very definitely become a big city. Indeed the existence of a community twice as large as the second largest Swiss city, and comprising as it does one-twelfth of the total population of Switzerland and half the population of the Canton of Zürich, is sometimes regarded today as a serious problem for the political and social structure of Switzerland.

THE ORIGIN AND DEVELOPMENT OF ZÜRICH

This becomes intelligible only when it is remembered that Switzerland enjoys a federative organization unlike any other on earth. The confederation is divided into 25 cantons which, in turn, are organized into 3,000 or more communes. The existence side by side of numerous public bodies with extensive autonomy and their organization into a free union of free communities is a noteworthy characteristic of the Swiss state. The task presented by the increase in size of Zürich was to find a means of ensuring the inner equilibrium of the whole, in spite of the rise to power of one part.

The origin of modern Zürich is significant. It resulted in 1893 from the fusion of twelve independent communes. Formerly, the city had consisted only of what is now known as the 'old city' (*Altstadt*) with about 28,000 inhabitants, then eleven outlying communes were added. These had grown from peasant communities during the nineteenth century into real suburbs, and together they possessed three times the population of the city they enclosed. The first incorporation was followed by a second in 1934, by which eight further suburban communes were added to the city. The importance of this second incorporation was considerably smaller than that of 1893; still the population of the city was further increased by roughly one-fifth.

What has given Zürich its present greatness and its present importance has been, more than anything else, its economic development. Although not the capital of the country, Zürich became during the nineteenth century more and more the centre of the Swiss economy. Some of the biggest manufacturing plants of the country are situated in Zürich: yet Zürich is not a typical industrial city. Numberless threads of the world-wide Swiss export trade have their meeting point here. Zürich contains most of the Swiss insurance companies and banks of international renown. It forms the nerve centre of the highly developed economy of the eastern part of Switzerland. To what extent the economic life of the country has become concentrated here can be seen by the fact that roughly one-sixth of the tax revenue of the confederation is raised directly or indirectly by the City of Zürich. And the importance of the city as a centre of culture is emphasized by the fact that Zürich is the home of the Federal Institute of Technology and of the largest university in Switzerland. Zürich enjoys a rich and varied artistic life.

Although Zürich exceeds the normal dimensions of a Swiss town by all standards, it has preserved—as will be shown in what follows—the characteristic political and constitutional make-up of a Swiss commune: namely far reaching autonomy and a pronounced democratization of public life.

THE MUNICIPAL CONSTITUTION

The national structure of Switzerland consists of three different entities, communes, cantons and the confederation. The communes are not, as they are in many other countries, purely local administrative bodies. They are public corporate bodies with limited autonomy. This applies to the City of Zürich, too, as can be demonstrated in three ways.

The organization of Zürich is determined in part by a law of the canton—in fact by the same law which in 1893 effected the fusion of the twelve communes. This is, however, merely a skeleton law. It enacts, for example, in contrast to the original picture of a Swiss commune, that the city shall be a representative referendum democracy. It is not a gathering of the whole electorate, periodically called together—the so-called communal assembly—which is the highest communal authority in most of the communes of German-speaking Switzerland. The exercise of supreme power is shared between the communal parliament, the so-called communal council, and the people themselves, who by means of ballot voting are invited to give their decision in certain questions. The powers, and especially the constitution and organization, of the city authorities, are not rigidly fixed by cantonal law. It is left to the city to decide for itself the details of its constitutional structure. It is for the commune alone to decide on the regulations for the organization of communal administration. In large measure, therefore, Zürich has been entrusted with the right to determine its own organization. It has made use of these powers by issuing a communal code, in reality a municipal constitution. All essential regulations concerning the organization of the city and the responsibilities of the municipal government are contained in the communal code at present in force, which was accepted by a general vote of the people on January 15, 1933. The code forms the basic law for the commune, just as the cantonal constitutions do for the cantons and the federal constitution does for the confederation. The regulations of the communal code have been supplemented and clarified by further rules, which correspond to the decrees of canton and confederation known merely as 'laws,' in contrast to 'constitution.'

ADMINISTRATIVE AUTONOMY

More important than the right to its own organization, is the possibility which the city has of carrying out the public responsibilities entrusted to it, independently within certain limits. This administrative autonomy can be seen in two fields:

(*a*) The responsibilities of the city consist in part in applying the laws promulgated by confederation and canton. The Swiss confederation, as distinct from the United States of America, does not always set up a federal body to apply federal laws. The confederation makes use of the cantons to a large extent to apply its laws and see that they are respected. The cantons for their part have to some extent delegated to the communes the duty of applying the federal law imposed on them by the confederation. Apart from this, the cantons have in many cases entrusted the communes with the application of cantonal laws. Although in all these cases the communes merely have an executive function, they still retain the possibility of independent action. The officers of the commune, who apply the laws of the confederation or of the canton, are the servants only of the commune, and are responsible solely to the commune. In this, the Swiss communal organization differs fundamentally from that of other countries. If the superior cantonal or federal authorities wish to lodge a complaint about the administrative activities of the commune, they cannot apply to an individual official. They must deal with the civic authorities, who in the matter in question represent the commune in its relationship with the state. This results in the commune having a say in the matter of how the appropriate cantonal and federal laws are applied to its people. The communes when applying the law pay particular attention to the special local conditions. There is no danger of legal decrees being applied after some uniform pattern, dictated by a central office. In dealing, too, with centrally organized matters, the principle of local self-government is maintained. This provides an essential guarantee for the right of freedom of the citizen. The individual can be sure that those measures which affect him most intimately and most acutely are enacted by authorities whom he knows and who also understand his personal circumstances.

(*b*) The city has not, however, merely the function of applying the law. It is also empowered to enact legislation—within relatively narrow limits. For example, town planning, in particular, is based on regulations issued by the communes themselves. The canton has indeed issued a law about the construction of private buildings. But this law contains only the minimum of regulations. The communes have the power to issue further restrictive regulations, going beyond those of the canton. In practice, these powers have continually grown in importance. The modern tendency in town planning to restrict or even prohibit building in certain regions and to create industrial zones in other areas, lies within the competence of the city itself. It has in recent years made increasingly more use of these opportunities. The city possesses limited legislative powers also in several matters of the police. The same is true in the field of social services. The social measures which the City of Zürich has carried through on its own competence and on the basis of its own legislative regulations, have often anticipated later measures by the canton or confederation. Zürich's action in encouraging the modern co-operative society building programme is an example of this. Long before a federal or cantonal old age pension scheme existed, the City of Zürich had created one out of its own means. This was later enlarged and extended by a cantonal law. Hence it was the city which took the initiative in legislation within the canton, and out of a municipal institution there developed a cantonal instrument.

Zürich's independent position is also shown in matters of finance. The commune is free to incur such expenditure as may seem necessary to it for the fulfilment of its responsibilities. The right of inspection, which the cantonal authorities possess over the financial administration of the city, plays in practice only a minor role. It has the character of a general supervisory power, which hardly ever acts as an appreciable hindrance to the authorities of the commune. The yearly budget, for example, is settled by the commune entirely independently. It has not even to submit the estimates to a higher authority.

MUNICIPAL FINANCE

Here it must be emphasized that in Swiss law the budget does not have the same function as in most countries. It does not possess the status of a law but merely registers the income and expenditure which are determined by other decrees. But these decrees, which form the actual legal basis for the budgeted expenditure, are also passed independently by the city itself. Only in exceptional cases—usually when a citizen lodges a complaint before the higher supervisory authorities—can the canton interfere in a concrete instance in the finances of the city. At the time of the incorporation of modern Zürich in 1893, a special authority was appointed, a kind of expert commission such as no other commune has, whose function it is to supervise the finances of the city. In this institution can be seen something of the mistrust which the Canton of Zürich, at that time strongly rural, felt towards the growing city. In reality, this body has hardly ever come into the limelight. Its role is limited to making an annual report to the higher supervisory authority.

Of much greater importance are the general laws which the canton has enacted on debt redemption. Only to a very limited extent is it possible for the communes, and especially Zürich, to finance their expenditure by means of loans. For example, even in the case of unusual expenditure such as a big non-recurring building programme, the law demands that one-third of the expenditure be paid out of current income. The result of this is that the capital indebtedness of the City of Zürich remains exceedingly small. The debt not covered by investments is smaller than the ordinary revenue of the city for six months. Here too the cantonal regulations have a purely minimal character. The City of Zürich as a rule makes every effort to meet its expenditure out of current income, even beyond the amount prescribed by the canton.

Today, taxes make up the biggest part of the city's income. The tendency is to attach more and more importance to them as a source of municipal finance. Nearly three-fifths of the total income come from this source. The revenue from taxation is, however, only to a very small extent the result of special communal taxes. The only purely communal taxes are the three forms of land tax (change of ownership tax, property tax, and ground rent tax) which the city can raise itself, but which if levied on the basis of a decision by the city authorities, must comply with certain cantonal regulations. These purely communal taxes have represented in recent years only about 7 per cent of municipal tax revenue. By far the most important source of income for the city is the surtax levied by the commune on the cantonal taxes on income and capital. Cantonal law determines the taxation on capital and income of private individuals and corporate bodies; and the assessment of these taxes is also made under the direction of the canton. The commune has the right to determine the surtax rate. Every year in the budget debate, the communal council must determine what percentage of the cantonal taxes the city shall levy as a surtax. This tax-surcharge levied by the commune is invariably larger than the cantonal tax. It corresponds to an average taxation of roughly 4–4$\frac{1}{2}$ per cent of the city's earned income, at present a total of 1·5–2 thousand million Swiss francs yearly, and a taxation of the city's capital of roughly 0·18–0·2 per cent, at present 7–8 thousand million francs. These figures do not, of course, show the full burden of taxation. It is substantially higher because the canton and confederation, as well as the commune, levy taxes on capital and income. This system whereby the canton fixes the form of the taxes, but the commune helps determine how high they shall be, means that the city can at any time secure the income necessary to cover its varying expenditure, thus preserving its right to independence in financial matters.

It is undeniable that some degree of dependence on the canton does exist, because the latter determines the whole system of taxation for the city. How considerable this dependence is felt to be can be seen from recent developments. For some time, the amount of income exempt from taxation has gradually been increasing, and in this way the burden of taxation has been shifted more and more onto the higher income and capital-owning group. Income and capital are on the average substantially lower in the rural communes of the Canton of Zürich than in the city, and the main burden of cantonal taxation has shifted to the city. Today, the City of Zürich helps towards financing the rural districts, by means of the cantonal taxes it raises, which in part form the financial contributions which the canton returns to the poorer communes.

PUBLIC UTILITY SERVICES

Apart from taxes, the most important source of income for the city comes from productive communal undertakings. As in almost all the bigger Swiss communes, the supply of electricity, water, and gas has been a communal monopoly for decades. The communes were able to secure this monopoly because they had control of the public land—the streets and public spaces. They were in a position to decide, therefore, who should be licensed to supply electricity, water, and gas on public ground.

Besides these public supply undertakings, the municipal transport system (tram, bus, and trolley-bus) is also a communal concern. Its importance for municipal finances is by no means small.

Formerly, all these undertakings showed a surplus over and above the interest on capital, which was as high as 20–25 per cent of the yield from taxation. Today the position has changed somewhat. The net surplus from the productive undertakings is less than 15 per cent of the yield from taxation, and a further drop rather than a rise in earnings seems likely. This trend is largely due to the fact that the democratic system of running the public undertakings, over which the communal council and the citizens themselves exercise a constant control, prevents a more commercial policy being adopted. It is hardly possible to get the assent of a majority of the people to a rise in prices and fares. This has placed the public transport undertakings particularly in an ever more precarious position. It is probable that in future these will no longer be a source of income but will become a financial liability to the city. The remaining income of the city comes from interest on its own capital and property, and from dues as well as from subventions for special purposes from canton and confederation.

THE DEMOCRATIC ORGANIZATION OF THE CITY OF ZÜRICH

In scarcely any other community on earth does direct democracy go so far as in the communes of German-speaking Switzerland. An assembly of all electors domiciled in the commune meets periodically and decides on matters of policy for the commune. In Zürich, until the first incorporation of outlying communes, communal assemblies were held. It was only the growing size of the community and the practical difficulties which the holding of a civic assembly in these circumstances entailed, which led the larger Swiss towns to give up the communal assembly and set up in its place other institutions, namely the communal council and ballot voting. But it can still be seen how the present day statutory organization of the towns has evolved from the old assembly system of former times.

All the chief municipal officials are elected for a four-year term by the electorate. That applies to the communal council consisting of 125 members, as well as to the executive committee (*Stadtrat*), composed of nine members, headed by the mayor. In the same way the school authorities, the five regional school boards and the central school board with at present 43 members, as well as the teachers in the elementary schools, are elected directly by the people themselves; whereas the chief officials for the poor law authorities and the board of guardians are appointed by the communal council. The election of the communal council takes place according to an electoral system which assures all parties proportional representation. Other authorities are elected on a majority system; but in practice care is taken to see that the parties are represented in proportion to their strength.

THE REFERENDUM

The most striking expression of communal democracy in the City of Zürich is the far reaching right of the people to vote directly on important questions. People are called on to take a direct part in the affairs of the commune in three ways:

(*a*) All changes in the municipal statutes require the assent of the electorate. This is significant because the municipal statutes—the constitution of the City of Zürich—to a very real extent determine both the administrative organization as well as numerous important principles to be followed in discharging the responsibilities of local government. Important innovations can hardly be introduced without altering the municipal statutes. The people have reserved to themselves the right of assent in all fundamental questions of municipal policy. This partly explains why hardly any essential change has occurred in the structure of the commune, despite the fact that the political majority in the City of Zürich has changed twice (in 1928 in favour of the social democrats, and after the Second World War in favour of the parties of the centre).

(*b*) A further important part of the referendum is the so-called obligatory 'finance referendum.' All decisions of the commune involving a new capital outlay of more than one million francs, or an annually recurring expenditure of more than fr. 50,000, require the sanction of the electorate. In practice, most of the referenda in Zürich have to do with expenditure. Taking the average of recent years, the community has to vote on some five or six expenditure bills a year. A negative vote is infrequent; on the whole the people tend to vote for the motions proposed by the authorities. The reason for this is that in every municipal department all important parties are represented, and thus all political shades of opinion share the responsibility for public work. One thing, however, must not be overlooked: the finance referendum exercises a very strong preventive action. It ensures that the authorities only lay before the people expenditure bills which have undergone careful scrutiny and which are likely to receive the assent of the majority.

(c) The second instrument of direct referendum democracy in the communes is the so-called 'optional referendum.' Decisions of the communal council must be put to the general vote of the commune provided that this is demanded within twenty days by 2,000 members of the electorate, or by one-third of the members of the council itself. There are very few decisions of the communal council which cannot be placed before the electorate, but these include—a fact which may seem strange—the decision over the annual estimates and the annual fixing of the tax rate (i.e. the percentage of cantonal tax to be charged). These exceptions can readily be explained by the fact that no independent statutory importance at all attaches to the estimates. New expenditure is not decided by means of the budget. It must be approved by special bills of the communal council or of the electorate. In view of these special bills, however, the democratic right to collaboration is assured in very large measure. Just as the voter can make use of the referendum against a decision of the council, he can also demand the consideration of a particular measure, by means of a 'people's bill' (a type of motion). This right to a referendum or a people's bill is not often exercised in practice. These instruments are effective more by their mere existence than by direct use. The fact that they do exist is, nonetheless, very useful in ensuring a constant regard for the will of the electorate.

DEMOCRATIZATION OF THE MUNICIPAL ADMINISTRATION

The City of Zürich is today the second biggest employer in Switzerland. In its service are 10,000 teachers, officials, clerks and other workers. It requires no explanation to see that communal democracy, so important in Switzerland, must be very different in the city from what it is in the small rural communes. A solution was consciously sought which would make it possible to assure to an administrative body of great size and complexity the largest possible measure of democratic efficacy.

One method employed was to decentralize the administrative body. The municipal administration has been subjected to a certain amount of sub-division, especially in a practical sense. The executive committee (*Stadtrat*) is not responsible to the communal council as the supreme authority for all matters of administration. By its side, and to some extent on an equal footing, there exist a municipal school board, a board of guardians, and another board for the poor law.[1] They carry out the responsibilities of their own administrative work independently, and are

---

[1] The board of guardians and poor law board are appointed by the communal council.

directly answerable to the cantonal boards of control. The existence of different boards side by side is the present day expression of an earlier state of affairs, whereby, for example, the management of education and the poor law were not in the hands of a political body but of other communal associations. Today, these communal associations have become fused. Only church affairs (in the Canton of Zürich there is a state evangelical church) are still today in the hands of special communities, the city's twenty-two parishes.

Less marked than this form of constitutional decentralization is the functional territorial decentralization. Only in education is there a well marked territorial division of labour, in that the responsibilities for school affairs is left to five largely independent regional school boards in the city. The tendency today is to aim at a greater territorial decentralization in other administrative fields too. But in practice the realization of this ideal is extraordinarily difficult. More successful and more effective have been the efforts by means of town planning to emphasize the regional structure of the city more strongly, and so to stress the independence of the various city districts.

Another method of democratization of administration in Zürich has been to put the executive officers on boards together. It represents a deeply-rooted conception of Swiss public law, that executive powers should as far as possible lie in the hands of a board or group and not in the hands of individual officers. And so in principle it is the executive committee (*Stadtrat*) as a body which decides all important administrative business. Individual members of the board are entrusted with certain fields of activity, or departments. In this position they have certain decisions to take. But in many matters they are merely empowered to make proposals to the collective body of the board. Where individual members of the executive committee, as heads of the departments under their care, themselves decide certain matters, they are often assisted by expert commissions in a consultative capacity. The function of these commissions is not merely that of an expert; they are at the same time a means of democratizing the administration.

Finally, the position of the mayor does not correspond to that of the mayor in numerous cities abroad. Yet his importance in Zürich is greater than that of the president of the cantonal government or even of the federal government. In contrast to them he is chosen by the people for the whole four-year period of office as mayor. He does not merely have to perform his office in rotation with other members of the *Stadtrat*. Nor like the other members does he head an actual administrative department. Comparatively little administrative work comes his way. This emphasizes his role as a representative one. And at the same time he is able to devote himself to the more general interests of the commune.

The structure of the City of Zürich demonstrates, as do most political and constitutional bodies in Switzerland, the tendency to strike a balance between the opposites, to unite divergent wishes in a middle solution. A way has been sought to reconcile the direct communal democracy handed down from the past with the exigencies of a modern city. It is clear that this attempt has in general been successful, when one sees how energetically Zürich has solved the problems presented to it.

SUPPLEMENTARY STUDIES

Cologne

LORENZ FISCHER

PETER VAN HAUTEN

LORENZ FISCHER

Born 1895. Died October 28, 1956. Graduated at Munich University 1919–23. Dr. oec. publ. Lecturer in Economics at the Volkshochschule Marienbuchen 1927–29. Senior Secretary at the Statistisches Reichsamt, Berlin, 1929–45. Second director of the Statistical Office of the City of Bremen 1945–46. Director of the Statistical Office of the City of Cologne 1946–56. Lecturer in Statistics at Cologne University 1949–56.

Author of numerous publications, including: 'Über das Wachstum großer Städte' in *Städtestatistik in Verwaltung und Wissenschaft* (1950); 'Die dritte Welle' (Evacuations 1939 to 1950) in *Der Städtetag* (1951); *Untersuchungen von Straßenverkehrsunfällen nach ihren auslösenden Ursachen.* (Part I). 'Hypothesen nach Ergebnissen der Straßenverkehrsunfallstatistik in Köln 1952' in *Statistiche Mitteilungen der Stadt Köln* (1954); 'Die Stadt, in der wir leben, in *Festschrift zur Eröffnung des Kölner Hauses der Begegnung* (1955); 'Stadtgebiet und Stadtregion als statistische Größe' in 55. *Tagungsbericht des Verbandes Deutscher* Städtestatistiker (1955).

PETER VAN HAUTEN

Born 1897. Studied at Bonn and Cologne Universities. Dr. rer. pol. 1924. Statistical adviser at the Institute of Social Biology and assistant of Professor Schumpeter at Bonn University 1926–28. Senior Secretary at the Statistisches Reichsamt, Berlin, 1928–33. Entered municipal administration in 1934. Since 1939 head of department in the Statistical Office of the City of Cologne and since 1942 also second director of the same office.

Author of several publications, including: 'Der Einfluß der Wanderungsbewegung auf das Kölnische Volkstum' in *Volkstumspflege und Volkskunde* (1952); 'Besteht zwischen städtischer Binnen- und Außenwanderung ein deutlicher Zusammenhang?' in *Statistische Mitteilungen der Stadt Köln* (1954); 'Die Kölner Wehrmachtstoten aus zwei Weltkriegen' in *Statistische Mitteilungen der Stadt Köln* (1955); 'Die kommunalen Verkehrsbetriebe im Spiegel der Statistik' in *Kommunalwirtschaft* (1956).

Cologne

COLOGNE, the city of the cathedral and the centre of traffic, is completely thrown on its own resources. It is neither supported by a group of big cities with similar economic interests, nor does it have the additional attraction of being the capital of a German 'Land.' Its character is still, as it always has been, determined by its geographical situation.

POSITION AND IMPORTANCE OF COLOGNE

Just as London grew around the bridge across the Thames, so Cologne owes its foundation to its unique position on the arterial roads of the Rhine. As early as 38 B.C. a Roman military camp was established on the left bank of the Rhine in the territory of the Eburones,[1] who had been defeated by Caius Julius Caesar in 53 B.C. Later, Marcus Vipsanius Agrippa ordered the *oppidum Ubiorum*, also known as *ara Ubiorum*, to be built for the Teutonic tribes of the Ubians, who, about 12 B.C., had been forced to settle on the western bank of the Rhine. In A.D. 50 this settlement, under the new name of *Colonia Claudia Ara Agrippinensis* (C.C.A.A.), was granted the status of a colony with all the privileges of a Roman city. Here the representatives of Roman emperors, members of the imperial families, as well as emperors themselves and anti-emperors had their residences; among them were Agrippa, Germanicus, Vitellius, Trajan, Postumus, Constantine the Great, Silvanus. Moreover, the *ara Ubiorum* was intended to become the religious and cultural centre of the planned province of Greater Germania.[2] As this province was never established, this plan could not be carried out, but its very conception foreshadowed the cultural importance of the *colonia*, as it was soon to be called.

The decision of the Roman generals in selecting this site for a settlement was determined by the fact that here the Rhine valley with its natural roads running from south to north and the traffic flowing in the east-west direction intersected. The valley of the River Wupper offered the shortest and easiest entrance into Northern Germany. The first permanent bridge connecting Cologne with the right bank of the Rhine was built by

[1] For this and the subsequent data, cf. E. Kuphal, 'Köln,' in E. Keyser: *Rheinisches Städtebuch*.

[2] F. Fremersdorf, 'Geschichtlicher Überblick,' in *Köln 1900 Jahre Stadt*.

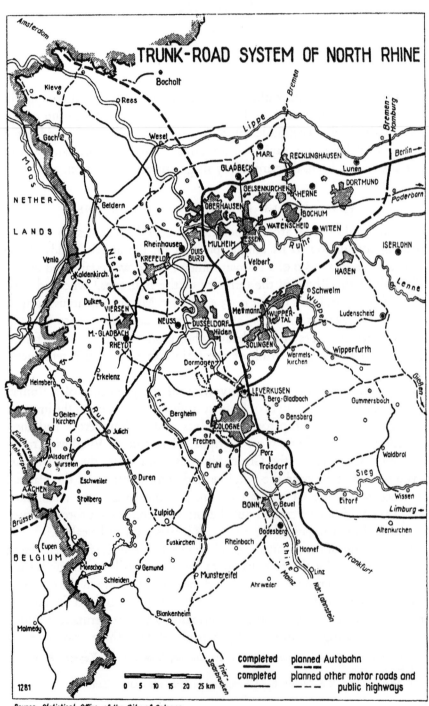

TRUNK-ROAD SYSTEM OF NORTH RHINE

	completed	planned Autobahn
	completed	planned other motor roads and
		public highways

Source. *Statistical Office of the City of Cologne*

Constantine in A.D. 310. It served the purposes of the Roman Empire, but during the migration of nations it was also used by conquering Teutonic tribes. Another point in favour of the site was the fact that it was easily accessible to ships from the open sea, an opportunity that was used not only by the Romans but also by the Normans who destroyed and ransacked Cologne in 881. In the Middle Ages, the city used its position to acquire the so-called *Stapelrecht* (stapling rights), i.e., the privilege of enforcing transhipment of all goods crossing the Cologne border between the Upper and the Lower Rhine. This law was in force as late as 1831. When it was repealed, the railways were coming into existence. Cologne soon became the first railway centre of Western Germany and, what is more, of Western Europe,[1] using again the time-honoured road along the Wupper.[2] At present the routes London–Cologne–Frankfurt–Vienna–Belgrade or London–Netherlands–Cologne–Basle–Italy intersect here with the routes Paris–Brussels (or Luxemburg)–Cologne–Berlin–Warsaw–Moscow or Paris–Cologne–Hamburg–Copenhagen–Stockholm.

When, in the twentieth century, motoring renewed the interest in, and importance of, roads and highways, Cologne soon acquired importance as an *Autobahn* centre, the facilities of which are still being improved. Before the Second World War, a large airport, which was used by about twenty-five airlines, was situated in the very outskirts of the city. The importance of the present airport, which is situated outside the city area, is steadily increasing.[3]

Cologne continued to be a cultural centre, as is testified by the establishment of an episcopal see (mentioned as early as 313).[4] Refounded in 625, it was elevated to the dignity of an archbishopric by Charlemagne in 785. In 1164 Reinald von Dassel had the relics of the Three Magi transferred to Cologne. The cathedral, which was begun in 1248 to house their shrine, was, in the course of time, to become one of the most inspiring sights of the world. The cathedral and the shrine of the Magi gave Cologne its spiritual centre. In 1248 the Dominicans and in 1260 the Franciscans began to conduct their *studium generale* in Cologne, thus establishing the two forerunners of the university (founded in 1388). Famous men, such as Albertus Magnus, Duns Scotus, Meister Eckehart, Suso, and Tauler taught here. Among their students was Thomas Aquinas. Some well-

[1] Th. Kraus: 'Köln, Grundlagen seines Lebens in der Nachkriegszeit,' in *Die Erde*, No. 1, Berlin, 1954.

[2] B. Kuske: *Die Großstadt Köln*.

[3] E. Rössger: *Entwicklung und Stand des nordrheinwestfälischen Luftverkehrsaufkommens*.

[4] E. Kuphal: 'Köln'; H. May and E. Meyer-Wurmbach, 'Das heilige Köln,' in *Köln 1900 Jahre Stadt*.

known schools of painters (Stephan Lochner, Bartel Bruyn) and a school of goldsmiths which was just as widely known, are evidence of the cultural importance of the city. It was this feature of Cologne history, its artistic accomplishments, that, after centuries of stagnation, inspired Cologne's spiritual rebirth in the Rhenish Renaissance[1] during the Napoleonic era. The length of time by which it preceded the economic revival is surprising. The economic strength gathered during the nineteenth and early twentieth centuries enabled Konrad Adenauer, then mayor of Cologne, to reopen the university at the expense of the city in 1919.[2] It had been closed by the French occupational forces in 1798, and up to the very abdication of the Hohenzollerns, the government of Prussia had refused to reopen it. Today it is the second largest university in the Federal Republic.

[1] J. Nadler: *Literaturgeschichte des Deutschen Volkes*, v. 2, Regensburg, 1938.

[2] Chr. Eckert: *Universität Köln*, Memoir, 1919; Chr. Eckert: *Die neue Universität*, 1921; F. Zinsser, *Universität Köln 1919–1929*.

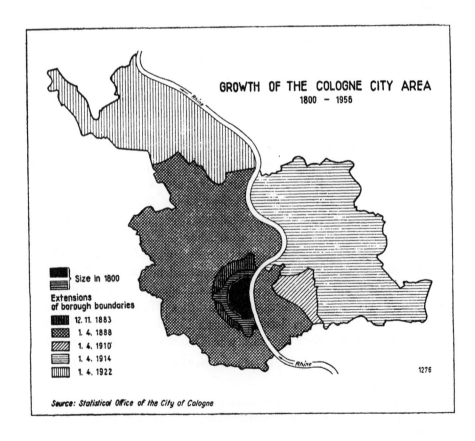

GROWTH OF THE COLOGNE CITY AREA
1800 – 1956

Size in 1800

Extensions of borough boundaries
- 12. 11. 1883
- 1. 4. 1888
- 1. 4. 1910
- 1. 4. 1914
- 1. 4. 1922

1276

Source: Statistical Office of the City of Cologne

THE GROWTH OF THE CITY AND ITS POPULATION
1800–1956

At the beginning of the nineteenth century Cologne had about 44,000 inhabitants, who lived in an area of 770 hectares.[1] For a long time the city's fortifications prevented any expansion, but in 1883 an order of the Prussian Cabinet permitted the city boundaries to be pushed out as far as the outer ramparts beyond the defence works, thus adding approximately 236 hectares of unbuilt sites to the city. Even then all organic growth was severely hampered by the fortifications, and it was only after 1888, the date of their final abandonment, that the city could begin to grow normally. The number of inhabitants had already risen steadily:[2] in 1817 it had reached and surpassed 50,000; 100,000 in 1852; 150,000 in 1883, and 175,000 in the beginning of 1888. It had thus already increased fourfold since the beginning of the century when, on April 1, 1888, a number of suburbs on the left bank, which had profited from, and shared in, the rapid growth of Cologne, were incorporated. At the same time the city, which up to then had been confined to the left bank of the Rhine, crossed the river by incorporating the town of Deutz and the commune of Poll. This addition of a total of 79,500 inhabitants raised the population beyond the quarter-million mark. By that time the area had grown to more than eleven times its original size.

Five years later, Cologne had a population of approximately 300,000, and after another ten years of approximately 400,000. In 1910 the defences of Deutz on the right bank were given up, making way for another extension of the city boundaries: this time Kalk, a town adjoining Deutz, and the commune of Vingst were added to the city, with 33,825 inhabitants in an area of not more than 599 hectares. This addition made the population shoot up beyond the half-million mark and Cologne became the all but largest city in Prussia.[3]

The acute housing shortage caused by the rapid growth of large factories, especially on the right bank, made an extension towards the east into the town of Merheim desirable, but this community was also claimed by the city of Mülheim on the Rhine, which, after a period of flourishing trade established by emigrants from Cologne, had also been industrialized during the nineteenth century. This clash of interests was solved on

---

[1] For these and the subsequent data, cf. E. Goebel: *Das Stadtgebiet von Köln.*

[2] Cf. *Statistische Jahrbücher der Stadt Köln*, 1948 ff.

[3] Goebel, *loc. cit.*, p. 21; see also *Statistisches Jahrbuch Deutscher Städte*, v. XVIII, census of 1910, p. 768, Breslau, 1912.

x*

April 1, 1914, by the incorporation of Mülheim as well as Merheim into Cologne, which added another 81,700 inhabitants and approximately 8,000 hectares to the city. After this, Cologne had the second largest area of all cities in Germany—it was only surpassed by the Administrative Union of Greater Berlin—and the third highest population.[1] It was thus a perfect example of that type of conglomeration of people which Weber has called the 'most remarkable phenomenon of the nineteenth century.'[2]

[1] Cf. *Statistisches Jahrbuch Deutscher Städte*, v. XXI, pp. 48–9, Breslau, 1916.
[2] A. F. Weber: *The Growth of Cities in the Nineteenth Century*, New York, 1899.

DEVELOPMENT OF THE POPULATION OF COLOGNE
1850 - 1955

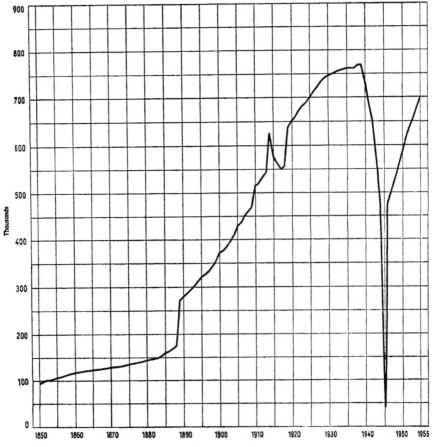

Thousands

Source: *Statistical Office of the City of Cologne*

1276

But even then Cologne was still a fortress. Not only was the city boundary on the left bank marked by defence works, but there was also a strong garrison stationed in the city itself. Rifle-ranges, drill-grounds and the harbour belonging to the corps of engineers prevented any expansion along the river towards the north. When, in fulfilment of the Treaty of Versailles, Cologne had all its forts and military installations removed, the way was free for the—up to now—last extension of the city boundaries, leading to the incorporation of Worringen, a large village of 7,600 inhabitants and of 5,400 hectares of excellent industrial and housing sites. On an area of 25,100 hectares[1] the city now numbered 667,100 inhabitants, a figure that, by 1925, had grown to 698,000,[2] by 1933 to 757,000, and by 1939 to 772,000. Thus only 125 years after the Congress of Vienna, Cologne had seventeen and a half times as many inhabitants as at the beginning of that period.

It was at this point of its development that the city was dealt the severest blows in its history. During the Second World War it lost 50,000 of its inhabitants: soldiers who either died in action or whose fate was never ascertained and civilians killed in the bombings. In 265 air-raids whole quarters of the town were reduced to dust and ashes. The population fled into the surrounding rural areas, so that in 1945, at the time of the occupation by the allied troops, not more than 40,000 people were living in Cologne.

But when the inhabitants began to return into the ruins of the town, they flowed back in enormous numbers: in July 1945 the 200,000, in November the 400,000 mark, was surpassed. The census of October 1946 showed a resident population of 491,400, that of September 1950, of 595,000, and at the end of October 1956 the city had again a population of 730,000, i.e. 94·4 per cent of the highest pre-war figure. The growth continues to be remarkably constant, and is, mainly because of the still considerable housing shortage, similar to the one of a hundred years ago. This growth is shared by the adjoining communities which are thus linked more and more with the city proper.

THE ADMINISTRATIVE DISTRICTS OF COLOGNE
PRINCIPLES OF REORGANIZATION

The big cities of modern times are complex entities whose constituent parts, though connected with the greater organism, form more or less

---

[1] Cf. *Statistisches Jahrbuch Deutscher Städte*, v. XXII, Leipzig, 1927.
[2] *Op. cit.*, p. 347.

independent units differing considerably in development, e.g. the decisive factors governing the life of the city centre may not apply at all in the suburbs. No two quarters are completely the same. All of them are closely connected with the city as a whole, but they also lead their own life. If the eventful history of Cologne is taken into account, the high percentage (58 per cent) of inhabitants who were born in the city,[1] comes as a surprise. This figure expresses the loyalty of Colognians towards their native city admirably, but what it cannot express is the fact that, in marked contrast to many other big cities, there continues to be a strong feeling of attachment to the various small quarters of the town, a feeling that, in a memoir addressed to the municipal council, is called 'the clan spirit of the various quarters.'[2] Some years before that, the city architect, referring to the city and its quarters, had coined the phrase: 'the Colognian federation of towns.'[3] Realizing this basic fact, the municipal council agreed to an administrative reorganization that would recognize it. Therefore, on January 1, 1955, the city was divided into eight large administrative districts (*Stadtteile*), comprising forty-eight smaller units (*Stadtbezirke*). Thus this reorganization creates two administrative levels: the larger districts contain several adjoining quarters, whose social structure may vary between that of a village and that of a middle-sized town. New administration buildings called 'consolidated branch offices' will be located near the centres of these larger districts. They will contain all those municipal offices that allow of decentralization. Three of these new administrative districts are comparable to middle-sized towns of 10,000 to 90,000 inhabitants, whereas the other five of them would correspond more closely to cities of 100,000 to 150,000 inhabitants. All these districts are large and unified enough to permit their inhabitants to feel first of all that they belong to 'Mülheim,' 'Ehrenfeld' or 'the old part of town.' This feeling of unity is continually being strengthened by the activities of numerous local clubs and societies, some of which may look back on centuries-old traditions, as in the case of guilds, while others may be of more recent origin, as e.g. the *Veedelszög*, i.e. special carnival parades organized by various quarters. Other unifying factors are the special local customs, like choosing the Queen of the May in some quarters and the celebration of kermises in others. All this goes to show that geographical situation and history had prepared the way for combining the quarters

[1] P. van Hauten: 'Die Gebürtigkeit der Kölner Bevölkerung,' in *Statistische Mitteilungen der Stadt Köln*, 1951.

[2] *Denkschrift über die Neueinteilung des Stadtgebietes*, Köln, January, 1953.

[3] R. Schwarz: *Das neue Köln, ein Vorentwurf*, Köln, 1950; cf. also for this, van Hauten: 'Köln als Städtebund,' in *Statistische Mitteilungen der Stadt Köln*, 1955.

into districts, and that all that was left to do was to follow their leads carefully, if necessary by consulting the population about their preferences.

The non-municipal authorities working in Cologne, such as state government offices, the Federal Revenue Administration, the Federal Mail, and the state police have already adopted the new district scheme. Other legal bodies with separate administrative functions will soon do the same. The new districts differ considerably in area and population. On an average there are 29 inhabitants per hectare in Cologne. The nucleus of the city, straddling the Rhine, heads the list with 91·6 inhabitants per hectare, though it was this part that suffered most from war damage and depopulation. In the southern quarter of the 'Old City' the residential parts number as many as 140 inhabitants per hectare, and in the quarter of Buchforst even 155·1 per hectare. These figures and those of the lowest density—Worringen with 2·6 inhabitants per hectare—show the wideness of the scale expressing the enormous structural differences of the Cologne districts that run the gamut of all the occupations a metropolis has to offer, from industry, transport and commerce to agriculture, vegetable-growing and livestock breeding.

THE AGGLOMERATION OF COLOGNE

Any data that consider the city area in isolation are apt to give an incorrect picture of the importance of a great city. Therefore, social statisticians in many countries, taking into account historical, methodological and administrative points of view, have examined the great cities as parts of agglomerations, conurbations, standard metropolitan areas, etc., a method that tends to produce a much more balanced impression. These attempts have been analysed by the editor in the first part of the present volume. At the moment, methodological preparations are being made with a view to examining agglomerations, to be defined according to objective criteria, in the next German census. As these studies have, however, not yet reached their final stage and as, above all, the list of essential criteria is far from complete, recent studies have made use of, and improved on,[1] the definitions of city agglomerations arrived at by purely statistical and geometrical methods developed in Germany at the turn of the century.[2]

[1] Cf. L. Fischer, 'Probleme der Abgrenzung städtischer Agglomerationen' in Report of the 55th Conference of the Association of German City Statisticians at Augsburg, 1955, and 'Sonderprobleme der Abgrenzung städtischer Grossagglomerationen in Nordrhein-Westfalen' by the same author, in *Sonderheft der Statistischen Mitteilungen der Stadt Köln*, 1955.

[2] von Hasse and Schott; cf. above all Schott: *Die großstädtischen Agglomerationen des Deutschen Reiches 1871–1910*, Breslau, 1912.

The first German[1] studies using the method of objective criteria, if adapted to the case of Cologne, would indicate a minimal ring of agglomeration around the city which includes the rural district of Cologne towards the west and fourteen neighbouring communities towards the east and north. The total area of this agglomeration—i.e. the city plus its ring—amounts to 941 square kilometres, which at the beginning of 1956 were inhabited by approximately 1,146,000 people, or 1,218 per square kilometre. The economic structure of Cologne and its ring of agglomeration is best reflected in the relative numbers of persons gainfully employed in agriculture, trade and public or other services. In 1950 agriculture and forestry accounted for 4 per cent, trade, commerce, industry and handicrafts for 77 per cent, public administration and public or other services for 19 per cent of the gainfully employed in the entire agglomeration. With 78 per cent in trade and commerce and 20 per cent in the services, the City of Cologne is only 1 per cent in each field above the average, whereas the ring of agglomeration, with 75 per cent and 17 per cent respectively, is not more than 2 per cent below it. These figures indicate a surprising similarity in the structures of the city and its surrounding ring. Even agriculture and forestry, which did not account for more than 2 per cent in the city, do not surpass 8 per cent in the ring, and though their share varies, of course, in some of the communities, it nowhere approaches anything like a limit that would point to an essential structural difference.

According to the geometrical definition of agglomerations, that of Cologne would include a circular area around the centre of the city, having a diameter of 40 kilometres. As this area would clash with the agglomerations of Bonn, Düsseldorf and the Wupper agglomeration of Wuppertal-Remscheid-Solingen, it could not be exactly circular in shape but would form a circle clipped by three secants. What remains is an area of 1,122 square kilometres which in the beginning of 1956 was inhabited by almost exactly 1,200,000 people, i.e. averaging 1,068 per square kilometre. As this includes the above-mentioned group of 1,146,000 inhabitants almost completely, the social structure is essentially the same as discussed above.

The constituent parts of the agglomeration of Cologne are closely interrelated, not only because of the well developed public transport system that facilitates commuting within this area, but also because many Colognians who had to leave the city during the air raids are still living

---

[1] For instance those by Boustedt: 'Die Stadtregionen in Bayern' in *Bayern in Zahlen*, 1953, and 'Methoden zur Abgrenzung von Stadt und Umland' in *Berichte zur Deutschen Landeskunde*, 1955.

in the adjoining communities. Many of the expellees from the eastern provinces of Germany, living in those communities, consider Cologne their obvious place of work. Moreover, the recession of the lignite mines towards the north helps to enlarge the potential working force of Cologne, which is thus becoming the place of employment for an ever-increasing number of people from the surrounding areas.

MUNICIPAL REPRESENTATION AND ADMINISTRATION

After Cologne had been occupied by the French in 1794, Ferdinand Franz Wallraf, the last Rector of the old University, in a letter to the Convention of Paris, defended the rights of the 'Senate of the Ubians which, according to the testimony of Caesar and Tacitus, had existed for 2,000 years'[1] and stressed that the principles of liberty, equality and fraternity which were then being preached in France, had already been practised in the Cologne city-republic, governed by its own citizens, for hundreds of years.[2] In 1735 the Lord Mayor, Johann Nikolaus Du Mont, expressed the same thoughts in a great speech which he made before the Paris Convention. However, history went its own inexorable way, undeterred by this proud sense of tradition in the ancient city.

The Napoleonic legislation of the year 1800 in France turned the civic communities (*Gemeinden*) into administrative districts. Thus the entire administration of the civic community was handed over to a *Maire* (Mayor), appointed by the government and subject to the control of the Prefect. To help the Mayor there was a local council with very limited powers. The French prefect system was also introduced into the German districts on the left bank of the Rhine which were under French occupation and of which Cologne was one. In contrast to this was the municipal constitution (*Städteordnung*) which was introduced into Prussia in 1808 by Stein and by virtue of which the towns were again recognized as independent communities, and the community spirit and interest in public affairs stimulated by the citizens' participation in administration. The towns were granted free self-administration of their own affairs and were only subject to state control in matters of absolute necessity. The citizens' active influence on the town administration was assured by the creation of an assembly of town representatives who were to be consulted in administrative matters. Thus the reform of Baron von Stein marked the

---

[1] Cf. E. Gothein: 'The Economic History of the City of Cologne' in *Die Stadt Köln* 1815–1915, vol. I, Cologne, 1916.

[2] Cf. also Nadler: *loc cit.*, p. 424.

beginning of civic self-administration in Germany.[1] Although Cologne became a Prussian town in 1815, the civic law along French lines remained valid until the proclamation in 1845 of the Prussian municipal constitutions for all towns and communities in the Rhine province. In 1856 the Rhenish Municipal Constitution was proclaimed and remained effective for almost eighty years. According to this constitution, the three-class electoral system prevailed, under which only tax-payers were granted the vote, the franchise was classified according to the amount of taxes paid, and women were not yet entitled to vote. In the Rhenish Municipal Constitution it was the duty of the town representatives to make all the more important decisions and to control municipal administration. The Lord Mayor, elected by the assembly of town representatives and established in office by the King, was an enfranchised member and chairman of the council of town representatives and, at the same time, the head of the entire civic administration. It was his task to put the resolutions of the town representatives into effect. As a result of this twofold function he had an extremely powerful position and could, indeed, exercise great influence over the fate of the town.

After the First World War the three-class electoral law was done away with and a general, equal franchise introduced—the vote being granted to all citizens who had completed their twentieth year and also extended to women. There was a complete shift when the German National Socialist Party came to power in 1933. The civic communities were subjected to the ideas of state authoritarianism. The Lord Mayors elected before 1933 were replaced by unconditional supporters of the National Socialist Party. The town representatives were deprived, for all practical purposes, of their rights and had merely advisory functions. It is true that the Decree of German Civic Constitutions issued in 1935 by the government of the Reich still appeared to subscribe to the spirit of self-administration. Yet in practice the so-called 'leader principle,' with its slogan 'authority over your inferiors, responsibility to your superiors,' always assured that power lay in the hands of one man.[2]

The power of the National Socialists collapsed with the general collapse of Germany in 1945. The British occupation authorities[3] did away with those civic regulations which had a definite national-socialist character

---

[1] P. Schoen: 'German Administrative Law' in *Enzyklopädie der Rechtswissenschaft*, vol. IV, Munich, Leipzig, Berlin, 1914.

[2] E. Becker: 'The Development of German Civic Communities and Civic Organizations and its Bearing on Present Day Civic Affairs' in H. Peters's *Handbuch der kommunalen Wissenschaft und Praxis*, Berlin, 1956, S. 103/4.

[3] Ibid., p. 106.

and introduced a new civic system on English lines. They made the honorary Lord Mayor chairman of the town council and the full-time Town Clerk chief of municipal administration so that the councillors and municipal administration each had separate leadership. The council of elected representatives once again became the body controlling the administration.

THE COUNCIL AND ITS COMMITTEES

Those regulations which were decided upon as a result of the Occupation Law were finally incorporated into the municipal constitution for the State of North-Rhine-Westphalia on October 28, 1952.[1] The number of councillors is regulated according to the number of inhabitants in the city. The Council of the City of Cologne consists of 66 members chosen by the citizens of Cologne. These councillors are elected by general, direct, and secret ballot for a period of four years. The Lord Mayor, who is chairman of the Council and represents the town in its external affairs, is chosen from among the council members and by them. According to the regulations of the Council's statute, he has three deputies who are also chosen from the council members. The first deputy of the Lord Mayor has the official title of Mayor. The period of office of the chief representative and his deputies is two years. They are eligible for re-election.

According to Section 4 of the Civic Constitution the Town Council has to draw up statutes within the framework of its activities. This was carried out in Cologne on June 24, 1954. In accordance with the regulations of the Civic Constitution the Council has to be convened at least once every two months. In Cologne the town representatives meet on an average once a month, according to the volume of business which has arisen for discussion. The meetings of the Council are public. The public may, however, be excluded from certain sessions dealing with special matters. The Council controls the administration. It can at any time demand reports on all civic matters from the Town Clerk, who is subject to its authority. The Council supervises the execution of its own resolutions and also that of its committees and the general conduct of administrative affairs. It is also entitled to inspect civic records.

The Council is responsible for all matters of administration, for the appointment of members and deputy members of committees, of the Chief Town Clerk and the heads of departments, and it also has the

[1] Report on the State of Law and Civic Regulations (*Gesetz und Verordnungsblatt für das Land NRW*), 1952, No. 57, pp. 23 ff. Cf. also E. Becker, *loc. cit.*, pp. 107 ff.

task of bestowing or withdrawing honorary citizenship. It is responsible for decisions concerning the alteration of municipal boundaries, and for the issue, alteration and cancellation of municipal statutes, especially those concerned with civic expenditure and public appointments. It is also responsible for fixing all public levies and private payments, and for the examination and approval of the yearly financial report. Without a resolution of the Council, no decisions can be made as to the disposal of municipal funds (apart from straightforward business expenditure for the purposes of administration), nor as to the establishment, enlargement, taking over or dissolution of public institutions and the town's commercial enterprises, nor as to the undertaking of new duties not provided for by law.

In plenary session the Council can appoint committees. In the municipal constitution provision is made for the appointment of a Chief Committee, a Finance Committee and a committee for the examination of financial reports.

It is the duty of the Chief Committee, which consists of fifteen members, to co-ordinate the work of all other committees. It makes decisions in those matters subject to Council resolution which allow of no delay. The Finance Committee prepares the financial statute and makes decisions concerning the execution of financial plans, in so far as other committees are not responsible. The committee for the examination of financial reports examines the yearly financial reports and co-operates for this purpose with the Public Accountant's Office, which is the body that deals with the examination and control of all municipal offices, including all municipal enterprises. The Public Accountant's Office is immediately responsible to the Council and subject to its decisions.

In Cologne there are, owing to the city's complex administrative needs, a further thirty-three committees, e.g. School Committee, Committees for Commerce and Harbours, Housing Committee, Cultural Committee, Youth Welfare Committee, and the working committees of the gas, electricity and water supply departments.

The Lord Mayor is chairman of the co-ordinating Chief Committee. The other committees elect a councillor from among their members as chairman. All advisory reports for the Council and its committees are prepared and put forward for discussion by the Chief Town Clerk. The minutes of the various committee meetings are submitted to him, so that he can always obtain any necessary information.

THE PRESENT MUNICIPAL ADMINISTRATION
AND ITS ORGANIZATION[1]

The executive power is in the hands of the Chief Town Clerk, who is responsible for the preparation and execution of the Council's resolutions. He must either be an authorized magistrate or have the qualifications of a higher Civil Servant, and he is elected to office for a period of twelve years, like his immediate colleagues, whose period of office is also fixed at twelve years by statute. He appoints, promotes and dismisses, in accordance with the municipal law and regulations, the officers, clerks and workmen; but so far as higher officers and principal clerks are concerned, he must obtain the approval of the Chief Committee. (The Council must approve the appointment of Heads of departments.) His general deputy is the Town Clerk. The scope of the municipal officers' duties is decided by the Council, which here follows the suggestions of the Chief Town Clerk. In order to ensure administrative efficiency, the Chief Town Clerk is bound to hold regular conferences with his officers. The Chief Town Clerk and his officers are present at Council meetings.

The municipal administration has to fulfil two quite different types of duties. On the one hand, there are the purely administrative affairs of the town itself within its own boundaries (in so far as they are not otherwise provided for by law) and for which it alone is entirely responsible. On the other hand, there are the duties deputed to it by the State government, which decides by law upon the extent and duration of such delegation. Concerning all other matters, article 28 of the Constitution of the Federal Republic provides the local authorities with full autonomous authority, within the framework of the law. The sphere of local self-government[2] includes, for example, the cleaning of streets, refuse disposal, the building of roads and paths, traffic, gas, electricity and water supplies, and cultural affairs. The duties delegated by the State to the municipal authorities include registration of persons and the Registrar's Office, housing, elementary schools, the supervision of State elections, statistical investigations and land disposal.[3]

Yet we can see in one further respect that the Cologne authorities have two distinct types of duties. For besides discharging the usual local government tasks, they have to deal with certain special duties arising from postwar problems; and these confront the municipal administration with

[1] Cf. here the last-mentioned sources.
[2] Cf. for this H. Peters: *Handbuch der kommunalen Wissenschaft und Praxis*, in various places.
[3] Cf. K. Mehnert and H. Schulte, *Deutschland, Jahrbuch 1953*.

extreme difficulties and considerable expense which can only be met by additional grants from the State government. The budgets of all local government departments show a constant expenditure for reconstruction activities. We shall examine this in more detail in the section dealing with finance.

The public is constantly kept in touch with the work of the municipal authorities by the Public Relations Office, which is in close contact with the press and the radio. In addition to this, the administrative report issued yearly by the Statistical Office and containing printed reports of the public meetings of the town council gives the public a clear picture of the difficulties and problems of the various municipal departments of which we shall now give a short description.

The duties of the Town Clerk encompass two large and important offices. The Real Estate Office has to ensure that the administration has sufficient sites at its disposal to carry through the town's reconstruction programme. In view of the requirements of modern town planning the Real Estate Office is confronted with a particularly difficult problem in the old central core of the city, with its abundance of dwarf-sized sites which have developed in the course of centuries.[1]

The Statistical Office is also under the direction of the Town Clerk. Owing to an historical development peculiar to German municipal government, it has come about that German municipal authorities are able to use the help of local statistical bureaux, independent of the Federal and State Statistical Bureaux. The Cologne Statistical Bureau, which has existed since 1883, is the centre for local and statistical investigation, a sociological research organ for the Council and the administration, and an advisory body for important administrative and planning programmes. Closely connected with the Statistical Bureau are the Inhabitants' Registration Office and the Electoral Office, the latter of which arranges all federal, state and local elections.

The field of duties of the General Administration, which is under the direction of the head of the Personnel Department, extends to arranging the division of administrative duties and deciding on the responsibilities and powers of the various departments, and to working for the simplest and most economical discharge of business and the personal care of all municipal staff. The department here co-operates above all with the Organization Department and the Personnel Department. The City Treasurer is responsible for the direction of municipal finance and taxes

---

[1] Cf. for this E. Sirp: 'The Quarter around the Greek Market in Cologne' in *Statistische Mitteilungen der Stadt Köln*, H. 1, Köln, 1956.

and has to draw up the yearly municipal budget and to arrange for the discharge of budget plans, which means that he has to be prepared for all new practical eventualities. The great responsibility of this post accounts for his great power in influencing the extent of municipal expenditure. He directs the treasury (together with the mortgage accounts), municipal revenue and the tax office. The supply of water, gas, and electricity is the task of the Public Works Department. The Department of Public Transport is in charge of all local tram and bus lines (which are municipal property) and arranges all business matters with into-town transport companies, e.g. the Cologne–Bonn Railways. The tasks of encouraging the growth of industry and of sponsoring frequent industrial fairs in Cologne, of increasing the efficiency of Cologne as a central point in national transport and of its four harbours, and finally of encouraging and furthering the tourist industry, falls to the lot of a very important department in Cologne, whose offices—the Office for the Encouragement of Commercial and Tourist Enterprises, the Municipal Tourist Office, and the Office of Harbour Administration—indicate their respective duties by their very names.

The Building Department is divided into three offices: one, the so-called *Tiefbau* office, is responsible for Cologne's bridges across the Rhine (the third city bridge is in the process of being built), for modern streets in the highly built-up city area, for the provision of largely one-way traffic facilities and of many miles of roads in residential areas. The second office, including the *Hochbau* office, the Building Supervision Office and the Residential Building Office, deals with all the business involved in building up the residential facilities of the city, from the first applications for building licences to the official inspection of the finished buildings; and to a great extent it is also responsible for settling questions of finance for building and approving of public subventions for such enterprises. Besides this it has to deal with the municipal building of administrative offices, schools, etc. The third office is the Town Planning Office which caters for the architectural future of the old metropolis, drawing up plans and deciding on new roadways. The Public Parks and Cemeteries Office looks after the extensive municipal gardens, parks and cemeteries. The Department of Social Administration has not only to deal with the usual social and youth welfare work but also with the care of war victims; there is, for example, the care and guidance of displaced persons and the numerous evacuees from Cologne and, in the Housing Office, the care of needy persons seeking accommodation. Finally there is a special office which deals with all applications from *Lastenausgleichsberechtigte* (victims of war damage, entitled to financial assistance from the State), and to whom 50 million marks a year, on an average, have been distributed after each application has been officially checked.

The Public Health Department, in addition to the usual public health activities, now lays especial stress on modern measures for the prevention of disease: it is very active in caring for the population (including school children and municipal workers) in various advisory and welfare offices, with considerable help from local doctors. The plan for building new municipal hospitals confronts the Hospital Department with many difficult tasks.

The Education Department, besides ensuring regular instruction at elementary, vocational and specialist schools and also a number of secondary modern and grammar schools, has to make sure that bombed-out schools find other suitable accommodation. It also has to make plans for future demands for school instruction from the ever-increasing population. The increase in *Kindergarten* schools and nurseries deserves special mention.

The Department for Cultural Affairs had, as a result of bomb-damage, to provide for the rebuilding of all museums, theatres and colleges, including the greater part of the University and University College Hospital, a programme which will take years to complete. Fortunately the art treasures and the contents of the city libraries remained safe. Some of the sub-sections of this department are the College of Music, the Gürzenich Orchestra, and the College of Administration. According to a new agreement the State of North-Rhine-Westphalia gives financial assistance to the University and its College Hospitals, as the city cannot bear this financial burden by itself. One should also mention the Municipal College for Evening Classes, the Historical Preservation Office, and an institution which is peculiar to Cologne, the Office for the Preservation and Encouragement of Cologne Traditions and Customs.

Among the divisions of the municipal administration, there are also the departments for Law and Public Security and also for Public Institutions (the individual offices are characterized by their names, e.g. the Law Office, the Office for Public Order, and the Office for Defence Expenditure) and also offices responsible for markets, slaughter houses, the disposal of refuse, public paths, and the bailiff's office.

THE POLITICAL LIFE OF THE CITY

Article 28 of the Constitution of the German Federal Republic of May 23, 1949, rules that in all states, country districts, and civic communities the people must elect political representatives by direct, free, equal and secret ballot. The parties, which may be freely founded and whose activities are quite independent, are guaranteed by Article 21 of the Constitution the right to participate freely in political life and to endeavour to further their own principles, in so far as their aims do not contravene democratic principles.

The parties so far founded in the Federal Republic were represented in the Cologne City Council of 1956. They are: the Christian Democratic Union (CDU); the German Social Democratic Party (SPD); and the Free Democratic Party (FDP).

The German Communist Party which was also represented in the City Council after 1945, declined in importance from year to year and by 1952 it failed to gain the minimum of 5 per cent of the total votes required for representation by the municipal electoral law. Other so-called *Splitter-Parteien* (very small parties) who also failed to attain the minimum of 5 per cent were the Free People's Party, the United German Block, the German Party, and the German Centrum Party.

According to the municipal electoral law of North-Rhine-Westphalia of 1956, all German nationals who have completed their twenty-first year and have been resident in the respective municipality for three months have the right to vote but may only be candidates for election after the completion of their twenty-fifth year. The City Council in Cologne consists of sixty-six members, half of whom are elected by direct ballot (on the principle of a relative majority) by thirty-three electoral districts. The other half of the representatives are taken from the reserve lists of those parties who have attained the afore-mentioned 5 per cent of the votes, so that the number of votes and the numbers of representatives of the various parties are in direct proportion. Men and women vote in separate polling booths.

Since the last war there have been four municipal elections; on October 13, 1946; October 17, 1948; November 9, 1952 and October 28, 1956. The number of persons entitled to vote increased from 317,914 (1946) to 521,791 (1956) as a result of the increase in population. That is to say, 72 per cent of the population were entitled to vote in the last election. On that occasion 341,387 persons voted, which is 65·4 per cent of those entitled to vote. This percentage was higher than in the municipal elections of 1952 and 1948, with 63·6 per cent and 54·7 per cent respectively; although it failed to reach the post-war record of the year 1946 with 73·6 per cent. Male voters exceeded female voters with a percentage of 68·2 per cent as compared with 63·3 per cent.

The valid votes cast on October 28, 1956, were divided as follows:

TABLE I

Party	Men	Women	Special Votes	Grand Total	in Per cent
SPD	80,937	72,656	1,702	155,295	46·0
CDU	54,247	83,281	4,019	141,547	41·9
FDP	12,296	11,938	256	24,490	7·2
FVP	5,978	4,811	88	10,877	3·2
GB/BHE	2,857	2,814	76	5,747	1·7
TOTAL	156,315	175,500	6,141	337,956[1]	100·0

The Social Democrats obtained an absolute majority in ten of the thirty-three wards, the Christian Democrats in two wards. The Free People's Party and the coalition of the German Block and the Refugee Party failed to obtain 5 per cent of the votes cast, so that they were not entitled to any seats from the reserve list. Of the total of 66 seats the Social Democrats obtained 32 seats (18 by direct election), the Christian Democrats 29 (15 by direct election) and the Free Democrats obtained 5 seats all from the reserve list. Thus the Social Democrats became for the first time the strongest party in the City Council. Compared with the election of November 1952 they gained 6 seats while the Christian Democrats lost 2 seats and so forfeited their former majority. The Free Democrats who had won 9 seats in 1952 lost 4 in 1956. The former Mayor, Theo Burauen, a member of the Social Democratic Party, was elected Lord Mayor. His deputy is Dr. Ernst Schwering, who held the position of Lord Mayor in the last five years and is a member of the Christian Democratic Party.

PROBLEMS OF RECONSTRUCTION AND EXPANSION SINCE 1945

The social and economic life of the city at the end of the Second World War presented a completely negative picture. In 1945 there were only 20,000 residential buildings left (including many which were partially damaged) from the pre-war total of 68,600 such buildings. The number of houses and flats was reduced to a quarter of the pre-war number. Of the 252,400 houses and flats officially recorded in 1939 only 63,000 were in a condition fit to live in. Single rooms had been reduced in number from 980,700 to 230,000, i.e. by 76·5 per cent. The schools of the town

[1] The difference between this figure and the total of 341,387 given above is accounted for by the exclusion of invalid votes in Table I.

had only 313 classrooms between them in 1945 as compared with 2,589 before the war. Of the 140 churches and chapels of the Roman Catholic and Protestant denominations, 81 were totally destroyed, 8 were 40–60 per cent destroyed, 12 were 15–40 per cent destroyed, 11 were 5–15 per cent destroyed, 21 were 5 per cent destroyed: only 7 were undamaged. There were no more theatres or museums. Of the 56 cinemas existing in the year 1939 there remained only 7 after the war. The stocks of books of the municipal libraries were reduced to one-seventh of their former total. In the municipal and private hospitals there remained only 1,627 beds of the pre-war number of 7,264. The municipal and suburban tramways had 1,150 trams and 79 omnibuses in 1938. After the war they had 208 trams and 23 buses. All five Rhine bridges were destroyed. Covered-in storage space in the harbours of Cologne stretched over an area of 125,328 square metres; after the bombing there remained 17,739 square metres. The rubble and debris were carefully estimated at the end of the war and were found to amount to 30 million cubic metres. By the end of September 1956, 14 million cubic metres of this debris had been cleared away.

Immediately after the war the municipal authorities had a plan drawn up for the reconstruction of the town—a plan which was to take into consideration both historical tradition and the requirements of modern life. New and healthy residential areas are being built up for the population which has increased by 20,000 persons per year on an average during the last five years. The number of residential buildings had risen to 52,400 by the end of September 1956, and flats and houses reached a total of 182,100. Thus 76·4 per cent and 72·1 per cent of the pre-war figures for dwelling-places had been restored. However, since the number of inhabitants has increased more rapidly than the provision of flats and houses, there is now an average of four people per dwelling-place or 1·1 persons per room, which is considerably more than the pre-war ratio of 3·06 or 0·79.

One of the most urgent tasks of the municipal authorities is the provision of new classrooms for schools. On May 15, 1956, 109 school buildings with 1,443 classrooms were again in use. Nevertheless, 51·1 per cent of pre-war classrooms are still missing. This general lack of rooms in schools affects the elementary schools in particular.

The rebuilding of the museums, whose art treasures were fortunately saved from the air raids, and of a new theatre, are in progress and near completion. The stocks of books in the municipal libraries have also almost reached pre-war figures with a total of 122,140 volumes at the end of September, 1956. The number of hospital beds has already passed the pre-war total with a figure of 7,580, so that there are 10·4 beds per 1,000 inhabitants. The extent of roadways and pavements is now 13 million square metres more than before the war as a result of the building of new roads. The municipal transport authorities have not yet been able to bring up the number of vehicles available to pre-war numbers but a considerable improvement has been made in transport conditions as a result of the introduction of very large modern vehicles. On November 1, 1956, 737 trams and 240 buses were once more in use.

Four of the Rhine bridges have been completely restored, one partially. The building of a completely new bridge was begun in the summer of 1956. The solution of the traffic problem is perhaps the most difficult task in hand, for the number of private cars and vehicles owned in Cologne has far exceeded pre-war figures (with a total of 73,287 on August 30, 1956): nowadays there is an average of one car per 10·1 inhabitants (as compared with one vehicle per 17·6 inhabitants in 1939). Tourist activity is also now greater than before the war, although Cologne was always a favourite centre for tourists. Indeed, one realizes how the life of Cologne is constantly expanding in scope and activity when one considers the very considerable increase in the consumption of water, electricity and gas: a few comparative statistics may serve to illustrate this point.

TABLE II

Year	Water (per million cubic metres)	Electricity (per million kilowatts)	Gas (per million cubic metres)
1939	43·6	330·2	89·7
1955	52·1	773·9	169·9

Yet it is not merely statistical evidence of this kind which gives a convincing picture of the prosperity of the town: a walk through the city's streets gives proof enough of the energy and vitality of the people of Cologne and their fine efforts at reconstruction.

THE ECONOMIC STRUCTURE OF THE CITY

Before the war, Cologne possessed highly productive industries and was, in fact, the most important centre of trade and commerce in Western Germany. The favourable geographical position of the town had made it a highly important central point in the country's transport. The complexity and variety of its commercial activity grew from the healthy and prosperous development of the old Hanseatic town. Yet the war inflicted such severe damage on Cologne's commerce that it seemed doubtful whether the city's resources of energy would be sufficient to rebuild a prosperous commercial and business system, which could form a basis of livelihood for the population. The old trades and handicrafts which, since the Middle Ages, had constantly been a most important economic element in the life of the town and which had greatly helped the development of related industries in the city,[1] were most severely affected by the war. In 1947, from among the approximately 7,500 owners of workshops and business concerns who were registered at the Chamber of Handicrafts, approximately 4,000 had entirely lost their workshops and business premises. According to investigations made after 1948, 43 per cent of industrial buildings and their machinery had been destroyed. Yet both these facts give a very incomplete picture of the total loss suffered: for there were an incalculable number of small concerns which either entirely disappeared in the destruction of the town or which had to move outside the city: and the damage inflicted on many other branches of commerce and industry was never actually registered. After the first new beginnings of commercial life following the currency reform of 1948, a period of intensive and successful business reconstruction took place. The census of business concerns taken on September 13, 1950, showed that there were once again approximately 32,000 workshops and business premises available (as compared with 42,000 in 1939) and that they employed 277,000 workers (cf. 327,000 in 1939). While the number of business concerns has diminished by one quarter since 1939 (from 42,000 to 32,000), the number of employees has decreased by only one-sixth (from 327,000 to 277,000).

A survey of the various business concerns and workshops shows that Cologne's economic character has not changed essentially. Now, as in pre-war years, local commerce and industry are dominated by commerce, finance, banking and insurance; also by manufacturing industries (those distinct from iron and metal industries) and the production of iron and metals and the manufacturing of such goods, each of which industries employs 20 per cent of all the city's workers. We must not forget the transport industry, which employs one-eighth of the working population. Thus we can see that Cologne has not only regained its former position in commerce and industry, but has been able to expand in certain directions. This is especially true of the insurance business: Cologne has become

---

[1] Cf. B. Kuske: *Die Großstadt Köln als wirtschaftlicher und sozialer Körper*, Cologne, 1928, pp. 37–8.

the West German centre of insurance, since many of the leading German insurance offices have moved into the city from Berlin. No less than fifteen insurance companies have now established their head offices in Cologne and the town is also the home of over seventy regional offices in the insurance world. The number of branches of commerce, characteristic of the town but of more than local importance, is considerable and includes many spheres of activity. First of all come banking and exchange activities, which have developed in Cologne through the centuries. It is true that the city has lost some of its former predominance in financial affairs since the stock exchange has been moved to Düsseldorf, though the goods exchange remains in Cologne; the State Central Bank for North-Rhine-Westphalia (an institution replacing the former Reichsbank) has also now been established in Düsseldorf, the capital of North-Rhine-Westphalia. Yet in the industrial sphere, the building of machinery and transport vehicles, the electro-technical and chemical industries, and the production of foodstuffs, sweets, alcohol and tobacco, have retained their very important positions. Cologne is also the headquarters of many important industrial and commercial organizations, the chief of which are the Federated German Industries, the General Federation of Housing Enterprises, and the Central Representative body of Commercial Representatives and Brokers.

Since 1950 there have been further flourishing developments in the city's economy which are, however, difficult to describe statistically because there has been no official 'stock-taking' since that time: and our view of current economic developments is limited to certain restricted facts. However, we find some indication of the present economic state of the town by consulting the figures of persons employed in Cologne, as given by the Labour Exchanges—including manual workers, clerks and office workers, officials and civil servants. Apart from self-employed persons, there were approximately 213,000 employees in Cologne in 1950. In 1955 290,000 such persons were registered at the Labour Exchanges. Thus the number of posts available in Cologne has increased by more than one-third in the space of five years.

All branches of Cologne's commerce and industry have shared in this process of expansion, though concrete figures are available only for industry and building. The building industry has naturally played an extremely vital part in the life of a town so severely damaged as Cologne. In 1950 20,000 persons were engaged in building, so that the industry has become, together with trade and commerce, the most important activity of the city, judged by the number of its employees. In June 1950 its business returns amounted to some 14·6 million marks and in the following years increased to almost three times that amount.

The monthly reports issued by local industry give the current returns for all concerns employing ten or more persons. In 1950 (when the last census of workshops and business premises was held), Cologne's industry brought in a total return of 1·5 billion marks, including returns from abroad amounting to 143·7 million marks, this being almost 10 per cent of the total sum. Five years later, the total return had increased by more than double the original amount and was now 3·5 billion marks. In this last figure are included returns from abroad, amounting to 546·8 million marks and reflecting how closely Cologne's industry is connected with foreign countries.

Cologne's permanent Industrial Fair is very important for the town's economy. It is, in fact, one of the most valuable and excellent commercial enterprises in Western Germany and is already of significance for the whole of Europe. The Spring and Autumn Fair (household goods, iron ware, textiles and clothing) and various special exhibitions help multitudes of interested customers and visitors from home and abroad to keep abreast of latest developments in the commercial and technical world and to make useful business connections.

FINANCE AND TAXATION

The unusually high rate of destruction in Cologne (over 70 per cent of all buildings) has given rise ever since 1945 to such a bewildering number of urgent problems that not even a 'peace-time' budget could cope with them. The city's financial position, owing to the fact that along with other German towns it is unable to profit from the biggest German taxes (income, corporation and turnover), bears three main characteristics: insufficient funds, a high degree of dependence on government grants, and an overstraining of the only local rates which keep pace with market trends, namely, tax on tradesmen's profits and capital.

TABLE III

GOVERNMENT GRANTS TO THE NORMAL CITY BUDGET

		Grants (million Dm)			
Fiscal year	Net expenditure (million DM)	general	special purposes	total	Grants as a % of net expenditure
1953–54	256	18	34	52	20·44
1954–55	270	17	44	61	22·78
1955–56	306	11	43	54	17·83

An average share of 20·2 per cent of current net expenditure must be considered as high and not particularly conducive to a sense of local responsibility, especially when during the last three years an average of 72 per cent of these grants have been given for precise purposes.

The share in the revenue from local rates borne by a limited number of town residents and legal bodies (companies) paying tradesman's profits tax has evolved since 1938–39 in the following way:

TABLE IV

		included in total local rates			
		Tax on tradesman's profits		Tax on real property	
	Total local rates	revenue in	as % of total	revenue in	as % of total
Fiscal year	(Million DM)	million DM	revenue	million DM	revenue
1938–39 (RM)	102	41	40·7	34	33·8
1951–52	76	48	62·8	19	25·2
1953–54	107	74	69·0	22	21·0
1955–56	137	100	73·0	24	17·7

Whereas in pre-war years the tax on tradesman's profits and the tax on real property yielded similar amounts, after the war in view of the difficult conditions pertaining to the tax on real property (lack of payment as a result of destruction of property, legal exemption for priority housing construction, cessation of rent payment, and so on) the revenue from this tax fell markedly. When we take into account the fact that the price index has on the average risen by 100 per cent since 1938 the picture becomes clear: tax on real property today yields only an ample third of the pre-war revenue. The main weight, therefore, has fallen with great severity on the tradesman's profits tax with the result that the so-called 'levy rates,' the extent of which are determined annually by the City Council and which have been somewhat increased, have been only slightly reduced in Cologne during the last few years (from 340 per cent to 330 per cent). An overall federal ruling decided upon by the Government and the *Länder* at the beginning of October, which takes into account the general nature of this problem, comes into force on January 1, 1957. It exonerates from payment of this tax wide sections of craftsmen, private traders and owners of small enterprises with monthly returns of up to 2,400 DM and reduces the scale of rates in the income brackets above this limit without, however, providing the municipalities compensation for the loss in revenue.

This disproportion between urgent tasks on the one hand and available financial resources on the other (revenue from the city's own duties and taxes, government monetary grants, income on property and redemption

TABLE V

NORMAL AND SPECIAL EXPENDITURE 1953–1955

Net Expenditure[1] under the Regular Budget	Fiscal Years 1953	1954 Million DM	1955	Total 1953–55	Share of clearing War Damage
1 General administration	12·2	12·4	12·6	37·2	0·8
2 Police and public order	9·8	8·2	9·0	27·0	—
3 Schools including	35·0	39·8	47·4	122·2	33·9
(a) Primary	11·9	14·4	17·7	44·0	17·3
(b) Secondary and Secondary modern	12·9	15·0	17·2	45·1	9·9
(c) Vocational	6·0	5·8	7·6	19·4	3·9
4 Cultural institutions including	23·4	24·9	26·7	75·0	5·1
(a) University, etc.	10·2	10·6	12·0	32·8	1·7
5 Social work including	38·0	41·2	45·9	125·1	2·0
(a) General social work and war damage relief	19·7	19·7	21·2	60·6	—
(b) Youth assistance	8·3	9·2	10·4	27·9	0·6
6 Health services including	16·6	19·4	20·0	56·0	4·9
(a) Hospitals	8·8	10·8	9·6	29·2	3·0
7 Construction and building including	39·3	39·0	44·9	123·2	24·3
(a) Roads, bridges, etc.	17·1	20·6	23·6	61·3	3·9
(b) Housing schemes	5·3	4·1	8·4	17·8	0·4
8 Public institutions	44·0	48·7	58·3	151·0	9·9
9 Economic enterprises	10·0	11·1	12·9	34·0	1·5
9 Levies and taxes including	27·7	24·9	28·3	80·9	0·8
(a) on real property	9·2	7·5	8·9	25·6	0·7
(b) special	7·2	6·5	8·1	21·8	—
TOTAL	256·0	269·6	306·0	831·6	83·2

[1] Excluding all expenditure that comes under the Acts concerning the equalization of war burdens and subsidized housing and also excluding mutual financial transfers between the administrative branches.

TABLE V—*continued*

Special Budget	Fiscal Years 1953	1954 Million DM	1955	Total 1953–55	Share of clearing War Damage
Total	23·0	63·5	86·5	173·0	68·1
including					
(a) General administration (Town Hall and offices)	1·0	4·1	6·3	11·4	11·4
(b) Theatre	0·1	1·1	4·8	6·0	5·8
(c) Other cultural expenditure (museums)	0·6	2·9	2·7	6·2	6·2
(d) Housing schemes	5·2	11·4	10·1	26·7	0·6
(e) Roads, streets, bridges, etc.	2·7	9·6	13·0	25·3	0·9
(f) Real property (general)	9·5	14·7	17·0	41·2	0·7
(g) Promotion of commerce	0·4	5·7	9·3	15·4	15·0
(h) Public transport and utilities	2·7	10·1	10·8	23·6	9·4

from loans and credit undertakings) explains why there is still a shortage of schools, hospital beds, tramcars and tramlines, omnibuses and trolley-buses, asphalt road surfaces, street lamps, etc., in Cologne, and also why ruins and heaps of debris are still to be found, although almost half of them have been cleared away. Meanwhile, with the growth of industry and trade, new urgent requirements are constantly thrusting themselves forward, such as the building of bridges and new thoroughfares, the extension of harbours, fairgrounds, etc., which do not take into account the city's financial position, demand instant attention and so delay the clearance of remaining war damage. Thus in 1954-55 the acquisition of sites for the construction of a new thoroughfare and new bridge ramps cost the normal and special budgets 15 million DM, while only 1 million DM were available for war damage clearance. The development of an old city heavily damaged by war into a modern centre of communications imposes a violent strain on its budget.

Table V shows the normal and special expenditure of the City of Cologne during 1953-55.

A survey of the expenditure for the various branches of administration taken over a period of three years so as to avoid fortuitous circumstances shows public services to be the essential activity of the municipal administration. Schools, welfare and youth assistance, public building and workers' settlements, each costing roughly similar amounts (120–125 million DM), take up approximately one-sixth of the normal (current) expenditure. Public utilities, i.e. 'day-to-day' things like sewerage, rubbish clearance,

75. COLOGNE

The Cathedral and main station are shown in the foreground. In the background is the river Rhine, with the Fair ground on the right and the suburb of Muelheim on the left.

[Photo: Aero-Lux

76. COLOGNE

In the foreground is the Rheinau-harbour and the left bank of the Rhine. On the opposite bank is the suburb of Deutz.

[*Photo: Aero-Lux*]

77. COLOGNE

The western part of the city, with ring-road and insurance buildings.

[*Photo: Aero-Lux*

78. COLOGNE

Public park at the Ebertplatz, with office buildings.

[Photo: Hartzenbusch

street cleaning and lighting, fire extinguishing equipment, as well as cemeteries, parks and gardens, and traffic improvements take up almost one-fifth of current expenditure.

Cultural amenities in the University, museums and theatres, required 75 million DM from city funds during the last three years. This includes considerable sums for the University clinics, so that the 56 million DM spent on health requirements do not represent the total expenditure in this important field.

The current expenditure for transport, power stations and waterworks owned by the City of Cologne are separated by law from the administrative budget and, according to commercial principles, come under a separate account. These enterprises therefore essentially come under the city's special budget together with investment expenditure (capital, credits, emergency funds).

The city acts as a so-called *Gewährsverband* for the City of Cologne Savings Bank, i.e. as a guarantor of investors' savings. This gives the city the right of supervision over the Savings Bank, whose accounts are kept separate from the normal budget of the city administration, which records only personnel expenses as permanent items.

The special budget shows a much more fortuitous appearance than the normal budget. So many outstanding tasks, such as the restoration of destroyed bridges, had by 1952 been successfully dealt with, that in 1953–55 only preliminary work for new projects (see above) appeared in the special budget. On the other hand the reconstruction of the Town Hall for a long time lagged behind the other administrative buildings, and the same was the case for the rebuilding of the theatre and the museums. The further clearance of war damage from the fair grounds (to facilitate traffic) required greater funds, particularly in the last two years; and similar needs were registered for the power stations and waterworks and above all for modernization and war damage clearance in the city's transport enterprises.

War damage clearance in the three years in question took up approximately 40 per cent of the special budget. From current expenditure, which pays mainly for the daily care of the community and the cost of personnel, exactly 10 per cent of the sum spent on war damage clearance during the three years was needed for material expenses, particularly for schools, public building and utilities, but also for cultural and health amenities.

However dissatisfying and restrictive the limitation of the city's funds, the re-emergence of its financial strength, when ten years ago it was almost totally destroyed, is a noteworthy fact. But should these gains continue

Y

to be diverted immoderately into government accounts, then the time is not far off when the pressure of urgent requirements will bring about a situation of conflict.

Town planning is the task of the *Städtebauamt* (Town Planning Office), which consists of a planning section and an operational section. The operational section deals with the legal side of planning control and sees that individual building applications do not conflict with over-all town planning.

The law applying to planning is the Reconstruction Act of the *Land* North-Rhine-Westphalia of April 29, 1952. On the basis of this law the City of Cologne Town Planning Office, in close collaboration with other competent city offices, outside authorities and departments affected, worked out from 1950 to 1954 an over-all plan, which acquired legal force in 1956. It shows the intended development of town planning for the municipality as a whole and is drawn up to meet the needs of a total population of 800,000. The over-all plan provides the guiding principles for the different operational plans. These show the areas for public and private use, the distribution of green-belt, traffic and building areas, transport installations, the main supply and drainage systems, the manner in which the building areas may be used and the extent of their use, the covering of building sites according to area and height and the distribution of building blocks. The over-all plan and the operational plans are subject to the decisions of the City Council.

The drawing-up of the over-all plan and the operational plans comes under the jurisdiction of the municipalities. The State only regulates the procedure and checks whether the plans agree with the over-all planning aims of the *Land*. The municipality is obliged under the Reconstruction Act to take into account the intentions of the *Land* in planning. The over-all plan and the operational plans require the approval of the Board of Control to become valid. The latter checks whether the operational plans agree with the aims of the over-all plan and are in accord with legal requirements.

For planning purposes the city is to be divided up into districts and quarters, as has already been done for administrative purposes. This division should correspond to the original and traditional structural units represented by the individual districts. It can be done in different ways: in the inner city there are new traffic routes forming the boundaries of the individual town quarters. Outside the thickly built-up inner city the town is divided

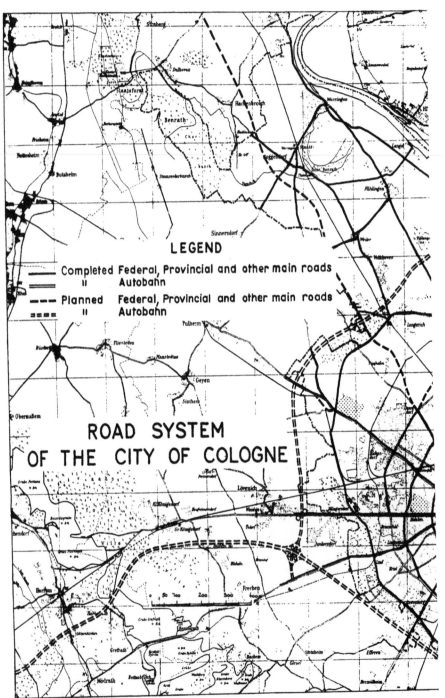

LEGEND

Completed Federal, Provincial and other main roads
      "   Autobahn

Planned   Federal, Provincial and other main roads
      "   Autobahn

ROAD SYSTEM
OF THE CITY OF COLOGNE

Source: Statistical Office of the City of Cologne

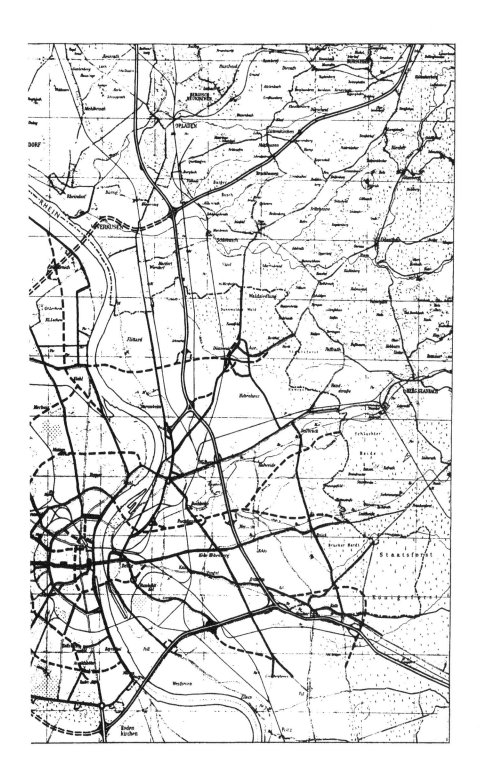

up by green areas and empty spaces. The Inner Green Belt which clearly separates the inner city from its suburbs is particularly valuable. In Ehrenfeld and Nippes, for example, there are central spaces which have small gardens, sports grounds, children's playgrounds, rest gardens, surrounded by built-up areas, which are thus divided up from within. Numerous new living quarters have been built or planned as self-contained units.

The reconstruction of destroyed bridges and the re-orientation of traffic routes is important for the life of the city. The proximity of the *Autobahn* is a very favourable circumstance. The town plan envisages the construction of eleven non-built-up roads to feed into the *Autobahn* and to lead right into the centre of the city. The most important traffic road in the inner city is the so-called 'collecting tangent' (*Sammeltangente*), that is, the road leading from the *Autobahn* in the west (boundary road—Grenz-strasse) over the planned North bridge, through the Inner Canal Street (*Innere Kanalstrasse*) to the Bonn circuit. It borders the inner city and collects all outgoing traffic. The inner city presents the most difficult problem. It is covered with a widely spread and well-distributed network of new traffic streets. The backbone of this system is the North–South Street. For east-west traffic there exists today only one main traffic street. The Gotenring bridge at present under construction provides another East–West connection and alleviates the excessive burden carried by the narrow Deutzer bridge. The North–South Street has been envisaged as a main artery for in-and-out traffic. The North–South Street is to have no shops and to provide no access to property so that it may properly fulfil its function. Subways and overhead crossings in the inner city are, for town planning and economic reasons, envisaged at only a few places, e.g. at the crossing of North–South Street and the East–West axis and at the bridge-heads of the new Gotenring bridge. In anticipation of a traffic increase of three times its present amount, about 17,000 vehicle parking places will be needed in the inner city. A considerable number are already there. Public parking places are being so planned that it will be possible at a later date to add further storeys to them. With the aid of the Garage Regulation of 1939 every builder is being required to provide parking space for any new building or building alteration. At the present time there are three two-storey garages under construction in the city area with a total parking capacity of approximately 1,200 vehicles. Parallel to the main shopping streets, streets for loading and unloading goods are either already in existence or being planned in order to keep the shops constantly supplied.

Public transport is receiving special attention. The trams wherever possible run on separate sections of road from the other traffic. There are no plans for removing the trams from the inner city. The moving of the main railway station (on account of its proximity to the Cathedral and traffic difficulties) which was planned after the war, is out of the question today owing to the cost.

For air traffic, Cologne is connected with Brussels and Bonn by means of a helicopter station which is centrally situated. For long distance air travel Cologne is favourably served by the aerodrome in Wahn which lies midway between Cologne and Bonn. Both are within quick reach and served by good roads.

The countryside preserved in the city area of Cologne is under planning protection. The green areas border on the Rhine and also on the inner and outer green belts. Intercommunications lead to the Rhenish woods of the nearby Bergisches Land.

A speciality of Cologne planning is the linking of traditional town planning values with modern requirements. Public opinion in Cologne desires the careful preservation of valuable old buildings, but is prepared to accept side by side with them the most modern innovations.

THE RELATION OF THE CITY ADMINISTRATION
TO THE LAND AND FEDERAL GOVERNMENTS

In the preamble and separate sections of the 1949 constitution of the German Federal Republic only the relation between the federal government and the *Länder* is mentioned, from which it follows by implication that local authorities come exclusively under the jurisdiction of the *Länder*. In accordance with this each *Land* in the Federal Republic has enacted its own municipal legislation. According to the North-Rhine-Westphalia municipal regulations the municipalities form the basis of the democratic state order. Article 28 of the Constitution grants municipalities the right to conduct the affairs of the local community within the framework of the law on their own responsibility (*Allzuständigkeit*). The responsibility of the Cologne city administration covers town planning, maintenance of cultural amenities, public assistance (required by law), supply of gas, water and electricity, local transport, road building, street cleaning and refuse clearance. Apart from these administrative responsibilities the city has certain duties to fulfil, the so-called *Auftragsangelegenheiten*, which are legally assigned to them by the state, such as registration with the police, registration of births, marriages and deaths, living accommodation, primary school education, the carrying out of elections. In certain similarly regularized matters, such as the annual making out of *Steuerkarten* (tax cards) for employees, the carrying out of the *Lastenausgleich* (equalization of burdens) for refugees and war casualties, and the drawing up of jurors' lists, the city administration becomes an agent of the state's financial or judicial authorities and is subject to the supervision of the competent departments or ministries of the *Land* government. The general state supervision of the City of Cologne is carried out by the president of the administrative district. The chief supervisory authority is the Minister of the Interior of the *Land* North-Rhine-Westphalia. The general supervision covers municipal administration as a whole. Its object is to make sure that local administration is pursued in conformity with the law. The supervisory authority may at any time enquire into the affairs of the municipality for this purpose.

Should the law appear to be infringed by decisions of the Council or by the *Oberstadtdirektor*, the supervisory authority can make use of its right of objection and if necessary overrule the decisions or arrangements to which it objects. In spite of the independence of the free cities—which, on an equal basis with the *Landkreise*, together form the *Landebene*—the supervisory authority has the right to intervene in order to compel fulfilment of tasks by the City Council as laid down by law. In an extreme case the Minister of the Interior may appoint a representative to carry out all or certain of the tasks of the municipality or, by means of a *Land* government decision, to dissolve the City Council. In such a case the city has the right of appeal to the administrative court (*Verwaltungsgericht*). Where the Minister acts in default of the municipal authority, he recovers the cost from the latter.

A REGIONAL AUTHORITY

Before the war the city and *Land* districts of the Rhine province were joined together in a single *Provincialverband* (provincial authority). In its place the law of May 12, 1953, has created the *Landschaftsverband Rheinland* as a regional local authority with the task of fulfilling requirements not confined to specific areas, particularly as regards road maintenance and welfare. Cologne is a member of this regional authority which, together with the (state) presidents of administrative districts, forms the intermediate organ between the *Land* government and the local authorities. The regional area embraces the northern part of the Rhineland. Its voting organ is the *Landschaft* council, whose members are elected by the individual member bodies. The *Landschaft* council supervises the work of the

Landschaft administration. Like the *Landtag* and the City Council, it also appoints special committees for different fields.

The *Landschaftsverband* is a joint board comprising 23 rural and 17 urban authorities, of which Cologne is one. It covers a territory of 12,520 square kilometres containing a population of 7,888,885 persons (in August, 1956). The powers it possesses are conferred by statute; it is a self-governing authority and is a legal corporation in public law.

The *Landschaftsverband* is empowered to deal with the following matters: (1) Health and welfare. In this sphere, the regional authority has established a regional centre for war victims and their dependants, and also a regional youth welfare office; (2) The maintenance of highways. These include the main roads of the region, the *Autobahn*, and national trunk roads; (3) Cultural services and amenities; (4) Certain clearing house and banking functions; (5) Participation in town and country planning; (6) Participation in public utility and transport undertakings; (7) The regional authority has the power of taking over the management and administration of the social insurance schemes for industrial injuries in agriculture, the local government service, and the fire brigade service.

The expenditure of the regional authority is met by fees and contributions paid in accordance with the provisions of tax legislation; by contributions from the member local authorities; and by financial grants from the *Land*. The City of Cologne is paying a contribution of 4·7 million DM in the financial year 1956. The regional authority cannot itself levy taxes. The Minister of the Interior for the *Land* is the supervising authority.

CONCLUSION

The city looks back on a history of about two thousand years during which its existence has been repeatedly threatened. It has survived all these dangers. The most recent catastrophe lies barely more than a decade back. A new chapter in Cologne's history began in the year 1945. Cologne rose again like the phoenix from its ashes. Again, as in earlier centuries, it had to rely on its own strength since it was given none of the impetus which the new capitals of the *Länder* seem to have had for the asking. The wounds of the Second World War are not yet completely healed. The outskirts of the town still show large gaps. The ruins of many houses are still visible. Many of the previous residents have not returned. Nevertheless, particularly since 1948, there has been a continuous upward development. There is lively activity in all parts of the communal organism. Within two or three years Cologne should at least have reached, if not exceeded, its pre-war level of population. The city is growing, and so, too, is the attraction which it has always exerted both at home and abroad. 'The upward trend has proved itself stable and has passed the most crucial tests.'[1] Council and administration are conscious of the historical mission of their city and have therefore adapted their plans for the future to the needs of modern times. In this way they hope, as far as is humanly foreseeable, to provide a foundation for the re-emergence and secure continuance of community life, so that coming generations will be able to complete what was begun anew during the past ten years.

[1] L. Fischer: 'On the Growth of Large Towns' in *Städtestatistik in Verwaltung und Wissenschaft*, edited on behalf of the association of German City Statisticians by Dr. Bernhard Mewes, Berlin, 1950.

Johannesburg

L. P. GREEN

B.Sc.(Econ.) Hons., London, 1939; Diploma in Education, Exeter, 1940; Diploma in Public Administration (Distinction), London, 1948; M.Sc.(Econ.) in Politics, London, 1949; Ph.D. in municipal government, Natal, 1954. War service 1939 to 1946. Enlisted in ranks and rose to Staff Captain, G.H.Q. (India). Lecturer in Politics and in Public and Municipal Administration, University of Natal, 1949–53. Engaged by Johannesburg City Council to draft municipal charter and prepare memoranda for submission to Transvaal Local Government Commission of Inquiry, 1953–54. Attached to Town Clerk's Department, Johannesburg, since 1954 to continue with memoranda, give evidence to Commission and undertake research into problems of city government.

Author of 'Public Monopolies,' in *Monopoly and Public Welfare* (University of Natal Press, 1952); 'The Administration of Municipal Housing in Durban,' in *The Durban Housing Survey* (1952); *History of Local Government in South Africa: An Introduction* (1957); and numerous papers and articles in *South African Journal of Economics*, Conference Proceedings of Institutes of Town Clerks and Municipal Treasurers and Accountants (S.A.), *South African Treasurer*, *Municipal Affairs* (S.A.), and other journals.

Johannesburg

NEARLY 6,000 feet above sea-level, Johannesburg is one of the highest great cities of the world and the metropolis of the Union of South Africa. It is surrounded by the rolling Transvaal veld that shelves away to the north and south from the hills of the Witwatersrand on which the city stands. Its climate is good. In winter, the daily sun quickly disperses night frosts. In summer, afternoon rains prevent the high temperatures to be expected only 1,800 miles south of the equator. Outside the city centre, the air is clear and bracing, suburban gardens are brilliant with colour, the streets green with trees and many of the houses are veritable mansions. Life is bustling, vigorous and cosmopolitan, and the visitor finds lacking only the mellowness of age and the charm of antiquity. For Johannesburg is wholly modern. Nevertheless, most of its political, governmental and planning problems are those of every metropolis. Some of them will be considered here, after a description of the city's social, economic and local government history, its place in South African life, and its municipal organization. And some solutions will be offered, before turning to consider that most complex of modern urban phenomena, the metropolitan or city region, of which Johannesburg, like so many other great cities, is already an established centre.[1]

THE GROWTH AND SIGNIFICANCE OF JOHANNESBURG

On September 8th, 1886, Paul Kruger, the President of the Transvaal South African Republic, signed a proclamation declaring the farms of Langlaagte and Randjeslaagte a public gold digging as from the following month. The biggest gold rush in history began. From every part of the world a stream of people converged on an area of bare, treeless veld vaguely described as being '35 miles from Heidelberg . . . and situated on the Witwatersrand Range.' By the following December, when the township of Johannesburg was established on Randjeslaagte, the population had grown from 100 to almost 3,000 people. Today, 70 years later, the jurisdiction of the Johannesburg City Council extends over 120 square miles and an estimated concentration of 1,000,000 people. Of these, 350,000 are of European descent and 650,000 non-European or mixed.

---

[1] The views expressed in this paper are solely those of the author. The Johannesburg City Council is neither responsible for them nor necessarily in agreement with them.

Johannesburg is thus doubly unique among the great cities of the world. It is not yet older than many of its citizens, and it has been built on gold. Nature did not dictate that men should live on Randjeslaagte for the usual geographical reasons, and neither commerce nor strategy determined its location. It is a city set on a range of hills to which water, food, coal, wood, iron and all the other necessities of life had to be brought by road, rail and pipe-line. But it was not yet 14 years old when it caused the Anglo–Boer War that still bestrides national politics, and it began to wrest the metropolitan leadership of Southern Africa from Cape Town long before it reached man's estate. It has transformed the country's balance of population, wealth and power, its international standing, and the structure of its social and economic life. It now accounts for one-twelfth of the Union's population in only one-four thousandth part of its area.

All this has come to pass because of gold. The largest gold-mining industry in the world centres on Johannesburg. In 1954, over 13 million ounces of gold were produced in South Africa to the value of over £165 million. The Witwatersrand mines accounted for 12 million of those ounces to the value of £148 million, and they are now beginning to produce uranium concentrates that will eventually raise a further £30 million a year. Since mining on this scale could never have been undertaken without appropriate manufacturing and commercial ancillaries, it is not surprising that Johannesburg has become the Union's leading industrial centre, the hub of probably the largest concentration of manufacturing in the southern hemisphere, and the commercial capital of Southern Africa. Quite naturally, too, it has for long been the financial centre and capital market of a sub-continent; in fact, the Johannesburg Stock Exchange began in 1887 and thus antedates the Chamber of Commerce founded in 1890 and the Chamber of Industries formed in 1914. As a result, Johannesburg today accounts for just over two-fifths of the aggregate national industrial output, and the total labour force employed in mining and manufacture in the metropolitan region is estimated at 500,000 persons, including over 400,000 non-Europeans.

The national road, rail and air transport routes now pivot on Johannesburg and graphically illustrate its economic dominion over every other town and city except, perhaps, Cape Town, 1,000 miles away. It bids fair to become the country's cultural and educational centre in spite of its youth, boasting already the largest public library, founded in 1889, the largest university, incorporated in 1922, the leading orchestra, a growing art gallery and a rapidly maturing theatre life that originated in 1887. Nevertheless, unlike so many other great cities it is no national capital—legislative, administrative or judicial—and unlike Paris, Rio de Janeiro or Cairo it has never held the political destinies of a nation in its hands. The accident of its birth has rendered it a cosmopolitan but predominantly English-speaking island in what is traditionally the most Anglophobe province of South Africa and, through the weighting of the electoral system, the country's political centre of gravity remains in the platteland or rural areas. Johannesburg is the fulcrum of national politics and not the lever.

A sequence of three chapters has so far been written in the municipal history of South Africa. The first, from 1786 to 1840, was compiled at the Cape of Good Hope; the second from 1854 to 1908, in Natal; the third from 1903 onwards, in the Transvaal, and largely by Johannesburg.

Urban local government began in Cape Town in 1786 with an appointed Municipal Committee of three Dutch East India Company officials and three burghers as an off-shoot of the Colonial High Court. Ten years later, the first British occupation, that lasted until 1803, eliminated the official side of the Committee and transformed it into an appointed Burgher Senate. This remained in control of municipal affairs until 1827; except for an experimental period between 1803 and the second British occupation beginning in 1806, when the Colony was returned to the Batavian Republic which toyed with the idea of an elected town council. The Senate was abolished in 1827 because of judicial reforms, and in 1836 a Municipal Ordinance introduced locally representative and responsible municipal boards of commissioners throughout the Colony, although Cape Town itself had to wait until 1840. These boards were ruled by public meetings and were similar to some English parish governments to be found in towns such as Liverpool before 1835.

The initiative in local government now tended to pass to Natal. The 1836 Ordinance was copied in this new Colony eleven years later, but contained so many elements of direct government as to render it ineffective in but the smallest of towns. In consequence, a committee of inquiry was established whose recommendations led to a thorough reform in 1854 that grafted important features of the English Municipal Corporations Act of 1835 upon the Cape system. Incorporation, the mayor, councillors, committees, town clerk and local government servants now made their début and municipal powers were significantly extended. The success of this reform soon had repercussions in the Cape, beginning with Cape Town itself in 1867, and led to strong municipal governments in Pietermaritzburg and Durban whose experiments in public health and

Native administration at the end of the century paved the way for Union legislation affecting all municipalities.

When Johannesburg was born, therefore, there was already a well-founded system of urban self-government in South Africa whose basic principles enjoyed the established authority of political history and tradition. In fact, the South African Republic had already introduced the Cape system of 1836 to its few urban centres north of the Vaal River; the Natal system had prematurely appeared for a short while in 1880 during the first British annexation of the Transvaal; and the Voortrekkers had brought an old rural local government system northwards from the Cape when founding the inland republics, and had usually made it elected rather than appointed. Nevertheless, the overwhelming influx of people, turmoil and excitement attending the birth of a boom town, the antagonisms naturally engendered between gold-seeking Uitlanders and pastoral Boers, and the subsequent war, for a time prevented the handing on of the local self-government tradition to Johannesburg. For sixteen years its administration was in a state of tutelage, and it was not until 1903 that it was able to follow in the footsteps of the Cape and Natal and tread the path of municipal self-government.

THE BEGINNINGS OF LOCAL GOVERNMENT

Like most vast organizations, the Johannesburg City Council began in a very small way. It started in November, 1886, as a mere Diggers' Committee of nine elected members under the *ex officio* chairmanship of a mining commissioner appointed by the South African Republic with general controlling powers. After numerous local complaints, the Committee was replaced in December, 1887, by a Gezondheids Comité or Sanitary Board headed by the commissioner and consisting of five elected members and two other Government nominees. The Board was reconstituted two years later, when a Government commissioner became its head, the elected membership was increased to twelve, and its area was fixed at some five square miles. It lasted until 1897, although its powers remained wholly inadequate. Its organization was rudimentary, and the Republic preferred to augment State revenues by granting franchises or concessions to the highest bidders for the supply of water, gas, electricity, public transport and market facilities.

The Sanitary Board's end was a result of the Jameson Raid. In the last days of 1895, Dr. Jameson crossed the western Transvaal border with some 500 armed men and made for Johannesburg. On January 2nd, 1896, they were captured, Cecil Rhodes' dream of an African empire was shattered, and the stage was set for the Anglo-Boer War. In February, a draft Bill for the abolition of the Gezondheids Comité and its replacement by a Stadsraad was gazetted. In the circumstances, it is not surprising that Johannesburg was to enjoy even less local self government than in the past. While the Government commissioner had ceased to be chairman of the Sanitary Board, a town council was now to be created with a salaried official at its head responsible to the Central Government and called the burgemeester or mayor. He was to be a much more independent and powerful figure than the commissioner, being appointed for five years by the State President as the chief executive officer of the council and chairman of its deliberations. He was to be assisted by two and later four schepenen or aldermen elected by the council, and might be removed only by the President himself. In addition, the non-Burghers or Uitlanders were to lose their voting rights.

The Bill was eventually modified to permit half of the council's twelve members to be elected by the non-Burghers, but few other changes were made before the new Stadsraad or Town Council was legally constituted in December, 1897. It governed the town for just over two years. Yet, although its jurisdiction was extended to about nine square miles, and although it boasted nine departments and later increased its membership to twenty-four, the Council's powers were little more effective than the Sanitary Board's. They but extended to sanitation, fire, market, street, and gas and electricity services, and to the making of local regulations or by-laws. As constituted, the Council was incapable of generally serving the community it governed, which had already expanded to nearly 102,000 people in 1895 owning property valued on a capital basis at £5½ million, and it was soon embroiled in the three year war that began in October 1899.

At the end of May 1900, British forces entered Johannesburg and the town's local government passed into the hands of an acting burgemeester or mayor appointed by the military governor. He remained in control until the following May, when twelve prominent citizens were nominated as members of a new town council. Paradoxically, this nominated council was the prologue to fifty years' municipal progress and the foundation of the present organization. It was created by proclamation after a draft municipal law had been outlined for the Transvaal by the acting mayor. This followed exclusively the provisions of the municipal laws of the Cape and Natal; it itself drew upon municipal legislation of Cape Town, Port Elizabeth and East London in the Cape, and the latest laws of Natal;

and it thus built directly upon an established tradition of local self-government.

Between 1901 and the end of 1903, when an elected town council first met, the nominated council was busy creating a municipal organization appropriate to the undertaking of much wider functions than the Stadsraad had ever enjoyed. It was guided by a certain amount of experience. Some of its members and officials had served the Stadsraad and other municipalities in Natal and the Cape; the acting town clerk, Lionel Curtis, had been private secretary to the Chairman of the London County Council; and questionnaires on organization were sent to both Durban and Cape Town. It was also guided by the provisions of the proclamation. This outlined a structure consisting of a nominated governing council with a chairman and deputy chairman that met weekly in public; a system of standing committees appointed by the council for the better management of its affairs, including an obligatory finance committee; and departmental officers, including a town clerk, treasurer, town engineer and such other officers as the council might desire and the military governor approve to undertake the town's daily administration. And it drew up domestic regulations within this framework based on those of the London County Council.

Largely as a result of the nominated council's representations, the municipal area was extended to 83 square miles, powers were obtained to levy an annual rate of up to 3d. in the pound upon the market or capital value of land and buildings, to borrow money subject only to central approval, and to license certain trades and road vehicles. In addition, extensive municipal functions were added to those of the Stadsraad. The council was given the exclusive right to operate public transport, extended powers of control over public health including sewerage and drainage works, the right to control the retail supply of water, a rather vague power to supervise Native locations and Asiatic bazaars, and rudimentary town-planning controls over the development of new townships or estates within its boundaries. The central government continued to run the police and education services, its approval was needed for all by-laws and loans, and it gave no comprehensive power to the council to undertake works or services for the general good of the municipality. But it permitted the nominated council to elect its own chairman and deputy chairman in 1902, and its constant objective was the early realization of municipal self-government with a minimum of central supervision.

When the elected town council took office at the end of 1903, therefore, it was presented with a well-articulated and multi-purpose civic administra-

79. JOHANNESBURG

From the south-west, with the Hillbrow–Berea Ridge in the background cutting off the northern suburbs from the central business district.

[*Photo: South African Railways*]

80. JOHANNESBURG
Joubert Park.

[Photo: South African Railways

81. JOHANNESBURG

Flats at Clarendon Circle, Hillbrow.

[*Photo: South African Railways*

82. JOHANNESBURG

Meadowlands Native Township constructed by the Natives Resettlement Board.

tion that was both locally representative and responsible. It consisted of a council of 30 councillors elected for three years, who annually elected their own mayor and deputy mayor; a system of five standing committees of management concerned with finance, general purposes, public health, public works, and tramways and parks; and fifteen executive departments and divisions including the town clerk's, town treasurer's, town engineer's, public health, assizer's, fire, and light and power departments. The municipal population had grown to 109,452 persons and the total rateable value to £36½ million. The gross annual income was now £390,000, of which rates accounted for £265,000, and the net loan debt had risen to £486,000.

During the next fifty years, this organization was to be severely tested by a remarkable growth in municipal population and problems unmatched elsewhere in South Africa. Its response naturally made it a pioneer in many respects and it became the main author of the third chapter of the Union's local government history, whose contribution can be conveniently summarized by a description of the city's organization and functions today as compared with the position in 1903.

THE JOHANNESBURG CITY COUNCIL:

MUNICIPAL POWERS AND FUNCTIONS

Johannesburg was granted city status in 1928 but this has brought no special privileges of government. Its powers and functions flow almost entirely from the general municipal legislation of the central and provincial governments applying to all urban centres of any consequence. Under the South Africa Act of 1909, power to legislate for local government was given to each of the four provincial administrations that replaced the old colonial governments. These consist of an Administrator appointed by the Governor-General of the Union, a provincial council elected by the parliamentary voters of the province, and an executive committee of four members elected by the council by proportional representation. The Administrator is president and member of the executive committee, which is constitutionally accountable to no one. At the same time, the central government was not divested of the power to legislate for local authorities itself, and has in fact legislated for them in regard to public health, housing, slums, Native affairs, licensing and the zoning of land use according to race. Its agent in these matters is usually the Administrator, who is not then answerable to the executive committee, and Johannesburg is thus subject to a dual control that often overlaps and is sometimes confused.

Nevertheless, the city has expanded its services at a very great rate. The increasing magnitude of its civic task, and of its response, is shown by the following rough indices. In 1905 the municipality's estimated population was 170,000 persons, its net loan debt was £5 million and the annual revenue turnover £743,000. In 1935 these figures had risen to 430,000 persons, £8 million and £4 million respectively. In 1955 the population was estimated to be 1,000,000 persons, the net loan debt had risen to almost £40 million, and the revenue turnover for the financial year ending in June 1956 was estimated at £18¼ million. The annual budget has reached a record total of £28 million on both revenue and capital accounts.

The council employs nearly 25,000 men and women of all races in the operation of the largest transport concern in South Africa, apart from the South African Railways; the largest electricity undertaking after the national Electricity Supply Commission; the largest wholesale and retail market; the largest urban non-European affairs organization; the largest public library; the largest housing agency; one of the largest civil engineering concerns; public health and welfare organizations second only to those of the Union; probably the largest buying agency outside that of the central government's; the largest fire department; the largest single licensing body; one of the largest land and estate administrations; and even a municipal farm.

The actual size of these undertakings can be gauged from current statistics. The public transport budget allows for an expenditure of £2¼ million during the financial year, and the electricity budget for over £4 million. The annual sales turnover in the produce market is now £6 million; non-European affairs, including housing, account for an annual expenditure of £2 million; the library has an annual circulation of 2¼ million books (in 1956); housing will absorb almost £2 million, including £1½ million for Natives; and the city engineer's department has over 10,000 employees. Health and social welfare services are costing £1½ million a year, and central purchases of stores and materials on revenue account amount to more than £3½ million annually.

The council also operates a municipal airport, a busy abattoir and livestock market, an extensive water undertaking and the largest gas works in South Africa. It owns a very fine art gallery, an Africana museum, a well-equipped zoo, 6,000 acres of parks and open spaces, a golf course, swimming baths, bowling greens and tennis courts. It runs a fever hospital and owns municipal cemeteries; and it controls urban development within the municipal boundaries by means of a comprehensive town-planning scheme. Practically the only local services that it does not undertake are police and education. But it employs mobile and point-duty traffic police at a cost of over £30,000 a year; it subsidizes the University of the Witwatersrand (where it recently founded a Chair of Local Government and Public Administration), it assists other educational and cultural bodies with grants-in-aid, and it runs its own nursery schools.

To finance these undertakings, the city currently expects to derive over £3½ million from a rate of 4½d. in the pound on land values, £4 million from service charges, £9½ million from its trading undertakings (which will contribute almost £1 million to general revenues), £660,000 from licences and Native registration fees, £800,000 from rents, £250,000 from traffic and other fines, and some £200,000 from government subsidies and refunds (including the Native services levy). The council will finance more than £1 million of its proposed capital expenditure from its own resources, over £6½ million from its consolidated loans fund, £800,000 from the Native services levy fund and over £1 million from government housing loans.

THE COUNCIL AND ITS COMMITTEES

At the centre of this multiple organization is the city council, consisting of forty-two councillors elected by parliamentary voters on the basis of one for each ward for three years and retiring together. The council elects its own mayor and deputy mayor from among its members to serve for one year as the city's civic and social heads; meets ordinarily once a month under the mayor as chairman to debate and resolve on the important issues of local government; and annually appoints some nine standing committees of from five to twelve members each for the better management of particular divisions of its business. These committees have matters referred and delegated to them by the council (in the latter case taking decisions on its behalf), report monthly to the council, elect their own chairmen and vice-chairmen, and work in close collaboration with the heads of the municipal departments and their senior officials.

Thus, the organization is shaped like a wheel. The council is the hub and is mainly concerned with the settlement of policy; the committees are the spokes, formulate policy for the council's decision, supervise the execution of decided policy by the departments, and exercise certain delegated powers of decision themselves. The departments form the rim of the wheel, assist the committees to formulate policy, and then carry it out as directed by the council or its committees. Council meetings are formal debating occasions held in the council chamber once a month. They are attended by the town clerk, city treasurer and other senior officials, and are open to press and public alike. Committee meetings are much less formal round-table discussions held in committee rooms, from which both the press and the public are excluded. They are each looked after by a committee clerk who records the minutes of their proceedings and sees that decisions and instructions are communicated to all concerned; and they are attended by the town clerk or his representative, appropriate departmental heads and other senior officials, who take part in discussions but have no vote. Every councillor must be a member of at least one committee, and may attend the meetings of any other committee with the right to speak but not to vote.

Since almost the whole field of civic administration is covered by the several committees, which are now the workshops of municipal government, they are very important links between the council and its departments; and as municipal business has expanded in scope and variety, the task of keeping them in step has become more and more vital. For this reason, a finance committee (which is the only statutory or obligatory committee) has become recognized as essential and has developed a strong financial control since 1903. In addition, a general purposes committee, mainly consisting of the chairmen of the other committees, has remained a feature of the municipal organization since the days of the nominated council, and an establishment committee to which all staff matters are referred was set up in 1954. In 1955 there were nine standing committees, viz:—

General Purposes	Non-European Affairs and Housing
Finance	Public Amenities
Establishment	Public Utilities
Health and Social Affairs	Works and Traffic
Licensing and Town Planning	

These committees establish certain sub-committees to investigate and report on matters referred to them. They meet ordinarily once a month in the morning or afternoon, but often have one or two additional special meetings. The Council does not normally create special committees to deal with particular problems, but it has twice appointed a special housing committee in recent years to speed up Native housing.

The committees supervise the work of twenty-two separate departments or divisions, each under departmental heads or officers of equivalent status who meet every month as a departmental committee under the chairmanship of the town clerk to consider and report on major departmental and civic proposals. This committee was first created in 1929 under the chairmanship of the chairman of the General Purposes Committee, but was reconstituted on its present lines in 1937 when a technical sub-committee was also established. The sub-committee now consists of the town clerk (as chairman), the city treasurer and the city engineer, together with departmental heads affected or interested. It meets more often than the departmental committee, and it considers and reports on similar subjects and any matters referred to it by the council's committees.

THE COUNCIL'S DEPARTMENTS AND STAFF

The Council's twenty-two departments or divisions are as follow:—

Abattoir and Livestock Market
   Department
Art Gallery
City Engineer's Department
City Health Department
City Social Affairs Department
City Treasurer's Department
City Valuation Department
Director of Housing
Electricity Department
Fire Department
Gas Department

Licensing Department
Market Department
Non-European Affairs Department
Organization and Methods Division
Parks and Recreation Department
Public Library and Africana Museum
Rand Airport
Staff Board
Town Clerk's Department
Traffic Department
Tramways Department

These departments have many branches or sections, such as the stores and buying branch, water branch, estates branch, cleansing branch, building survey branch, town planning office, sewer branch, ambulance section, Native employment bureau, Native influx control office, medical division, sanitation division, legal section, staff control office and municipal reference library.

The work of the technical departments is self-explanatory, but it should be stated that the director of housing is concerned only with the building of non-European houses, and that the Health Department administers the European housing schemes. The Social Affairs Department has undertaken family welfare work, homes for the aged, community social centres, children's play centres, food distribution services, a communal restaurant, a central register of social welfare organizations, sheltered employment projects, hostels, youth services, welfare supervision in parks, and poor relief; but some of these functions are now transferred to the Union Department of Social Welfare. The Non-European Affairs Department administers both Coloured and Native housing schemes and hostels, Native beerhalls, tearooms and restaurants, seven non-European townships and an emergency camp; and it controls the influx and employment of Natives throughout the municipal area. The Town Clerk's Department—with its legal and estates section, committee clerks, records office, duplicating section, typists' pool, translation section, despatch section, and public information and research services—is a central office through which all official correspondence, departmental reports and committee and Council agendas, minutes and decisions, are channelled, distributed and despatched.

The staff consists of 3,500 salaried and over 4,000 daily paid European employees, and more than 17,000 non-European clerks, labourers, messengers and nurses. The European members are organized in eleven trade unions of which the largest is the Johannesburg Municipal Employees' Association, they have operated sick benefit societies since 1908, and they run a sports club. On the council's side, their affairs are a matter for the Staff Board of three, and an organization and methods division. The Staff Board was appointed in 1950 to advise on all promotions, appointments, salary rates and the creation of new positions; and, within the limits of council policy and the terms of conciliation agreements and arbitrators' awards, it has gradually been given power to dispose of many routine items and lower-grade appointments and promotions. It is now completing a comprehensive scheme of minimum qualifications for all positions and grades, and it works in close co-operation with the organization and methods division. This division was set up in 1953 to advise departments on organizing new work, to undertake a planned review of existing work, to carry out a continuous organization and methods survey in all departments, and to make special investigations and studies as required. In practice, the town clerk has a particular interest in the operations of both bodies.

With these developments, the council is justified in claiming that it now has adequate and satisfactory machinery to encourage efficiency, and to control appointments, the creation and classification of positions and the determination of salary grades. In fact, its present personnel problems mainly arise out of its relations with the trade unions under complicating

Union legislation, and usually concern demands for general increases in wages. But these problems should soon be met by the establishment of a permanent arbitration tribunal for the settlement of local authority labour disputes throughout South Africa, in place of the existing *ad hoc* system of awards and agreements.

POLITICAL PROBLEMS

Johannesburg has not escaped the strife of party politics. Of the three phases in its political history, the first, from 1903 to 1911, was preparatory when political differences remained largely personal and people were occupied with laying the foundations of local and national government in a spirit of unity. The second, from 1912 to 1947, witnessed the rise and fall of a Labour Party, the elevation of political differences to a local party level, and the development of strong Labour and United Party caucuses. The third, from 1948 onwards, has seen the invasion of the municipal field by the National Party, the elevation of political differences to the national level, and conflict between the United Party Council and the Nationalist-dominated central and provincial governments.

One result of this conflict has been a dispute between the council and the Minister of Native Affairs over the removal of Natives from the city's western areas. Originally the council's own slum clearance scheme, it is now the duty of a Natives Resettlement Board appointed by the Minister to carry out government policy. Another result has been a recent but isolated intrusion of party polemics at a public reception following the induction of the mayor, whose office is by custom non-political. A third result is that, because of the present balance of political forces, many Johannesburg citizens do not expect differential treatment from the province although every other circumstance demands it—but they may be wrong, as provincial controls have recently been relaxed in certain directions.

Undoubtedly, this alignment of local and national parties has other serious disadvantages for the city's government, and it is condemned in many quarters. But it is inevitable. Today's urgent national problems of political, economic and communal life are but the large-scale projection of the political, economic and communal problems of our cities, and they are bound to become the centres of political controversy. Johannesburg cannot change its role as the fulcrum of national politics in South Africa, and no party can afford to withdraw from its government unless it has lost all prospect of a parliamentary majority.

The caucus system is as inevitable as the party system. The formal organization of the Johannesburg City Council has already been described, but the informal is equally important. Twenty-eight of the forty-two councillors belong to the United Party and meet as a caucus group to control the votes of its members in municipal affairs. The group meets regularly under the party leader, has its own agenda and party whip, and effects its control by the ultimate threat of dismissal from the party. In practice, it decides each year who shall be nominated for the positions of mayor, deputy mayor, and chairman, vice-chairman and members of the various committees. The eight Nationalists form a similar caucus under their leader, and occasionally walk out *en bloc* from council meetings.

The Johannesburg local government officer is thus faced with two parallel mechanisms: the formal structure of the council and its committees consisting of members of both parties, and the informal machinery that controls those members, distributes functions between them and determines the balance of power between the various committees within the limits of their terms of reference. He may find his well-considered recommendations rejected by a committee because the majority party caucus has already taken a decision without the benefit of his personal advice. Whatever his formal status and powers may be, he will find them to vary unofficially with the status and powers within the party of the chairman of the committee to which he is mainly responsible.

Yet the majority-party caucus is a co-ordinating element in the city's government that can give it direction and leadership, and impart stability to its administration. Its abolition would leave a void that would probably be filled by a single rather than collective personality; and no party can really dispense with it. In short, the system has come to stay, and the only realistic solution to the problems it raises is to attempt to incorporate it into the official municipal structure.

The most logical solution might be to enhance the co-ordinating and policy-planning powers of the General Purposes Committee, which consists mainly of the chairmen of other committees, and to deny to the opposition party and independent members any seat on it. This move would offend a basic municipal principle that gives every councillor a share in committee business; but it would certainly recognize political realities, end the conflict between the formal and informal organizations, and openly concentrate responsibility for the planning and co-ordination of policy where it really lies—in the majority party.

CONSTITUTIONAL PROBLEMS

The constitutional fathers of 1903 built upon a tradition that, for all its complex history, had originated in forms of direct local government. Indirect, representative government did not eliminate all traces of that origin. Thus, when they were faced with the problem of turning a representative body into an effective administrative organ, they solved it by dividing the work of government between committees of the council. Later on, a judicial decision confirmed that every councillor had the right to be a member of at least one committee.

The city's constitution has thus two pillars of self-government: the elected council and the committee system. Nevertheless, a Transvaal commission of inquiry now investigating the structure of local government throughout the province has condemned the committee system root and branch. It admits certain advantages—that the system permits specialization, provides small enough bodies to get things done, and gives every councillor a chance of sharing in municipal management. But it finds that committees are directly responsible for splitting the municipal organization into water-tight compartments; for encouraging councillors to interfere with officials; for relegating the full council to little more than a rubber stamp; for taking up too much of the voluntary councillors' time; and for bringing about a lack of interest in municipal affairs on the part of the electorate.

The commission's answer for Johannesburg is a single executive committee of five salaried members elected by proportional representation by the council for three years, and possibly one or two subordinate *ad hoc* committees with advisory and certain executive powers. It believes that this solution will counteract the tendency towards party political divisions and the strengthening of the caucus system, and that it will resuscitate council meetings.

Some of these recommendations will be considered later, but it is at once clear that proportional representation on a single executive cannot reverse an historical political trend. In fact, the caucus would even more easily control the municipal machine, and the commission's stated objective is in any case some form of cabinet government. If that is so, then the answer already given to the party political problem is more logical and practical. It would threaten the second pillar of local self-government, but only in so far as minority parties are concerned; the opposition would have a vested interest in keeping the council chamber alive as a debating forum; and the resulting constitution would be a natural evolution from the past in tune with present circumstances.

Inconsistently enough, the commission also states that the electorate's acknowledged civic apathy and ignorance is an inherent defect of universal suffrage, to be accepted as inevitable by the framers of democratic constitutions. Here it certainly underestimates the possibilities of education in citizenship at adolescent and adult levels, of publicity by means of lectures, films, exhibitions, the radio and the press, and of fostering public relations by means of trained officers. The commission ignores the tendency for the percentage of voters to drop as municipalities grow, and the possibilities of extending contacts between the council and the community by co-opting non-councillors on to standing committees, and by representing economic, professional and other associations on advisory committees. Above all, it fails to recognize how far the citizen's degree of civic interest and sense of responsibility depend upon the extent of the powers entrusted to him through his representatives.

As already stated, Johannesburg's powers and functions are no greater than those of any other Transvaal municipality, and it is subject to identical higher controls. The city council has recently submitted to the commission a memorandum in favour of freeing Johannesburg's government from the ties which bind it so closely to the province, and of granting it definite home-rule powers by means of a municipal charter. It believes that the charter should confer two kinds of powers; those appropriate to the government of a metropolitan city, and those required for experimental purposes. At present, the former are mainly limited in their exercise. The latter can be provided for in either of two ways—either provincial procedure can be established for the annual or biennial extension of particular powers at the city's request, or the charter can include a general clause authorizing the undertaking of any activity that is in the interests of the local community, and is not subject to any provincial or parliamentary statute. In the latter case, a financial limitation could apply, such as a defined maximum annual expenditure linked to rate income.

However, these powers are asked for by representatives of only two-fifths of the city's million inhabitants. The non-European majority has no vote, and they raise a most difficult problem. The radicals' solution of electing non-European representatives to the city council tends to ignore the general political climate, the racial barriers, the transitional and unstable state of urban non-European society, and the great variations in its members' cultural and educational attainments. The reactionaries' solution of complete separation in the social, economic and political spheres tends to ignore the fact that four-fifths of the industrial labour force of the Witwatersrand is non-European. Since there are already elected Native advisory boards in the Native townships that act as links between the council and the non-European population, the best solution may well lie in the direction of their development into subordinate local authorities. An Urban Bantu Authorities Bill has in fact been published which would create elected local government bodies in these townships, but it has a number of drawbacks that might prove fatal in practice. Not the least is a divided control in which a city or town council is the Minister of Native Affairs' junior partner.

There is certainly a need for much more research into the social organization and class structure of the urban non-European community before firm and effective proposals can really be made for their future local government. For example, a strong middle class has always been the backbone of municipal self-government, but it is doubtful if Johannesburg's non-Europeans have yet evolved more than a rudimentary class structure on European lines.

ORGANIZATIONAL PROBLEMS

In spite of the commission's findings, most of the organizational problems of the city's government do not spring from the committee system. They arise from the need progressively to divide the labour of government as it grows in volume and variety. Whatever the form of city government, its work must be divided according to major services to form departmental blocks to which money, material and men are allocated; and these blocks must be sub-divided into branches and sections. There are two basic reasons for this division in breadth and depth. The one is technical: the main groups of services must be separated out and sub-divided in order to bring special skills and techniques to their performance. The other is human: the basic working groups must be restricted to an effective size where effectiveness depends as much on the capacities, needs and teamwork of the employees as on the capacity of management.

A large municipality such as Johannesburg will thus build up a system of departments each more or less self-contained, each concerned with a major service and its ancillaries, each needing appropriate techniques, and each made up of many working groups, sections and divisions. If these groups are to be kept to an effective size, and if special skills are to be applied to new duties, the groups, sections and divisions must multiply as municipal business grows, and the number of departments must also tend to multiply. Organizational growth thus develops an inherent tendency to disintegrate, which is strengthened by the need to delegate responsibilities to and within departments as they expand in response to municipal evolution. In fact, decentralization becomes as imperative as specialization; but both lead to fragmentation.

The system of committees in Johannesburg has certainly encouraged fragmentation of the departmental system, but is itself the result of the same process of specialization and decentralization. In consequence, both the spokes and the rim of the organizational wheel have threatened to fly off its hub as the city has expanded municipal services, and remedies have been sought in the sphere of co-ordination. On the one hand, financial control has been brought to a fine art by means of the annual budget and financial regulations. On the other hand, administrative co-ordination has been developed in the General Purposes Committee, in the Town Clerk's Office, by means of the Departmental Committee and Technical Sub-Committee, and through domestic regulations, the O. and M. Division, and the Staff Board. But there is still room for improvement.

As far as the committee system is concerned, the Finance Committee has been unable to confine itself to purely financial aspects in spite of its terms of reference. The General Purposes Committee has tended to lose its co-ordinating function in a welter of general purposes, whereas its prime responsibility should be the general oversight and review of policy and machinery on behalf of the council with a view to giving direction and coherence to municipal development. But it can hardly undertake this task without reference to financial considerations. Of course, the Finance Committee can always be asked for its views, but in Johannesburg there seems to be no good reason why the two functions of financial control and administrative co-ordination should not be combined within the committee system by means of a strong co-ordinating and finance committee.

This committee might well consist of the chairmen of the other committees, together with two or three ordinary members; and it could set up two sub-committees, one for establishment and O. and M. matters, and one to deal with accounts and routine financial items. The Establishment Committee would thus be abolished and the number of committees reduced to seven. The formal organization would be brought into line with the informal caucus system, as previously suggested. And the greater streamlining of the committee system would lessen the demands on councillors' time, and generally speed up municipal business.

The commission's recommendation of a single executive committee of five salaried members is not a better solution to the problem of fragmentation. Apart from its political and constitutional disadvantages, it would still have to divide the labour of government amongst its members; or set up sub-committees; or consider a different group of services each day in rotation. As the commission foresees, it would have to exercise all the powers of the existing committees to undertake this work, but, as it seems to admit, the work might in fact be too much for a single executive in Johannesburg, no matter how it divided its business.

As far as the departmental system is concerned, the town clerk is recognized in standing orders as the chief administrative officer and he acts as a co-ordinating *primus inter pares* relative to other heads of departments. But his status has always depended very much on his personality, his salary today is the same as the city treasurer's and city engineer's, and his function has never been adequately defined.

Actually, much of it concerns the establishment and maintenance of a good system of communication throughout the organization. This is firstly a matter of good structure, so that the network of communications connecting the main points of the organization is well devised and maintained. Secondly, it involves drawing up and revising a comprehensive, published code of standing orders that maps this structure and lays down the functions and procedure of the council, its committees and departments. Thirdly, it involves the main aspects of staff management or establishment work. In addition, the machinery and activity of an organization cannot be co-ordinated effectively without the systematic ordering of operations, and it will still lack harmony and logic if the policy to be executed is inconsistent.

All this implies that the town clerk as a co-ordinating chief administrative officer should be responsible for the maintenance of the communication system (including personnel administration), the business of organization and methods, and the co-ordination of policy and machinery. But this responsibility has never been fully recognized in Johannesburg. Ideally, besides being chairman of the Departmental Committee and Technical Sub-Committee to ensure the co-ordination of policy proposals before the committee stage, the town clerk should have at his disposal O. and M., establishment, legal, secretarial, records, research, and information and public relations services; and his status should be officially confirmed in other than standing orders.

So far, he has no trained public relations or information officer, the Staff Board and O. and M. Division remain independent bodies; he is not supported by statute or special council resolution, and he is too burdened with ordinary departmental duties to undertake what should be his main responsibility. It follows that his deputy should be made the *de facto* head of the Town Clerk's Department, so that he himself could devote more time to the larger problems of municipal government. He could then become a general adviser to the council on administrative, organizational, constitutional and even civic matters. His status could be strengthened by making him responsible to the suggested co-ordinating and finance committee, and by defining his duties and authority in general terms by committee or council resolution upon his appointment. His salary could be raised above that of any other officer, if only by a token amount.

Such a development of the town clerk's function would be in line with recent experiments in some of the larger American cities such as New York, Los Angeles and Philadelphia, and with the recommendations of the O. and M. inquiry into the organization of the Coventry City Council in England. But it would not go so far as the proposal of the Transvaal commission of inquiry to set up a principal officer very similar to the city manager or Californian chief administrative officer. This overlooks the fact that no American city with well over half a million inhabitants has a city manager, and that it is rare for a single individual to be responsible for the running of a large-scale multiple business undertaking. A group can more competently grapple with its economic, technical, governmental and human problems.

FINANCIAL PROBLEMS

Johannesburg's financial position compares favourably with that of many other great cities, and its independence from government grants is unusual. Moreover, the annual assessment rate has remained the same for the last four years. But services could have been expanded and improved if more money had been available, and rates will have to rise in the future if additional sources of revenue are not made available.

In South Africa generally, a gap between municipal needs and means appeared in 1931 and 1932, the years of depression. A growth of communal services and equipment but not financial resources, subsequently widened this gap and continued to widen it after 1946. The main reasons were the far greater use of the motor-car; rapid industrialization, with its consequent influx of Natives into the urban areas; demands for expansion and higher standards in personal and environmental health services; new standards in town planning that added to the costs of municipal servicing; and the development of social welfare and recreational services.

As municipal needs thus multiplied, municipal revenues remained stationary as a percentage of the national income, being 3·1 per cent in 1929, 3·4 per cent in 1938 and 3 per cent in 1948. In 1949 expenditure on housing, public health, social welfare and Native administration accounted for 27·5 per cent of the total annual expenditure of nineteen major municipalities including Johannesburg, but central subsidies in respect of these services amounted to only 6 per cent of their total income. Since new sources of revenue had not been made available, the strain had to be taken by existing sources not designed to support the new demands. The alternatives were increased government grants or the loss of services to central and provincial authorities, but neither of these was a palatable solution.

One early method of easing the strain was to increase trading profits, which rose as a percentage of municipal revenues from 24·5 per cent to 25·7 per cent between 1929 and 1938. But in 1948 they were down to only 7·4 per cent, and they are likely to fall even further because the present capital shortage entails the financing of improvements in trading undertakings from their revenues, and most undertakings have substantial development programmes that cannot be delayed. The municipalities have thus intensified their demands for a greater share of the non-national tax field, which is monopolized by the province except for the local rate.

Johannesburg has been no exception to the general rule. Between 1948 and 1953 its trading profits remained at some 7 per cent of its municipal revenues; government grants increased and together with the new Native services levy, which is a form of subsidy, finally equalled trading profits in importance; and assessment rates, and charges for sanitary and allied services in lieu of rates, continued to be relied on very heavily. In 1948 they accounted for 63·5 per cent of the city's revenues; in 1953 they still accounted for 54·8 per cent.

Clearly, new financial means must soon be found if the city's government is to continue to satisfy the growing demands of modern urban life. A great city needs stable sources of revenue which will not fluctuate too greatly in good and bad times; sources of revenue that will be expansive in the long run, and thus permit a gradual extension and raising of standards of services; and sources of revenue that will permit it to undertake experiments in municipal government. Above all, it must maintain a budgetary independence of higher authorities.

The simplest way of ensuring a city's independence is to give it some kind of property tax. Provided this does not have to bear too much of the burden of maintaining and expanding local services, it is quite a good tax. It usually avoids double taxation and it can be stable, productive, reasonably just, and fairly simple and certain in incidence. But if it carries too heavy a load, and if too much property is exempted, the tax will become inequitable, as it is no longer properly related to personal wealth and tends to be regressive. Moreover, it discriminates against house-dwellers in favour of flat-dwellers, and its incidence on owners and tenants changes with changing economic conditions. The higher the rate, the more important these defects become.

Johannesburg already has such a tax, and it has so far been sufficient to ensure budgetary independence. But it cannot be raised much further without magnifying its defects; it can hardly support municipal experiments, and it is evidently not expansive enough in the long run to permit needed extensions of services. The council has thus submitted to the commission of inquiry that the Transvaal municipalities should be given a two-thirds share in the provincial motor vehicle tax, a half share in the provincial entertainment tax, and a one-third share in the provincial betting, totalisator and auction sales taxes. Provincial revenues would be reduced by only 6·54 per cent, but Johannesburg would enjoy the major share of a transfer of some £2½ million a year.

Nevertheless, the city would still need further taxes for experimental purposes. The most appropriate are to be found amongst business, sales and personal or local income taxes, and its charter should at least permit the controlled levying of one or more of these. Provincial revenues would not suffer relatively from such changes. Central and provincial revenues rose from 13·2 to 19 per cent of the national income between 1929 and 1952, and the Union could easily increase its existing 50 per cent subsidy to the province, which has never enjoyed budgetary independence and has no particular need for it.

PLANNING AND DEVELOPMENT PROBLEMS

The city's need for increased revenue arises from an economic revolution that manifests itself in housing, traffic, transport, migratory and social problems. All these are aspects of town planning and development.

Johannesburg was originally laid out like any other Transvaal mining town. There was little expectation of permanent settlement once the surface mines had been exhausted, and it was only in 1890 that the possibilities of deep mining were recognized and the city's future assured. Thus, what is now the central business district was designed on a grid-iron pattern with small plots or stands, narrow roads, no open spaces and many right-angled intersections. Moreover, unlike most other South African municipalities, which were originally granted many square miles of Crown land, Johannesburg has had to buy space for its parks and gardens. And, in spite of the imaginative extension of its jurisdiction in 1903 and the control over new township development then received, the council enjoyed no positive town-planning powers until 1931. Five years later, its first provisional town planning scheme was prepared.

The planning consultants reported that, because it had never been planned, the developed part of the municipal area consisted of an un-correlated aggregate of separate townships which unnecessarily increased the costs of municipal services. Places of scenic beauty had been lost to the public, and the desire to provide the largest possible number of corner stands had resulted in small blocks, numerous intersections, and a conse-quent slowing down and constant interruption of traffic. The original grid-iron pattern and its subsequent extensions were also condemned from commercial and aesthetic standpoints, and the consultants added that the general adoption of this pattern throughout South Africa not only resulted in a loss of urban character and individuality, but did little to foster the civic spirit of urban populations.

The road system had many defects, including wrong widths and bad siting; uncoordinated main traffic routes, and lack of connections between the chief radiating arteries; narrow central streets and main roads; numerous cross-streets checking traffic on the main arteries; and a badly sited railway that cut Johannesburg in two and severed its northern residential area from its business area. Land-use zoning had not been attempted and industrial factories, business premises, shops and dwelling-houses were intermingled regardless of their effect on each other. But the consultants believed that the city would re-build itself in the next fifty years so that, if zoning regulations were introduced forthwith, non-conforming uses would gradually disappear and it would eventually assume an orderly and satisfying appearance. They suggested the adoption of seven types of zoning with appropriate density and height regulations; zoned 2,000 acres for agricultural purposes; and set aside a large area for non-European development, with surrounding open spaces or tree belts.

In addition, the consultants criticized the planning ordinance because it permitted no control of areas outside the municipal boundaries. Since the city was restricted in the south by the gold mines, it had to grow towards the north, east and west. While expansion along the east-west ridge would eventually come under the control of neighbouring munici-palities, there were no local authorities to the north. This problem was soon taken up nationally by a government committee that reported in

z.

1940 in favour of proper control over all peri-urban development in the interests of public health. The province thereupon established an appointed Peri-Urban Areas Health Board in 1943 with planning powers, whose main efforts have been confined to the north and south of Johannesburg's boundaries.

For the municipal area itself, a pre-war town planning scheme was eventually agreed upon and finally proclaimed in 1946. This now controls all development by the zoning of land use, except in the case of mining property; provides for a new civic centre with an imposing situation overlooking the down-town area; and assures the city's orderly growth for many years to come. Few modifications to the scheme have been needed so far, but the question of racial zoning is now being considered by a government board and will affect certain zoning boundaries.

In the meantime, arrangements with the South African Railways have attacked the problem of the northern approaches to the city's centre, and new bridges and access roads now span the railway. But the growth in motor transport has been so rapid that the council called in American traffic engineers in 1954 to design a new highway system. Their published plans involve £33 million, including a £23 million expressway system; but the alternative may be a slow strangulation that is already well advanced in and around the central business district at peak hours. On the other hand, some people fear that the expressways will merely pour more traffic into the inadequate central street system and aggravate the city's public transport difficulties by encouraging the use of private cars.

Traffic is not the only planning and development problem. Housing is possibly an even more formidable task, as it is largely a matter of quickly providing a decent roof for the hundreds of thousands of non-European inhabitants of shack and shanty areas in and about the municipality. A departmental survey of the western areas of Johannesburg in 1950, which consisted of six townships with a population of 77,000 non-Europeans, described general conditions of suspicion, squalor, drunkenness, crime, delinquency and filth of which many European citizens had hardly any knowledge. Yet all these places lay within the municipal boundaries.

Today, there is a four-pronged attack on Native housing. The influx of Natives into the urban area is controlled by a permit system that is effectively stemming the human tide. The council has appointed a director of housing who, in one year and using Native labour, has built 3,050 houses at a cost of from £220 to £260 each. The Natives Resettlement Board is rehousing the thousands of Natives it is removing from the western areas to a new outside township. And the council is commencing vast 'site-and-service' schemes under the aegis of the Minister of Native Affairs that will enable Natives to build temporary huts on some 50,000 serviced sites, on the understanding that the council will begin replacing them with permanent dwellings in five years' time.

The financial burden of this attack is very great, especially as most Natives cannot really afford economic rents, yet there is every reason to believe that it can be sustained with the help of government funds and subsidies—provided the root cause of the housing problem is also tackled. However, the council resolved in 1949 that no further extraordinary steps should be taken to encourage the expansion of industry in the city, and the Minister of Native Affairs has now decided as an interim measure to oppose all further applications for the development of industrial townships on the Witwatersrand.

Whatever its effects on other towns and cities, this decision should at least help Johannesburg to catch up on its housing backlog, and give breathing-space for the formulation of future plans on a regional rather than town or city basis. Unfortunately, the machinery for the regional planning of land use is still in its infancy and no adequate organization has so far been created for controlling the development of the metropolitan region centring on Johannesburg—a matter to be considered more fully later in this chapter.

RELATIONS WITH HIGHER AUTHORITIES

The dual control exercised by the central and provincial authorities over Johannesburg's government has already been mentioned. In theory it is certainly onerous, but in practice many controls are nominal and the province has recently relaxed its supervision in important instances. Nevertheless, at present it has at its disposal the following nine kinds of control. First, there is the power to legislate in regard to local government. This is the basic means, as the city council may exercise only those powers granted to it by the province and state, which may both prescribe any condition and procedure they think fit. Second, there is the Administrator's confirmation of by-laws, which extends as far as the standing orders governing the domestic conduct and machinery of the council.

Third, there is the Administrator's power to approve the council's administrative schemes and particular acts. For example, it may not begin to build sewage or water works, or works for the supply of heat, light or power, without his consent; and it cannot sell electrical or gas appliances to consumers, or finance their purchases by more than £100 each, without his general approval of such schemes. It must submit every town-planning scheme to the Administrator who, after referring it to the Provincial Townships Board, may reject or approve it with any modifications he thinks fit. It cannot raise loans from the Provincial Housing Loans Fund unless its housing schemes are approved by the Administrator; and sub-economic schemes must in any case receive his approval, although if they concern Native housing they must be passed by some eight other bodies, including central government departments. And the council must obtain the Administrator's approval or consent to at least seventeen specific acts, including contracts with its own officials, the establishment of recreation grounds, the letting of such grounds and the advancement of loans to sporting bodies, the substantial alienation or acquisition of immovable property, financial assistance to any national or public cause, the permanent closing of public places, and the performance of functions incidental to those permitted by ordinance.

Fourth, the province has wide powers of financial control. It prescribes the method of valuation for assessment purposes and confines the rate to a maximum of 7d. in the pound. It enumerates the council's possible sources of revenue, obliges it to appoint a finance committee, outlines its accounting and budgetary systems, details its procedure in regard to borrowing, approves its depreciation regulations, and prohibits borrowings without the Administrator's approval of their amounts and conditions. Fifthly, the audit of municipal accounts must be carried out by officers appointed by the Administrator, and they may surcharge persons responsible for unauthorized payments or deficiencies or losses due to negligence or misconduct.

Sixth, the province plays a part in the appointment and dismissal of the council's senior officials. For example, the council is obliged by legislation to appoint a town clerk and a medical officer of health; special conditions attach to the dismissal of these officers and the treasurer and city and electrical engineers; and their removal or any reduction in their salaries must be approved by the Administrator, except for the medical officer of health, who is protected by the Minister of Health.

Seventh, the Administrator has powers of compulsion and action in default in health and financial matters, although these are most unlikely to be applied to the city council. Eighth, either the Administrator himself or a board established by him may exercise an appellate function in connection with the council's activities, as in the case of town planning. Ninth, the Administrator may at any time appoint a commission of inquiry into any matter concerning the city council and its activities.

At least the first three of these kinds of control might well be relaxed, especially in connection with the closing of public places, the establishment of recreation grounds, the purchasing of land, the establishment of water, electricity and gas works, the promulgation of by-laws, and the limitation on the performance of incidental functions. Moreover, although they may often be little more than nominal in practice, many more controls could in fact be modified or abolished in the interests of municipal autonomy and of administrative relief at both local and provincial levels. While such controls may be necessary in the case of smaller, inexperienced municipalities, most of them are inappropriate in the case of a city whose annual budget vies with that of the province, and whose officers are sometimes more expert and usually more permanent than the provincial civil servants. Furthermore, their incidence is heightened by central controls in the spheres of Native housing and administration, public health, racial zoning and capital borrowings.

Leaving aside the question of central controls, if, as the recent trend in administrative practice seems to recognize, there are good reasons for relaxing provincial controls in the case of Johannesburg, and if many of these should continue to apply to smaller municipalities, some way should be found of differentiating by legislation between the city and other local authorities. The best method is the municipal charter that will grant certain additional powers of local government but concentrate on removing the present provincial leading-strings, nominal or otherwise. A second method is a special chapter for Johannesburg in the Local Government Ordinance.

Neither of these would affect a further external problem, which concerns the relations between Johannesburg and its neighbours within the metropolitan region.

THE METROPOLITAN REGION

The most striking feature of modern urbanization is the rise of the metropolitan or city region. It is the result of two revolutions, the one industrial and the other technical, and its two basic characteristics are therefore economic and technological.

Intensive study of the economies of metropolitan regions shows that their economic functions may be classed as either basic or non-basic. The basic functions are those performed for markets mainly outside the region, such as mining or steel-making, where the product is exported to earn regional imports rather than consumed locally within the region. The non-basic functions are those performed for markets mainly within the region, such as the running of departmental and provision stores, whose supplies are not normally exported. The wealth and prosperity of the region ultimately depend upon its basic economic functions. Its boundaries depend very much upon the service areas of its non-basic economic functions.

The result of these economic activities is no mere massing of factories, warehouses, shops, banks or network of roads and railways. People make a city. A metropolitan region is characterized by a highly organized society adapted to its environment and able to satisfy people's cultural as well as economic needs. But this society could never have evolved without the assistance of technical advances accompanying the industrial revolution, and its structure is not simply the result of the fact that towns and cities are the chief places of work. It is also the result of universal changes in means of transport and communication.

The motor-car and electric railways have extended the potential urban radius to some 15 miles from the work centre, and the area of possible residence to some 700 square miles. Other technical advances, such as the telephone, radio, electric power transmission, water reticulation, septic tanks and hydraulic wells have also helped to spread the urban population outwards over administrative boundaries into suburbs, towns, townships and villages surrounding the major city centres. Division of labour has established particular relationships between these satellites and the central cities, so that each tends to undertake special functions for the others and all together develop into an urban complex.

In the result, most metropolitan regions today seem to have two main boundaries and one subsidiary boundary. At the centre is the city hub or central business district, with its great buildings, soaring land values and congested traffic. Outside the subsidiary boundary of this hub is the rest of the metropolitan zone proper embracing a population in daily contact with the hub and whose limits determine the first of the two main boundaries. Beyond the metropolitan zone proper lies an extensive and expanding zone embracing more remote areas whose inhabitants, although not in daily contact, are economically dependent on the hub for many specialized services. Its boundary, in itself an indeterminate and shifting zone, marks the confines of the metropolitan region and the limit of the hub's regional attraction.

So far, this metropolitan concept has not been applied to the Witwatersrand and a great deal of social science research is needed before valid, practical conclusions can be drawn. But investigations by the Union's

Natural Resources Development Council into the area known as the Southern Transvaal have indirectly provided information about possible metropolitan regions, and a Johannesburg metropolitan region covering most of the Witwatersrand can certainly be distinguished.

The Southern Transvaal, that centres on Johannesburg and stretches from Pretoria in the north to Vereeniging 73 miles to the south, and from Randfontein in the west to Nigel 50 miles to the east, is the heart of a wider human-use region. This region's boundaries extend in a broad arc of some 200 miles from Witbank in the east through Pretoria to Klerksdorp in the west. It then runs 70 miles south to Odendaalsrus and the gold-fields of the Orange Free State, before returning some 210 miles in a north-easterly direction to Witbank. It is a mining region whose basic industries produce gold and uranium, coal, iron and steel. And the N.R.D.C.'s investigations into its heart, as a preliminary to devising a means of controlling its land use, show that the Southern Transvaal may be divided into three main zones.

In the north, there is a zone centring on Pretoria, which today has a population of less than 300,000, but which the Natural Resources Development Council estimates may have a population of almost 2,000,000 in fifty years' time. In the centre, there is a zone extending for 45 miles from Krugersdorp in the west to Springs in the east and following the contours of the Witwatersrand reef, which already has a population of well over 1,500,000. In the south there is a younger zone centring on Vereeniging but expanding rapidly along the valley of the Vaal River. Here are three possible metropolitan regions, and the most important is the second that centres on Johannesburg.

Already, there are pointers to the main boundaries of this region. To the east of Johannesburg, there is no extensive break in a line of residential development beginning at its municipal boundary and continuing through the municipalities of Germiston, Boksburg, Benoni, Brakpan and Springs. These are all well-established urban centres, of which Boksburg may mark the boundary of the inner metropolitan zone. To the west of Johannesburg lie Florida, Roodepoort-Maraisburg and Krugersdorp, in that order. In the last few years an ever-increasing number of city people have been heading westwards and it is already estimated that 15,000 Europeans travel to work in Johannesburg from the West Rand every day, more than 2,500 by road and the rest by electric railway. Florida, which is 25 minutes by rail from Johannesburg, seems to mark the boundary of the inner metropolitan zone on this side of the city, but it may be extending to Roodepoort.

The area bordering the northern municipal boundary of Johannesburg is already well developed as a suburban dormitory under the control of the Peri-Urban Areas Health Board. To the south, the board has planning schemes extending to the rural centres of Walkerville and Meyerton some 20 miles away, where many smallholders combine daily work in the city with market gardening. The southern obstacle of the gold mines is thus being over-run, and new townships have in fact sprung up between the mines and the city's southern limits.

In the result, the probable inner metropolitan zone covers the area of two city councils (Johannesburg and Germiston), four town councils, two village councils and much of the area of the Peri-Urban Areas Health Board, particularly to the north of Johannesburg. The outer metropolitan zone may well incorporate the area of at least six town councils and the sparsely populated countryside between them, including part of the area of the Peri-Urban Areas Health Board mainly to the south of Johannesburg.

It is true that no precise estimate of the region's maturity can yet be made, but a reasonable inference can be drawn from the following facts— the existence of fifteen local authorities with a total population nearing 2,000,000 people; the production of £148 million worth of gold annually; an income almost three times as great derived from manufacture alone; a fleet of some 170,000 motor-vehicles; and an average income per head amongst the European population estimated by a world planning authority as possibly greater than that of the United States of America. All these facts point to the arrival of a prosperous metropolitan society on the Witwatersrand nearing viability, and to the need to consider its future local government without further delay.

THE FUTURE GOVERNMENT OF THE REGION

The many possible forms of government applicable to a metropolitan region may be classed under two general headings: (a) better means of co-ordinating the efforts of existing local authorities, or of undertaking particular services, without interfering with the present framework of local government; (b) fundamental changes in the local government structure of the metropolitan region.

The first category includes four possibilities: (1) the granting of extra-territorial jurisdiction to existing local authorities in such specific matters as town planning, public health, public transport, and the supply of heat, light and power; (2) the creation, for similar purposes, of *ad hoc* joint boards or committees representative of the local authorities concerned;

(3) the establishment of regional public utility corporations representative of local authority and other corporate consumers; (4) the establishment of special authorities that are neither self-governing nor representative of local authorities, but which do not entail changes in the existing local government units. Johannesburg has experienced all of these means of co-ordination in regard, for instance, to public transport, the Rand airport, the Rand Water Board, and the Peri-Urban Areas Health Board respectively.

The second category includes a further four possibilities: (1) the annexation of local authorities contiguous to the central city, or their consolidation with it to form a single new authority; (2) the merger of special and *ad hoc* authorities with the central city; (3) the establishment of a loose association of all local authorities within a particular area for advisory and planning purposes; (4) the establishment of some kind of two-tier system of local government. Johannesburg has already been advised by its Departmental Committee and planning consultants that its boundaries should not be extended because of topographical, technical, financial and planning limitations. It has taken over a number of local utility companies from time to time, and the Resettlement Board will eventually hand over to the council and be disbanded. And it has belonged to the Advisory Council of Reef Municipalities since its formation in 1938. This is a voluntary association of local authorities on the Witwatersrand that has been of great use for the mutual exchange of views and consideration of provincial and central policies affecting local government, although it is not established by statute and is not a federal body similar to the one created, for instance, in Berlin in 1911.

With annexation ruled out, the city's experience thus covers almost all but the last possibility, that is, some kind of two-tier system. It leads to the belief that this is probably the correct answer to the problem of metropolitan Johannesburg's future government, since all other answers have been responses to various stages in the region's development. Now it has come of age, a new response is needed—as in London, Manchester, Toronto, Montreal, and Miami, Florida, which have all set up, or have been recommended to set up, two-tier systems of local government.

It is true that many practical problems will be involved in creating a two-tier structure. Should only the inner metropolitan zone be involved, consisting of nine authorities? Will sufficient able councillors be found to serve the regional as well as lower-tier local authorities, and how should they be elected? Will local political differences prove a serious stumbling-block? Will the region be financially viable, and what revenue

z*

resources will it enjoy? Can the boundaries of either tier of authorities be closely related to the economic optimum service areas of the various local government functions? Above all, can local communities of interest be preserved?

The future constitution of a metropolitan region such as Johannesburg's cannot be discussed without reference to these and similar questions, and much detailed research is needed. Indices must be devised to discover the inner and outer metropolitan boundaries, and the relationships between the various cities and towns. The comparative costs of government in the existing local authorities must be investigated, the quality of services compared, and some answer arrived at to the question of whether there is at present a waste and duplication of effort. The optimum size of the various local government services must also be considered. In some cases, as perhaps transport, the optimum point may not be critical and the size of the unit may be variable over a fairly wide range; in others, such as water and sewage, indivisible economic factors may complicate matters; in others, especially personal services, the optimum size will always remain relatively small.

Because of the complex problems involved, a tentative two-tier system of local government for the Johannesburg metropolitan region can hardly be proposed at this stage. It can only be stated that the function of the upper tier regional authority should be to co-ordinate rather than undertake local government services as, for example, in the Manchester plan of 1947. It is true that co-ordination may sometimes be possible only by placing a service under the direct management of the regional local authority, and that sometimes technical, economic or financial considerations may demand regionalization to achieve optimum conditions; but the principal role of a regional local authority should be to plan rather than to execute. It should undertake the administration of services only where there are very strong reasons for doing so.

Even so, a distinction must be drawn between different planning authorities and regions for different purposes. In a new industrial country, the drawing up of guide plans to facilitate control over the balanced development of extensive regions of human use, and over the distribution of population and industry within those regions, is a matter for an authority distinct from those engaged in town planning or even metropolitan planning and control.

As far as the Witwatersrand is concerned, if a regional council were to be created for the Johannesburg metropolitan region, it should undertake metropolitan planning and control within the aegis of a regional planning board concerned with the planning of land use for at least the Southern Transvaal. This planning board could be a small, expertly staffed Government body with members drawn from central, provincial and local government departments, and agencies such as the N.R.D.C. It could be assisted by an advisory committee of representatives nominated by the local and regional authorities and the industrial and commercial associations concerned, before which all proposals would be laid by the board for consideration and report back prior to their inclusion in any plan.

The board's function would be to draw up guide plans of future land use in the Southern Transvaal, including industrial, agricultural, commercial, residential and recreational uses. It would have no executive authority and would plan, but not control, development.

Within the framework of such guide plans, the metropolitan regional council could be empowered to draw up comprehensive plans for the controlled development of town and country within its boundaries, subject to central or provincial approval. The planning and development of specific towns would be left to the lower tier local authorities concerned, subject to the overall control of the regional council, which would thus replace the present Provincial Townships Board within the metropolitan boundaries. Planning and development outside the boundaries of lower-tier authorities but within the metropolitan area could be undertaken by the regional council itself, and it might be given the power to act in default should a local authority fail to draw up plans or control development for insufficient reason.

Tokyo and Osaka

MASAMICHI ROYAMA

MASAMICHI ROYAMA

Graduated from the University of Tokyo, 1920. Assistant Professor of Public Administration at the University of Tokyo, 1923–28; Professor of Public Administration and Political Science at the University of Tokyo, 1928–39. Member of the House of Representatives, 1941–45; President of Ochanomizu University since 1953. President of Japanese Society for Public Administration, 1950–54; President of Japanese Association for the study of Public Utilities, 1949–56. Member of Governmental Committee on Reorganization of Local Government; Member of Governmental Committee on Development and Planning of Land and Water Resources.

Author of: *Chiho Gyosei Ron*, 1937; *Eikoku Chiho Gyosei No Kenkyu*, 1949; *Chiho Seido No Kaikaku*, 1953.

Tokyo and Osaka

IT IS needless to say that Tokyo, the capital, is the political centre of Japan. It has, moreover, become the greatest metropolis of economic as well as cultural activities in Japan. It has a population of over 8,000,000, which increases at the rate of some 300,000 a year. The metropolitan city of Tokyo may well be compared with a gigantic sphinx, yet it keeps on growing day after day without ever having a break. Tokyo is on an important cross-road of international communication and is a focal point in the exchange of culture in East Asia. It is also believed that the day is not far off when Tokyo will again begin to play an important part in the politics of the Far East.

The City of Osaka is situated geographically in the centre of Japan and ranks second after Tokyo. The area of the city covers 205 square kilometres, containing a population numbering 2,550,000. From olden days the city has been active as a commercial and financial centre and, since modern industrialization started in Japan, it has rapidly become one of the biggest industrial cities in the Far East.

THE GROWTH OF THE METROPOLITAN COMMUNITY

The western Pacific embraces a beautiful chain of several islands called Japan, the main one of which is Honshu. Tokyo, the capital, is situated in about the middle of Honshu and faces the Bay of Tokyo. Although a local authority, the metropolitan city of Tokyo is neither a simple prefecture nor a mere city. It is, so to speak, a great entity and is composed of three parts—the urban area which is divided into scores of wards, the country area which contains farm and mountain villages, and the several islets which are scattered to the south of Tokyo Bay. The total area of Tokyo is 2,031·17 square kilometres, of which wards occupy 578·65 square kilometres; mountain and farm villages 804·39 square kilometres, and islets 299·26 square kilometres. Its geographic location extends from 138°53' to 142°26' East longitude and from 24°14' to 35°53' North latitude. Its climate is temperate and the average temperature in 1954 was 14·8 deg. C.

The brief history of Tokyo is as follows. During the feudal age when Tokyo was called Edo, Dokan Ohta (1432–86), a feudal lord, discovered that Edo was strategically important from the viewpoint of land and water traffic and constructed the castle of Edo in 1457. The building of the castle gave an impetus to the construction of the fortress-town and led eventually to the later development of Edo and Tokyo. Edo, however, declined in prosperity for a short time after Dokan Ohta passed away, but it received a new impetus when Iyeyasu Tokugawa (1799–1867) settled there and drew up a large-scale town plan. Throughout the days of Tokugawa's reign, Edo, the Shogunate capital, developed as the political and economic centre of Japan. At the height of its prosperity, Edo had a population of 1,400,000, which must have been the largest for any city in the world at that time.

With the ending of the feudal era in 1868, the capital of Japan was transferred from Kyoto to Edo, which was renamed Tokyo. With the subsequent progress of Japan as a modern nation, Tokyo continued to develop as the political, economic and cultural centre of Japan. In the field of town planning, Western ideas and methods have been introduced and Tokyo is gradually taking shape as a modern city in the true sense of the word.

Tokyo has so far been visited by two great disasters which resulted in heavy casualties—one was the great earthquake and fire of 1923 and the other was the air raids during the Second World War. In the first disaster, Tokyo had about 300,000 houses burnt with about 1,500,000 victims (which was 60 per cent of the then urban population of Tokyo). It was most fortunate for Tokyo at that time that great sympathy and assistance were extended to the city not only from within the country but also from many foreign countries, as a result of which reconstruction work under the planning of the central government was completed in seven years and an entirely new modern city was born. The greatest of all the changes resulting from this calamity was the marked development of suburban districts. During the Russo-Japanese War of 1904–5 and especially during the First World War, the overflowing urban population had already begun to find an outlet into suburbs. The urbanization of surrounding towns and villages was further spurred by the earthquake of 1923. As a result, in 1932 as many as 84 towns and villages in the districts adjoining Tokyo were integrated with the metropolitan city.

The casualties caused by air raids during the Second World War were far greater than those of the 1923 disaster. Deaths totalled about 100,000; wounded about 130,000; missing about 7,000; and the total number of persons whose homes were destroyed by air raids about 3,000,000. It is surprising to note that, within ten years after the end of the war, many of

83. TOKYO

The Diet building from the rear. The water in the background is a part of the moat of the Imperial Palace Plaza.

84. TOKYO

The Marunouchi Block, where a large number of Japan's large corporations have their head offices. The moat on the right is a part of the Imperial Palace Plaza.

85. TOKYO
A general view of the central area.

86. TOKYO

The wholesale markets and wharves.

the facilities of Tokyo have been rehabilitated to the pre-war level or even above that standard.

Nevertheless, the city management of Tokyo is confronted with a new and great difficulty: namely, how to solve the social problems brought about by the enormous increase in its population.

The following is an outline of present-day Tokyo as seen from the viewpoint of political, economic and cultural activities.

Tokyo is the capital of Japan, where the three departments of legislature, judicature and administration of the government are located; it is, therefore, the fountainhead of all political activities. In view of the retarded progress of industrialization, the Japanese Government adopted a policy of vigorously protecting and extending assistance to various industries, which eventually resulted in the establishment of a close relationship between politics and industry. This has made Tokyo, which is situated in an advantageous geographical location, the greatest economic centre in Japan. Tokyo has thus become the largest consumer city as well as the greatest manufacturing centre in Japan. In 1953 there were in Tokyo 42,000 factories employing 670,000 employees. The amount of goods shipped out of Tokyo in that year was 820,000 million yen. According to the statistics of 1952, 47 per cent of the entire foreign trade of Japan is centred in Tokyo. The national income in 1952 was 4,893,900 million yen, whereas the earnings of the citizens of Tokyo amounted to 552,600 million yen, or 11 per cent of the total national income. Compared with the *per capita* national income of 57,245 yen, that of the citizens of Tokyo amounted to 79,746 yen or 39 per cent more than the former.

In addition, Tokyo boasts of being the educational and cultural centre of Japan. The concentration of educational institutions and facilities in Tokyo is especially great in colleges and universities. Out of the 72 state universities in Japan, Tokyo has 11, while 58 out of 120 private universities and 61 out of 184 junior colleges are also situated in Tokyo. About 200,000 out of some 410,000 university students, i.e., 50 per cent of the total enrolled number, attend universities in Tokyo. Besides higher education, there is a remarkable concentration in Tokyo of the great undertakings engaged in mass communication such as newspapers, broadcasting and publications. In one sense, it may well be said that the present-day culture of Japan is produced in Tokyo and is disseminated throughout the nation.

These activities of the metropolis extend far beyond the administrative boundary of the city into much wider areas of adjacent prefectures. Tokyo has on hand a number of important problems awaiting solution and readjustment from the standpoint of administration and the public interests of community life, such as the construction of dwelling houses, improvement of police services, amelioration of transport facilities for commuters, integration of factory areas between Tokyo and Yokohama, transportation of commodities, and improvements of the central wholesale market. These problems are created by the growth of the enormous metropolitan community in and around the City of Tokyo.

In spite of the difference in its constitutional status as a local authority, the same phenomena relating to the growth of a metropolitan community can be observed in the case of Osaka.

Roughly speaking, the topography of the City of Osaka is similar to that of Tokyo, but it is more level, and most of the city is located about three metres above the sea level. The River Yodo which runs through the city empties into the Inland Sea, at the mouth of which stands the port of Osaka, built and operated by the Osaka municipality. This port serves as the hub of domestic as well as overseas trade. Numerous branches of the River Yodo glide through the city together with many canals, furnishing abundant transportation facilities to the citizens of Osaka. These water facilities have significantly assisted the economic and cultural development of the area.

On the other hand, due to the fact that Osaka stands on the deltaic area of the River Yodo, the ground of the city is comparatively weak, and this situation has been worsened owing to the drawing of water in great quantities for industrial use. Tremendous subsidence of the ground has already occurred. It is no wonder that the installation of a city water supply system for industrial use, the construction of breakwaters to prevent inundation caused by high tides, and the provision of sewerage in the low-lying lands, have come to occupy a position of importance in the municipal administration of Osaka.

From olden days the City of Osaka has been called the centre of commercial and industrial activities, while Tokyo was the political centre. In 1954 there were 150,000 industrial units in the city employing 1,150,000 citizens. The number of firms engaging in wholesale trades was 20,000, which is 15 per cent of the national figure. Osaka occupies the first place in the country for the volume of business transactions, among which the wholesale trades handling textile products and other personal articles amount to 1,211,300 million yen annually, or 50 per cent of the entire domestic trade.

CHANGES IN POPULATION

It is estimated that Tokyo had a population of about 1,000,000 at the time of the Meiji Restoration of 1868. After the Russo-Japanese War (1904–5) and the First World War, its population began to show a rapid increase and, in spite of the great earthquake and fire of 1923, it reached the figure of 7,800,000 by 1942.

In August 1945 Tokyo was literally reduced to ashes and had only 3,060,000 inhabitants. The loss of overseas territories due to the defeat in the last war and the resultant repatriation of several millions of Japanese, and the surplus labour force in farm villages and other small local cities and towns again contributed to the sudden concentration of population in Tokyo, which far surpassed the record of pre-war days. The census taken in October 1955 revealed that Tokyo's population exceeded the pre-war peak and reached 8,037,084 (about 9 per cent of the entire population of Japan). This amounts to an annual increase of 434,000 during the ten years since the end of the last war.

The following points may be noted concerning this population increase in the post-war period: (1) The social increase due to the difference between moving in and moving out of the city amounts to between 70 and 80 per cent of the population increase; (2) the overwhelming portion of this social increase is made up of labourers, the majority of whom are without a fixed job and these put heavy pressure on the labour market of Tokyo; (3) about 70 per cent of those who moved in had no ability to become tax-payers in their previous places of residence. In consequence, the greater number of them have been putting a heavy burden on the administrative and financial activities of Tokyo from the viewpoint not only of taxation but also housing and social security; (4) 87 per cent of the entire population of Tokyo is found in the ward area. Compared with the distribution in 1940, a considerable decrease was seen in the central part of the city, and twelve wards showed a decline in 1955. Five of these twelve wards have less than 70 per cent of the population they had in 1940. On the other hand, a very high rate of increase was seen in the suburbs of the outer wards and their environs and several of them showed an increase of from 200 to 400 per cent. The same phenomenon is also seen in the case of neighbouring prefectures. The majority of the increased population are the commuters and their families. These facts go to prove the powerful attraction of Tokyo, which penetrates deep into all the environs beyond the present administrative boundary. With regard to the mobile population, it is found that about 450.000 commuters and students (both day and evening)

TABLE I

POPULATION GROWTH OF TOKYO (1920–55)

Date	Entire Area of Tokyo Metropolis	Ward Area
1920	3,699,428	3,358,196
1925	4,485,144	4,109,113
1930	5,408,678	4,986,913
1935	6,369,919	5,895,882
1940	7,354,971	6,778,804
1945	3,488,284	2,777,010
1950	6,277,500	5,385,071
1955	8,037,084	6,969,104

TABLE II

CAUSES OF POPULATION GROWTH OF TOKYO (1946–55)

Term	Population Increase	CONTENTS			
		Net in Migration	Rate	Natural Increase	Rate
1946	670,831	641,750	96	29,081	4
1947	516,350	416,479	81	99,871	19
1948	407,415	295,123	72	112,292	28
1949	484,837	367,377	76	117,460	24
1950	400,596	303,963	76	96,633	24
1951	384,732	297,766	77	86,966	23
1952	333,871	247,443	74	86,428	26
1953	349,966	269,488	77	80,478	23
1954	283,672	203,736	70	79,936	30
1955	262,335	177,716	68	84,619	32

come into Tokyo from nearby prefectures every day, whereas some 150,000 go out of Tokyo to places as far away as 50 to 70 kilometres.

In Osaka, the modern municipal system was first established in 1889 when the city had a population of 470,000 in an area of 15 square kilometres. However, the rapid development of the Japanese economy caused Osaka to expand by leaps and bounds, and the city's boundaries were enlarged in 1897, in 1925, and again in 1955 to its present size of 205 square kilometres. The population of the city reached its maximum in 1940, when it numbered 3,250,000, but this gradually declined as the result of the war. Air raids damaged one-third of the city and by the end of the war in 1945 the population of the city had decreased to 1,030,000. Many industries in the city received deadly blows.

However, life has now been brought back into the Japanese economy, and this has created boom conditions in Osaka's industries. According to the 1955 census, the city has a resident population of 2,550,000, with a density of 12,417 per square kilometre, the highest of any city in Japan.

The daily floating population of Osaka is quite large. Those who have jobs in the city but live outside the administrative boundary number about 370,000, while those who go out of the city in the daytime are approximately 50,000. The daytime population is thus 2,870,000.

GOVERNMENT AND ADMINISTRATION

1. *Tokyo Metropolitan Government*

For the purpose of local administrative organization, Japan is divided into major authorities of 42 *Ken*, 2 *Fu*, 1 *Do* and 1 *To*, and there are minor authorities in 496 cities, 1,867 towns and 2,418 villages. They are all provided for in the basic Local Autonomy Law. Tokyo is also one of these local bodies, but its special character is recognized in its status of *To*, which means metropolis in Japanese. Tokyo as a *To* has the compound structure of a great city and a prefecture.

The local self-government system in the modern sense of the term was started in Japan when the municipal government of cities, towns and villages was introduced in 1888, followed by the prefectural system and the county system in 1890. This local self-government system, however, was modelled on that of Prussia and was characterized by strong centralized authoritarian rule and limitation of the autonomy of local bodies. Although the prefectures were local bodies, they were governed by prefectural governors who were centrally appointed officials. The governors enjoyed strong authority over the prefectures themselves and also exercised powerful supervisory authority over cities, towns and villages.

Such centralization was the basic characteristic of the local administrative system in Japan since the Restoration, but considerable changes have taken place during the following fifty years. For instance, with the growth of political parties in national politics democratic principles became gradually realized in the structure of local government, and with the introduction of universal manhood suffrage in 1925 local autonomy increased and supervision by the central government diminished.

When the local self-government system was first established in 1889, the city of Tokyo was placed within and under the Tokyo Prefecture. However, a special municipal system was applied to the three cities of Tokyo, Osaka and Kyoto and the duties of the mayors and deputy mayors of these cities were exercised by the centrally appointed prefectural governors and secretaries (as in the case of the Department of the Seine and Paris). This was done in view of the political importance of these large cities, and especially because Tokyo was the capital of the nation. The special municipal system was abolished in 1898 owing to the desire of the citizens for a greater measure of autonomy. The general municipal system enjoyed by other cities in Japan was later applied to the city of Tokyo, without taking into consideration its peculiarities as a great city.

In line with the movement demanding a special status (similar to that of a county borough in the United Kingdom), attempts at special legislation for the city of Tokyo were made on several occasions by the Government. All their efforts, however, came to nought because no agreement was reached between the House of Peers and the Government on the one hand, and the House of Representatives and the City of Tokyo on the other, concerning the area of Tokyo, the method of appointing the head of the city, and the autonomy of wards. In particular, there was an acute difference of opinion on whether the head of the municipality should be officially appointed or elected directly by popular vote. The expansion of the municipal area of Tokyo in 1932 accelerated the campaign for the enactment of a special municipal system but it again failed to bear fruit. This conflict between the Tokyo Prefecture and the City of Tokyo continued until the day of the outbreak of the Pacific War. The establishment of the Metropolitan Government system in 1943 finally put an end to a struggle which had lasted for the previous fifty years.

Under the Metropolitan Government system both the Tokyo Prefecture and the City of Tokyo were abolished, and the whole area was taken over by the newly established Metropolitan Government of Tokyo. However, the former system of centralization which was seen in the appointment of the head of the municipality and limitation of the authority of the municipal council has been further intensified. The most important factor contributing to the establishment of the Metropolitan Government system was the national emergency caused by the war, which was capitalized by the Central Government. In the same year of 1943, reorganization took place of the entire local government system (cities, towns and villages). These two changes contributed greatly to a reversal of the democratic tendency which had until then been growing in Japanese local government.

The new Constitution which came into being after the Second World War expressly protects the local autonomy, and the establishment of the Local Autonomy Law has reorganized Japan's local government so far as the system itself is concerned. The centralized authoritarian rule has been abolished and the rule by the Central Government has been changed to that of local self-government. The Metropolitan Government system of Tokyo, which had retained a status similar to the prefectural government, has also come to assume various democratic characteristics such as the election by popular vote of the chief executive, extension of powers of the council, and the autonomous status of the special wards,[1] etc.

The Metropolitan Government of Tokyo is the major unit of local government in the Tokyo metropolis, which comprises 23 wards, 8 cities, 22 towns and 17 villages. The difference between Tokyo metropolis and other prefectures lies in the fact that the metropolis combines the powers of a prefecture and those of a large city.

The governmental organization of the metropolis follows the formula of the presidential system in the United States of America. Both the Metropolitan Council, which is the legislative organ, and the Metropolitan Governor, who is the head of the executive, are directly elected by popular vote. The Council expresses the will of the local communities and the executive department carries out administration based upon the decisions of the council. The Council and the Executive stand on an equal footing and maintain the so-called system of checks and balances.

The Metropolitan Council consists of 120 members, including the Chairman and Vice-Chairman. The members are directly elected and their tenure is four years. The Chairman and Vice-Chairman are elected from among the members of the Assembly. The metropolitan area is divided into 32 electoral districts, of which 23 are in the urban areas and 9 in the rural areas. The number of councillors elected by each electoral district depends on the number of voters.

The major functions of the local authorities which must be decided by the Metropolitan Council include (1) the enactment of by-laws or the alteration or abolition thereof; (2) the determination of the estimated annual revenue and expenditure; (3) the approval of a report of the final accounts; and (4) matters relating to the levy and collection of local taxes, rents, fees, etc. The council possesses a wide range of authority, which includes, besides the power of decision, approval of the appointment of personnel, examination of papers of the executive department, and the power to investigate the affairs of the local authorities, etc.

The regular meetings of the council are held four times a year. The right to convene the council is in the hands of the Governor, but an extraordinary session may also be called by the members of the council (at the request of more than a quarter of the quorum). There are 16 standing committees of the council. They are established for each division of the administrative departments. Each committee is charged with the duty to

[1] The position of the special wards is described on pp. 731–33.

investigate the business of the department concerned and also to examine bills and petitions. For the purpose of examining matters which are specially referred by the decision of the Council, *ad hoc* committees may be established.

The Governor, in his capacity as the executive head, represents the metropolis. Besides carrying out the business of the Metropolitan Government, he also disposes of matters delegated to him by the Central Government in accordance with laws and ordinances. For the purpose of dealing with a wide range of business, the following departments are maintained: (1) General Affairs, (2) Financial, (3) Tax, (4) Public Welfare, (5) Sanitation, (6) Labour, (7) Economic Affairs, (8) Works, (9) Building, (10) Port and Harbour, (11) Public Service, (12) External Affairs, (13) Public Cleansing, (14) Chief Accountant, (15) Fire Protection, (16) Transportation, and (17) Waterworks.

The Governor is assisted by two Deputy Governors, a Chief Accounting Officer, the Directors of Bureaux and a large staff. The staff, including teachers and those engaged in various undertakings, numbered 136,044 in January 1956.

There are, in addition, the following independent organs: (1) Education Board, (2) Election Board, (3) Auditors, (4) Personnel Commission, (5) Public Safety Board, (6) Local Labour Relations Board, (7) Expropriation Commission, (8) Fisheries Adjustment Commission, (9) Fresh Water Fisheries Control Commission, and (10) Valuation of Assessment Commission. These administrative commissions exercise independent powers of their own and function separately from the Governor.

2. *Municipal Administration of Osaka*

The City of Osaka was established as a self-governing local unit in 1889. Although a council was set up then because Osaka was an important district from the viewpoint of the national government, a mayor was not installed as the executive officer, and the office of the mayor was carried out by the governor of the prefecture, namely the governor of Osaka-fu. In 1898 a mayor was appointed as in other cities, towns, and villages, and the mayors at that time were elected by the municipal council.

In 1947 the principle of election of the mayor by popular vote was adopted. Under the existing system the machinery of municipal administration in the City of Osaka is different from that of the Metropolitan Government of Tokyo, because its legal status is only that of a city within the Osaka Prefecture.

Each city, town, or village is provided with a local council. The number of its members is fixed in proportion to population, but cities, towns, or villages may decrease the number of members on their local councils by their own by-laws. In the case of Osaka, the size of the council specified by the general law is 96, but the city has reduced the number of members to 73. The term of office is four years and the members are elected directly by the inhabitants. Each councillor in the City of Osaka receives 27,500 yen a month as remuneration and 16,500 yen as compensation for expenses. It is usual for each member of a local council to follow an occupation of his own. However, in recent times, the office of councillor in big cities such as Osaka has become complicated and onerous, and there is a general tendency for councillors to give all their time to the work of the council.

The council votes the budget submitted by the mayor once each fiscal year, recognizes the settlement of accounts, approves regulations, and decides other important matters. Moreover, it is authorized to demand the audit and inspection of administrative procedures. Osaka City Council has nine standing committees, dealing with Finance and General Affairs, Economy, Education, Welfare, Labour, Construction, Transportation, Water, and Fire Protection. These committees all participate in municipal administration in their specific fields.

The chief executive organ of the municipality is the mayor, who directs the municipal administration and represents the city. The mayor is elected directly by the inhabitants of the city, and his term of office is four years. Moreover, the mayor submits the budget and draft regulations to the council, and supervises and executes all the municipal affairs except those falling under the jurisdiction of the administrative commissions.[1] The mayor also supervises and executes business of a national nature delegated to him with full powers by law.

As assistants the mayor has three deputy-mayors, a treasurer, numerous clerks, engineers and other employees. The deputy-mayors assist the mayor in supervising the work done by clerks, and act on his behalf in case of incapacity. The treasurer supervises the receipts and disbursements of the municipality. The mayor appoints the deputy-mayors and the treasurer with the agreement of the city council, and their term of office is also for four years.

The following sections and bureaux are established for the purposes of disposing of administrative affairs falling within their jurisdiction.

The City of Osaka is divided into 22 wards, with a ward office for each. These wards are not self-governing bodies as in the case of the special

[1] See below, p. 731.

wards in Tokyo, but are administrative districts with the staff headed by the chief of the ward acting as the auxiliary staff for the mayor. The total number of staff of the wards is 3,494.

The ward office disposes of affairs directly connected with the daily lives of the citizens living in the ward, such as assessment and collection of municipal taxes, census registration, registration of inhabitants, and allocation of staple food. These offices are playing a big part in creating closer ties between the mayor and the citizens, and they enable the citizens to get direct contact with the municipal administration.

Besides these administrative organs, the mayor has under his jurisdiction the management of a university which is composed of faculties of economics, law, commerce, science and engineering, medicine, and home economics. The teaching and research staff numbers 704.

The executive organization is as follows:—

Name of Office	No. of Staff	Matters dealt with (as of April 1, 1956)
Secretariat	80	Secretarial work, ceremonies, social affairs
Liaison Section	4	Liaison with external persons, etc.
Public Hearing Section	21	Hearing of citizens' opinions, publication of municipal policies
Accounting Division	39	Receipt and reimbursement of cash, accounting of articles and auxiliary affairs for treasurer, settlement of accounts
Personnel Section	66	Appointment and dismissal of personnel, salaries and wages, welfare
Administrative Bureau	99	Archives, regulations, statistics, survey on municipal administration
Financial Bureau	141	Budget, taxation
Social Relief Bureau	759	Public assistance, children's welfare, industrial training centres, nurseries
Economic Bureau	514	Guidance and training media and small-sized enterprises, central wholesale market of foodstuffs, slaughter-houses, sight-seeing, international trade fairs
Sanitation Bureau	967	Local hygiene, public hygiene, hospitals
Public Cleansing Bureau	1,543	City cleaning, cemeteries, burial, cremation
Planning Bureau	357	City planning, readjustment of town-lots
Public Works Bureau	392	Administration of real estate owned by municipality, construction and supervision of municipal dwelling houses and buildings
Port and Harbour Bureau	637	Administration of municipal ports
Fire Protection Bureau	2,506	Fire fighting and fire prevention
Transportation Bureau	9,629	Operation of municipal streetcars, subways, buses
Water Works Bureau	1,899	Operation of water works

The City of Osaka formerly engaged in the distribution of electricity as a municipal enterprise, but in 1942 this was transferred to a governmental corporation established for the purpose of the war-time economy. After the end of the war the corporation was dissolved and electricity supply was handed over to private enterprises in nine regions, each of which has been given a monopoly. The City of Osaka did not succeed in regaining responsibility for the supply of electricity. In Japan, municipal enterprise in the field of gas and electricity undertakings has been permitted only in the formative period when they acted as pioneers.

As executive organs of the municipality there are, besides the mayor, a number of administrative commissions such as the Board of Education, Election Board, Personnel Commission, Auditing Commission, etc. They are almost identical with those which exist in the Metropolitan Government of Tokyo and other big cities.

PROBLEM OF THE SPECIAL WARDS IN TOKYO

The existence of special wards is a unique feature in the local government system of Japan and is found only in the Metropolitan Government of Tokyo. The history of the wards goes back to 1878 when 15 wards were established in conformity with the law concerning county, ward, town and village organization. Although the wards did not have the status of a legal entity, each ward had a ward office and a headman who was appointed by the national government. Ward councils were established in the following year, 1879, and the members were elected by popular vote. These wards became local authorities as a result of the City Organization Law of 1889. They retained that status even when the Metropolitan Government was established in 1943.

The Local Autonomy Law of 1947 enhanced the status of the special wards and conceded to them, in principle, the same powers as those enjoyed by a city. Considerable restrictions were, however, placed upon the special wards by metropolitan by-laws, particularly in regard to their powers relating to personal and property rights so that the systematic and co-ordinated development of all the wards in the entire area of Tokyo might be achieved. This led to a struggle for power between the Metropolitan Government and the special wards. An amending law of 1952 emphasized the co-ordination of the relations between the Metropolitan Government and the wards, and laid it down that special wards are local bodies under the Metropolitan Government. Thus, the special wards became subordinate local authorities which are authorized to deal with matters coming within the scope either of the powers conferred on them by legislation and ordinances or of the powers delegated to them by the metropolitan by-laws. Their independent legal status was recognized and each ward still maintains its ward council as in the past. The headman of a ward is, however, no longer elected by popular vote but is appointed by the ward council with the concurrence of the Metropolitan Governor.

There is a great difference in the matters dealt with by special wards and those administered by ordinary cities. The following matters are handled by the Metropolitan Government: (1) comprehensive planning and matters requiring liaison and co-ordination; (2) matters requiring a high degree of expert technique; (3) matters requiring uniform handling; and (4) matters liable to be run in an inefficient way and wasteful if they are handled separately by each ward. A limited power is granted to wards with respect only to matters which have a close bearing upon the daily life of the inhabitants in the ward, and matters which need handling in accordance with local conditions, all of which are specified. The more important of these matters are as follow: (1) establishment and maintenance of primary and secondary schools and kindergartens; (2) construction of special ward roads; (3) establishment and management of small parks, playgrounds and children's recreation grounds; (4) establishment and management of libraries, community centres and public halls; (5) planting of roadside trees and establishment of street lighting and the management thereof; (6) street scavenging; (7) establishment and management of municipal pawnshops, clinics, public lavatories and retail markets; (8) management of public ditches; and (9) issue of identification papers.

In order to conduct these affairs, the wards are given power by the metropolitan by-laws to levy taxes based upon the provisions of the Local Tax Law and they can also collect ward taxes. Besides the above, wards have other revenues in the form of grants-in-aid, rents and fees.

The special wards, which number 23, vary greatly in size and resources. For instance, Setagaya Ward, a residential area, is 62 square kilometres in area with a population of 530,000 and a budget amounting to 1,500 million yen. On the other hand, Chiyoda Ward, which is situated in the civic centre, is only 11·7 square kilometres in area with a resident population of 120,000 (although it has a very large population in the daytime) and a budget of 400 million yen. Each of the wards has gone through its own peculiar process of development. The wards which lack modern facilities are mostly those in the grip of financial difficulties due to sudden increase in population. It becomes necessary, therefore, to make a financial adjustment between the Metropolitan Government and each ward on the one hand, and among the wards themselves on the other. The Metropolitan Government, taking into consideration the various views proferred by each ward, makes this adjustment by virtue of the metropolitan by-laws. Each ward enjoys a limited right of self-government and is subject to different administrative conditions. It is an important duty of the Metropolitan Government, therefore, to raise the level of administration in the wards as a whole to a higher level and to improve uniformly the welfare of the citizens.

All the problems pertaining to special wards lead in the end to the question of the proper relationship between the Metropolitan Government and the wards themselves, or of how to define the character of wards. The solution of these problems may be sought from the standpoint of the history of the wards, social realities, the evils of large-scale municipal administration, and the right balance between democratic decentralization of authority and the efficient centralization of power. If the citizens participate more fully in municipal government through taking more interest in the wards, and if wards are utilized for giving people a civic training, then the democratic value of the special wards will be further increased. From the standpoint of the metropolitan community as a whole, however, the problem of the special wards in Tokyo must be considered in the light of the more efficient and better co-ordinated administration of the capital city.

LOCAL POLITICS AND PARTIES

In the Council of Tokyo Metropolitan Government, as in the National Diet, the two party system has been established. Of the 120 members of the Metropolitan Council, 83 belong to the conservative Liberal-Democratic Party. It stands in an overwhelmingly superior position to the opposing Socialist Party, which has returned only 34 members to the council. Besides these major parties, there are two Communists and one Independent. It cannot be said, however, that the party composition of the Metropolitan Council proves that the citizens are overwhelmingly in favour of the Conservatives, because the election of the Metropolitan Governor, which took place at the same time as that of the Metropolitan Council, turned out to be exclusively a contest between the present Governor, who is a conservative Independent member supported by the Liberal-Democratic Party, and a reformist member supported by the Socialist Party. The former received 1,309,000 votes while the latter got 1,191,000 votes. The fact that the former thus won by a narrow margin is a clear proof of the above statement. It is also of great interest to note that a striking discord is shown in the following table between the political loyalties of the members of the Diet, the Metropolitan Council, and the ward councils, who have all been returned by the same Tokyo citizens.

TABLE III

POLITICAL PARTY REPRESENTATION OF TOKYO VOTERS (1955)

Political Party	Members of House of Representatives (Feb. 1955 election)		Members of Metropolitan Council (April 1955 election)		Members of Ward Councils (April 1955 election)	
	Number	per cent	Number	per cent	Number	per cent
Liberal-Democrats	15	56	83	69	507	54
Socialists	12	44	34	28	94	10
Communists	—	—	2	2	41	5
Minor parties	—	—	—	—	13	1
Independents	—	—	1	1	283	30
TOTAL	27	100	120	100	938	100

According to this table, the percentage of support for the Socialist Party declined more and more in the case of local elections for the Metropolitan Council and the ward councils compared with that of the House of Representatives. On the other hand, the Liberal-Democrats received greater support in the Metropolitan Council elections than in the other elections. When we consider that the 30 per cent of Independents in the ward councils is made up mostly of conservatives, it becomes easy to see that the superiority of the conservatives is more predominant in local assemblies than in the National Diet. Although the presence of Independents in the House of Representatives and the Metropolitan Council is insignificant, they occupy nearly a third of the seats in the ward councils. This goes to show that there is still a strong belief that local politics should not be under the influence of political parties. At the same time, it also shows that the local organization of political parties is not yet fully developed. On the other hand, it may be pointed out that boss-ridden politics have established a strong foothold in league with the dormant civic consciousness of the citizens. This tendency is more conspicuous in farm villages than in the large cities in Japan.

Each political party in the Metropolitan Council is closely affiliated with the corresponding national political party. Both the former Prime Minister and the Secretary-General of the Socialist Party were at one time members of the Metropolitan Council. Many former members of the Metropolitan Council are seen today among the members of the National Diet.

There are no serious conflicts between the two major political parties in the Metropolitan Council such as the one which exists in the National Diet over the question of the revision of the Constitution or that of rearmament. This may be attributed to the fact that there is no big problem in the local politics of large cities in Japan capable of causing a violent controversy between the conservative and reformist parties. An example of the above statement is the deliberation of the draft of the Capital Area Readjustment Law which is receiving suprapartisan support. In the Metropolitan Council, only questions pertaining to local affairs in Tokyo are debated and political discussions concerning national affairs seldom take place. The same applies to election campaigns for the Metropolitan Governor.

The Osaka Municipal Council is composed of 73 members, 47 of whom belong to conservative parties, 23 to the Socialist Party, and the remaining three to the Communist Party. These parties are local organizations of the national political parties, and in principle stand for a common platform with the policies of their national parties. But they are not controlled by the central parties in connection with municipal administration except in the case of the Communist Party. This is due to the fact that in the case of municipal politics, their sphere of control over local politics in accordance with the policies of the party is very limited.

There are many cases where the members of the council approach the headquarters of the parties to which they belong in order to promote the interests of the citizens by influencing Government policies to their advantage. For instance, at the time of the campaign for the establishment of a special city system, equivalent to the county borough system in England, or at the time of the campaign in 1954 for continuation of the city police, the members of the Osaka Council combined for the purposes of the campaign without regard to the wrangling of factions or party differences.

The principal policies of the city are decided by the council, but actually the policies assume a definite shape when the annual budget bill, which is prepared and submitted by the mayor, passes the council. The council has authority to modify the budget bill, and in the City of Osaka successive mayors have been very co-operative and have held the will of the council in respect. In consequence, in not a single instance has it been necessary for the council to modify the budget. The council is also authorized to pass a resolution of lack of confidence in the mayor, if more than two-thirds of the members are present and there is a majority of not less than three-quarters in favour of the resolution. On the other hand, the mayor may adjourn the council in order to appeal to the citizens, but this has never occurred in the history of the city.

The citizens who have the right to vote for election of the mayor and the members of the council may demand, by means of a petition signed by a certain number of citizens, the recall of the councillors, the dissolution of the council, the making, revision or abolition of regulations, or the carrying out of audit inspections. But such actions have never so far taken place in Osaka.

FINANCE

1. *Tokyo Metropolitan Finance*

The general account budget (1956 fiscal year) of the Tokyo Metropolitan Government is 97,301 million yen. Compared with that of the Central Government of 1,034,923 million yen for the fiscal year 1956 (April 1, 1956–March 31, 1957) and that of the entire mass of local government expenditure of 1,045,670 million yen, the budget of the Tokyo metropolis amounts to a little less than 10 per cent each of the national budget and of the total local finance. This proportion coincides with the relative size of Tokyo's population, which is also a little less than 10 per cent of the total national figure.

The general account budget of the Tokyo metropolis covers expenditure relating to education, police, fire protection, social welfare, health and hygiene, civil engineering, ports and harbours, unemployment measures, housing, etc. Apart from the above budget, the Tokyo metropolis has three special accounts called Municipal Enterprises Special Accounts, which cover works connected with street-cars, buses, trolley-buses, waterworks and sewerage. These special accounts amounted to 29,125 million yen for the fiscal year of 1956. There are in addition three more special accounts amounting to 8,985 million yen to provide for the fund for maternity and child welfare loans, races (for horses, bicycles, motor-cars and motor-boats) and office supplies. The total budget of all these accounts of the Tokyo metropolis reached 135,413 million yen in 1956.

According to the table opposite, the largest of all the revenues is that of the metropolitan tax, which provides 53 per cent of the total amount. Next comes the grant from the national treasury which provides 19 per cent. Judged by the standard of local authorities in Japan, the financial position of Tokyo metropolis is considered quite autonomous and independent of the national treasury.

87. OSAKA

The Castle of Osaka, built in 1584. The tower was reconstructed by the city in 1929 to commemorate the enthronement of the Emperor.

88. OSAKA

The river Yodo flows through the centre of the city. It is divided into two tributaries which join together again at the lower reaches of the river. The City Hall, composed of three blocks, is seen on the tongue of land between the tributaries. [Photo: *Asahi Press*]

TABLE IV

THE TOKYO METROPOLITAN GOVERNMENT

TOTAL ESTIMATES OF REVENUE FOR 1956

	Unit: 1,000 *yen*	*per cent*
General account	97,301,805	
Metropolitan tax	51,670,281	53
Municipal enterprises and property income	2,624,955	3
Local redistribution tax	584,760	1
Fees and dues	5,074,590	5
Grant from National Treasury	18,895,700	19
Metropolitan loans	6,260,000	6
Money transferred	2,449,119	3
Contribution from special wards	1,257,647	1
Balance brought forward	1,946,158	2
Others	6,538,595	7
Special account	8,985,191	
Child and maternal welfare loan fund	68,839	
Races	7,391,198	
Office supplies	1,525,154	
Municipal enterprises special accounts	29,125,944	
Transportation	13,325,944	
Water-works	12,130,000	
Sewerage	3,670,000	
TOTAL	135,412,940	

Under the centralized tax system of pre-war days, the surtax on the national tax was the principal source of local revenue. As a result of the introduction of democratic local autonomy in post-war days, independent local taxes took the place of surtaxes. Based upon the recommendation of the United States Mission led by Dr. Shoup, who visited Japan in 1949, a complete reorganization of the local tax system took place. This has greatly increased the independent sources of local revenue and has established the local finance equalization system, whereby grants are given by the Central Government to those local authorities which are poor in revenue resources.

Tax revenues of the Tokyo metropolis, with the municipal property tax and business tax as the main sources, began to increase year by year until a partial reform of the tax system was enforced, the consequence of

AA

which has recently slowed down the rate of increase in revenue. If such a tendency continues for long, it may become very difficult to cope with the ever-increasing financial need which is being brought about by the growth of population.

The Metropolitan Government of Tokyo does not receive the equalization grant (now called the local grant tax) because its tax revenue is high. However, the amount paid by the national treasury in respect of certain functions or services imposed by national laws provides 19 per cent of the total revenue. The principal items coming under this head are the grants for public assistance and compulsory education.

Another big source of revenue for the Tokyo metropolis are the metropolitan loans. At present, in order to issue bonds, the approval of the Central Government must be obtained. The approval given in the past few years was extremely restricted. For this reason the progress of various construction works, such as the building of municipal dwelling houses and schools, the repair of roads and rivers, and block planning, has been very slow. It must be recognized that the Tokyo metropolis has done its best and spent a large sum of money out of its own resources during the past ten years for the rehabilitation of the metropolis out of the ruins caused by air raids. Compared, however, with the reconstruction work after the great earthquake and fire of 1923, which progressed very rapidly because Japan could float foreign loans in the United Kingdom and the United States, the present rehabilitation has been very slow due to the fact that Japan could not obtain foreign loans in the post-war days, nor could she raise enough domestic loans.

Nevertheless, it is a notable fact that the Central Government has recently changed its policy concerning approval of the issue of bonds. Bond issues will not hereafter be restricted to the small amount of 1·0–1·5 per cent of the revenue which has been the rule in the past. It is imperative for the metropolis to float sufficient loans if the various construction works required by the rehabilitation plan are to be carried out.

Table V shows that on the expenditure side education is the largest item and takes 25·5 per cent. It is followed by the expenses for the Metropolitan Government amounting to 12·5 per cent, most of which is to cover personnel expenditure on the 24,400 regular members of the staff. Next in order of magnitude come expenses for the police, public welfare, housing (which aims mostly at building municipal dwelling houses), public works (which includes roads, rivers and town planning), and unemployment counter-measures.

As stated below, when the development of the greater capital area

TABLE V

THE TOKYO METROPOLITAN GOVERNMENT

ESTIMATED EXPENDITURE ON GENERAL ACCOUNT FOR 1956

	Unit: 1,000 yen	per cent
Metropolitan Council	200,252	0·2
Metropolitan Government	12,192,458	12·5
Personnel Commission	36,019	—
Police	11,805,338	12·1
Fire protection	3,404,371	3·5
Education	24,792,721	25·5
Public works	6,053,741	6·2
Port and harbour	1,933,039	2·0
Housing	6,896,445	7·1
Public welfare	7,829,391	8·0
Asylums	243,272	0·3
Unemployment counter-measures	4,982,542	5·1
Health and hygiene	2,221,102	2·3
Public cleansing	2,394,970	2·5
Forestry	1,012,252	1·0
Commerce	2,099,788	2·2
Central wholesale market	646,281	0·7
Statistics	41,883	—
Election	112,457	0·1
Tax collection	827,878	0·9
Metropolitan loans	4,164,742	4·3
Disbursement loans	1,751,262	1·8
Reserves	100,000	0·1
Financial grants to special wards	1,257,647	1·3
Natural disaster relief fund	301,958	0·3
TOTAL	97,301,809	100·0

begins in the near future, the Tokyo metropolis must readjust and improve the various facilities in the metropolis. For this purpose, the second three-year capital construction plan has already begun, but it is estimated that as much as 50,000 million yen is required to complete the programme. To find the sources from which to get this huge sum will be one of the largest problems facing the Metropolitan Government of Tokyo in the days to come.

2. *Osaka Municipal Finance*

As to the municipal finance of Osaka, it is important to bear in mind that in Japan the tax system of the cities is somewhat different from that of the prefectures, and it differs also from that of metropolitan Tokyo. The principal taxes for the cities are the municipal resident tax and the real property tax, which together provide over two-thirds of all the tax revenues. The combined amount of grants from the national equalization fund and from the Prefecture of Osaka provide 18 per cent of the total income, which is slightly smaller than in the case of Tokyo Metropolitan Government.

TABLE VI

THE CITY OF OSAKA

TOTAL REVENUE FOR 1955

	Unit: 1,000 yen	per cent
1. General account	25,711,550	64·7
Municipal tax	12,780,876	32·2
Resident tax	3,297,545	
Real property tax	6,604,435	
Electricity and gas tax	1,591,308	
Other taxes	1,270,140	
Taxes by former laws	17,448	
Rents and fees	2,185,335	5·5
Grants from national and prefecture funds	5,000,246	12·6
Revenue from construction work carried out on behalf of other bodies	504,970	1·3
Income from disposals of real property	337,062	0·8
Revenue from municipal enterprises	1,017,880	2·6
Municipal loans	2,476,000	6·2
Other income	1,409,181	3·5
2. Special account	14,033,812	35·3
Transportation	9,454,625	23·8
Waterworks	3,532,703	8·9
Municipal loan	1,046,484	2·6
TOTAL	39,745,362	100·0

The following table shows the allocation of funds on General Account to various functions and services in the municipal administration of the City of Osaka.

TABLE VII

THE CITY OF OSAKA

EXPENDITURE OF GENERAL ACCOUNT FOR 1955

	Unit: 1,000 yen	per cent
Council expenses	87,198	0·2
Administrative expenses	2,328,615	5·9
Public works	5,126,096	12·8
Housing	1,993,815	5·1
Education	3,055,601	7·7
Public hygiene	1,156,510	2·9
Sanitary cleansing	1,389,254	3·5
Labour and employment	3,117,425	7·8
Industry and trade	600,452	1·5
Port and harbour	1,659,593	4·2
Police[1]	828,616	2·1
Fire-fighting	936,341	2·4
Election	89,209	0·2
Tax collection	220,307	0·5
Principal and interest of loans	1,871,119	4·7
Municipal enterprise	898,640	2·3
Others	352,759	0·9
TOTAL	25,711,550	64·7

PLANNING THE METROPOLITAN REGION

1. *Tokyo Metropolitan Region*

In the old days of Japan, both the capital of Nara and Kyoto were built into magnificent capitals by grandiose planning. It has been the earnest and constant wish not only of Tokyo citizens but also of the entire Japanese people to make Tokyo into a beautiful and modern capital where people can live in peace and also find work. The great earthquake and fire of September 1, 1923, completely reduced the cultural installations of the past years to ashes in a matter of a few hours. Again, the Second World War turned the greater part of the city into desolate ruins. The rehabilitation plan was hastily drawn up in accordance with the City Planning Law of 1919 and also the Special City Planning Law of 1946.

[1] Until the time when the municipal police system was transferred to the prefecture police in 1955, police expenses were borne by the city amounting to the sum of 3,297,267,459 yen in 1954, but at present the sum for the police was reduced to the expenses for the temporary Public Safety Commission.

It aimed roughly at (1) distribution of population, (2) land utilization, (3) public services (such as land adjustment, road improvement, water-works, sewerage system, etc.).

These planning measures were not adequate to cope with the ruined conditions of the metropolis, where political, economic and cultural activities began to find their way with the marked tendency to concentration of population after the end of the war.

It was for this reason that the Capital Construction Law was enacted in 1950 for the purpose of drawing up a fundamental plan to accomplish the great work of reconstructing the national capital.

This law embodied the following three major aims. In the first place, it declared anew that the metropolis would be constructed not only as a mere local authority but also as the centre of the nation's politics, economics and culture worthy of the name of capital. Secondly, it pointed out that, in constructing the nation's capital, it must be so planned and constructed as to be able to execute more efficiently every pivotal activity required in connection with domestic as well as international relations. Thirdly, it demanded that the Government and the whole nation should recognize it to be a matter of national importance and that they should participate and co-operate positively in the work and make every effort for its satisfactory completion.

In accordance with this law, a Capital Construction Committee was established in the Central Government in 1950. This committee is composed of nine members who are appointed by the Prime Minister. At present, the Minister of Public Works is the chairman of the committee and the members are composed of one member each from the House of Representatives and the Upper House, four men of learning and experience, the Governor of the Metropolitan Government and one member of the Metropolitan Council. As soon as the committee was appointed, it made an estimate of the future population of the capital and decided upon fifteen special items requiring immediate attention, such as road works, low-lying land, parks and green belts. Based upon this Capital Construction Plan, various individual plans have been published for each of these fifteen special items. For example, the item of road works is divided into individual plans for street widening, road repairs, road reconstruction, bridge reconstruction, open spaces, provision of parking spaces, etc. Furthermore, a five-year emergency construction plan for the metropolis has been adopted in order to execute these individual plans. This five-year plan was to start in 1952, but it required a huge expenditure of 116,500 million yen. This gigantic plan, therefore, was too heavy to

be shouldered by the metropolis and only 28 per cent of the entire programme has been completed in three years since 1952.

The slow progress of this five-year plan led to a demand for reconsideration of the scheme and the Capital Construction Committee was obliged
to draft a revised three-year plan. It eliminated the wishful figures included
in the five-year plan and put emphasis upon many priority projects. It
drew up a plan to execute the following eight projects during the three
years starting from 1956: construction of fire-proof houses, expansion of
compulsory educational facilities, readjustment of waterworks, and of
land and roads, expansion of sewerage system, repair and construction
of the port and harbour, and creation of parks and green belts. However,
the Capital Construction Committee recognized long before that one of
the most serious defects in the Capital Construction Law lay in the fact
that it applied only to the ward areas of the Tokyo metropolis and that
the capital construction work could be carried out only within the metropolitan area. The wider area which may be called the metropolitan region,
comprising the outlying places having a close social and economic
connection with the Tokyo metropolitan administrative area, should be
included in the plan for the metropolis. In considering the management
of the problems of a huge city, such an area and satellite towns must be
treated as one with their principal central city. Therefore, the necessity
of drawing up a wider plan covering such an area has become an acute
and important problem. For this reason, it becomes necessary to state
here the conception of the capital area.

The Capital Construction Committee decided to name as the capital
area the area consisting of the metropolis and the outer suburban area
which is closely connected with it (an area extending about 50 kilometres
in radius with Tokyo Station as its centre). The committee began in
January 1954 the investigation and analysis of past changes in population
and industries, in order to consider the possibility of readjusting the future
distribution of population and industries. As a result, it decided upon the
idea of the capital area in June 1955 and then set out to prepare a plan
which would hold in check the population and industries of the ward
areas in Tokyo beyond a certain limit in order to prevent the metropolis
from becoming excessively large. The committee also wanted to make
it possible for the capital zone to develop in a coherent and orderly manner
by fostering satellite towns in the capital's outer suburban area.

The overgrown City of Tokyo already has a population of 7,000,000
in its ward areas, and this is increasing at the rate of some 300,000 a year.
The evils of this excessive population growth have shown themselves in

various places and are making municipal activities extremely difficult. These evils are as follows:—

(1) Excessive concentration of population and industries; (2) uncontrolled urban development (uncontrolled surface development, the improper use of land and, especially, the lowering of efficiency due to excessive concentration in the heart of the metropolis); (3) the degradation of residential environments (arising from such causes as the mingling of dwelling houses and factories, shortage of houses, and the development of over-crowded areas); (4) lack of open spaces such as parks and green belts, and the loss of the natural amenities; (5) the deterioration of traffic conditions (congestion of commuting traffic, increase in the time and money spent in the journey to work, the stagnation of surface traffic); (6) shortage of public services (such as water-works, sewage system, facilities for disposing of dirt, markets, educational facilities, etc.); and (7) housing problem (shortage of 420,000 houses and the need for fire-proof houses).

In order to solve these problems, the following basic policy has been adopted:

(*a*) To alleviate the excessive concentration of population and industries in the capital, it is necessary in the first place to establish an integrated development plan on a national scale and to promote the development of provincial towns so that the inhabitants of such towns may be enabled to settle down in those places. For the sake of convenience, however, the present state of affairs is considered likely to continue for some time to come. Based upon this conception, a great regional plan is to be established to cover the capital area and systematic encouragement should be given to its fulfilment.

(*b*) This great regional plan will have the following basic policy:

The growth of the urban area of the capital is to be held in check to remain at a certain limit and its readjustment will be carried out. The population and industries which cannot be absorbed into this area are to be encouraged to move into the satellite towns to be fostered in the outer suburban area. Moreover, in order to protect the urban area of the capital from the danger of being linked with satellite towns, rural zones such as farm land and green belts will be maintained between them.

The metropolis of Tokyo is the territory comprised within a circle having a radius of 100 kilometres. This is divided into the following three areas in accordance with the basic policy set out above:

(1) *Inner urban area.* This is an area where the commuter can reach the heart of the metropolis in not more than one hour. It extends over 719 square kilometres. People from other areas tend to settle in this area, but

the regional plan provides that only the population and industries which are already settled in the inner urban area and are difficult to relocate elsewhere will be permitted to remain there or to increase. In order to rationalize the population density and land utilization, efforts will be made to have more high buildings, to secure as much open space as possible, and to improve public facilities so that the evils associated with great cities may be eradicated.

(2) *Green belt area*. This area surrounds the inner urban area. The belt has a width of 10 kilometres. Housing development or industrial construction is prohibited in the green belt.

(3) *Outer suburban area*. This area adjoins the outer side of the green belt area and is situated at a distance of about 25 kilometres from the heart of the metropolis. In order to absorb the population which might otherwise move into the central part of the metropolis, satellite towns will be constructed on the nucleus of existing towns. In encouraging the building of satellite towns, vigorous efforts will be made to attract various industries.

It is scarcely necessary to add that the basic idea of this capital area plan must be accompanied by a plan for the redistribution of population. However, a serious problem exists with respect to the scale and growth of population and its composition. The present population of 8,480,000 in the ward areas of Tokyo metropolis is expected to rise to 11,815,000 by 1975. There will thus be an addition of 3,335,000 persons for whom provision must be made. The question is how to distribute this large addition in the outer suburban area.

First of all, the industrial composition of the outer suburban area must be estimated and the great number of surplus people must be sent out of the ward areas to the places best suited to their abilities. The secret of decentralization of population is to make manufacturing industry (which has the greatest power of absorbing population) settle permanently in the areas outside the ward areas. If the population in 1975 is divided into the above three areas according to the population distribution plan, the inner urban area will receive 10,900,000 (8,500,000 for the ward area and 2,400,000 for neighbouring towns); the suburban green belt area, 1,450,000; and the outer suburban area, 5,700,000.

In order to establish and execute the capital plan, it was necessary to have stronger legislation than the Capital Construction Law with respect to the area to be included and the set-up of the planning authority. In 1956 the Capital Area Readjustment Law was passed by the National Diet. This law is an intensified Capital Construction Law which has been integrated with the Satellite Readjustment Promotion Law and the Industries

AA*

Readjustment Law which had been drafted by the Capital Construction Committee. It authorizes the Capital Area Readjustment Committee to investigate, formulate, propagate and promote matters concerning the Capital Area Readjustment plan. This committee has been established as an external office of the Prime Minister's office and is composed of the chairman, who is a minister, and four members (two of whom serve full time) appointed by the Prime Minister. A secretariat has been appointed to serve the committee. In addition, there is the Capital Area Readjustment Council acting as an advisory body for the Readjustment Committee. This council is composed of not more than 45 members. It contains four members of the House of Representatives, two members of the Upper House, not more than ten members of the staff of the administrative agencies concerned, all the prefectoral governors, not more than 16 presidents of the councils concerned, and not more than 13 men of learning and experience.

The Capital Area Readjustment Law embodies provisions intended to widen the scope of planning with respect to areas; to improve the carrying out of the readjustment plan and the effectiveness of the committee; and to provide more satisfactory financial measures. The weakness of the financial measures for the capital construction works is regarded as having been the chief reason why the Capital Construction Law failed to produce effective results. In view of this failure, the working expenses required for the Capital Readjustment Plan are to be included in the budget of the committee from the fiscal year 1957. In addition, the new law provides for government grants-in-aid, the transfer of national property, and loan facilities.

Thus, the Capital Area Readjustment[1] started with hopes for its success. There is much apprehension, however, about this type of regional planning, particularly as regards the co-ordination of various administrative departments of the Central Government in Tokyo and the neighbouring prefectures. It is feared also that the success of this kind of planning administration might lead to an attempt on the part of the Central Government to establish regional government of a bureaucratic type.

2. Osaka City Planning

The planning of Osaka comes under the City Planning Law of 1919, under which the planning of cities is supposed to be carried out, together

---

[1] The word 'Readjustment' used in this chapter in connection with the planning of Tokyo has a comprehensive meaning. It includes development, improvement, rehabilitation, co-ordination and consolidation. Hence it also includes the distribution of population and the relocation of industry.—*Editor*.

with the laws pertaining to land readjustment and building construction. However, this law is based on the conception that the work of city planning forms part of a national project, and the decision of the competent minister is final in designating planning areas and in establishing plans and projects. Only the execution of projects is delegated to the mayors of cities, towns or villages.

Moreover, although the unbuilt areas surrounding cities are to be included in the city planning areas, it is extremely difficult to do so if they encroach on other cities, towns or villages. Therefore, the area of the Osaka city plan has been limited to the territory comprised within the municipal boundary of Osaka.

The City of Osaka occupies a strategic position in the central industrial zone of Japan, and in consequence it suffers from excessive density of population, crowded buildings, and congested traffic. Many difficulties need to be overcome in order to straighten out these adverse circumstances. It is for this purpose that the municipality of Osaka operates public transport services such as trams, subways, and buses; has constructed public dwelling houses; carried out readjustment of lands, construction of roads, water and sewerage, the installation of a water system for industrial use, the building of ports and harbours, and the raising of the ground by piling up earth to counteract subsidence.

But the City of Osaka suffers from difficulties common to all big cities, and in addition there are some problems to be solved which are peculiar to the city owing to its situation in a big urban area surrounded by many satellite towns.

The number of persons coming into the City of Osaka in the daytime amounts to approximately 500,000, including the commuters, who are believed to exceed 370,000. Administrative boundaries nominally separate Osaka and the surrounding cities, towns and villages; but the satellites and the City of Osaka can be seen as forming in reality one undivided unit when one sees the unbroken mass of houses stretching over the whole area. The transportation facilities serving these places also form a closely knit network.

Consequently, many projects mapped out for the city are not necessarily limited to the area of Osaka city, but exert an influence on the surrounding cities and towns. For instance, ports and harbours in Osaka have a close connection with the port of Kobe which is located not far away. Coordination of these matters must be achieved in the near future by means of a consolidated plan.

As for the water supply, very many cities are relying on the River Yodo as their source of water and with the active demand for water brought about by an increase in the population and by demands for its industrial use, the intake from the river is exceeding its capacity, quite apart from the fact that the water is getting polluted by factories along the banks. For these reasons the development of a new water source, involving the direct utilization of Lake Biwa which lies in another prefecture, has recently come into the limelight. Shortage of dwelling houses is another of the problems from which the municipality of Osaka is suffering. Although it is trying its best to cope with the situation by constructing houses, it is far from satisfying the citizens, and suitable sites in the city no longer exist. The city is therefore compelled to look for suitable sites in the out-lying suburbs. This makes it necessary to have a comprehensive plan in order to ensure that offices and dwelling houses are located in a well-arranged manner.

As already explained, projects pertaining to city planning have come to affect surrounding cities, towns, and villages over a wide area; and a well co-ordinated plan for an expanding city in the industrial zone of the country is greatly needed. There are, however, several prefectures in this area, such as Osaka, Hyogo, and Kyoto, and there are many cities crowded in the area, so that unified planning and administration for the region is far from being realized.

Due to the fact that the City of Osaka is situated at the mouth of the River Yodo, which runs from Lake Biwa, it has a close connection with the lake and the waterways of the Yodo; and as regards river conservancy and water utilization, this is a matter of great concern to the municipality of Osaka. However, this waterway lies in a research area designated by the Government in accordance with the national Land Development Law; the planning and execution of the projects concerned with this area are entrusted to the Construction Bureau in the Kinki District, to local authorities on the spot, and to the prefectural governments concerned. The City of Osaka has no direct access to the project, which might jeopardize the development of the city. It is expected, therefore, that a regional planning scheme similar to the Capital Area Readjustment Law will be applied to the metropolitan area of Osaka in future.

RELATIONSHIP WITH CENTRAL GOVERNMENT

It is a difficult task to define the relationship between the State and local authorities, nor is it easy to determine the precise degree of participation by the State in the affairs of local authorities. This difficulty becomes more complicated if the local authority happens to be administering the nation's capital.

Japan's local administration in pre-war days was extremely centralized and followed the principle of bureaucratic administration. Too much emphasis was placed upon administrative control and a powerful degree of supervision was the order of the day. Furthermore, in regard to a large number of matters the permission or approval of the Government had to be obtained before action was taken, so that the State was a powerful guardian and supervisor of the local authorities.

Local administration in post-war days, however, has been considerably decentralized and the form of supervision by the Central Government has changed from administrative control to legislative and judicial controls. It has also changed in character from authoritative control to non-authoritative guidance, assistance and advice. This means that the autonomy of local public bodies is more highly respected and the status of local government administration has been raised.

This state of affairs exists generally in the case of local authorities, but it applies more or less in the same degree to the Tokyo metropolis, because although the Tokyo metropolis is the nation's capital, its status is merely that of a local authority so far as the constitution and the Local Autonomy Law are concerned. Therefore, the relationship between the State and the Tokyo metropolis is exactly the same as that subsisting between the State and other local authorities.

However, in connection with the metropolis the question has been raised as to the form and the degree of concern which the State should display in the affairs of the metropolis. This question came to the surface at the time when the Capital Construction Law and the Capital Area Readjustment Law were enacted. There are some people who insist that the capital is the centre and symbol of the State, and that it is therefore natural for the State to take a more active part in the affairs of the capital than in those of other large cities. Some people go so far as to declare that, in view of the special character of the metropolis, its autonomy should be greatly limited or it should not enjoy any autonomy at all. However, the citizens of Tokyo have, since the Meiji era, consistently continued to insist that Tokyo can advance and develop only through the exercise of autonomy. Dr. Charles A. Beard in his great work, *The Administration and Politics of Tokyo*,[1] published in 1923, wrote as follows:

'Now Japan seems to be midway between the ancient days when a wise emperor, like Kwammu, could plan, lay out, and construct a beautiful

[1] Charles A. Beard: *The Administration and Politics of Tokyo*, 1923, p. 139.

capital on his own motion and the modern days when public improvements must spring, in part at least, from popular desires and interest. The affairs of the city of Tokyo, for example, are no longer directed by the Shogun or solely by Imperial officers. At the very time when staggering problems of municipal administration were thrown upon the city, the power to plan and act was divided between the Imperial Government and a certain portion of the male citizens in Tokyo to whom the suffrage was granted. As in western countries, industrial cities grew up so rapidly in Japan that a wholly new type of urban civilization was created before the people became aware of the problems involved in the revolutionary changes. There is no single Imperial officer who can now control all the factors in the situation and bring immediate order out of the chaos.'

Even so long as thirty years ago, Dr. Beard made it very clear that the affairs of Tokyo could not be conducted by a bureaucratic system. This is more true at present when the democratic principle of decentralization has been thoroughly established. The State will surely be more concerned in the construction and management of Tokyo than in that of other local authorities. Tokyo also will find it necessary, in view of its national and international character, to adopt a special type of management for its affairs. Again, since the increase in Tokyo's population is caused mainly by the migration of people from other prefectures, and in view of the fact that Tokyo suffered more from the war damage than other cities, Tokyo should be in a position, on that account alone, to request the State for special and closer co-operation. Such questions as floating loans, joint defrayment of expenditure or national grants-in-aid must be considered within the limit proper for their special characteristics. Thus, the satisfactory development of the metropolis can only be assured if the State abandons bureaucratic dogma on the one hand and Tokyo wipes out its narrow-minded autonomy-consciousness on the other, and both of them observe the principle of mutual co-operation.

In connection with the relationship with Central Government, it is important to notice the fact that a big city such as Osaka is under dual control by the Central Government and the prefectural government, owing to being placed in the same category as the middle-size and smaller cities, towns and villages. The critical attitude of the citizens of Osaka towards this irrational position springs from the fact that a big city requires to be administered by methods and organizations beyond the limits of those existing in the prefectures, and also from the fact that the government of big cities exerts an influence over wide surrounding areas. The prefectural government, which is but a local body, is said to over-reach its abilities and exercise unnecessary formal control over the city administration, thereby hindering the smooth operation of duties and wasting money.

For example, it is not infrequent that in a big city, because a city and a prefecture are both local public bodies of the same nature, there is competition in building schools, clubs, and other facilities, the result being that at times unnecessary facilities are set up or that similar facilities are irrationally provided under different policies. There is, for example, a duplication of facilities in Osaka city for the collection of taxes. This is a phenomenon which frequently occurs under dual administration.

In order to do away with the disadvantages resulting from dual administration and control found in the government of Japan's major cities, there has been a movement during the past forty years or more for a special system for such cities: namely, to place the five cities of Osaka, Kyoto, Nagoya, Yokohama, and Kobe, whose populations are over one million in number, outside the jurisdiction of the prefectural government and to confer upon them the powers of both prefectural and municipal governments. This movement for a special system for big cities is the result of a demand for local authorities to be given greater administrative powers in the case of those municipalities whose economic and administrative capacity is equal to or exceeds that of the prefectural governments. The proposal has the support of taxpayers, who wish to eradicate the expense resulting from the dual administrative set-up.

This reform has been embodied in a Bill which has been before the National Diet on several occasions, but was rejected each time by the Upper House, which favours administrative centralization and bureaucracy. Nevertheless, in 1947 the way was open for the first time for its implementation when the law on local autonomy was passed with the object of establishing a democratic local government system in Japan.

Under the provisions of this law a system applying to special cities was established, whereby specially designated cities with a population of half a million or more stood outside the jurisdiction of prefectural governments, and were given the combined functions of prefecture and city, town, or village.

It was thus that the five major cities mentioned above put in a demand to the National Diet to be designated as special cities. Bills to achieve this were introduced in 1947 and again in 1952. But the results were abortive owing to violent opposition on the part of the prefectural governments concerned. These political conflicts and struggles between big cities and prefectures led the Central Government in 1952 to set up a body called the Committee for the Investigation of Local Government, composed of scholars, Diet members, representatives of local authorities and Government officials. The committee issued its first report in 1953, stressing those matters requiring immediate attention. To implement this report, the Government introduced a Bill into the Diet in 1956 which purports to do away with the rights of prefectural governors to issue permits and sanctions to the mayors of major cities, and to eradicate to a small extent the irrationalities seen at the present time in the major cities by transferring to a large extent the administration of the prefectures to the respective cities.

FUTURE OF TOKYO AND OSAKA

Both Tokyo and Osaka are improving the administration of their services in order to increase the welfare of the inhabitants within the regional community. Nevertheless, in order to make it possible for them to carry out efficient and rational city management, there is need for a new structure of administration which, for the sake of convenience, may be called 'regional government.' In other words, it means an administration which would cover a great area radiating from a large pivotal city and extending far beyond its administrative boundary.

In view of the fact that the administration of a large city has come to assume the character of a regional area and that co-operation among the local public bodies within this region has been growing as a matter of necessity, problems of regional planning and metropolitan self-government must be studied. It is to this end that new legislation such as the Capital Area Readjustment Law has been enacted and that much interest is shown by all the parties concerned in the future application of this law.

Select Bibliography

Select Bibliography

CLASSIFIED BY CITIES

AMSTERDAM

Brugmans, H. *Geschiedenis van Amsterdam*. Amsterdam, 1930. 8 vols.

D'Ailly, A. E. *Zeven eeuwen Amsterdam*. Amsterdam, 1944. 6 vols.

Doers, J. C. van der, and others. *Ons Amsterdam*. Amsterdam, 1950.

Mijksenaar, P. J. *Amsterdam, verleden, heden, toekomst*. Amsterdam, 1951.

Kranenburg, R. *Het Nederlands staatsrecht*. 6th ed. Haarlem, 1947. 2 vols.

Pot, C. W. van der. *Handboek van het Nederlandse staatsrecht*, 4th ed. Zwolle, 1950.

Peelje, G. A. van. *Administratief recht*. s'Gravenhage, 1927.

Pot, C. W. van der, and others. *Nederlandsch bestuursrecht*, 1932.

Kranenburg, R. *Inleiding in het Nederlandsch administratief recht*. Haarlem, 1941.

Oppenheim, J. *Het Nederlandsch gemeenterecht*. 5th ed., by C. W. van der Pot. Haarlem, 1928–30. 3 vols.

Bool, J. *De gemeentewet*. Zwolle, 1930.

Loenen, J. W. A. C. van. *De geemeentewet en haar toepassing*. 2nd. ed. Alphen aan de Rijn, 1934–38. 2 vols.

Vos, H. *De gemeentewet*. 4th ed. s'Gravenhage, 1933–38.

Sikkes, P., and Zadel, A. *Beknopt leerboek voor het gemeenterecht*. 6th ed. Alphen aan de Rijn, 1950.

Steinmetz, B. J. F. *Handboek der Nederlandsche overheidsfinancien*. Amsterdam, 1949.

Nap, N. A. *De wet betreffende de noodvoorziening gemeente-financiën 1948* (*De voorzieningen van het rijk met betrekking tot de gemeente-financiën*, vol. 7). Alphen aan de Rijn, 1949.

Prinsen, M. J. 'De functie van de gemeente in het huidige staatsbestel, gezien in het licht van de finansiële verhouding tussen rijk en gemeenten,' in *Tractatus tributarii, opstellen op gelastinggebied aangeboden Prof. Dr. P. J. A. Adriani*.

BOMBAY

Venkatarangaiya, M. *Beginnings of local taxation in the Madras Presidency*. Bombay, 1928.

Venkatarangaiya, M. *Development of local boards in the Madras Presidency*. Bombay, 1938.

Shah, K. T., and Bahadurji. *Indian Municipalities*. Bombay, 1925.

Cambridge History of India. Vol. 6, Ch. 28.

India. Resolutions of the Government of India on local self-government, 1882, 1915, 1918.

India. Memorandum presented to the Indian Statutory (Simon) Commission by the Government of India, 1929.
Indian Statutory Commission. Report. Vol. 1, Pt. 4, Ch. 4.

Wacha, D. E. *Rise and growth of Bombay municipal government*. 1913.
Masani, R. P. *Evolution of local self-government in Bombay*. 1929.
Bombay. Municipal Commissioner. Administration report for 1948–49.
Bombay. Outline of the master plan for Greater Bombay.
Modak, N. V. *Note on Greater Bombay*.
Bombay, Municipal Finances Committee. Report, 1948.
City of Bombay Municipal Act, 1888, as modified up to July 1950.
Bombay. The Greater Bombay scheme: Report of the Housing Panel, 1946.

Goode, W. S. *Municipal Calcutta*. Edinburgh, 1916.
Calcutta. Corporation of Calcutta Investigation Commission. Report, 1949–50. 2 vols.
Calcutta Municipal Act, 1923, as modified up to July 1950.

BUENOS AIRES

Bercaitz, M. A. *Procedimiento administrativo municipal*. Buenos Aires, 1946.
Baulina, A. V. *El gobierno municipal*. Cordoba, 1941.
Bucich Escobar, I. *Buenos Aires-Ciudad*. Buenos Aires, 1936.
Carranza, A. *La cuestion capital*. Buenos Aires, 1926–32. 5 vols.
Carril, B. del. *Buenos Aires frente al pais*. Buenos Aires, 1944.
Estrada, J. M. *Curso de derecho constitucional*. Buenos Aires, 1927. 3 vols.
Gonzalez Calderon, J. A. *Derecho constitucional argentino*. 3rd ed. Buenos Aires, 1931. 3 vols.
Greca, A. *Derecho Ciencia de la administración municipal*. Vol. 4. 2nd ed. Santa Fe, 1943.
Korn Villafañe, A. *Derecho municipal y provincial*. Buenos Aires, 1936–39.
Korn Villafañe, A. *Derecho público político*. Buenos Aires, 1936–39. 2 vols.
Macdonald, A. F. 'The City of Buenos Aires,' in his *Government of the Argentine Republic*. New York, Crowell, 1942.
Zabala, R., and Gandia, E. de. *Historia de la ciudad de Buenos Aires*. Buenos Aires, 1936–37. 2 vols.
Zavalia, C. *Tratado de derecho municipal*. Buenos Aires, 1941.
Primera Reunion Nacional de Municipios, Buenos Aires, 1945. Memoria. Buenos Aires, 1945.
Gómez Forgues, M. I. 'El regimen municipal en la capital federal.' *Revista de la Facultad de Derecho y Ciencias Sociales*. 3rd series. 4. Jan.–April 1941, pp. 135-172.
Zavalia, C. 'El gobierno de la ciudad de Buenos Aires.' *Revista de Derecho y Administración Municipal*. February 1940.
Works by Professor Rafael Bielsa:
Derecho administrativo, 4 vols. 4th ed. Buenos Aires, 1947.
Principios de derecho administrativo, 2nd ed. Buenos Aires, 1949.
Ideas generales sobre lo contenciosoadministrativo. Buenos Aires, 1936.

El estado de necesidad con particular referencia al derecho constitucional y al derecho administrativo. Rosario, 1940.

Sobre el recurso jerárquico. Principios generales y examen del mismo, 2nd ed. Buenos Aires, 1940.

La protección constitucional y el recurso extraordinario. Jurisdicción de la Corte Suprema. Buenos Aires, 1936.

El orden político y las garantías jurisdiccionales. Buenos Aires, 1943.

El problema de la descentralización administrativa. Buenos Aires, 1935.

Ciencia de la administración. Rosario, 1937.

El estadista y el pueblo. Buenos Aires, 1945.

Principios de régimen municipal, 2nd ed. Buenos Aires, 1940.

Relaciones del Código civil con el derecho administrativo. Buenos Aires, 1923.

Perfiles de juristas y políticos (Sarmiento, Costa, Ihering, Marshall); Rosario, 1939.

La jurisdicción contenciosoadministrativa. Rosario, 1949.

Alguños aspectos de la función pública. Santa Fe, 1941.

Los planes de estudio de derecho. Su unidad e integridad. Buenos Aires, 1950.

Los conceptos jurídicos ye su terminología. Rosario, 1946.

CHICAGO

Breese, G. W. *The daytime population of the central business district of Chicago.* Chicago, University of Chicago Press, 1949.

Burnham, D. H. *Plan of Chicago.* Chicago, Commercial Club of Chicago, 1949.

Chicago Plan Commission. *Rebuilding old Chicago.* Chicago, 1941.

Chicago Plan Commission. *Industrial and commercial background for planning Chicago.* Chicago, 1942.

Goode, J. P. *The geographic background of Chicago.* Chicago, University of Chicago Press, 1928.

Gosnell, H. F. *Machine politics: Chicago model.* Chicago, University of Chicago Press, 1937.

Lepawsky, A. *Home rule for metropolitan Chicago.* Chicago, University of Chicago Press, 1935.

Merriam, C. E. *Chicago: a more intimate view of urban politics.* New York, Macmillan, 1929.

Merriam, C. E., and others. *The government of the metropolitan region of Chicago.* Chicago, University of Chicago Press, 1933.

Pierce, B. L. *A history of Chicago.* New York, Knopf, 1937–40.

Pierce, W. H. 'Chicago: unfinished anomaly' in Allen, R. S. *Our fair city.* New York, Vanguard Press, 1947.

Smith, T. V., and White, L. D. *Chicago: an experiment in social science research.* Chicago, University of Chicago Press, 1929.

Smith, T. V., and White, L. D. *A decade of social science research.* Chicago, University of Chicago Press, 1940.

Walker, R. A. 'Chicago: planning in evolution' in *The Planning function in urban government.* Chicago, University of Chicago Press, 1950.

Wirth, L., and Bernert, E. H., eds. *Local community fact book of Chicago.* Chicago, University of Chicago Press, 1949.

U.S.A. Urbanism Committee. Our cities: their role in the national economy; report. Washington, 1937.

Frankel, S., and Alexander, H. 'Arvey of Illinois: new style of political boss,' *Collier's*, **124**, July 23, 1949, pp. 9–11.

Hepner, A. 'Call me Jake,' *New Republic*, **116**, March 24, 1947, pp. 20–23.

Lepawsky, A. 'Chicago: metropolis in the making,' *National Municipal Review*, April 1941.

Madison, R. 'Letter from Chicago,' *New Republic*, **112**, April 23, 1945, pp. 549–51.

Rubin, V. 'You've gotta be a boss,' *Colliers'*, **116**, August 25, 1945, p. 20.

COPENHAGEN

Aakjaer, S., and others, eds. *København før og nu*. København, Hassings Forlag, 1947–50. 6 vols.

Bruun, C. *Kjøbenhavn: en illustreret skildring af dets historie, mindesmaerker og institutioner*. Kjøbenhavn, Philipsen, 1887–1901. 3 vols.

Christensen, V. *København; Kristian VIII og Frederik VII tid*. København, Gad, 1912.

Dahl, F. *Københavns bystyre gennem 300 år*. vol. 1, 1648–1858. København, Munksgaard, 1943.

Holm, A. *Københavns kommunes forfatning*. København, Gydendalske Bokhandel, 1938.

Holm, A., and Johansen, K. *København 1840–1940: det Københavnske bysamfund og kommunens økonomi*. København, Nyt Nordisk Forlag, 1941.

Jørgensen, H. C., and Ipsen, K. *Laerebog i Københavns kommunes økonomi til brug ved undervisningen af Københavns kommunes assistentaspiranter*. København, 1949.

Jørgensen, H., ed. *København fra boplads til storby*. København, Hirschsprung, 1948. 2 vols.

Nielsen, O. *Kiøbenhavns historie og beskrivelse*. Kjøbenhavn, Gad, 1877–92. 6 vols.

Ramsing, H. U. *Københavns historie og topografi i middel alderen*. København, Munksgaard, 1940. 3 vols.

Ramsing, H. U. *Københavns ejendomme 1377–1728: oversigt over skøder og adkomster*. København, Munksgaard, 1943. 4 vols.

Trap, J. P. *København og Frederiksberg*. 4th ed. København, Gad, 1929.

Copenhagen. Stadsingeniørens Direktorat. København: de indlemmende distrikter: byplanmaessig udvikling, 1901–41. 1942.

Copenhagen. Stadsingeniørens Direktorat. København fra bispetid til borgertid: byplanmaessig udvikling til 1840. 1947.

Copenhagen. Stadsingeniørens Direktorat. Københavns gamle bydel: bebyggelse, befolkning, erhvert, trafik. 1947.

Historiske Meddelelser om København, udg. af Københavns Kommunalbestyrelse. Quarterly, 1907– .

Københavns Borgerrepraesentanteres Forhandlinger 1840–41—1841– .

Københavns Kommunalkalender; udg. af Københavns Statistiske Kontor
 Annual. 1931- .
Samling af Bestemmelser vedrørenda Københavns Kommune; udg. pa
 Kommunalbestyrelsens Foranstaltning. Annual, 1607-1863, 1897- .
Statistik Årbog for København, Frederiksberg og. Gentofte Kommune,
 1919- .

LONDON

Rasmussen, S. E. *London: the unique city.* 3rd ed. London, Cape, 1948.
Ormsby, H. *London on the Thames.* 2nd ed. London, Sifton Praed, 1928.
Sinclair, R. *Metropolitan man.* London, Allen & Unwin, 1937.
Gibbon, Sir G., and Bell, R. *History of the London County Council, 1889–*
 1939. London, Macmillan, 1939.
Morrison, H. *How London is governed.* Rev. ed. London, People's Univer-
 sities Press, 1949.
Harris, Sir P. *London and its government.* London, Dent, 1931.
Robson, W. A. *The government and misgovernment of London.* 2nd ed.
 London, Allen & Unwin, 1948.
Barker, B. *Labour in London: a study in municipal achievement.* London,
 Routledge, 1946.
London. Corporation. *The corporation of London,* 1950.
London County Council. Administrative County of London financial
 abstract 1938–39—1947–48. 1950.
London County Council. London housing statistics 1948–49. 1949.
London County Council. Statistical abstract for London, Vol. 31, 1939–48,
 with 1949 figures where available. 1950.
London County Council. Statistics of metropolitan boroughs 1950–51. 1952.
London County Council. *The youngest county.* 1951.

Forshaw, J. H., and Abercrombie, Sir L. P. *The County of London plan.*
 London, Macmillan, 1943.
London. Corporation. Improvements and Town Planning Committee.
 The city of London, a record of destruction and survival; the proposals
 for reconstruction as incorporated in the final report of the planning consultants,
 C. H. Holden and W. G. Holford. London, Architectural Press, 1951.
London County Council. Administrative County of London development
 plan, London, 1951. 2 vols.
United Kingdom. Ministry of Town and Country Planning. Greater
 London plan, 1944; by L. Abercrombie, 1945.
United Kingdom. Ministry of Town and Country Planning. Advisory
 Committee for London Regional Planning. Report. 1946.
United Kingdom. Ministry of Town and Country Planning. London
 Planning Administration Committee. Report. 1949.

LOS ANGELES

Bemis, G., and Basche, N. *Los Angeles county as an agency of municipal*
 government. Los Angeles, Haynes Foundation, 1947.

California University. Bureau of Governmental Research. Studies in Local Government.

No. 4. *Intergovernmental cooperation in the Los Angeles area,* by R. M. Ketcham, 1940.

No. 7. *Intergovernmental cooperation in pre-protection in the Los Angeles area,* by J. R. Donoghue, 1943.

No. 8. *Intergovernmental cooperation in public personnel administration in the Los Angeles area,* by J. N. Jamison, 1944.

No. 9. *Coordinated public planning in the Los Angeles region,* by J. M. Jamison, 1948.

No. 10. *Los Angeles County Chief Administrative Officer: ten years experience,* by A. Holtzman, 1948.

No. 11. *Cooperative health administration in metropolitan Los Angeles,* by M. G. Morden and R. Bigger, 1949.

No. 12. *Cooperative administration of property taxes in Los Angeles County,* by J. E. Swanson and others, 1949.

Cottrell, E. A., and others. *Metropolitan Los Angeles: a study in integration.* Vol. 1. *Characteristics of the Metropolis.* Vol. 2. *How the Cities Grew.* Los Angeles. Haynes Foundation, 1952.

Crouch, W. W., and McHenry, D. E. *California government, politics and administration.* 2nd ed. Berkeley, University of California Press, 1949.

Jones, H. L., and Wilcox, R. F. *Metropolitan Los Angeles: its governments.* Los Angeles, Haynes Foundation, 1949.

Kidner, F. L., and Neff, P. *An economic survey of the Los Angeles area.* Los Angeles, Haynes Foundation, 1945.

McWilliams, C. *Southern California country.* New York, Duell, Sloan & Pearce, 1946.

Nadeau, R. *The water seekers.* Garden City, Doubleday, 1950.

Robbins, G. W., and Tilton, L. D., eds. *Los Angeles: preface to a master plan.* Los Angeles, Pacific Southwest Academy, 1941.

Rush, J. A. *The city–county consolidated.* Los Angeles, The Author, 1941.

Scott, M. *Metropolitan Los Angeles: one community.* Los Angeles, Haynes Foundation, 1949.

Shevky, E., and Williams, M. *Social areas of Los Angeles.* Berkeley, University of California Press, 1949.

California. Legislature. Assembly Interim Committee on State and Local Taxation. *The borough system of government for metropolitan areas.* Sacramento, 1951.

Cottrell, E. A. 'Problems of local government reorganization,' *Western Political Quarterly,* 2, December 1949, pp. 599–609.

Crouch, W. W. 'Extraterritorial powers of cities as factors in California metropolitan government,' *American Political Science Review,* 21, April 1937, pp. 286–91.

Peppin, J. C. 'Municipal home rule in California.' *California Law Review,* 30, November 1941, pp. 1–45.

Stewart, F. M., and Ketcham, R. 'Intergovernmental contracts in California,' *Public Administration Review,* 1, Spring 1941, pp. 242–8.

MANCHESTER

Nicholas, R. *Manchester and district regional planning proposals.* Norwich, Jarrold, 1943.

Nicholas, R. *City of Manchester Plan.* Norwich, Jarrold, 1945.

Simon, E. D. *A city council from within.* London, Longmans, 1926.

Simon, E. D. *The rebuilding of Manchester.* London, Longmans, 1935.

Simon, S. D. *A century of city government: Manchester 1838–1938.* London, Allen & Unwin, 1938.

Simon, S. D. *How the Manchester Education Committee works.* Manchester, Manchester University Press, 1934.

MONTREAL AND TORONTO

Cassidy, H. M. *Public health and welfare organization in Canada.* 1945.

Buck, A. E. *Financing Canadian government.* 1949.

Leacock, S. *Montreal.* 1945.

Masters, D. C. *The rise of Toronto, 1850–90.* 1947.

Wright, F., ed. *The borough system of government for Greater Montreal,* 1947.

Crawford, K. G. *Local government in Canada.* (Mimeographed.) 1949.

Civic Advisory Council of Toronto. Committee on Metropolitan Problems. First report, sections 1–3. 1949–50.

Canada. Civil Service Commission. *The career basis of the Canadian civil service.* 1949. Unpublished.

Canada. Privy Council Office. *Commissions of Enquiry in Canada, 1867–1949.* 1950. Unpublished.

Taylor, G. 'Topographical control in the Toronto region.' *Canadian Journal of Economics and Political Science,* **2,** November 1936, pp. 493–511.

Callard, K. 'The present system of local government in Canada: some problems of status, area, population and resources.' *Canadian Journal of Economics and Political Science,* **17,** May 1951, pp. 204–17.

Crouch, W. W. 'Metropolitan Government in Toronto,' *Public Administration Review,* **14,** Spring 1954, pp. 85–95.

MOSCOW

Simon, E. D., and others. *Moscow in the making.* London, Longmans, 1937.

Ranerskii, I. *Moskva industrial'naia.* (Moscow Industry). Moscow, Moskovskii Rabochii, 1947. (In Russian.)

Webb, S. and B. *Soviet communism: a new civilization?* London, Longmans, 1935.

Towster, J. *Political power in the U.S.S.R., 1917–1947.* New York, Oxford University Press, 1948.

Dobb, M. H. *Soviet economic development since 1917.* London, Routledge, 1948.

Bienstock, G., and others. *Management in Russian industry and agriculture.* London, Oxford University Press, 1944.

Sloan, P. A. *How the Soviet state is run.* London, Lawrence & Wishart, 1941.

Maynard, H. J. *The Russian peasants and other studies.* London, Gollancz, 1947.

Karpinsky, V. A. *The social and state structure of the U.S.S.R.* Moscow, Foreign Languages Publishing House, 1948.

Besclovskii. *Course in the economics and planning of communal economy.* Moscow, 1945. (In Russian.)

Law on the Five Year Plan for the Rehabilitation and Development of the National Economy of the U.S.S.R., 1946–1950. London, Soviet News, 1946.

Gorodskoe Khoziaistvo Moskvy. (Moscow City Economy.) (Monthly published in Russian.)

Soviet News, issued by Press Department of the Soviet Embassy in London.

NEW YORK

Allen, R. S., ed. *Our fair city.* New York, Vanguard Press, 1947.

Black, G. A. *The history of municipal ownership of land on Manhattan Island to the beginning of sales by Commissioners of the Sinking Fund in 1844.* New York, Colombia University Press, 1897.

Booth, M. L. *History of the City of New York from its earliest settlement to the present time.* New York, Clark and Meeker, 1859.

Duffus, R. L. *Mastering a metropolis: planning the future of the New York region.* New York, Harper, 1930.

Federal Writers' Project of the Works Progress Administration in New York City. *New York city guide.* New York, Random House, 1939.

Flynn, E. J. *You're the boss.* New York, Viking Press, 1947.

Franklin, J. *La Guardia: a biography.* New York, Modern Age Books, 1937.

Irving, W. *Knickerbocker history of New York.* New York, Doubleday, 1928.

James, M. *The metropolitan life: a study in business growth.* New York, Viking Press, 1947.

Levy, F. N. *Art in New York: a guide to things worth seeing.* 5th ed. New York, Municipal Art Society, 1935.

Limpus, L. M., and Leyson, B. W. *This man La Guardia.* New York, Dutton, 1938.

McBain, H. L. *The law and practice of municipal home rule.* New York, Columbia University Press, 1916.

Myers, G. *The history of Tammany Hall.* 2nd ed. New York, Boni, 1917.

Pound, A. *The golden earth: the story of Manhattan's landed wealth.* New York, Macmillan, 1935.

Rankin, R. B. *Guide to municipal government—City of New York.* 5th ed. Brooklyn, Eagle Publishing Co., 1942.

Riis, J. *How the other half lives.* New York, Scribners, 1890.

—— *The making of an American.* New York, Macmillan, 1919.

—— *New York, past and present, its history and landmarks, 1524–1939: one hundred views.* New York, New York Historical Society, 1939.

Rodgers, C., and Rankin, R. B. *New York, the world's capital city.* New York, Harper, 1948.

Stoddar, L. *Master of Manhattan: the life of Richard Croker*. New York, Longmans, 1931.

Werner, M. R. *Tammany Hall*. New York, Doubleday, 1928.

Regional plan of New York and environs. 1929. 10 vols.

New York (City) Mayor. Annual report.

New York (City). City Planning Commission. Annual report.

Citizens Budget Commission. *Fiscal facts concerning the City of New York*. Annual.

New York (State). Chamber of Commerce. *A guide-book to the City of New York with historical, descriptive and statistical facts*. 1947.

New York (City). Department of Parks. Collection of miscellaneous reports concerning various projects connected with the work of Robert Moses. 1934–39. 4 vols.

New York (City). Department of Parks. *12 years of progress, 1934–45*.

New York (State). Constitutional Convention Committee. *New York: city government, functions and problems*, Vol. 5, 1938.

United Nations. Secretary-General. *Report to the General Assembly on the permanent headquarters of the United Nations, July 1947. (A/311)*.

World Trade Corporation. *Report to the City of New York on surveys of plan and program for the improvement of waterfront and world trade facilities, October 16, 1947*.

PARIS

Bernheim. *Le conseil municipal de Paris de 1789 à nos jours*. Thesis, Paris, 1937.

Chapman, B. *Introduction to French local government*. London, Allen & Unwin, 1953.

Chevalier, L. *La formation de la population parisienne au XIX siècle*. Paris, Presses Universitaires de France, 1950.

Chombart de Lauwe, P. H. *L'espace social dans la région parisienne*. Paris, 1951.

Cottez, J. *L'organisation de la préfecture de police*. Thesis, Paris, 1944.

Dubech, L., and Espezel, P. *Histoire de Paris*. Paris, 1931. 2 vols.

Felix, M. *Le régime administratif du Département de la Seine et de la Ville de Paris*. 3rd ed. Paris, Rousseau, 1946.

Georges, P., and others. *Etudes sur le banlieu de Paris*. Paris, Colin, 1950.

Lainville, R. *Le budget communal*. Paris, Godde, 1950.

Laurent, R. *Paris, sa vie municipale*. Paris, Godde, 1938.

Lelandais. *Le régime administratif du Département de la Seine et de la Ville de Paris*. Paris, Imprimerie Municipale, 1946.

Marabuto, P. *Les partis politiques et les mouvements sociaux dans la IV République*. Paris, 1948.

Vezien, P. *Les services publics industriels dans la région parisienne*. Thesis, Paris, 1940.

Organisation et attributions des services de la Préfecture de Paris, 1946.

Organisation et attributions des services du Département de la Seine et de la Ville de Paris. Paris, Imprimerie Municipale, 1946.

Attributions et pouvoirs du Conseil Municipal de Paris avant et après les decrets lois des 21 avril at 13 juin 1939, completés par l'ordonnance du 13 avril, 1945. Paris, Imprimerie Municipale, 1947.

Attributions et pouvoirs de Conseil Général de la Seine avant ct après les decrets lois des 21 avril at 13 juin, 1939, completés par l'ordonnance du 13 avril, 1945. Paris, Imprimerie Municipale, 1947.

Conseil Municipal de Paris. Règlement intérieur. Paris, Imprimerie Municipale, 1945.

Conseil Général de la Seine. Règlement intérieur. Paris, Imprimerie Municipale, 1946.

Paris. Bulletin municipal officiel de la Ville de Paris et annexe au recueil, des actes administratifs de la Préfecture de la Seine et de la Préfecture de Police.

Belmas. 'Le préfet de police,' *Revue du Droit Public*, **51,** 1934, pp. 373-405.

Goguel, F. 'Structure sociale et opinions politiques à Paris d'après les élections du 17 juin 1951,' *Revue Française de Science Politique*, **1,** 1951, pp. 326-33.

RIO DE JANEIRO

Araujo Gois, H. *O saneamento da baixada fluminense*. Rio de Janeiro, 1934 and 1939.

Backheuser, E. *A faixa litoranea do Brasil meridional*. Rio de Janeiro, 1918.

Barbosa, R. *Comentarios a constituicao federal de 1891*. vol. 5, São Paulo, 1934.

Biard. *Deux années au Brésil*. Paris, 1862.

Carneiro, J. F. '8,500,000 km²,' *Catolicismo, revolucao e reacao*. Rio de Janeiro, 1947.

Carvalho, D. de. *Historia de Cidade do Rio de Janeiro*.

Coelho, N. *A capital federal*. Rio de Janeiro, 1915.

Costa, N. *Historia da Cidade do Rio de Janeiro*. Rio de Janeiro, 1933.

Cruls, G. *Aparencia do Rio de Janeiro*. Rio de Janeiro, 1949. 2 vols.

Edmundo, L. *O Rio de Janeiro no tempo dos vice-reis*. 2nd ed. Rio de Janeiro, 1935.

Edmundo, L. *O Rio de Janeiro do meu tempo*. Rio de Janeiro, 1939. 3 vols.

Edmundo, L. *A Corte de D. Joao IV no Rio de Janeiro*. 3 vols.

Etienne, G. *A reconstrucao do Rio de Janeiro*. Rio de Janeiro. 1904.

Figueira de Almeida. *Historia fluminense*. Rio de Janeiro, 1939.

Fleiuss, M. *Historia administrativa do Brasil*. São Paulo, 1935.

Fleiuss, M. *Historia da Cidade do Rio de Janeiro*. São Paulo, 1928.

Freire, F. *Historia da Cidade do Rio de Janeiro*. Rio de Janeiro, 1912-14.

Grilo, H. *Problemas economicos do Distrito Federal*. Rio de Janeiro, 1947.

Ktzinger, A. M. *Resenha historica de Cidade do Rio de Janeiro*.

Lamego, A. R. *O homem e a Guanabara*. Rio de Janeiro, 1948.

Macedo, J. M. de. *Um passeio pela Cidade do Rio de Janeiro*. 2nd ed.

Macedo, R. *Efemeridas cariocas*. Rio de Janeiro, 1942.

Macedo, R. *Curiosidades cariocas*. Rio de Janeiro, 1943.

Magalhaes Correia. *O Sertao Carioca*. Rio de Janeiro, 1936.

Mello Barreto Filho and Hermeto Lima. *Historia de policia do Rio de Janeiro*. Rio de Janeiro, 1939–42. 2 vols.

Morais, L. J., and others. *Geologia e petrologia do Distrito Federal e imediacoes Ouro Preto*. 1935.

Noronha Santos. *Meios de transporte no Rio de Janeiro*. Rio de Janeiro, 1934.

Ribeyrolles, C. *Brasil pitoresco* (translation). São Paulo, 1941. 2 vols.

Rio, J. do. *Religioes do Rio*. 2nd ed. Rio de Janeiro, 1950.

Rugendas, J. M. *Viagem pitoresca atraves do Brasil* (translation). São Paulo, 1941.

Schlichthorst, C. *O Rio de Janeiro como e* (translation). Rio de Janeiro, 1943.

Seidler, C. *Dos anos no Brasil* (translation). São Paulo, 1941.

Taunay, A. de. *Rio de Janeiro de antanho*. São Paulo, 1942.

Teixeira, A. W. *Estrutura politica e direcao administratica do Distrito Federal*. Rio de Janeiro, 1950.

Teixeira, A. W. *Tributos do Distrito Federal*. Rio de Janeiro, 1950.

Vivaldo Coaraci. *O Rio de Janeiro no seculo XVII*. Rio de Janeiro, 1944.

Brazil. *Decreto lei no. No.* 3770 de 28/10/41: *Estatuto dos funcionarios publicos civis de prefeitura do Distrio Federal*. Rio de Janeiro, 1947.

Brazil. *Lei no.* 217 de 15/1/48: *lei organica do Distrito Federal*. Rio de Janeiro, 1948.

Ebling, F. K. 'Metropolitano,' *Revista do Engenharia*, no. 92, 1944.

ROME

Rome. Consiglio Comunale. Estratto verbale delle deliberazioni del Consiglio. (Bilancio preventivo 1950.) Roma, 1951.

Ferraris. *La capitale e il suo ordinamento*. Torino, 1912.

Vuoli. *L'ordinamento amministrativo della città di Roma*. Milano, 1927.

Testa, V. Articles in *Il Tempo*, currently.

Ceroni, G. Articles in *Messaggero di Roma*, currently.

'Roma.' Article in *Enciclopedia Italiana* (full bibliography).

Rome. Istituto di Studi Romani. *Storia di Roma*. In progress.

Bollettino statistico, in *Capitolium* published by the Rome municipality.

Rome. *Annuario statistico*.

Maroi. 'La popolazione industriale e commerciale della città di Roma,' *Roma*, 1934.

Rebecchini. *Les problèmes de l'urbanisme à Rome* (communication au 21ème Congrès des Capitales, Lisbonne, Octobre 1950.)

Reed, H. H. 'Rome, the third sack,' *Architectural Review*, 1950, February, pp. 91–110.

Camera di Commercio, Industria e Agricoltura di Roma. *Indici della ricostruzione e caratteristiche economiche di Roma e provincia*. Roma, 1953.

STOCKHOLM

Ahlmann, H. W. *Stockholms inre differentiering*, 1934.

Boalt, G. *Skolutbildning och skolresultat för barn ur olika samhällsgrupper i Stockholm*, 1947.

Dahlgren, E. W., ed. *Stockholm. Sveriges hufvadstad*. vol. 1. Stockholm, 1897.

Dahlström, E. *Trivsel i Söderort*, 1951.

Olsson, J. *Sveriges Kommunalstyrelse*. Stockholm, 1935.

Olsson, J. *Kommunal självstyrelse i Sverige*. Stockholm, 1950.

Sundberg, H. G. F., and Berglund, H., eds. *Kommunal författningshandbok*. 3rd ed. Stockholm, 1948.

Stockholm of today, 1939.

Stockholm, sa vie, ses habitants, ses œuvres sociales, 1939.

Stockholm ancien et moderne. Exposition organisée par le Musée de la ville de Stockholm, 1949.

Stockholm Stadsplanekontor. Generalplan för Stockholm, 1952.

William-Olsson, W. *Huvuddragen av Stockholms stads geografiska utveckling, 1850–1930*, 1934.

William-Olsson, W. *Stockholms framida utveckling*, 1941.

Stockholm. Stadskansli. Kommunal författningshandbok för Stockholm. Annual.

Kommunal författningssamling for Stockholm. Periodical.

Statistisk Årsbok för Stockholms stad.

Stockholm. Stadskansli. Statistisk Årsbok för Stockholms stad.

Stockholm. Stadskansli. Stockholm's Kommunal-Kalendar. Annual.

Guinchard, J. 'Administrative organisation of the City of Stockholm.' *Statistik Manadsskrift*. **25**, 1930. No. 10.

Robson, W. A. 'My investigation of city government in Stockholm,' *Municipal Journal*, **45**, 2268.

SYDNEY

Bland, F. A. *City government by Commission: Sydney's first experiment*. Sydney, Royal Australian Historical Society, 1928.

Bland, F. A. *Government in Australia*. Sydney, N.S.W. Government Printer, 1944.

Bertie, C. H. *Early history of the Sydney Municipal Council*. Sydney, City Council, 1911.

Greig, A. W. *Notes on the early history of local government in Victoria*. Melbourne, University Press, 1925.

New South Wales. *New South Waes: the mother colony of the Australias*. Sydney, Government Printing Office, 1896.

Australia. Annual yearbook of the Commonwealth of Australia.

New South Wales. Annual yearbook of the New South Wales Government.

Local Government Association of New South Wales. Conference Reports. Sydney.

Local government in the postwar world. 1945.

Local government finance. 1946.

Local government structure. 1947

Decentralisation. 1948.

Shire and Municipal Record, published monthly by the Local Government Association of N.S.W.

Australian Municipal Journal, published by the Municipal Association of Victoria.

Australia. Department of Local Government. Annual report.

Annual reports of the various statutory corporations.

Sydney. City Council. Annual report (up to the 1920's).

Sydney. City Council. Vade mecum (up to 1939).

WELLINGTON

Mulgan, A. *The city of the Strait.* Wellington, A. H. and A. W. Reed, 1939.

Wellington City. Year book.

Cox, J. 'A town planning exhibition,' *Landfall*, 1948, June.

New Zealand. Local Government Committee. Report. 1945.

New Zealand. Department of Internal Affairs. *Local government in New Zealand*, ed. by F. B. Stephens. 1949.

New Zealand. Official year book.

In preparation:

Official history of the city of Wellington.

Fitzgerald, L. J. *Recent local body elections in the Wellington district.* M.A. thesis, Victoria University College.

ZÜRICH

Schmid, W., ed. *Zürich. Stadt und Land.* Bern, Hallwag, 1938.

Hürlimann, M. *Zürich.* Zürich, Atlantis Verlag, 1948.

Hafner, K. *Zürcher Bürger- und Heimatbuch.* 2nd ed. Zürich, Verlag der Erziehungsdirektion, 1940.

Huber, M. *Heimat und Tradition: gesammelte Aufsätze.* Zürich. Atlantis Verlag, 1947.

Clerc, J. L. *Perspectives cavalières de Zurich.* Neuchâtel.

Zürich. Stadtrat. Zürich: Geschichte, Kultur, Wirtschaft. Zürich, Fretz, 1933.

Statistisches Jahrbuch der Stadt Zürich; hg. vom Statistischen Amt der Stadt Zürich, 1905.

Aeppli, H. *Das zürcherische Gesetz über das Gemeindewesen vom 6. Juni 1926; Verfassungsbestimmungen, Gemeindegesetz, Bürgerrechtsverordnung, Rechnungsverordnung.* 4th ed. Zürich, Polygraphischer Verlag, 1933.

Klöti, E. *Die Zürcher Stadtverfassung; kantonale Verfassungsbestimmungen, Zuteilungsgesetz von 1891, Vereinigungsgesetz von 1931, Gemeindeordnung.* Zürich, Polygraphischer Verlag, 1934.

Klöti, E. *Kommunalpolitik in der Stadt Zürich.* Bern, Föderativverband des Personals öffentlicher Verwaltungen und Betriebe, 1930.

Lüchinger, A. *Gemeindehaushalt der Stadt Zürich.* Zürich, Volkshochschule, 1949.

Schmidt, H. *Die Ausgaben der Stadt Zürich von 1893–1936.* (Dissertation der Universität Zürich.) Affoltern, 1939.

Heer, G. H. *Kleine Stadtbürgerkunde.* Zürich, Neue Zürcher Zeitung, 1950.

Zürich. Stadtkanzlei. *Geschichte der Zürcher Stadtvereinigung von 1893.* Zürich, 1919.

Zürich. Tagblatt der Stadt Zürich. *50 Jahre Zürcher Stadtvereinigung, 1893–1943.* Zürich, 1943.

Peter, J. 'Die städtischen Finanzen in den Kriegsjahren 1939–45.' *Zürcher Statistische Nachrichten,* 1946, Heft 2.

Oechsli, W. *History of Switzerland, 1499–1914.* Cambridge, University Press, 1922.

Hottinger, M. D. *The stories of Basel, Berne and Zurich.* London, Dent, 1933.

Dändliker, K. *Geschichte der Stadt und des Kantons Zürich.* Zürich, Schulthess, 1908–12. 3 vol.

Largiader, A. *Geschichte von Stadt und Landschaft Zürich.* Erlenbach, Rentsch, 1945. 2 vol.

Kläui, P. *Zürich: Geschichte der Stadt und des Bezirks.* Zollikon, Bosch, 1948.

Vogt, E. *Der Lindenhof in Zürich: zwölf Jahrhunderte Stadtgeschichte auf Grund der Ausgrabungen 1937–38.* Zürich, Orell Füssli, 1948.

Zurlinden, S. *Hundert Jahre Bilder aus der Geschichte der Stadt Zürich in der Zeit von 1814–1914.* 2 vols. Zürich, Berichthaus, 1914–15.

Kläui, P., and Imhof, E. *Atlas zur Geschichte des Kantons Zürich.* Zürich, Orell Füssli, 1951.

Ermatinger, E. *Dichtung und Geistesleben der deutschen Schweiz.* München, Beck, 1933.

Wehrli, M., ed. *Das geistige Zürich im 18. Jahrhundert: Texte und Dokumente von Gotthard Heidegger bis Heinrich Pestalozzi.* Zürich, Atlantis Verlag, 1943.

Zücher, R. *Die künstlerische Kultur im Kanton Zürich: ein geschichtlicher Überblick.* Zürich, Atlantis Verlag, 1943.

Escher, K., and others. *Die Kunstdenkmäler des Kantons Zürich.* Vols. 4–5. *Die Stadt Zürich.* Basel, Birkhäuser, 1939–49.

Hürlimann, M., and Jaeckle, E. *Werke öffentlicher Kunst in Zürich: neue Wandmalerei und Plastik.* Zürich, Atlantis Verlag, 1939.

Das Schweizerische Landesmuseum, 1898–1948: Kunst, Handwerk und Geschichte. Zürich, Atlantis Verlag, 1948.

Zürich. Official Information Office. *Guide to art in Zurich. Zurich in history and art.* Zürich, Verkehrsverein, 1945.

Senti, A. 'Förderung von Literatur, Kunst und allgemeiner Kultur durch die Stadt Zürich.' *Zürcher Statistische Nachrichten*, 1943, Heft 3, pp. 209–34; 1934, Heft 2, pp. 139–224.

Hürlimann, M. *E. T. H., die Eidgenössische Technische Hochschule in Zürich.* Zürich, Atlantis Verlag, 1945.

Amann, J. *Die stadtzürcherischen Schulen und ihre Fürsorgeeinrichtungen.* Zürich, Schulamt der Stadt Zürich, 1939.

Wehrle, A., ed. *Turnen und Sport in Zürich.* Zürich, Aschmann und Scheller, 1942.

Zürich (Canton) Erziehungsrat. *Die zürcherischen Schulen seit der Regeneration der 1830er Jahre: Festschrift zur Jahrhundertfeier.* Zürich, Verlag der Erziehungsdirektion, 1933–38. 3 vols.

Zürcher Volkswirtschaftliche Gesellschaft. *Zürichs Volks- und Staatswirtschaft.* Zürich, Girsberger, 1928.

Hermann, E. *Ein Jahrhundert Zürich und die Entwicklung seiner Firmen.* Zürich, Müller, then A B C Druckerei und Verlags-A.G., 1946–47. 2 vols.

Kläui, P. *Zürich: Geschichte der Stadt und des Bezirks.* Zollikon, Bosch, 1948.

Jäger, E. *Der Personennahverkehr der Stadt Zürich.* Zürich, Girsberger, 1946.

Weisz, L. *Studien zur Handels- und Industriegeschichte der Schweiz.* Zürich, Neue Zürcher Zeitung, 1938–40. 2 vols.

Weisz, L. *Die zürcherische Exportindustrie: ihre Entstehung und Entwicklung.* Zürich, Neue Zürcher Zeitung, 1936.

Haeb, F., ed. *Aus der Geschichte der Zürcher Arbeiterbewegung.* Zürich, Genossenschaftsdruckerei, 1948.

Richard, E. *Kaufmännische Gesellschaft Zürich und Zürcher Handelskammer,* 1873–1923. Zürich, Handelskammer, 1924. 2 vols.

150 Jahre Neue Zürcher Zeitung, 1780–1930. Zürich, 1930.

Zürich. Statistisches Amt Statistik der Stadt Zürich. Heft 54. Zürichs Wirtschaftsstruktur nach den Betriebszählungen, 1946. Heft 56. Zürichs Bevölkerung nach den eidgenössischen Volkszählungen bis 1941, 1949.

Guth, H. 'Die Pendelwanderung im Kanton Zürich 1941,' *Zürcher Wirtschaftsbilder,* I, 1945, no. 5/6, pp. 135–76.

Walther, P. *Zur Geographie der Stadt Zürich: der Siedlungsgrundriss in seiner Entwicklung und Abhängigkeit von den natürlichen Faktoren.* Zürich, Orell Füssli, 1927.

Real, W. H. *Stadtplanung: Möglichkeiten für die Aufstellung von Richtlinien am Beispiel der Verhältnisse in der Stadt Zürich.* Bern, Buri, 1950.

Scotoni, E. *Die Sanierung der Zürcher Altstadt: eine ökonomische Untersuchung.* (Dissertation Univ. Zürich.) Zürich, 1944.

Zürich. Hochbauamt. *Der soziale Wohnungsbau und seine Förderung in Zürich, 1942–1947. Les colonies d'habitation et leur développement a Zürich: Social housing and its development in Zürich.* 2nd ed. Erlenbach-Zürich, Verlag für Architektur, 1948.

Winkler, 'Das Stadtbild Zürichs im Wandel der Eingemeindungen.' *Mitteilungen der Geographisch-Ethnographischen Gesellschaft, Zürich,* **39,** 1938/39, pp. 111–66.

Peter, J. 'Die Wohnungspolitik der Stadt Zürich.' *Zürcher Statistische Nachrichten,* 1946, Heft 1.

SUPPLEMENTARY STUDIES

COLOGNE

Baedeker, K. *Köln und Umgebung.* Hamburg, Baedeker, 1954.

Göbel, E. 'Das Stadtgebiet von Köln und seine Entwicklung.' In: *Statistische Mitteilungen der Stadt Köln,* Sonderheft, 2. Auflage, 1948.

Keyser, E. *Rheinisches Städtebuch.* Stuttgart, Kohlhammer, 1956.

Köhler, H. *Köln. Natürliche Grundlagen des Werdens einer Großstadt.* Berlin, 1941.

Kuske, B. *Die Großstadt Köln als wirtschaftlicher und sozialer Körper.* Köln, O. Müller, 1928.

Rößger, E. *Entwicklung und Stand des Nordrhein-Westfälischen Luftverkehrs-aufkommens unter besonderer Berücksichtigung des Köln-Bonner Raumes.* Köln, 1956.

Zinsser, F. *Universität Köln,* 1919–1929. Köln, 1929.

Stadt Köln. *Die Stadt Cöln im ersten Jahrhundert unter Preußischer Herrschaft.* 1815–1915. 2 Bände, Cöln, Neubner, 1916.

Stadt Köln. *Statistische Mitteilungen der Stadt Köln. Herausgegeben im Auftrag des Oberstadtdirektors von Direktor Dr. Lorenz Fischer,* Zeitschrift seit 1946.

Stadt Köln. *Das neue Köln. Ein Vorentwurf.* Köln, Bachem, 1950.

Stadt Köln. *Köln 1900 Jahre Stadt. Aus Anlaß der stadtgeschichtlichen Austellung Mai–August 1950.* Köln, 1950.

Stadt Köln. *Verwaltungsberichte der Stadt Köln. Herausgegeben im Auftrag des Oberstadtdirektors vom Statistischen Amt.* Erscheinen jährlich.

JOHANNESBURG

Chilvers, H. A. *Out of the crucible.* London, Cassell, 1929.

Curtis, L. *With Milner in South Africa.* Oxford, Blackwell, 1951.

Gray, J. *Payable gold.* South Africa (Johannesburg), Central News Agency, 1937.

Green, L. P. *History of local government in South Africa: an introduction.* Cape Town and Johannesburg, Juta, 1957.

Holmes, I. Q. *Local government finance in South Africa.* Durban, Butterworth, 1949.

Jacobson, D. *Fifty golden years of the Rand,* 1886–1936. London, Faber and Faber, 1936.

Johannesburg. City Council. Official Guide, 2nd ed. Cape Town, Beerman, 1956.

Maud, Sir J. P. R. *City government: the Johannesburg experiment.* Oxford Oxford University Press, 1938.

Merry, T. G., ed. *Golden city diamond jubilee souvenir.* Johannesburg, Central News Agency, 1946.

Seventy golden years. Johannesburg, Felstar Publishing Co., 1956.

Johannesburg. City (Town) Council. Mayor's minute (annual), 1904 onwards.

Johannesburg. City Council. Memoranda submitted to the Transvaal Local Government Commission of Inquiry, 1953–1956.

Johannesburg. City (Town) Council. Vade mecum (annual), 1931 onwards.

Johannesburg. Provisional Town Planning Scheme. Explanatory report by the regional planning consultants and preliminary statement of proposals for development. March, 1936.

Johannesburg. Stadsraad. Burgomaster's Report, 1897.

Johannesburg. Town Council. Deputy Chairman's minute, 1901–1903.

Transvaal. Local Government Commission (Hofmeyr) Report: T.P.3, 1915.

Transvaal. Local Government Commission (Marais): First Interim Report, T.P.4, 1954.

Transvaal. Local Government Commission (Marais): Second Interim Report. May 16, 1955.

Transvaal. Local Government Commission (Stallard) Report: T.P.1, 1922.

Transvaal. Municipal Commission (Tucker) Report: T.G.12, 1909.

Institute of Municipal Treasurers and Accountants S.A. (Inc.). Annual conference proceedings. 1930 onwards.

Institute of Town Clerks of Southern Africa. Annual conference proceedings. 1948 onwards.

Institution of Municipal Engineers, South African District. Annual journal. 1948 onwards.

Municipal Affairs, Official Local Government Publications (monthly), Pretoria, 1935 onwards.

Official South African Municipal Year Book, S.A.A.M.E., Pretoria, 1910 onwards.

South African Municipal Magazine (monthly), Johannesburg, 1917 onwards.

South African Treasurer (monthly), official journal of the I.M.T.A., S.A. (Inc.), Johannesburg, 1928 onwards.

Union of South Africa. Official Year Book. Government Printer, Pretoria, 1910 onwards.

TOKYO AND OSAKA

Editorial Group for Manuscripts of the late Hiroshi Ikeda (IKEDA HIROSHI IKO KANKO KAI), A collection of articles of Hiroshi Ikeda on municipal problems (IKEDA HIROSHI TOSHI RONSHU), 1940.

Kanamaru, S. *Explanation of the Local Autonomy Law* (CHIHO JICHI HO SEIGI), 1948.

National Association of City Mayors (ZENKOKU SHICHO KAI), *The Municipal Year Book* (TOSHI NENKAN), 1956.

Osaka Municipal Research Institute. Reports. No. 6 (OSAKA SHISEI KENKYUJO HOKOKU, DAI ROKU SHU), *A study of the present status of city finance* (TOSHI ZAISEI NO JITTAI CHOSA), 1952.

Osaka Municipal Research Institute. Reports. No. 11 (OSAKA SHISEI KENKYUJO HOKOKU, DAI JUISSHU), *A study of the present status of city, town and village finance* (SHI CHO SON ZAISEI NO JITTAI CHOSA), 1954.

Osaka Municipal Research Institute. Reports. No. 14, No. 15 (OSAKA SHISEI KENKYUJO HOKOKU, DAI JUYON SHU, DAI JUGO SHU), *A study of actual status of metropolitan living area* (DAITOSHI SEIKATSUKEN NO JISSHO KENKYU), 1955.

Osaka Municipal Research Institute. Research Commission for Metropolitan Administration. Report. (DAITOSHI GYOSEI CHOSA IINKAI HOKOKU), 1950.

Osaka Municipal Research Institute. Research Commission for Metropolitan Administration. Report. (DAITOSHI GYOSEI CHOSA IINKAI. HOKOKU), 1951.

Royama, M. *On local administration* (CHIHO GYOSEI RON), 1937.

Seki, H. *Theory and practice of municipal policy* (TOSHI SEISAKU NO RIRON TO JISSAI), 1951.

Tokyo Institute for Municipal Research. A collection of articles in commemoration of the fiftieth anniversary of the establishment of local autonomy in Japan (JICHI-SEI HAPPU 50-NEN KINEN RONBUN SHU), 1938.

Tokyo Institute for Municipal Research (TOKYO SHISEI CHOSA KAI), Fifty-year history of local autonomy (CHIHO JICHI 50-NEN SHI), 1940.

Japan. Ministry of Construction. Tokyo Metropolitan Construction Commission. Report. (SHUTO KENSETSU IINKAI. HOKOKU). 1952, 1953, 1954, 1955.

Joint Office of Five Big Cities (GODAISHI KYODO JIMUKYOKU). Report on the survey of Tokyo metropolitan administration (TOKYO TOSEI CHOSA HOKOKUSHO), Vols. I & II, 1949.

Tokyo Institute for Municipal Research. Collection of Dr. Beard's speeches (BEARD HAKASE KOEN SHU), 1928.

Tokyo Metropolitan Government. Outline of Tokyo's government, 1954 (TOSEI GAIYO), 1955.

Tokyo Metropolitan Government. Ten-year history of Tokyo's government (TOSEI JUNEN SHI), 1954.

Tokyo Metropolitan Government. *Tokyo* (in English), 1955.

Osaka City. *City of Osaka* (in English), 1955.

Osaka City. *Development of Osaka City* (OSAKA SHI NO HATTEN).

Osaka City. *History of Osaka City* from 1926 (SHOWA OSAKA SHI SHI), 8 Vols. from 1951 to 1954.

Osaka City. *Municipal Year Book*, Annual.

Honjo, E. 'Sixty years of the administration of Osaka City (OSAKA SHISEI ROKUJUNEN)', Study of Municipal Research (TOSHI MONDAI KENKYU, a magazine of Municipal Research Association Osaka City), 1949.

Osaka City. Statistical Year Book of Osaka City (OSAKA SHI TOKEI SHO).

INDEX

PERSONS AND PLACES

BB*

INDEX OF SUBJECTS

For Product Safety Concerns and Information please contact our EU
representative GPSR@taylorandfrancis.com Taylor & Francis Verlag GmbH,
Kaufingerstraße 24, 80331 München, Germany

Printed and bound by CPI Group (UK) Ltd, Croydon, CR0 4YY
01/05/2025
01858512-0001